Mathematical Structure of Syntactic Merge

Linguistic Inquiry Monographs
Samuel Jay Keyser

A complete list of books published in the Linguistic Inquiry Monographs series appears at the back of this book.

Mathematical Structure of Syntactic Merge

An Algebraic Model for Generative Linguistics

Matilde Marcolli, Noam Chomsky, and Robert C. Berwick

The MIT Press
Cambridge, Massachusetts
London, England

The MIT Press
Massachusetts Institute of Technology
77 Massachusetts Avenue, Cambridge, MA 02139
mitpress.mit.edu

The MIT Press would like to thank the anonymous peer reviewers who provided comments on drafts of this book. The generous work of academic experts is essential for establishing the authority and quality of our publications. We acknowledge with gratitude the contributions of these otherwise uncredited readers.

This book was set in Times New Roman by Matilde Marcolli and Robert C. Berwick. Printed and bound in the United States of America.

Library of Congress Cataloging-in-Publication Data is available.

ISBN: 978-0-262-55252-3

10 9 8 7 6 5 4 3 2

EU product safety and compliance information contact is: mitp-eu-gpsr@mit.edu

τὸν φρονεῖν βροτοὺς ὁδώσαντα, τὸν πάθει μάθος
θέντα κυρίως ἔχειν. στάζει δ' ἔν θ' ὕπνῳ πρὸ
καρδίας μνησιπήμων πόνος: καὶ παρ' ἄκοντας
ἦλθε σωφρονεῖν.

Aeschylus, Agamemnon, 176-183

Contents

Series Foreword

We are pleased to present the eighty-ninth volume in the series Linguistic Inquiry Monographs. These monographs present new and original research beyond the scope of the article. We hope they will benefit our field by bringing to it perspectives that will stimulate further research and insight.

Originally published in limited edition, the *Linguistic Inquiry Monographs* are now more widely available. This change is due to the great interest engendered by the series and by the needs of a growing readership. The editors thank the readers for their support and welcome suggestions about future directions for the series.

Samuel Jay Keyser
for the Editorial Board

List of Figures

Editor's Note

Chapters 1 and 2 of this monograph are the joint work of Matilde Marcolli, Noam Chomsky, and Robert Berwick. These chapters have appeared elsewhere (`https://arxiv.org/abs/2305.18278` and `https://arxiv.org/abs/2306.10270`).

On May 29, 2023, Marcolli outlined chapter 3 to Chomsky in an email communication. Subsequent conversations followed between Berwick and Chomsky on June 15 and 16, 2023.

As a result of unforeseen circumstances that precluded Chomsky's further participation, it was subsequently developed and completed by Marcolli and Berwick.

While the authors of chapter 3 have tried to remain as faithful as possible to these initial interactions, Chomsky was unable to review and approve a final version. Its authors take full responsibility for any errors, inaccuracies, or possible misrepresentation that might have occurred.

Samuel Jay Keyser, editor

0 Minimalism and Merge: Introduction

Since 1993, followed by Chomsky's 1995 monograph [22], the generative enterprise for the study of language has developed in a direction that its originator, Noam Chomsky, has termed the "Minimalist Program." Minimalism focuses on the design of the human language faculty as a computational and cognitive process, aiming to see how far one can reduce the number of assumptions required to account for human language syntax in terms of its variation, acquisition, evolution, and use. Its key idea is that human-language syntax rests on a *single* structure building device called *Merge*, currently known as *free symmetric Merge*, for reasons that will become apparent in what follows. In Minimalism, Merge builds sentence syntactic structure—a finite computational process that generates an infinity of possible outputs. In previous, more familiar accounts, this computational process was carried out by some sort of generative grammar with an (often large and language-specific) set of rules, such as a transformational grammar; one aim over the past 60 years has been to reduce this rule set to be as small as possible for learnability and evolvability reasons. It is evident that any such computational process for human language requires *some* kind of device to assemble smaller parts into larger ones, and, as we shall see, the new theory also requires a way to examine and take apart the structures that it assembles in order to accommodate what fell under earlier accounts as transformational movement. We will return to an explicit example shortly.

To serve as a reference standard for the detailed theory of the current conception of Merge, in this monograph we follow what is laid out in the recent publication by Chomsky and colleagues, *Merge and the Strong Minimalist Thesis*, from November 2023 [40]. Sometimes we will cite this reference in the text simply as *Elements*. According to this account ([40], p.1), Merge generates an "infinite array of hierarchically structured expressions interpretable at a 'meaning' interface and available for externalization at a 'form' (sound/sign) interface, the so-called basic property of language."

The Merge framework thus assumes a tripartite division of human language into (1) a central syntactic computation, Merge, and then two interface channels: (2) a syntax–semantics interface, also known as the interface to the Conceptual–Intensional (CI) system; and (3) an interface to the Sensory–Motor (SM) system, where the hierarchical structure of syntax gets flattened into a form suitable for output or input (speech/sign or parsing), also known as "Externalization." Figure 0.1 sketches the basic picture in cartoon like form. The display at the top half of the diagram indicates how Merge first selects items from a lexicon of words and syntactic features, combining them in a kind of computational scratchpad called a *workspace*. Successive Merge actions lead to a sequence of modified workspaces, freely generating syntactic objects, which are then fed to either Externalization or the syntax–semantics interface. (Many of these freely generated objects will, of course, be ill-formed and thus never result in any intelligible outputs at these interfaces.) The bottom half of this diagram displays a slightly more elaborated version of the same tripartite subdivision, indicating that the hierarchical structures that a free symmetric Merge generates, binary rooted trees, are *nonplanar*—that is, they have no left-to-right structure but instead have their leaves unordered, like Alexander Calder mobiles. Externalization (depicted on the bottom left) yields the actual word ordering found in a particular language, a *planar embedding* of the unordered structures. On the right, we show one view of the syntax–semantics interface, reflecting what we adopt here.

In this monograph, our primary focus will be on Merge, in chapter 1, but that will lead in turn to important considerations of both Externalization and the syntax–semantics interface, in chapters 1 and 3. In chapter 2, we discuss this more recent formulation of Minimalism in comparison with earlier versions.

0.1 Why This Monograph?

There has been a rapid development and significant modification to the Minimalist Program over the past nearly three decades, leading to a flourishing linguistic field that includes, among much other work, more than a dozen monographs in this "Linguistic Inquiry Monograph" series alone. However, this same rapid development and change came with a proliferation of new terminology and sometimes unfamiliar notions, such as "workspace," "Minimal Search," "Minimal Yield," "Resource Restriction," and "No Tampering.". This new machinery and assumptions not only have led to vigorous debate on the details about Minimalism, but have also prodded some researchers to criticize

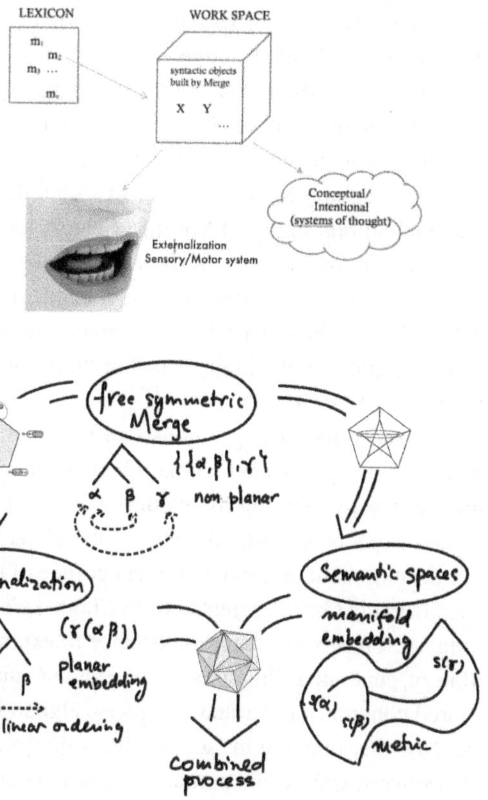

Figure 0.1
The Sensory–Motor (SM) and Conceptual–Intensional (CI) interfaces and the geometric model we will describe in Chapter 3. (The top part of the image is reproduced from [40] with permission.)

the entire Minimalist Program itself as vague, contradictory, or even unscientific; see [107] and [91], among several other sources.

Given this context, the central goal of this monograph is to present a precise, mathematical formalization of Chomsky's recent theory of Merge and its supporting assumptions as laid out in *Elements*, which at a single stroke furnishes a new way to explore its important linguistic implications clearly, while also laying to rest any fears that this framework itself might be formally incoherent. It is not. On the contrary, in the remainder of this monograph, we show that it can be described as a very particular kind of highly structured algebra. In a way, this result should not be surprising at all. Language, and in particular human-language syntax, is all about *structure*, while mathematics, especially algebra, is really also the study of structure.

Beyond this, the most important result of this investigation is the high level of coherence and consistency provided by this algebraic structure: many of the properties of the attendant linguistic theory —the generative process for syntactic objects, the necessity for workspaces in which this generation operates, the requirement and optimality of Minimal Search, and so forth—simply *fall into place* as a direct consequence of the algebraic structure. More strikingly, they follow by *necessity* from the basic mathematical assumptions of the model, in the sense that all these aspects of the model are required for intrinsic structural reasons, and the algebraic structure would break down if these aspects were modified arbitrarily.

The notion that a rigid and strongly constrained algebraic structure (ultimately underwritten by simple mathematical rules) dictates by intrinsic necessity the fundamental laws governing the natural world is a fundamental tenet of modern science. Especially in the domain of the physical sciences, it has guided the entire development of the magnificent edifice of modern theoretical physics. It is the "Miracle Creed" referred to in Chomsky's [34], [35]. It is in our view a striking result of significant importance to explicitly discover that, within the setting of generative linguistics, one can encounter the same level of highly structured coherence governed by precise algebraic conditions, as in modern physics. Not only that, but in fact the algebraic structures involved are exactly those that govern similar phenomena within the context of theoretical physics.

Such uniformity is also part of a basic tenet of modern science: that similar processes arising in different contexts will be governed by the same mathematical structures describing the same fundamental laws. As Galileo famously stated it, the universe is a book that is written in the language of mathematics.[1] Or, as more recently put by Richard Feynman in chapter 2 of his *Lectures on Physics* [64], "the aim is to see *complete nature* as different aspects of *one* set of phenomena" (our emphasis).

This algebraic formulation of Merge and Minimalism is the result of a collaboration between Noam Chomsky, a mathematician and theoretical physicist (Matilde Marcolli), and a computer scientist and computational linguist (Robert Berwick). It introduces a methodology that is of interest in its own right, taking an innovative tack that is quite distinct from the usual and more

[1] "questo grandissimo libro che continuamente ci sta aperto innanzi agli occhi, io dico l'universo. [...] Egli è scritto in lingua matematica," G. Galilei, *Il Saggiatore*, 1623.

familiar formal language theory or axiomatic approaches. So far as we know, it is entirely novel, distinct from any previous attempt to fully formalize Chomsky's theory, including computer implementations, although there have been other attempts to draw on algebraic methods to analyze Minimalism, usually in a very different way.[2] In a certain sense, this monograph brings up to date the algebraic analysis of context-free languages pioneered in the early 1960s by Chomsky and Schüztenberger [39], along with attendant techniques, introducing algebraic tools that have already been successful at tackling similar problems in the description of generative processes in the context of theoretical physics.

0.2 Key Ideas

In the remainder of this introduction, we provide some intuitions behind our algebraic formulation and how Merge and its associated machinery can be turned into mathematics, while following both the abstract and linguistic examples presented in *Elements* [40]. We will not repeat all the details of the linguistic theory of Merge, but refer the reader to *Elements* to follow along with the remainder of this monograph. We have also provided a separate chapter at the end of this monograph, chapter 4, which serves as a glossary regarding algebraic concepts that might be less familiar to some of the readers.

Further, our goal in this introduction will not be to cover the entire algebraic formalization of this framework, but rather to illustrate only in part the general approach that we take and how we arrive at a few of the key results regarding the algebra of Merge; how Minimal Search works in the theory by acting on what is called a Workspace; and the general algebraic character of our results. This turns out to crucially rely on an underlying formalism known in mathematics as *Hopf algebras*.

Why Hopf algebras? Hopf algebras are designed to describe general composition and decomposition operations for various types of mathematical objects, which typically are combinatorial, structured objects. That already suggests that they might be useful to capture language's combinatorial structure. In the case that we will be discussing here, the combinatorial objects are the "syntactic objects" (SOs) and the "workspace" that Merge constructs. The composition and decomposition operations (algebraically, what are known as the *product* and *coproduct*) will look like what is displayed in Figure 0.2, with the

[2] See, e.g., [47] for a similar use of the algebraic structure of magma, developed independently for comparing language and actions; and [192], [105], and [69], among others, for Minimalist computer implementations.

product assembling syntactic objects into Workspaces and the corresponding coproduct extracting what are called in current theory “accessible terms” from a workspace for further action by Merge.

A bit more detail about this composition/decomposition process is in order here. As the left side of Figure 0.2 shows, the product has *two* channels that serve as input and *one* that serves as output. In contrast, as shown on the right side of the diagram, the coproduct has one channel of input (a single syntactic object) and two channels of output (the two parts of the decomposition), where we use a *sum* of the outputs as a bookkeeping device to keep track of all the corresponding possibilities for the extraction of accessible terms (here showing one hierarchical object decomposed seven ways).

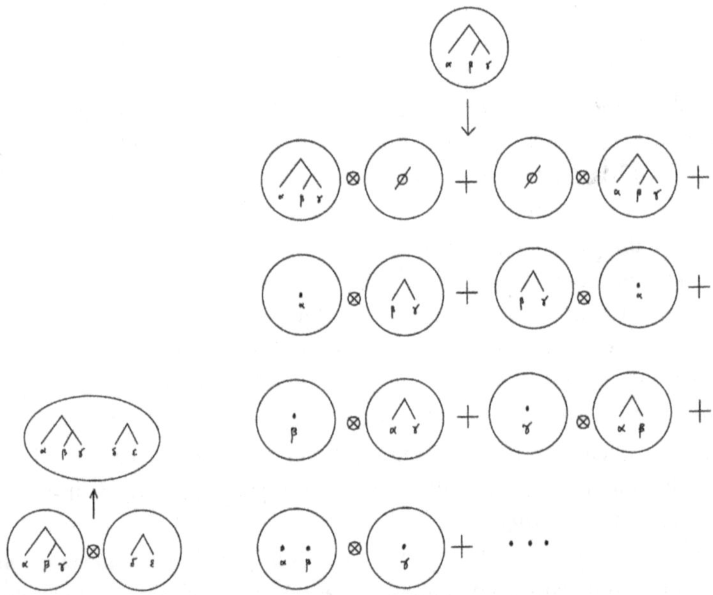

Figure 0.2
Assembling workspaces and extracting accessible terms as product and coproduct in a Hopf algebra.

To begin, we align the discussion that follows with the account in [40] that provides the definition of Merge as the single syntactic “structure building” operation. In the current Minimalist framework, sentences are built by successive applications of Merge (a derivation) starting from some initial set of syntactic features and a lexicon, but these operations take place only within a kind of “computational scratchpad,” called a *workspace* (WS), which represents available computational resources. In the *Elements* reference, the notation used for this is $WS = [P_1, \ldots, P_n]$, where each P_i is itself a syntactic object—either a lexical item or a syntactic object constructed by a previous application of Merge.

Readers familiar with Minimalism may recall that the "simplest" intuition behind this notion of Merge is that it is "just" binary set formation; for instance, merging two lexical items like *the* and *apple* yields the (unordered, binary) set, $\{the, apple\}$—a new syntactic object. Given the workspace idea, then "all" Merge does is act on a workspace W and output a new, modified workspace W', with the newly constructed set added. More generally, if we start with the workspace $[the, apple, ate]$, then the first application of Merge could put together *the* and *apple*, yielding a new syntactic object $\{the, apple\}$ and a new workspace $[\{the, apple\}, ate]$, while a second Merge application combining *ate* with the previously formed syntactic object yields

$$[\{ate, \{the, apple\}\}];$$

namely, the workspace consisting of a single syntactic object $\{ate, \{the, apple\}\}$. Importantly, recall that these set-notated syntactic objects themselves have no "left to right" linear order—the Merge operation that assembles syntactic objects is *nonassociative* and *commutative*. Rather, following Figure 0.1, linear order, as seen in any particular human language, is imposed by the Externalization process, the spell-out yielding speech/sign, or their inverse (parsing).

All this can be translated into mathematical form in a straightforward way. We introduce syntactic objects in section 1.1.2 of chapter 1. We start with a set $\mathcal{SO}_0$ of lexical items and recursively form the set $\mathcal{SO}$ of syntactic objects through repeated applications of a single commutative, nonassociative operation $\mathfrak{M}$, the binary Merge operation

$$\mathfrak{M}(\alpha,\beta) = \{\alpha,\beta\}.$$

This yields what mathematicians call the "free magma" generated by the set $\mathcal{SO}_0$.[3] More precisely, we write this generative process, where we construct the set of all possible Syntactic Objects over this set, $\mathcal{SO}$, by repeated application of Merge operating freely, as:

$$\mathcal{SO} = \text{Magma}_{na,c}(\mathcal{SO}_0, \mathfrak{M}),$$

[3] See [47] for a formulation of Merge as binary set formation using a similar algebraic approach. They note as we do: "The strong compositionality of language requires a particular formalism, a magma, to describe the algebraic structure corresponding to the set of hierarchical structures underlying sentences."

where the subscripts indicate that the magma is nonassociative and commutative; see section 1.1.2, and especially Definition 1.1.1 and equation (1.1.2). We also show that the set SO of syntactic objects obtained in this way is canonically identified with the set $\mathfrak{T}_{SO_0}$ of *abstract binary rooted trees*, with leaves labeled by the set SO_0. Here, *abstract* means that the trees are *nonplanar*—as we noted previously, they look like Calder mobiles—and mathematically, we say that they do not have an assigned planar embedding, reflecting the fact that there is no left-to-right ordering (e.g., word ordering) of the leaves of the trees, as required by Strong Minimalist Thesis (SMT) theory (which we also refer to as New Minimalism). A *planar embedding* (and hence word order) is imposed by the process of Externalization, an important matter to which we return in detail in later chapters.

If syntactic objects are nonplanar trees, then what about workspaces? In section 1.2, where we introduce the notion of a workspace, we use the notation $\mathfrak{F}_{SO_0}$ for the *set* of workspaces, mathematically identified with the set of *binary forests* whose connected components are trees in $\mathfrak{T}_{SO_0}$. This is equivalent to the linguistic description of workspaces as *multisets* of syntactic objects. The multiset possibility means that there can be repetitions of the same syntactic object in a workspace.

In the linguistic theory, this repetition is necessary because Merge—and in particular what is called Internal Merge—needs this kind of forest structure to locate elements of the syntactic objects called *accessible terms* by Minimal Search in order to form copies (what replaces movement in older accounts). We show that workspaces are in fact *necessary* for structural mathematical reasons. Our notation for a workspace, $F = T_1 \sqcup \cdots \sqcup T_n$, corresponds to the notation $WS = [P_1, \ldots, P_n]$ used in [40] or the notation $WS = [\mathrm{SO}_1, \ldots, \mathrm{SO}_n]$ used in Chomsky's more recent work (e.g., [29] and [31]).

With our notation, one can view the disjoint union operation $\sqcup$ that collects the syntactic objects $T_i \in \mathfrak{T}_{SO_0}$ to form the forest $F \in \mathfrak{F}_{SO_0}$ as a multiplication operation that works in combination with another comultiplication operation that disassembles workspaces into their accessible terms. The necessity for workspaces is then dictated by the algebraic coherence conditions on these two operations. As we discuss later in this book, the use of workspaces is one of the *key differences* between the earlier versions of Minimalism and Merge described in [22], as opposed to later versions, as in *Bare Phrase Structure* [23] (see also [26] and [28]); the algebra makes this difference completely clear.

With this algebra in hand, we can next turn to the important concept of *Merge acting on a workspace to yield a new workspace*—a derivation. Here, we will need not just syntactic objects and workspaces, but also the accessible terms of syntactic objects: the substructures accessible for further computation via Merge.

Why do we need this? Merge builds structure, but then it also must take it apart to find accessible terms to manipulate, to reproduce the effects of movement via copying. For this, we need the algebra associated with products and coproducts and have these work hand in hand, so these two operations must be *compatible*. As suggested previously, this is something like multiplication: a familiar multiplicative product takes two terms and operates on them to produce a single output, so one writes a product operation (on a vector space $\mathcal{V}$) as a map $m : \mathcal{V} \otimes \mathcal{V} \to \mathcal{V}$, with the "left and right channels" of the tensor product space $\mathcal{V} \otimes \mathcal{V}$ representing the two inputs to the multiplication $x \otimes y \mapsto m(x, y)$.

Here (and elsewhere), we use vector spaces as a convenient bookkeeping device to compile a list of all the decomposition possibilities–all the accessible terms–as a formal sum. Here's how it works: a coproduct takes one term and can find all the ways of splitting it apart, so one writes a coproduct (on a vector space $\mathcal{V}$) as a map $\Delta : \mathcal{V} \to \mathcal{V} \otimes \mathcal{V}$. Note that while it may seem unnecessary to consider vector spaces when defining a product that might just as well be defined simply on the elements of a set, it becomes *necessary* in the case of coproducts, since in general, there is no unique way of decomposing an object, but rather a list of possibilities that one can write as a sum. Hence, our approach requires a vector space structure: this is no serious issue, as any set X has an associated vector space $\mathcal{V}(X)$ for which the elements of the set X form a linear basis; namely, $\mathcal{V}(X)$ consists of all the formal linear combinations of elements of X. We adopt this simple and commonly used construction and associate with the set of all workspaces the vector space of formal linear combinations (i.e., formal sums) of them.

When it comes to syntactic objects and workspaces, a simple way of splitting apart the workspaces would be to again separate them into the individual syntactic objects that compose them. In algebraic terms, this type of coproduct operation would correspond to assigning the coproduct of each tree to be $\Delta(T) = T \otimes 1 + 1 \otimes T$ and, for a forest, $F = \sqcup_i T_i$ taking $\Delta(F) = \prod_i \Delta(T_i)$. These operations of product and coproduct *do* in fact satisfy compatibility and yield a *Hopf algebra*.[4]

However, this is not a very interesting coproduct: defined in this way, $\Delta(F)$ simply produces partitions of the workspace. From the point of view of the linguistic theory we need, with only these partitions available, we would lose

[4] The algebra $\mathbb{Q}[x]$ with the coproduct $\Delta(x) = x \otimes 1 + 1 \otimes x$ is known as the Hopf algebra of the additive group $\mathbb{G}_a$.

completely Internal Merge, which is required to replicate the copy theory of movement that we have assumed. There is, however, a much better and more natural choice of a coproduct operation, which still satisfies the required algebraic compatibility with the product that assembles workspaces and admits a formulation of Internal Merge.

Instead of just separating a workspace into its constituent syntactic objects, this coproduct can also access deeper substructures than just the connected components of a workspace. This coproduct not only disassembles the workspace into its connected components, but also disassembles each component (each syntactic object) into its *accessible terms*. This means replacing the simple $\Delta(T) = T \otimes 1 + 1 \otimes T$ with the more elaborate

$$\Delta(T) = \sum_{\underline{v}} F_{\underline{v}} \otimes (T/F_{\underline{v}}),$$

where the terms $F_{\underline{v}}$ are *subforests* (collections of disjoint accessible terms) inside T and the corresponding term $T/F_{\underline{v}}$ is what remains of T if the accessible terms in $F_{\underline{v}}$ are removed (pruned out).

The sum is taken over all such choices of subforests, and it includes the previous terms $T \otimes 1 + 1 \otimes T$ as parts of the sum. This extraction of accessible terms is what makes it possible for Merge to act by combining two of these terms into a new syntactic object and assembling the new workspace from the remaining terms of the disassembled one. In our formulation, the accessible terms of a syntactic object T are the full subtrees T_v, with a root at one of the vertices. A precise mathematical formulation of this operation will be presented in section 1.3 in chapter 1.

With this algebraic structure in hand, one can now account for both External Merge, which combines two components of the workspace, and Internal Merge, which needs to extract accessible terms within a single component. For instance, using the same example of section 3.3.2 of *Elements* [40], we can have a workspace of the form WS = {{was, {eaten, {the, apple}}}, and Internal Merge extracts the accessible term {the, apple} and produces a new workspace WS' = {{the, apple}, {was, {eaten, ~~{the, apple}~~}}}. In our setting, the pair consisting of the extracted accessible term {the, apple} and the remaining term {was, {eaten, ~~{the, apple}~~} (with the ultimately canceled deeper copy of the same accessible term) are exactly what we have in the two channels of the coproduct $\Delta(T)$ described previously.

In this way, we obtain a description of the action of Merge on workspaces that first uses the Hopf algebra coproduct to disassemble the workspace and produce these types of pairs, which are then deposited in the two channels of the coproduct output:

$$\{\text{the, apple}\} \otimes \{\text{eaten}, \sout{\{\text{the, apple}\}}\} .$$

We then use free symmetric $\mathfrak{M}$ to reassemble the new workspaces using the product operation of the Hopf algebra. This automatically also takes care of maintaining the distinction between *repetitions* in the workspace and *copies* produced by the Internal Merge, so it incorporates what is referred to in [40] as the FormCopy function. (This will be discussed in more detail in section 1.3.)

Given this algebraic formulation of Merge, we find that a number of desirable linguistic properties then follow easily as a direct consequence of the tightly constrained algebraic structure. The following subsection highlights the main linguistic results in the remainder of this monograph.

0.3 Summary of Main Linguistic Results

In chapter 1, we introduce the main aspects of the algebraic formulation of Merge and how it works, given the SMT, as outlined previously: we provide a mathematical formulation of syntactic objects, workspaces, the action of Merge on a workspace to yield a new workspace, Minimal Search, Minimal Yield, Resource Restriction and the like. We also propose an algebraic model for the Externalization procedure that follows the syntactic structure formation implemented by free symmetric Merge.

In particular, we provide rigorous mathematical proofs for several important properties outlined in *Elements* [40], which shows that *these properties are directly dictated by the intrinsic necessity of consistency of the algebraic structure*. In particular, we prove the following:

- Workspaces are *necessary* for the consistency of the product and coproduct operations of the Hopf algebra.
- A Merge operation with higher arity than binary would necessarily both overgenerate and undergenerate compared to binary Merge, and is therefore empirically ruled out in favor of binary Merge.
- Merge acting on a workspace satisfies what is referred to in *Elements* [40] as the computationally-motivated "Third Factor Principle" of *Resource Restriction* and *Minimal Yield*.
- Minimal Search is indeed *minimal*, in the sense that it optimizes a cost function expressed in terms of a grading (ranking) of the coproduct, selecting as the optimal leading terms what amounts to Internal Merge and External Merge, while eliminating other possibly linguistically undesirable forms of Merge such as Sideward and Countercyclic Merge.

- Merge is Markovian, in the sense that it determines a Hopf algebra Markov chain. While Sideward Merge is subdominant and negligible in the structure formation process (as shown by Minimal Search), it plays a role in the Markovian property by ensuring a strong connectedness condition that makes the Perron–Frobenius theory applicable to the construction of the transition matrices of the Hopf algebra Markov chain.
- Phase Theory, syntactic head, labeling algorithm, and other aspects needed for semantic interpretation, like FormSet and Theta Theory, also have a natural formulation in our setting involving properties of the Hopf algebra coproduct and the notion of *operad*, another fundamental algebraic structure.
- The core computational structure of free symmetric Merge is the same sort of structure that governs the most basic example of combinatorial Dyson–Schwinger (DS) equations in physics, the equations that recursively solve the least action problem in the physics of fundamental interactions.

Given our formalization of Merge, we can then begin to examine, at least in part, how to analyze two interfaces, Externalization and the Conceptual–Intensional system, in the form of a syntax–semantics interface that we keep as theory-neutral as possible.

In chapter 1, we also compare our model of Externalization to other types of "linearization" algorithms, such as Richard Kayne's Linear Correspondence Axiom (LCA). In our model, what is referred to in linguistics as "linearization" corresponds to imposing a choice of planar embedding on abstract binary rooted trees. To this end, we introduce a precise mathematical definition of a *head function* that abstracts the main properties of syntactic heads as formulated by Chomsky in *Bare Phrase Structure* [23]. Using this formulation, we prove that the LCA is *not* equivalent to Externalization in general.

In chapter 2, we explore the consequences of our algebraic formulation to see whether it can be used to tell us something about the differences between linguistic theories regarding Merge. We carry out a comparative analysis, at the level of the algebraic structures, between the formulation of Minimalism in terms of free symmetric Merge, as in the SMT presented in [40] and in chapter 1 of this monograph, as opposed to older formulations of Minimalism that used features and planar trees. To be specific, we choose for comparison here Edward P. Stabler's Computational Minimalism [192], simply because it is very clearly formalized and widely known due to its relationship to formal languages. We prove the following:

- In this alternative account of Minimalism, Internal and External Merge *cannot* be seen as cases of the same operation, unlike in our setting and the more recent accounts by Chomsky and others.

- Feature checking in Stabler's Minimalism introduces a significant level of additional complexity in the algebraic structure, compared to the current theory. More precisely, the absence of workspaces and the fact that Merge operations are partially defined on domains dictated by label-matching conditions force the use of a different and algebraically more complex Hopf algebra and the selection of a further structure in the form of a family of right-ideal coideals in this Hopf algebra, which keep track of the feature checking problem. These structures are a partial substitute for the Hopf algebra quotients (like those that would normally be associated with DS equations), and their presence is due to the combined effects of the three main factors where Stabler's Minimalism differs from Chomsky's more recent accounts: planar trees, the absence of workspaces, and feature checking.
- This family of right-ideal coideals in the Hopf algebra account for the computational complexity of feature checking in Stabler's Minimalism. We illustrate this by providing a provable, explicit estimate of the growth of the feature checking problem in a chain of derivations when Merge is applied repeatedly. This result shows that, under iterations of Merge, the problems with partially defined operations and domain checking compound, resulting in a rapid growth in the computational complexity. This result agrees with what has been found in axiomatic treatments of this older version of Minimalism, as described, for instance, in [105].

In chapter 3, we advance and discuss a mathematical model for the syntax–semantics interface based on our algebraic theory for syntax. The basic structure that we consider for a syntax–semantics interface consists of a computational process for syntax (given by free symmetric Merge) combined with a relational-topological structure on semantic spaces (realized in various possible models and aimed to be as general and theory-neutral as possible). We show that the problem of assigning semantic values to syntactic structures in a way that satisfies consistency over substructures can be addressed using a mathematical procedure used in theoretical physics, and known as *Bogolyubov preparation* and *Birkhoff factorization*.

To quickly get a sense of how this mechanism works, consider the following very simple example that we discuss in section 3.2.5 of chapter 3. Consider the sentences "France is a republic," "France is hexagonal," and "France is a hexagonal republic". The first two have an immediate and clear semantic interpretation, while the third appears odd in view of the semantic awkwardness of the expression "hexagonal republic", assuming that a polygonal shape is not usually regarded as a property of a form of governance. However, the sentence "France is a hexagonal republic" can be modeled by a syntactic object of the form $T = \{a, \{b, \{c, d\}\}\}$, with a, b, c, d the lexical items *France, is, hexagonal, republic*, respectively. The Hopf algebra coproduct extracts accessible terms from this object and yields terms of the form $T_v \otimes T/T_v$ in the resulting decomposition. These terms include, for example,

$$c \otimes \{a, \{b, d\}\} \quad \text{and} \quad d \otimes \{a, \{b, c\}\},$$

which reproduce as the quotient structures $\{a, \{b, d\}\}$ and $\{a, \{b, c\}\}$ the two sentences "France is a republic" and "France is hexagonal," which have straightforward semantic interpretations, but it also contains terms such as

$$\{c, d\} \otimes \{a, b\}.$$

These terms track the precise place (the extracted term $\{c, d\}$ "hexagonal republic") where the assignment of semantic values runs into problems. The mathematical Birkhoff factorization construction described in chapter 3 (and in additional detail in section 4.6 in chapter 4) is designed to incorporate this type of consistency checking across substructures into a recursively defined function that modifies an initial assignment of semantic values to track its behavior over all substructures. In this way, the interpretation of the substructure *hexagonal republic* can be properly located as the source of the problem of coherent semantic values.

This technique again originates in the context of theoretical physics, where a very similar problem of assigning meaningful physical values to the combinatorial products of a generative process (Feynman graphs) is solved using the same method.

Overall, again we find that once the algebraic structure is in place, many desirable properties can be directly derived by way of structural necessity. We use this to obtain several results, depending on various choices of the model for the semantic spaces that we consider, all within the same algebraic formalism:

- An algorithm for checking the consistency of assignments of semantic values over syntactic substructures (e.g., 0/1 truth value type assignments).
- A simple computational model that can be implemented in a vector space semantics (as in the currently natural language processing method of word embeddings, the simplest being just the familiar example of cosine similarity), as well as in Viterbi (probabilistic) semiring parsing.
- By incorporating rectified linear unit (ReLU) thresholds, a *geometric formulation* related to neural network models in terms of hyperplane arrangements (see [92]).
- A proof that this model avoids known problems that arise in other tensor-product models for language semantics, such as [188]. A more general

comparison with other models, including these tensor models, tree-adjoining grammars (TAGs), and other models inspired by physics, is outlined in section 2.5 in chapter 2.

In chapter 3, we then discuss how the two processes of Externalization (the interface with the Sensory–Motor system) and our syntax–semantics interface (the interface with the Conceptual–Intensional system) can interact and combine. This again can be expressed in mathematical terms, and it is described by a *geometric* model based on moduli spaces of trees with metric structures, as well as the combinatorial object known as the *associahedron.* We can fold into this geometric model various aspects of linearization: Externalization, syntactic parameters, Kayne's LCA, and the like. This model suggests a novel approach for a mathematical theory of syntactic parameters and the geometry of the space of "possible languages," although here we shall put much of this aside for future inquiry.

Chapter 3 then turns to Paul Pietroski's compositional semantics from the perspective of our algebraic model. We prove that, in our model, the following properties hold:

- Merge suffices to determine Pietroski's *Combine* operation.
- Pietroski's account of *predicate saturation* also follows from Merge and our proposed mathematical structure (expressed in terms of the mathematical notion of an *operad*) of the magma of syntactic objects.
- Asymmetrical Pair-Merge is not needed to describe adjunctions and their invisibility to syntax.
- The generative power of adjunctions, among other possible syntactic structures, can be measured in terms of codimensions of strata in polytopes arising from the Tamari order on planar trees.

Next, in section 3.8, we provide a series of explicit examples drawn from *Elements* to illustrate how we can use our model to analyze language-specific properties such as the following:

- Thematic role assignment and the distinction between External Merge and Internal Merge (a dichotomy referred to in section 5.2 of [40]). This dichotomy refers to a segregation of External Merge and Internal Merge in semantics: in *Elements* only External Merge assigns thematic roles.
- Double-object constructions in terms of binary branching, which turns out to be forced by the algebra.

- Obligatory control, as in section 5.3 of *Elements*, is obtained in terms of restrictions to diagonals (modeling the FormCopy operation) in the operad of syntactic objects, and an induced form of the Hopf algebra coproduct.

In section 3.9, we consider instead the Heim–Kratzer (HK) model of semantics [93]. We show that it is possible to *topologize* HK semantics, in order to make it compatible with our model of the syntax–semantics interface. This can be done by assigning to the inductive construction of types in HK semantics and the associated sets of functions between types, an additional structure of topological spaces, also built inductively. Although the inductive construction does not fully preserve all the properties of these topological spaces, it suffices to develop a good formalism for both Boolean and Viterbi parsing. We also show that the compatibility with our formalism is even better if one considers an extension of HK semantics where the $\{0, 1\}$ truth values are replaced by fuzzy $[0, 1]$-values representing confidence levels.

At the conclusion of Chapter 3, we analyze the current form of neural network transformer architectures and attention modules used in the setting of large language models (LLMs). The main results we obtain are the following:

- We prove the compatibility of LLMs with generative linguistics in general, contrary to some existing statements in the literature. This is because we can prove that attention modules satisfy the same algebraic structure as our basic model of syntax–semantic interface, demonstrating that LLMs do not in any way "invalidate" generative linguistic models.
- We show that LLMs can be seen as partial solvers of an *inverse problem* of reconstructing syntactic structure from the image of syntax inside semantics through the syntax–semantics interface. Consequently, recent results that have recovered syntactic structure from LLMs (at least in part) are, in fact, completely to be expected; although full recovery of syntax might turn out to be very difficult computationally. See [134] for a recent example of the possible detection of certain syntactic relations in LLMs.

0.4 Acknowledgment

We especially would like to thank Riny Huijbregts for his many comments and suggestions, for his frequent helpful feedback and advice, and for always being available to answer our many questions. We also benefited from comments, questions, and conversations with many other people, including Kurusch Ebrahimi-Fard, Martin Everaert, Sandiway Fong, Danny Fox, Jeffrey Heinz, Norbert Hornstein, Tim Hunter, Richard Kayne, Minhyong Kim, Joachim Kock, Dominique Manchon, Jack Morava, Paul Pietroski, Jon Rawski,

Barry Schein, Juan Pablo Vigneaux, and Yiannis Vlassopoulos. The first author received support for this work from the National Science Foundation grant DMS-2104330, the FQXi Foundation grants FQXi-RFP-1 804 and FQXi-RFP-CPW-2014, SVCF grant 2020-224047, and the Center for Evolutionary Science at the California Institute of Technology.

1 The Mathematical Structure of Syntactic Merge

1.1 A Mathematical Model of Syntactic Merge

We consider here the formulation of Minimalism as presented in [28], [29], and [31], and in the *Elements* text [40], with a fundamental free symmetric Merge operation of binary set formation. We refer the reader especially to *Elements* for a detailed account of the underlying linguistic theory. Our analysis here will primarily focus on the underlying mathematical structure of that theory.

Our formulation here of syntactic objects and of the action of Merge on workspaces is the same as described by Chomsky in [28] and [29]. What we make explicit here are the following aspects of this account:

- The algebraic structure (magma) of syntactic objects, used implicitly in [28], [29], [31]
- The structural aspects of the sequence of workspaces constructed during derivations, namely product and coproduct operations, also used implicitly in [28] and [29] in the description of the Merge action
- How to write the Merge action in such a way that the role of these algebraic structures manifestly appears

We stress again that in order to make it easier for the readers to align our notation and terminology here with [28], [29], [31], and especially with [40], we outline in section 1.1.1 the precise matching of notation between this monograph and the "*Merge and the Strong Minimalist Thesis*" book [40] in the Cambridge *Elements* series.

1.1.1 Notational Conventions

Since in this monograph we discuss the mathematical structure of Merge, we have chosen a notation that is standard in mathematics for the various objects that we are considering. This is necessary in order to articulate mathematical arguments where the terminology and notation is already well established and the linguistic terminology and usage sometimes has been less clear.

An obvious drawback of this choice is that mathematical notation does not always agree with the notation that is well established in generative linguistics. Consequently, in this subsection we outline the basic notational translations between this text and the *Elements* text [40], which should be regarded as the main companion text, where the *same theory* is introduced in a purely linguistic setting.

- **Syntactic objects**: In section 1.1.2, where we introduce syntactic objects, we use the notation $\mathcal{SO}_0$ for the *set of lexical items and syntactic features* and both the notation $\mathcal{SO}$ and $\mathfrak{T}_{\mathcal{SO}_0}$ for the *set (magma) of syntactic objects*. As will become clear in the following, we will use the first notation when we emphasize the algebraic structure of the set of syntactic objects and the second notation when we emphasize the identification with the set of binary rooted trees. The first notation, $\mathcal{SO}$, follows [29], where the individual syntactic objects are denoted by $\mathrm{SO}_i, \mathrm{SO}_j \in \mathcal{SO}$. The second notation, $\mathfrak{T}_{\mathcal{SO}_0}$, is the standard mathematical notation for the set (magma) of abstract binary rooted trees with leaves labeled by the set $\mathcal{SO}_0$. By "binary" rooted trees we will always mean "full binary" (no non-branching vertices) unless otherwise stated. The two sets $\mathcal{SO}$ and $\mathfrak{T}_{\mathcal{SO}_0}$ can be identified, so we mostly use the notation $T \in \mathfrak{T}_{\mathcal{SO}_0}$ for syntactic objects. Also, the set notation such as $\{\alpha, \{\beta, \gamma\}\}$ is used in [40], as well as in the previous literature such as [28], [29], and [31], for syntactic objects. In mathematics, this notation and the tree notation are used equivalently:

  $\{\alpha, \{\beta, \gamma\}\} =$ 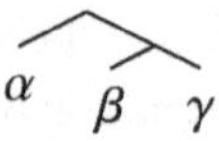

  with the important caveat that these are *not* planar trees, so that, for instance,

  $\{\alpha, \{\beta, \gamma\}\} =$ 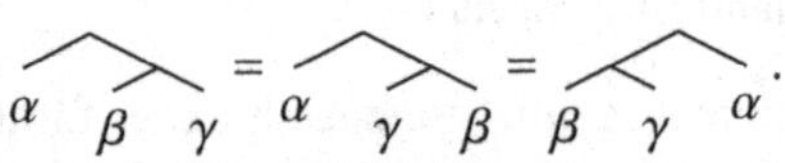

  We do mention explicitly throughout the text when it happens that trees are abstract (not planarly embedded) $T \in \mathfrak{T}_{\mathcal{SO}_0}$ (the case of syntactic objects), and when they come endowed with a planar embedding $T^{\pi} \in \mathfrak{T}^{pl}_{\mathcal{SO}_0}$. The latter case never occurs in core Merge, but it does occur in Externalization and in other "linearization" procedures (*planarization*, in mathematical terminology).

- **Workspaces**: In section 1.2, where we introduce the notion of a workspace, we use the notation $\mathfrak{F}_{\mathcal{SO}_0}$ for the set of workspaces, mathematically identified with the set of *binary forests* whose connected components are trees in $\mathfrak{T}_{SO_0}$. This is equivalent to the linguistic description of workspaces as multisets of syntactic objects. Our notation $F = T_1 \sqcup \cdots \sqcup T_n$ for a workspace corresponds to the notation $WS = [P_1, \ldots, P_n]$ used in [40] or the notation $WS = [\mathrm{SO}_1, \ldots, \mathrm{SO}_n]$ used in [29] and [31]. Note that we use the term *workspaces* here because while at any one point in a derivation there is a single workspace, a sequence of workspaces is constructed in the course of a derivation. As we explain later in this chapter, Merge is a transformation of the set of workspaces, which means that it takes one workspace as its input and it outputs a new workspace. In the course of a derivation, one has several applications of Merge; hence, an initial workspace is transformed by the first application of Merge into the next one, which is in turn transformed by a second Merge operation, and so on.

- **Accessible terms**: These are the parts of a workspace that are accessible for computation; namely, they can be targeted by the action of Merge. They are defined in section 3.3.2 of *Elements* [40] (see equation (14) in *Elements* [40]), where one distinguishes between "terms" and "members" of a workspace. In our terminology, the accessible terms of a syntactic object T are the subtrees T_v, with v a non-root vertex of T and T_v denoting the subtree below v in T, see Definition 1.2.2 for a more precise discussion. The description of accessible terms of the workspace in [40] corresponds here to our (1.2.4). So, as in the example that follows (14) in [40], if we have a workspace of the form $T \sqcup \mathfrak{M}(T_1, T_2)$ (or $[a, \{b, c\}]$ in the notation of [40]), then the syntactic objects T and $T' = \mathfrak{M}(T_1, T_2)$ (i.e., a and $\{b, c\}$) are "members" in the terminology of [40] (*connected components* in mathematical terminology), while the syntactic objects T_1 and T_2 (i.e., b and c) are *accessible terms*.

- **Action on workspaces**: In [40], the notation (see equations (10) and (12) of [40])

$$\mathrm{Merge}(P_1, P_2, WS) = WS' = [\{P_1, P_2\}, \ldots]$$

is used for the action of Merge on workspaces, which we discuss in section 1.3. We write this same action equivalently as a map $\mathfrak{M}_{S,S'} : \mathfrak{F}_{SO_0} \to \mathfrak{F}_{SO_0}$, for a pair S, S' of syntactic objects, as:

$$\mathfrak{M}_{S,S'}(F) = F' = \mathfrak{M}(T_v, T_{v'}) \sqcup T/T_v \sqcup T'/T_{v'} \sqcup \hat{F}\,.$$

Here, $F = T \sqcup T' \sqcup \hat{F}$ and T, T' are the components (members) containing accessible terms that match the objects S, S', and $\hat{F}$ are all the remaining (unchanged) components. In this case

$$\mathfrak{M}(T_v, T_{v'}) = \{T_v, T_{v'}\} =$$ 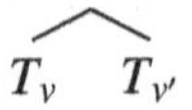

matches the term $\{P_1, P_2\}$ in the notation of [40] while our $T/T_v \sqcup T'/T_{v'} \sqcup \hat{F}$ is just subsumed into the ... in $[\{P_1, P_2\}, \ldots]$ of *Elements* [40]. We will provide more details on this way of writing the action of Merge in section 1.3.
In fact, in equation (10) of *Elements* [40], the more general form of a hypothetical m-ary Merge is described as:

$$\text{Merge}(P_1, \ldots, P_m, WS) = WS' = [\{P_1, \ldots, P_m\}, \ldots] .$$

The operation written here as $(P_1, \ldots, P_m) \mapsto \{P_1, \ldots, P_m\}$ is the same one that we write as

$$\mathcal{B}(T_1 \sqcup \cdots \sqcup T_m) = \qquad (1.1.1)$$ 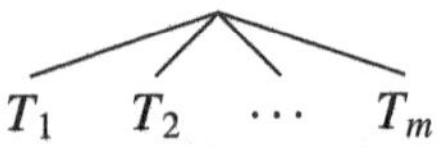

in (1.3.5) in Definition 1.3.2. We follow the notation $\mathcal{B}$ for this operation, because it is commonly adopted in the mathematical physics literature. We will discuss the case of a hypothetical m-ary Merge in section 1.11, and the FormSet operation, which also involves, in a different way, the grafting operation (1.1.1) in section 1.16.

We now outline more precisely how the various objects listed above (syntactic objects, workspaces, accessible terms, and the action of Merge) should be understood from a mathematical perspective.

1.1.2 Syntactic Objects and the Merge Magma

As in [28] and [29], one considers, as the starting point in the construction of the set of *syntactic objects* $\mathcal{SO}$, an initial set, which we denote by $\mathcal{SO}_0$, consisting of lexical items and syntactic features.

Definition 1.1.1. *The set* $\mathcal{SO}$ *of syntactic objects is the free, non-associative, commutative magma over the set* $\mathcal{SO}_0$,

$$\mathcal{SO} = \text{Magma}_{na,c}(\mathcal{SO}_0, \mathfrak{M}) , \qquad (1.1.2)$$

with the binary Merge operation:

$$\mathfrak{M}(\alpha, \beta) = \{\alpha, \beta\} . \qquad (1.1.3)$$

This means that, as described by Chomsky in [28] and [29], the set $\mathcal{SO}$ is obtained from an initial set $\mathcal{SO}_0$ through iterations of the Merge operation (1.1.3). This procedure generates elements of $\mathcal{SO}$ of the form $\{\alpha, \beta\}$ or $\{\alpha, \{\beta, \gamma\}\}$, for

$\alpha,\beta,\gamma \in \mathcal{SO}_0$, and so on. The Merge operation (1.1.3) acts on the set $\mathcal{SO}$, yielding the structure of nonassociative, commutative magma.[5]

Remark 1.1.2. The description of the set $\mathcal{SO}$ of syntactic objects given in Definition 1.1.1 provides an identification

$$\mathcal{SO} \simeq \mathfrak{T}_{\mathcal{SO}_0} \tag{1.1.4}$$

of the set of syntactic objects with the set of binary, nonplanar, rooted trees, with leaves labeled by elements of $\mathcal{SO}_0$.

By *nonplanar* we mean that we regard objects $T \in \mathfrak{T}_{\mathcal{SO}_0}$ as abstract trees, without fixing a choice of a planar embedding. This implies that there is no choice of a linear ordering on the leaves of such trees. As in [28] and [29], word order (i.e., the linearly ordered externalized form of sentences) is considered part of the Externalization process, not of the core computational mechanism of syntax given by Merge.

1.1.3 On the Use of the Tree Notation

In mathematics, there are two different *notations* that can be used for the objects $T \in \mathfrak{T}_{\mathcal{SO}_0}$, the nonplanar (also called abstract) binary rooted trees with leaves labeled by elements of the set $\mathcal{SO}_0$. One *notation*, referred to as the "set notation," represents these objects as balanced bracketed expressions such as

$$\{\alpha, \{\beta, \gamma\}\},$$

with $\alpha,\beta,\gamma \in \mathcal{SO}_0$. The other *notation*, referred to as the "tree notation," represents *the same objects* graphically as trees, such as

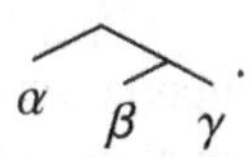

in this example, where no planar embedding of the tree is assigned, the tree represented here is *the same* as (for example)

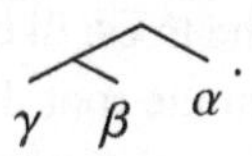

These two choices, set or tree notation, are *completely equivalent*, in the sense that the set of balanced bracketed expressions described here is *canonically isomorphic* to the set of nonplanar binary rooted trees; therefore, these two

[5] The existence of magma structures like these in generative linguistics was also recently observed independently in [47], in a somewhat different context.

notations are, for all mathematical purposes, *completely indistinguishable*. In mathematics, set notation is very rarely used because it would be too cumbersome and unnecessarily complicated to write any proof in that notation. So tree notation is the standard default. This is the default that we will follow in this book, for exactly the same reasons.

In the linguistic literature on Merge and the Strong Minimalist Thesis (SMT), in particular, for instance, in Chomsky's [29], it is pointed out that tree notation may be confusing and that the use of set notation is preferable.

There are indeed two main reasons why tree notation may be confusing. One of them is very simple and the second one is more subtle, and we will discuss here how to avoid both problems. Indeed, the problems are correctly identified in these warnings in [29] and elsewhere, but we argue here that both issues are better addressed, not in terms of a change of notation since the two notations describe the same mathematical objects, but in terms of which *algebraic structures* one considers for this set of objects.

The two main potential sources of confusion when using tree notation for syntactic objects are the following

1. *Planar versus abstract trees*: When one uses set notation, there is a completely clear and immediate way of distinguishing planar and nonplanar (abstract) trees. Namely one uses the $\{\, ,\, \}$ bracket (as in unordered sets) in the case of abstract trees, where no planar structure is assigned; and the $(\, ,\,)$ bracket (as in lists) for planar trees, where the set of leaves is ordered. When using tree notation, one is forced to draw the tree on the page, hence using a planar embedding, while having to remember that the embedding is not part of the data.
2. *Operations on trees*: A more important caveat in Chomsky's [29] and elsewhere, about the use of tree notation refers to the fact that drawing trees may suggest putative "operations" on trees that might not correspond to viable linguistic operations (especially referring to operations that act via insertions in lower levels of the trees, like Countercyclic and Late Merge, or grafting of trees away from the root, like Parallel Merge). As we discuss in section 1.1.3.2, such insertion operations do exist as mathematical operations on trees, but they belong to *different algebraic structures* on the set of trees, and a direct analysis of the corresponding algebraic structures reveals whether these are genuine extensions of Merge and whether they are compatible with the basic structure of Merge in SMT. Thus, whether such operations do occur is a consequence of their algebraic structure, regardless of what notation is used to represent trees.

We discuss the first, simpler, issue in section 1.1.3.1, and the second, more subtle issue in section 1.1.3.2.

1.1.3.1 Planarity and lists versus sets In the formulation given, in Definition 1.1.1 and Remark 1.1.2, we identify syntactic objects with *nonplanar* binary rooted finite trees with leaves labeled by the set $\mathcal{SO}_0$, where "nonplanar" means that no choice of a planar embedding is taken for the tree. These are often also referred to as "abstract trees." This is the usual mathematical description of the elements of the free, nonassociative, commutative magma on a given set.

While *planar trees* (trees with a choice of a planar embedding) and *abstract trees* might at first appear to be similar mathematical objects, their combinatorial properties are very different, and this accounts for several significant differences, in linguistics, between older forms of Minimalism and the newer form that we discuss in this chapter and monograph. This is discussed more explicitly in Chapter 2, based on our paper [139].

In the linguistics literature, the passage from planar trees in the older versions of Minimalism to abstract trees is usually discussed using the terminology *sets* to refer to the abstract trees as elements of the free, nonassociative, commutative magma. The reason for the use of this terminology is that in dropping the planar structure, one replaces an identification of the set of leaves with parenthesized *lists* (*ordered sets*, often referred to in the linguistics literature as "strings") with just sets (in fact, more precisely *multisets*). In order to avoid the conflict of terminology between sets and multisets, we prefer here to follow the standard mathematical terminology and refer to the syntactic objects as abstract binary rooted trees (with no assigned planar embedding).

It is important to note that, because all the trees are binary, the clash of terminology between sets and multisets is very mild when one considers syntactic objects. Indeed, since trees are binary rather than n-ary, with some $n \geq 3$, the only repetitions of labels that give rise to multisets can be on two consecutive ones, so there is an unambiguous way of labeling the same objects by sets. For example, a multiset of the form $\{\{a, a\}, b, \{c, d\}\}$ can be written equivalently as the set $\{\{a\}, b, \{c, d\}\}$, with the convention that a set with the form $\{a\}$ stands for the abstract tree 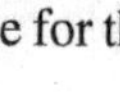
.

However, even with binary trees, the clash of terminology between sets and multisets becomes much more problematic when it comes to describing *workspaces*, as we will see in section 1.2. These are genuinely multisets that *do not have* an equivalent description as sets. Hence, the mathematically correct notion to use for them is *binary forests* (disjoint unions of a finite collection of abstract binary rooted trees), rather than *sets*. Indeed, forests are multisets where the same tree (the same syntactic object) may appear more than once, what in linguistics is called *repetitions*. This is to be expected, as the same syntactic object may be used repeatedly, in different ways, in the course of a derivation. This is another of the reasons why we will not be using the "sets" terminology that is more common in the linguistics literature, and we prefer to adopt the mathematical notation of trees (with no planar structure) and forests.

1.1.3.2 Algebra and operations on trees As we discussed, the more subtle and more serious concern expressed in [29] about the use of the tree notation for syntactic objects lies in the fact that several types of mathematically well defined operations on trees are possible, and are suggested by the geometric visualizations of trees, which do not necessarily correspond to viable linguistic operations. One should note, however, that the change of notation from "tree notation" to "set notation" may be suggestive of which operations are preferable, in the sense that it can make certain operations more transparent and others more cumbersome to write, but since the underlying objects are still the same in both notations, the corresponding operations are still there and well defined. We want to argue here that one can in fact explicitly analyze the kind of algebraic structure that various types of operations on trees satisfy, and compare them with the algebraic structure describing the free symmetric Merge operation in SMT, and on the basis of this comparison, we argue about the relevance and compatibility of such operations within the mathematical structure of the linguistic model.

Some of these operations have been adapted to linguistic use: for example, some forms of Countercyclic Merge or Late Merge that involve operations that graft and grow trees via insertions at non-root vertices. Operations of this form are also considered in tree-adjoining grammars. However, these other proposed forms of Merge, involving this type of insertion operations on trees, have been criticized in [29] and elsewhere on linguistic grounds (as violations of the "extension condition"), and are not seen as part of the Merge formulation in the SMT. We will be analyzing these operations explicitly in section 1.7.

The main important point here is that the set $\mathfrak{T}_{SO_0}$ of abstract binary rooted trees, as well as the associated set $\mathfrak{F}_{SO_0}$ of forests that we will be discussing in section 1.2 and subsequently, *are not just sets*. They carry algebraic structures. For example, we have already seen that $\mathfrak{T}_{SO_0}$ is a free nonassociative commutative magma. Additional algebraic structures will be described in section 1.2 and subsequently, involving the notion of Hopf algebra. One of our main goals will be to show that *this algebraic structure* is responsible for the form and properties of the free symmetric Merge and its action on workspaces within the framework of the SMT.

Then the key point, regarding several possible other types of operations on trees, such as those involving insertions and grafting at internal vertices, is that they can be defined, and *they have their own algebraic structure*. The relation between different algebraic properties of different types of tree operations determines whether they are mutually compatible and whether some of these operations are genuine extensions of Merge or are in fact obtainable from the usual Merge operation.

Thus, we will follow the principle that the issues of concern are independent of the choice of *notation* with which one represents the syntactic objects, and depends entirely on which *algebraic structure* is considered. We will return to this point several times throughout the book, especially in section 1.7 with the discussion of the insertion operations that were the main reason of concern in the comments on the "tree notation" in [29].

Another related concern expressed in [29] is the fact that the tree notation might erroneously give the impression that "there has to be something at the root of the tree", but we hope that the reader will understand that there is absolutely no mathematical ground for this impression, as the magma that generates the objects of $\mathfrak{T}_{SO_0}$ creates *no* root labels, and in fact no labels at any non-leaf vertices. We will be discussing in section 1.15 how one can introduce labeling algorithms and subdomains of the set $\mathfrak{T}_{SO_0}$ on which labelling can happen, but the syntactic objects in $SO = \mathfrak{T}_{SO_0}$, by definition, have nothing at all attached to the root, nor there is any reason to expect them to do so.

1.2 Workspaces: Product and Coproduct

We next introduce workspaces, as in [28] and [29], and the action of Merge on workspaces. We first introduce workspaces with a bialgebra structure related to the combination of workspaces and the extraction of accessible terms with the cancellation of copies.

Definition 1.2.1. *Workspaces are nonempty finite (multi)sets of syntactic objects. The identification* (1.1.4) *between syntactic objects and binary, nonplanar, rooted trees, with leaves labeled by elements of SO_0, induces an identification,*

$$\mathcal{WS} \simeq \mathfrak{F}_{SO_0}, \tag{1.2.1}$$

between the set $\mathcal{WS}$ of all workspaces and the set $\mathfrak{F}_{SO_0}$ of binary nonplanar forests (disjoint unions of binary, nonplanar, rooted trees) with leaf labels in SO_0.

Note that, with this definition of workspaces, we allow the presence of repeated copies of the same syntactic object in a workspace, since a forest can have multiple connected components that are isomorphic to the same tree. This is needed for the two operations "combination of workspaces" and "extraction of accessible terms" described next to be well defined, as the result of these two operations can produce repeated copies of the same tree, even when starting with a forest that has none.

Definition 1.2.2. *Given a binary, nonplanar, rooted tree $T \in \mathfrak{T}_{SO_0}$, let $V_{int}(T)$ denote the set of all internal (non-root) vertices of T. For $v \in V_{int}(T)$, let $T_v \subset T$ denote the subtree consisting of v and all its descendants. Let $L_v = L(T_v)$ be the set of leaves of T_v. The set of accessible terms of T is given by*

$$Acc(T) = \{L_v = L(T_v) \mid v \in V_{int}(T)\}, \tag{1.2.2}$$

while we write

$$Acc'(T) = \{L_v = L(T_v) \mid v \in V(T)\} = \{T\} \cup Acc(T). \tag{1.2.3}$$

We can identify, when convenient, $Acc(T)$ with the set of T_v (rather than the sets of leaves $L(T_v)$) and also with the set $V_{int}(T)$ (of corresponding root vertices). For a workspace given by a forest $F = \sqcup_a T_a \in \mathfrak{F}_{SO_0}$, the set of accessible terms is

$$\alpha(F) := Acc(F) = \bigcup_a Acc(T_a). \tag{1.2.4}$$

Thus, we have the total number of vertices of the forest given by the sum

$$\#V(F) = b_0(F) + \#Acc(F), \tag{1.2.5}$$

where $b_0(F)$ is the number of connected components (trees) of the forest F, the number of members of the workspace. We define the size of a workspace F by

$$\sigma(F) := \#Acc'(F) = \#V(F) = b_0(F) + \#Acc(F). \tag{1.2.6}$$

Namely, the number of syntactic objects plus the total number of accessible terms.

Remark 1.2.3. In the linguistics literature, one sometimes prefers to define the number of accessible terms in a workspace F as $\sigma(F)$ of (1.2.6), rather than as $\alpha(F)$ of (1.2.4), also counting the individual components (members) as part of the accessible terms of the workspace.

The size $\sigma(F)$ of the workspace, defined as in (1.2.6) is consistent with [28], [29] and agrees with the definition of size used in [67]. As pointed out by Riny Huijbregts, it may be preferable to consider the effect of Merge on workspaces in terms of the counting of accessible terms $\alpha(F)$, rather than the size $\sigma(F)$. We will discuss and compare the effect on various size-counting in section 1.6.2.

In this section, we describe the *set of all possible workspaces* and the algebraic structure that it carries. In section 1.3, we will introduce the action of Merge on the set of workspaces, as in [40]–namely, an operation that takes *a workspace* as its input and gives *a new workspace* as its output.

The set of workspaces is endowed with two operations: (1) a *product operation* that combines workspaces by taking their union is simply given by the disjoint union on the set of forests (a commutative and associative product, with unit given by the empty forest); and (2) a *coproduct operation*, which provides all the possible extractions of accessible terms. In order to be able to consider all accessible terms simultaneously, one considers, instead of the set $\mathfrak{F}_{SO_0}$, as before, a space comprised of the formal linear combinations of elements in this set. Namely, we denote as $\mathcal{V}(\mathfrak{F}_{SO_0})$ the $\mathbb{Q}$-vector space (or the $\mathbb{Z}$-module) freely generated by elements of $\mathfrak{F}_{SO_0}$, formal linear combinations of binary nonplanar forests with rational (or integer) coefficients, so one can sum over all the possible extractions of accessible terms (see (1.2.8) later in this chapter).

This construction that assigns to a set B the vector space $\mathcal{V}(B)$ with basis B (namely, the vector space of formal linear combinations of elements in B), is standard in mathematics. The vector space operations of linear combinations (sum and scalar multiplication) just act on the coefficients of such linear combinations. The advantage in the case of composition/decomposition operations on the elements of B lies in being able to handle the case where there are multiple possibilities for either composition or decomposition by combining all the different possibilities into a single element given by a formal sum. In a case like ours, where composition is single valued but decomposition is multivalued, the product is a map $m : B \times B \to B$ (extended bilinearly to $m : \mathcal{V}(B) \times \mathcal{V}(B) \to \mathcal{V}(B)$), while the coproduct Δ assigns to elements of B a sum of elements in the vector space $\mathcal{V}(B) \otimes \mathcal{V}(B)$ (extended by linearity to $\Delta : \mathcal{V}(B) \to \mathcal{V}(B) \otimes \mathcal{V}(B)$). The tensor products account for the fact that product multiplication has two inputs and the coproduct has two outputs, the two parts of a decomposition.

The construction of the coproduct is designed to extract from a workspace the accessible terms that are needed for Merge computation. For every accessible term T_v of a syntactic object T, we want $\Delta(T)$ to contain a term of the form $T_v \otimes X$, with X representing what is left of T when the accessible term T_v is extracted. The suitable notion of "what is left" of a combinatorial object when a subobject is extracted should then be describable in terms of a suitable cancellation (i.e., a quotient operation). There are different possibilities for how to construct a quotient T/T_v, illustrated in Figure 1.1. We describe them here and compare their properties in section 1.2.1.

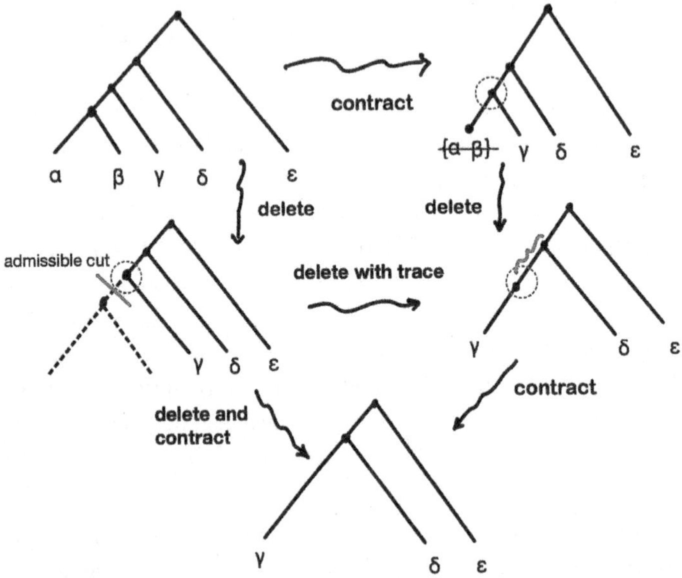

Figure 1.1
Different ways of taking the quotient of a binary rooted tree by a subtree.

The usual way of defining the quotient of T by T_v, in the context of Hopf algebras of rooted trees in mathematics and theoretical physics, is to *contract* (shrink) the entire tree T_v to a single vertex, so the root vertex of T_v becomes a leaf. We write this quotient with the notation $T/^cT_v$ to stress that it is obtained by *contraction*. With this definition of the quotient, in particular, one has $T/^cT = \bullet$, which is the tree consisting of a single root vertex. We extend this definition to the case where we extract a disjoint collection of accessible terms $F_{\underline{v}} = T_{v_1} \sqcup \cdots \sqcup T_{v_n}$ from T.

Definition 1.2.4. Let $T \in \mathfrak{T}_{SO_0}$ be a syntactic object, and let $F_{\underline{v}} = T_{v_1} \sqcup \cdots \sqcup T_{v_n}$ be a forest $F_{\underline{v}} \in \mathfrak{F}_{SO_0}$ where the components T_{v_i} are nonintersecting accessible terms of T. We define $T/^cF_{\underline{v}}$ as the binary rooted tree obtained from T by contracting each T_{v_i} to its root vertex v_i, which becomes a leaf of $T/^cF_{\underline{v}}$. The leaf v_i carries a label of the form $\not{F}_{\underline{v}}$. We call this label the *trace* of T_v.

For example, consider syntactic objects of the form

$T_1 =$ 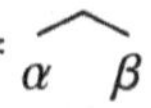

and $T_2 =$ 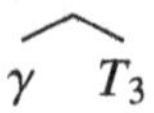

with $T_3 =$ 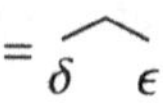

and consider the forest $F_{1,3} = T_1 \sqcup T_3$ with T_1 and T_3 being accessible terms of the syntactic object

$T = \mathfrak{M}(T_1, T_2) =$ 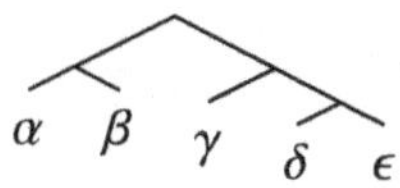

.

The quotient $T/^c F_{1,3}$, in the sense of Definition 1.2.4, is given by

$T/^c F_{1,3} =$ 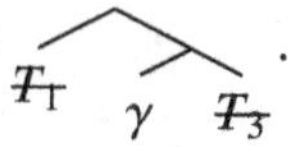

. (1.2.7)

The other way that one can define the remainder of the extraction of an accessible term T_v (or a collection $F_{\underline{v}}$ of accessible terms) from a syntactic object T is by deleting the extracted term. Simple deletion leaves a tree that is no longer necessarily binary, but a unique maximal binary tree is defined by it (obtained via the contraction of some edges), which one can take as the final result of the quotient operation. We use the notation $T/^d F_{\underline{v}}$ for this quotient to stress the fact that it is obtained by *deletion* and to distinguish it from the quotient $T/^c F_{\underline{v}}$ of Definition 1.2.4. When it is clear from the context which quotient is used, we will drop the notation and just write $T/F_{\underline{v}}$ in either case.

Definition 1.2.5. *Given a rooted binary tree (with no assigned planar embedding) $T \in \mathfrak{T}_{SO_0}$ and a collection of disjoint accessible terms $F_{\underline{v}} = T_{v_1} \sqcup \cdots \sqcup T_{v_n}$ in T, consider the rooted binary tree $T \setminus F_{\underline{v}}$, obtained by removing all the trees T_{v_i} from T. There is then a unique maximal rooted binary tree that can be obtained from this complement $T \setminus F_{\underline{v}}$ via the contraction of some edges. That resulting rooted binary tree is what we call the "quotient $T/^d F_{\underline{v}}$."*

In the example (1.2.7), the quotient $T/^d F_{1,3}$ consists of the single lexical item γ. With this definition, we obtain $T/^d T = \emptyset$.

There is an intermediate step that more clearly explains the relation between the quotients by rephrasing the extraction of accessible terms using the notion of "admissible cuts," as in Definition 1.2.6.

Definition 1.2.6. Let $T \in \mathfrak{T}_{SO_0}$ be a syntactic object. An admissible cut C on T is the removal of a collection of edges of T, with the property that no two of them lie on the same path from the root to one of the leaves. An admissible cut separates T into a forest $\pi_C(T)$ consisting of all the subtrees disconnected from the root by the cut, and a tree $\rho_C(T)$ that is the (not necessarily binary) rooted tree comprising the remaining component that still contains the root vertex. We use the notation $T/^{\rho}\pi_C(T) := \rho_C(T)$.

An example of an admissible cut and the corresponding subforest $\pi_C(T) = F_{\underline{v}}$ is illustrated in Figure 1.2. Here one sees that the remaining tree $\rho_C(T)$ is not binary, because it also contains non-branching vertices.

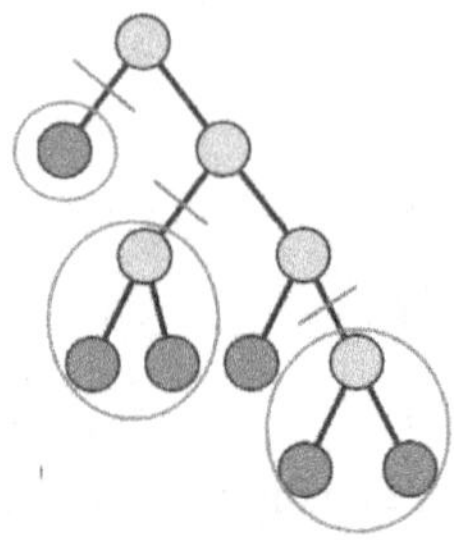

Figure 1.2
A subforest of accessible terms in a syntactic object and the corresponding admissible cut.

The following observation follows directly from this definition of "admissible cut".

Lemma 1.2.7. *Given a syntactic object $T \in \mathfrak{T}_{SO_0}$ the subforests $F_{\underline{v}} \subset T$,*

$$F_{\underline{v}} = T_{v_1} \sqcup \cdots \sqcup T_{v_n},$$

where $T_{v_i} \subset T$ are all accessible terms, disjoint in T, are in bijective correspondence with the admissible cuts of T, with $\pi_C(T) = F_{\underline{v}}$. The quotient operation $T/^d F_{\underline{v}}$ of Definition 1.2.5, then, is the unique maximal binary rooted tree determined by the nonbinary tree $\rho_C(T)$ of the admissible cut with $\pi_C(T) = F_{\underline{v}}$.

We therefore have three slightly different choices of a remainder term, when a collection $F_{\underline{v}}$ of accessible terms is extracted from a syntactic object T:

1. The contraction quotient $T/^c F_{\underline{v}}$;
2. The remainder term $T/^{\rho} F_{\underline{v}} = \rho_C(T)$, corresponding to the admissible cut C with $\pi_C(T) = F_{\underline{v}}$;
3. The deletion quotient $T/^d F_{\underline{v}}$.

These three cases are exactly the three possibilities illustrated in Figure 1.1. In linguistic terms, we should think of these possibilities in the following way, in the case where the extracted term is a single accessible term T_v. The first case corresponds to the FormCopy operation described in [40], where one *copy* is the extracted accessible term T_v and the other is the *deeper copy* that remains visible in $T/^c T_v$ only as a label of the leaf obtained by contracting the subtree. The second case (with the admissible cut made of a single cut edge) has the extracted accessible term $T_v = \pi_C(T)$ and a *trace* represented by a vertex in $\rho_C(T)$. In the third case, the deeper copy is cancelled (while the extracted term T_v remains) and the trace is not explicitly carried over in the resulting syntactic object.

The relation between these three forms of remainder term of the extraction of an accessible term accounts for the fact that the deeper copy (the first case) that remains as a trace (the second case) and is not explicitly visible (in the syntactic object resulting from the third case) is still interpreted at the semantic interface. Thus, in terms of the diagram of Figure 0.1 in the introduction (which we will return to in Chapter 3), we should think of $T/^c F_{\underline{v}}$ as the form that goes to the Conceptual-Intensional (semantic) interface for interpretation, $T/^d F_{\underline{v}}$ as the form that is used in Externalization, and $T/^\rho F_{\underline{v}}$ as the form that the combined process accesses. The extraction of accessible terms and construction of the corresponding remaining objects can be expressed in the form of a coproduct operation, as in Definition 1.2.8.

Definition 1.2.8. Corresponding to these three choices, we can form decompositions of a syntactic object T in the form of a coproduct. For $\omega \in \{c, d, \rho\}$, take

$$\Delta^\omega(T) := T \otimes 1 + 1 \otimes T + \sum_{\underline{v}} F_{\underline{v}} \otimes T/^\omega F_{\underline{v}}, \tag{1.2.8}$$

where the sum is over all the subforests $F_{\underline{v}} \subset T$ consisting of disjoint accessible terms of T. We can absorb the *primitive part*

$$P(T) := T \otimes 1 + 1 \otimes T \tag{1.2.9}$$

into the sum, with the first term corresponding to the full tree T seen as an accessible term of the workspace and the last term as the extraction of the empty subforest. The coproduct (1.2.8) is extended to forests $F = \sqcup_a T_a$ in the form

$$\Delta^\omega(F) = \sqcup_a \Delta(T_a),$$

or equivalently

$$\Delta^\omega(F) = F \otimes 1 + 1 \otimes F + \sum_{F_{\underline{v}} \subset F} F_{\underline{v}} \otimes F/^\omega F_{\underline{v}}.$$

We write (1.2.8) as a sum:

$$\Delta^\omega(F) = \sum_{n \geq 0} \Delta^\omega_{(n)}(F), \tag{1.2.10}$$

where the terms $\Delta^{\omega}_{(n)}$ involve the extraction of a subforest given by a collection of n accessible terms in $\mathrm{Acc}'(F)$. It is also customary to use the notation

$$\tilde{\Delta}^{\omega}(F) := \Delta^{\omega}(F) - (F \otimes 1 + 1 \otimes F) = \sum_{F_{\underline{v}} \subset F} F_{\underline{v}} \otimes F/^{\omega} F_{\underline{v}} \tag{1.2.11}$$

for the non-primitive part of the coproduct.

While only the terms $\Delta^{\omega}_{(1)}(F)$ and $\Delta^{\omega}_{(2)}(F)$ are used in the action of Merge (see section 1.3), all the higher terms $\Delta^{\omega}_{(n)}(F)$ are required for the algebraic properties of the coproduct (see section 1.2.1) and will play a role in another linguistic operation that is a necessary part of the Minimalist model, called FormSet, which we will discuss in section 1.16.

Remark 1.2.9. As we explain in more detail later in this chapter, the coproduct Δ^c is well defined as given in (1.2.8) and is in fact the restriction to nonplanar binary rooted tree of the well-studied coproduct of the Connes–Kreimer Hopf algebra of Feynman graphs [45]. The restriction to trees of the Connes–Kreimer Hopf algebra was already considered by [19]. The coproduct Δ^ρ requires a larger Hopf algebra, as the right channel of the coproduct can contain rooted trees that are not binary. In fact, it is the coproduct of the other Hopf algebra used in the Connes–Kreimer theory of renormalization [45], the Hopf algebra of rooted trees (not necessarily binary) with the coproduct given by admissible cuts. The relation between the restriction to trees of the Hopf algebra of Feynman graphs and the Hopf algebra of rooted trees is discussed in detail in [19]. Finally, the coproduct Δ^d, which performs the "cancellation of deeper copies" in linguistic terms, satisfies a weaker relation where the lists of terms for the coassociativity relation match, but only up to certain multiplicities in some of the terms. We will not discuss explicitly how to use multiplicities on vertices and edges to correct this counting, as this will be analyzed in the forthcoming [146]; rather, we will work with the relation between the quotients $T/^c F_{\underline{v}}$, $T/^\rho F_{\underline{v}}$, and $T/^d F_{\underline{v}}$ described previously and the coassociativity property of Δ^ρ and Δ^c.

Before discussing further the algebraic properties of the coproducts (1.2.8), we can look at a simple example. To see what (1.2.10) means, consider the case of equation (17) of *Elements* [40], where we have a workspace of the form WS = [{eaten, {the, apple}}], or in our notation, $F = T$ with

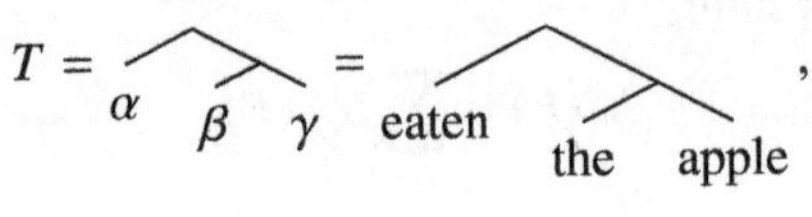

with α ="eat(en)", β ="the", and γ ="apple" $\in \mathcal{SO}_0$. Keep in mind that this is a nonplanar tree, so

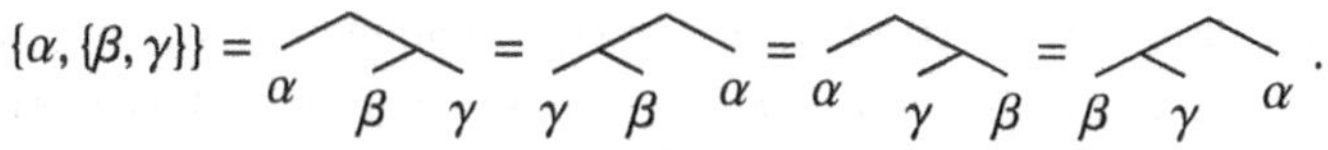

Then the coproduct produces a list of accessible terms in the left channel, accompanied by the complementary term in the right channel, in the following way:

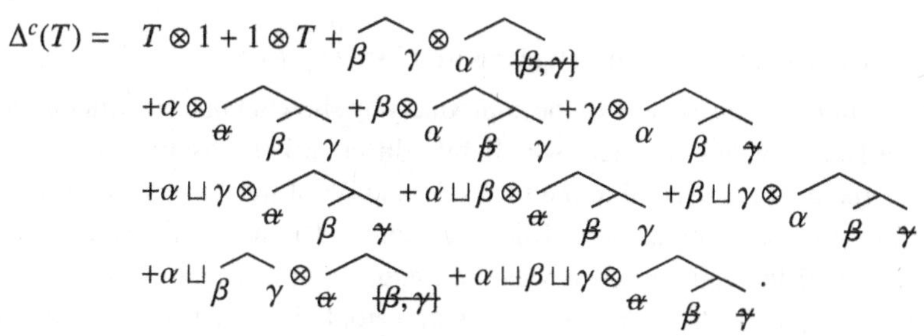

Meanwhile, in the deletion form this reads as

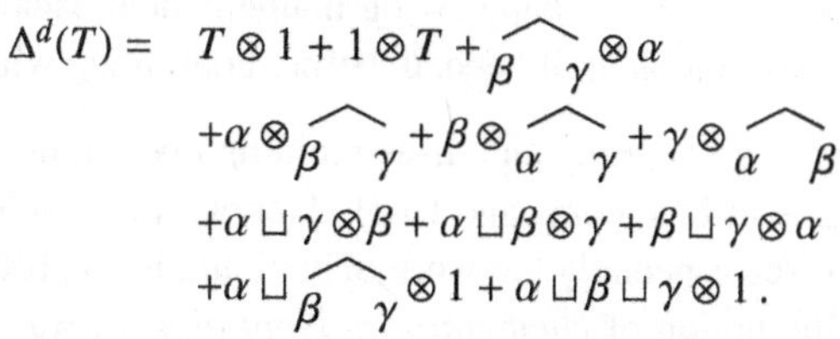

The non-primitive terms in the first two lines correspond to the terms in $\Delta_{(1)}$; the remaining non-primitive terms are in $\Delta_{(2)}$, except the last one, which is in $\Delta_{(3)}$. All these terms are there for "structural reasons"–namely, to ensure good behavior of the coproduct under iteration (we will discuss the coassociativity in Lemmas 1.2.10 and 1.2.12).

We see that in each term of these sums, where the formal sum is just a way of handling the list of possibilities, the left channel of the coproduct exhibits a selection of one or more of the possible accessible terms of the workspace and the corresponding right channel shows what remains of the workspace if those accessible terms are removed. So the coproduct takes a workspace and "disassembles" it into all its possible constituent parts that are available for computation (in the left channel of the coproduct output), while at the same time keeping track (in the right channel) of what would remain of the rest of the workspace if those accessible terms are used (i.e., the cancellation of the deeper copy).

So, in all expressions like this, the sum should be read as a list of alternative possibilities, and each tensor product term as the pair:

- a possible choice of accessible terms for computation
- the corresponding effect produced by this choice on the rest of the workspace.

As we discuss in section 1.3, this list of possibilities is the input to Merge, which uses it to produce a new workspace. Summarized in a short slogan, the coproduct presents all the possible items available for the Merge computation.

1.2.1 Combinatorial Hopf Algebras and Workspaces

We conclude this discussion of the coproducts by showing that the workspaces form a Hopf algebra. In fact, two slightly different but closely related algebraic structures correspond to the forms Δ^c and Δ^d of the coproduct (with the instrumental use of Δ^ρ as part of the argument). The relation between these two forms of the coproduct encodes the linguistic operation of "cancellation of deeper copies." The general definition of a Hopf algebra, as well as the corresponding conditions on product and coproduct and their compatibility, are summarized in section 4.2 of Chapter 4 on mathematical background, which we invite the readers to look at, if needed, before continuing with this section.

The description of assembling and disassembling operations on combinatorial objects in terms of bialgebras and Hopf algebras was developed by Gian-Carlo Rota [178]; see especially the work of Joni and Rota [106] and Schmitt [185]. The specific notion of *combinatorial Hopf algebra* was introduced by Loday and Ronco [126]. For a general approachable introduction to composition and decomposition operations described by Hopf algebras, see [12].

In this section, we demonstrate that workspaces with the composition and decomposition operations described in the previous section result in Hopf algebras that fit into a well-studied class called *combinatorial Hopf algebras*. These are Hopf algebras with the following properties:

1. They are *graded*, in the sense that the underlying vector space $\mathcal{V} = \oplus_{k\geq 0}\mathcal{V}_k$ is graded, with multiplication and comultiplication compatible with the grading, in the sense that degrees add when elements are multiplied:
$$\cdot : \mathcal{V}_k \otimes \mathcal{V}_\ell \to \mathcal{V}_{k+\ell}\,,$$
and the coproduct $\Delta : \mathcal{V}_k \to \oplus_{j=0}^{k}\mathcal{V}_j \otimes \mathcal{V}_{k-j}$ has a primitive part $P(x) = x\otimes 1 + 1\otimes x$ of degree $\deg(x)$, while all the remaining (nonprimitive) terms are of lower degree:
$$\Delta(x) = x \otimes 1 + 1 \otimes x + \sum x' \otimes x''$$

with $\deg(x') < \deg(x)$, $\deg(x'') < \deg(x)$; that is,

$$\tilde{\Delta} = \Delta - P : \mathcal{V}_k \to \oplus_{j=1}^{k-1} \mathcal{V}_j \otimes \mathcal{V}_{k-j} .$$

2. The homogeneous components $\mathcal{V}_k$ of the underlying vector space $\mathcal{V}$ are spanned by sets B_k of combinatorial objects (e.g., permutations, sets of partitions, sets of finite graphs, of (abstract or planar) rooted trees, of matroids, etc.). The sets B_k are finite for each fixed k, with k specifying a measure of the "size" of the objects.
3. The coproduct describes decomposition operations on the given combinatorial objects, with the terms $x' \otimes x''$ representing, for instance, pairs of a subobject and a quotient object.
4. They are connected; namely, the degree zero part $\mathcal{V}_0$ consists of just scalars (with the only object of size zero spanning this one-dimensional space).
5. As a consequence of the graded connected property, the bialgebra structure entirely suffices to determine inductively an antipode, which takes the form

$$S(x) = -x - \sum S(x') \cdot x'', \tag{1.2.12}$$

for $\Delta(x) = x \otimes 1 + 1 \otimes x + \sum x' \otimes x''$, with x', x'' of lower degree, so one obtains a Hopf algebra structure.

The Hopf algebra formed by workspaces is a combinatorial Hopf algebra in the sense described here.

The following statement is a specialization to binary forests of sections 4.1 and 4.2 of [19].

Lemma 1.2.10. *Let $\mathcal{V}^c(\mathfrak{F}_{SO_0})$ denote the vector space (over $\mathbb{Q}$) spanned by the workspaces $F \in \mathfrak{F}_{SO_0}$, endowed with the product given by the disjoint union $\sqcup$ and the coproduct Δ^c of* (1.2.8). *The space $\mathcal{V}^c(\mathfrak{F}_{SO_0})$ is graded by the number of edges. Then $(\mathcal{V}^c(\mathfrak{F}_{SO_0}), \sqcup, \Delta^c)$ is a graded* bialgebra. *This induces a Hopf algebra structure on the complement in $\mathcal{V}^c(\mathfrak{F}_{SO_0})$ of the span of the lexical items and features.*

Proof. The multiplication given by disjoint union is both associative and commutative, in the case of *nonplanar* trees and forests, with the unit given by the empty forest. Coassociativity of the coproduct

$$(\mathrm{id} \otimes \Delta^c) \circ \Delta^c = (\Delta^c \otimes \mathrm{id}) \circ \Delta^c$$

follows by the well-known coassociativity of the coproduct of the Connes–Kreimer Hopf algebra of Feynman graphs of [45]. It suffices to verify it on trees T; the case of forests follows. In essence, in $(\mathrm{id}\otimes\Delta^c)\circ\Delta^c(T)$, the operation $\mathrm{id}\otimes\Delta^c$ extracts accessible terms from the contracted trees in the right channel of the range of $\Delta^c(T)$ and produces further contractions of these, leaving the previously extracted accessible terms of T unchanged, while in $(\Delta^c\otimes\mathrm{id})\circ\Delta^c(T)$, the operation $\Delta^c\otimes\mathrm{id}$ extracts accessible terms from the previously extracted accessible terms of T and produces corresponding contractions while leaving the previously contracted trees unchanged.

The reason why the two operations $(\mathrm{id}\otimes\Delta^c)\circ\Delta^c$ and $(\Delta^c\otimes\mathrm{id})\circ\Delta^c$ yield the same result lies in the fact that the accessible terms of accessible terms of T extracted by $(\Delta^c\otimes\mathrm{id})$ are themselves accessible terms of T, and the resulting contractions (of the accessible terms extracted by $\Delta^c(T)$) can themselves be seen as accessible terms produced by $(\mathrm{id}\otimes\Delta^c)$, as described in the proof (for more general graphs) in Theorem 1.27 of [46]. The number of edges gives a grading on $\mathcal{V}^c(\mathfrak{F}_{\mathcal{SO}_0})$, but the term in degree zero is not one-dimensional (as the connected condition for Hopf algebras would require) because the trees consisting of a single leaf identified with their labeling items in $\mathcal{SO}_0$ are also of degree zero. These single-point trees give rise to noninvertible group-like elements that do not have inverses, preventing the construction of the antipode S, so the resulting structure is only a bialgebra, not a Hopf algebra. It does induce a Hopf algebra, however, when passing to the quotient by the ideal generated by the elements $1-\alpha$, with 1 the product unit and $\alpha\in\mathcal{SO}_0$. This reduces the degree zero part to a one-dimensional space, and one obtains a graded connected bialgebra, and hence a Hopf algebra with the inductive construction (1.2.12) of the antipode; see section 4.2 of [19]. This quotient can be identified, as a vector space, with the subspace of $\mathcal{V}^c(\mathfrak{F}_{\mathcal{SO}_0})$ spanned by those forests that, if nonempty, have a nonempty set of edges. □

Lemma 1.2.11. *Let $\tilde{\mathfrak{F}}_{\mathcal{SO}_0}$ denote the set of all (not necessarily binary) forests with leaf labels in the set $\mathcal{SO}_0$ (including in this case the possibility of an empty label). The vector space $\mathcal{V}(\tilde{\mathfrak{F}}_{\mathcal{SO}_0})$ has the structure of commutative, associative algebra with the product $\sqcup$, and the coproduct Δ^ρ makes it a graded connected Hopf algebra (a combinatorial Hopf algebra). Let $\tilde{\mathfrak{F}}^{\leq n}_{\mathcal{SO}_0}$ denote the subspace spanned by "at most n-ary" forests (internal vertices of valence at most $n+1$). The coproduct Δ^ρ is restricted to the subalgebra $\mathcal{V}(\mathfrak{F}_{\mathcal{SO}_0}\subset\mathcal{V}(\tilde{\mathfrak{F}}_{\mathcal{SO}_0})$ as a map:*

$$\Delta^\rho:\mathcal{V}(\mathfrak{F}_{\mathcal{SO}_0})\to\mathcal{V}(\mathfrak{F}_{\mathcal{SO}_0})\otimes\mathcal{V}(\tilde{\mathfrak{F}}^{\leq 2}_{\mathcal{SO}_0}),\tag{1.2.13}$$

where the trees in the right channel of $\Delta^\rho(T)$ may contain internal vertices of valence 2.

Proof. The combinatorial Hopf algebra structure on $\mathcal{V}(\tilde{\mathfrak{F}}_{SO_0})$ determined by $\sqcup$ and Δ^ρ is the same as the Connes–Kreimer Hopf algebra of nonplanar (and not necessarily binary) rooted trees; see [45] and [19]. For $T \in \mathfrak{T}_{SO_0}$, in the non-primitive part of the coproduct $\tilde{\Delta}^\rho(T) = \sum_C \pi_C(T) \otimes \rho_C(T)$, the terms $\pi_C(T)$ in the left channel are all forests in $\mathfrak{F}_{SO_0}$ since deleting edges of an *admissible cut* implies that each resulting component is an accessible term of T (see Lemma 1.2.7). In the right channel of $\tilde{\Delta}^\rho(T)$, the source vertex of each of the edges in a cut C will become a vertex of valence 2 (if internal) or 1 (if the original root of T) in the term $\pi_C(T)$, which is therefore in $\tilde{\mathfrak{F}}_{SO_0}$ but not necessarily in $\mathfrak{F}_{SO_0}$. □

Consider then the coproduct Δ^d. In this case, we have the following described in Lemma 1.2.12.

Lemma 1.2.12. *With the coproduct Δ^d of the form* (1.2.8)*, for any $T \in \mathfrak{T}_{SO_0}$ (and hence for any $F \in \mathfrak{F}_{SO_0}$), the list of terms in $(1 \otimes \Delta^d) \circ \Delta^d(T)$ matches the corresponding list of terms in $(\Delta^d \otimes 1) \circ \Delta^d(T)$. Let $d(C, C') = \min_{e \in C, e' \in C''} d(e, e')$ be the distance between admissible cuts given by the shortest path in T between the cuts. The terms that correspond to pairs of admissible cuts with distance $d(C, C') \leq 1$ occur in $(1 \otimes \Delta^d) \circ \Delta^d(T)$ and $(\Delta^d \otimes 1) \circ \Delta^d(T)$ with different multiplicity.*

Proof. We have

$$(\Delta \otimes \mathrm{id}) \circ \Delta(T) = (\Delta \otimes \mathrm{id}) \sum_{\underline{w}} F_{\underline{w}} \otimes T/F_{\underline{w}} = \sum_{\underline{v},\underline{w}} F_{\underline{v}} \otimes (F_{\underline{w}}/F_{\underline{v}}) \otimes T/F_{\underline{w}},$$

where the sums are taken over subforests with the disjointness condition, in which the subforests $F_{\underline{v}} \subset F_{\underline{w}}$ consists of either full components of $F_{\underline{w}}$ or of subforests of the components. The first case gives terms of the form $F_{\underline{v}} \otimes F_{\underline{u}} \otimes T/F_{\underline{v},\underline{u}}$ for $\underline{w} = (\underline{v}, \underline{u})$. On the other hand, we have

$$(\mathrm{id} \otimes \Delta) \circ \Delta(T) = (\mathrm{id} \otimes \Delta) \sum_{\underline{v}} F_{\underline{v}} \otimes T/F_{\underline{v}} = \sum_{\underline{u},\underline{v}} F_{\underline{v}} \otimes (T/F_{\underline{v}})_{\underline{u}} \otimes (T/F_{\underline{v}})/(T/F_{\underline{v}})_{\underline{u}}.$$

We distinguish among these terms the case where the subtrees of T with the root at u_i are disjoint from the trees of $F_{\underline{v}}$, where we have $(T/F_{\underline{v}})/(T/F_{\underline{v}})_{\underline{u}} = T/F_{\underline{v},\underline{u}}$; and the remaining cases, where some vertices u_i in $\underline{u}$, as vertices of T, are above some vertices v_j of the components of $F_{\underline{v}}$, where the corresponding quotient is $(T/T_{v_j})/T_{u_i} = T/T_{u_i}$ and $(T/T_{v_j})_{u_i} = T_{u_i}/T_{v_j}$. Thus, we see that we obtain the same two types of terms with the same set of terms for both $(1 \otimes \Delta^d) \circ \Delta^d(T)$ and $(\Delta^d \otimes 1) \circ \Delta^d(T)$. Consider, then, pairs of admissible cuts with $d(C, C') \leq 1$. It suffices to consider the case where both cuts consist of a single edge. Figure 1.3 illustrates the sources of the difference in the counting of multiplicities. □

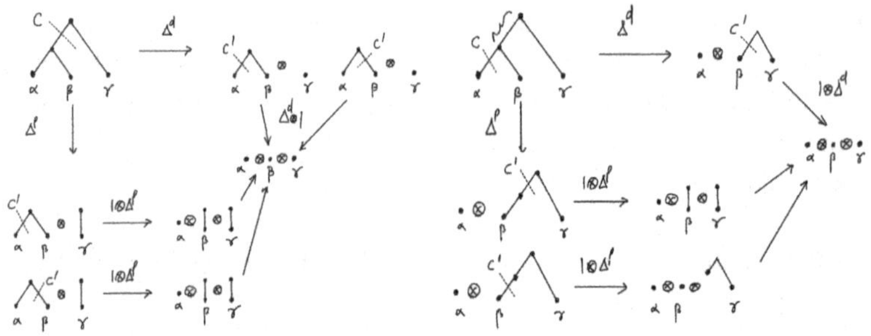

Figure 1.3
Multiplicities: cutting C first leaves two cuts C' in $(\Delta^d \otimes 1)\Delta^d(T)$, producing two identical terms, but only one in $(1 \otimes \Delta^d)\Delta^d(T)$.

This possible discrepancy of multiplicities in the counting of terms has no significant consequences in our setting, as we will see in section 1.3.

1.3 Action of Merge on Workspaces

We next introduce our formalization of the action of Merge on workspaces via an operator that performs a search for matching terms. This operator will be applied to the terms of the coproduct; that is, to the accessible terms that Merge is supposed to work on. In fact, as we have seen, the left-hand-side of the coproduct produces this list of accessible terms—the search runs over the list—while the right-hand-side of the coproduct keeps track of the corresponding cancellation of copies. We introduce the Merge action via some preliminary definitional steps.

1.3.1 Matching Terms

As we explained in section 1.2, the coproduct on workspaces produces the list of accessible terms available for computation. We introduce here an operator that inspects all the terms produced by the coproduct for matching terms with an assigned pair of syntactic objects.

Definition 1.3.1. Consider two syntactic objects, $S, S' \in \mathfrak{T}_{SO_0}$. Define a linear operator $\gamma_{S,S'} : \mathcal{V}(\mathfrak{F}_{SO_0}) \to \mathcal{V}(\mathfrak{F}_{SO_0})$ by setting

$$\gamma_{S,S'}(F) = \begin{cases} F & F = S \sqcup S' \\ 0 & \text{otherwise,} \end{cases} \tag{1.3.1}$$

extended by linearity. We then define

$$\delta_{S,S'} : \mathcal{V}(\mathfrak{F}_{SO_0}) \otimes \mathcal{V}(\mathfrak{F}_{SO_0}) \to \mathcal{V}(\mathfrak{F}_{SO_0}) \otimes \mathcal{V}(\mathfrak{F}_{SO_0})$$

as

$$\delta_{S,S'} = \gamma_{S,S'} \otimes \mathrm{id}\,. \tag{1.3.2}$$

Thus, we have

$$\delta_{S,S'}(S \sqcup S' \otimes F') = S \sqcup S' \otimes F'\,,$$

while all basis elements $F \otimes F'$ that are not of this form are mapped to zero.

Note that $\gamma_{S,S'}$ (hence $\delta_{S,S'}$) is a *linear operator*, but *not* an algebra homomorphism since $\gamma_{S,S'}(S \sqcup S') = S \sqcup S'$, while $\gamma_{S,S'}(S) = \gamma_{S,S'}(S') = 0$.

When applied to a term of the form $F_{\underline{v}} \otimes F/F_{\underline{v}}$, for $F = \sqcup_{i \in \mathcal{I}} T_i$, the operator $\delta_{S,S'}$ gives

$$\delta_{S,S'}(F_{\underline{v}} \otimes F/F_{\underline{v}}) = S \sqcup S' \otimes T_a/S \sqcup T_b/S' \sqcup F^{(a,b)}\,, \tag{1.3.3}$$

with $F^{(a,b)} = \sqcup_{i \neq a,b} T_i$, if there are indices $a, b \in \mathcal{I}$ such that $T_{a,v_a} \simeq S$, $T_{b,v_b} \simeq S'$, or

$$\delta_{S,S'}(F_{\underline{v}} \otimes F/F_{\underline{v}}) = S \sqcup S' \otimes T_a/(S \sqcup S') \sqcup F^{(a)}\,, \tag{1.3.4}$$

with $F^{(a)} = \sqcup_{i \neq a} T_i$, if for some $a \in \mathcal{I}$ we have two disjoint accessible terms $T_{a,v_a} \simeq S$ and $T_{a,w_a} \simeq S'$. If there is more than one choice of indices a, b for which matching pairs $T_{a,v_a} \simeq S$, $T_{b,v_b} \simeq S'$ exist, then on the right-hand-side of (1.3.3), one obtains the sum over all the possibilities. All other terms of the coproduct where no matching occurs are mapped to zero.

1.3.2 Grafting

Next, observe that the operation (1.1.3) on syntactic objects factors through a grafting operator $\mathfrak{B}$ on forests, familiar from physics, as defined next.

Definition 1.3.2. *Let $\mathfrak{T}^{\mathbb{N}}_{SO_0}$ denote the set of all n-ary finite rooted trees with arbitrary $n \in \mathbb{N}$, with no assigned planar structure and with leaves labeled by the set SO_0. Let $\mathfrak{F}^{\mathbb{N}}_{SO_0}$ be the set of finite forests with connected components in $\mathfrak{T}^{\mathbb{N}}_{SO_0}$. Let $\mathcal{V}(\mathfrak{T}^{\mathbb{N}}_{SO_0})$ and $\mathcal{V}(\mathfrak{F}^{\mathbb{N}}_{SO_0})$ denote the $\mathbb{Q}$-vector spaces spanned by these sets. The grafting operator $\mathfrak{B} : \mathcal{V}(\mathfrak{F}^{\mathbb{N}}_{SO_0}) \to \mathcal{V}(\mathfrak{T}^{\mathbb{N}}_{SO_0})$ is the linear operator* *defined on generators by*

$$\mathcal{B}(T_1 \sqcup T_2 \sqcup \cdots \sqcup T_N) = \underset{T_1 \quad T_2 \quad \cdots \quad T_N}{\bigwedge} , \tag{1.3.5}$$

with the convention that if $F = T$ *is a single tree, then* $\mathcal{B}(T) = T$*, and that* $\mathcal{B}(F \sqcup 1) = \mathcal{B}(F)$.

The grafting operator $\mathcal{B}$ is well-known in the mathematical formulation of perturbative quantum field theory, as it is the operator that defines the recursive structure of Dyson–Schwinger (DS) equations; see [5] and [65].

Lemma 1.3.3. *The Merge operator* $\mathfrak{M}$ *of* (1.1.3)*, namely the multiplication operation in the Magma* $\mathrm{Magma}_{na,c}(\mathcal{SO}_0, \mathfrak{M})$*, determines a bilinear operator* $\mathfrak{M} : \mathcal{V}(\mathfrak{T}_{\mathcal{SO}_0}) \otimes \mathcal{V}(\mathfrak{T}_{\mathcal{SO}_0}) \to \mathcal{V}(\mathfrak{T}_{\mathcal{SO}_0})$*, defined on generators as*

$$\mathfrak{M} : T \otimes T' \mapsto \mathfrak{M}(T, T') = \underset{T \quad T'}{\wedge} , \tag{1.3.6}$$

where we define $\mathfrak{M}(1, 1) = 1$ *and* $\mathfrak{M}(T, 1) = \mathfrak{M}(1, T) = T$*. This operator factors through the grafting operator* $\mathcal{B}$ *restricted to the range of multiplication* $\sqcup$*, in the form of the commutative diagram*

$$\begin{array}{ccc} \mathcal{V}(\mathfrak{T}_{\mathcal{SO}_0}) \otimes \mathcal{V}(\mathfrak{T}_{\mathcal{SO}_0}) & \xrightarrow{\quad \mathfrak{M} \quad} & \mathcal{V}(\mathfrak{T}_{\mathcal{SO}_0}) \, . \\ & {\scriptstyle \sqcup} \searrow \quad \nearrow {\scriptstyle \mathcal{B}} & \\ & \mathcal{V}(\mathfrak{F}_{\mathcal{SO}_0}) & \end{array}$$

Proof. Since trees and forests do not have an assigned planar structure, both $\mathfrak{M}$ and the operator $\mathcal{B}$ do not depend on the order of the trees. Moreover, the image of $\mathcal{V}(\mathfrak{T}_{\mathcal{SO}_0}) \otimes \mathcal{V}(\mathfrak{T}_{\mathcal{SO}_0})$ under $\sqcup$ consists of forests with two connected components, so their image under $\mathcal{B}$ is still a binary tree, which takes the form (1.3.6). □

1.3.3 The Merge Operators

The operation (1.3.3), in combination with the operation (1.1.3) on syntactic objects, and the bialgebra structure on workspaces, contributes to the definition of the action of Merge on workspaces as described in Chomsky's work ([28], [29], and [40]), that we can now define in the following way.

Definition 1.3.4. *Let* $\mathcal{B}$ *be the grafting operator of Definition 1.3.2, and consider the coproduct* $\Delta = \Delta^d$ *of* (1.2.8)*. The action of Merge on workspaces* *consists of a collection of operators:*

$$\{\mathfrak{M}_{S,S'}\}_{S,S'\in\mathfrak{T}_{SO_0}}, \quad \mathfrak{M}_{S,S'} : \mathcal{V}(\mathfrak{F}_{SO_0}) \to \mathcal{V}(\mathfrak{F}_{SO_0}),$$

parameterized by pairs S, S' of syntactic objects, which act on $F \in \mathcal{V}(\mathfrak{F}_{SO_0})$ by

$$\mathfrak{M}_{S,S'}(F) = \sqcup \circ (\mathcal{B} \otimes \mathrm{id}) \circ \delta_{S,S'} \circ \Delta(F). \tag{1.3.7}$$

Remark 1.3.5. There is a small difference between the definition of the operators $\mathfrak{M}_{S,S'}$ that we give in (1.3.7) and the definition followed by [28], [29], and *Elements* [40]. We show next that these descriptions agree in the case where matching elements for the pair S, S' are found among the accessible terms of the workspace. The difference is in the case of no matching terms. In the formulation of [28], [29], and [40] one defines

$$\mathfrak{M}_{S,S'} : \mathfrak{F}_{SO_0} \to \mathfrak{F}_{SO_0} \tag{1.3.8}$$

as maps on the set of workspaces (assigning to a workspace another new workspace). When no matching terms for S, S' exist, it is assumed that $\mathfrak{M}_{S,S'}$ acts as the identity map, leaving the workspace unchanged. This is the natural choice, but it has a drawback when it comes to algorithmic implementation: one can make any chain of Merge derivations infinitely long by repeatedly applying a loop consisting of the identity map (applying maps $\mathfrak{M}_{S,S'}$ with no matching terms). From this perspective, it appears more convenient to assume that, when no matching terms for S, S' exist, the derivation crashes and does not continue. This cannot be directly implemented if one writes $\mathfrak{M}_{S,S'}$ as a function of sets (1.3.8), but it can easily be expressed if one considers the induced map of vector spaces

$$\mathfrak{M}_{S,S'} : \mathcal{V}(\mathfrak{F}_{SO_0}) \to \mathcal{V}(\mathfrak{F}_{SO_0}) \tag{1.3.9}$$

as in Definition 1.3.4. In the vector space $\mathcal{V}(\mathfrak{F}_{SO_0})$, there is a natural way of expressing the idea that the derivation crashes: mapping to the zero-vector (which exists as an element of $\mathcal{V}(\mathfrak{F}_{SO_0})$, but not as an element of $\mathfrak{F}_{SO_0}$).

If one wishes to retain the original assumption that Merge acts as the identity (rather than the zero map) when no match exists for S, S', then one needs to modify the map (1.3.7) (which maps to the zero vector in this case). One can obtain such a modification, for example, by requiring that (1.3.7) holds when $\delta_{S,S'}(\Delta(F)) \neq 0$ and one sets $\mathfrak{M}_{S,S'}(F) = F$ otherwise, or by a modification of the maps $\delta_{S,S'}$ with the same effect. This expresses the requirement that if no matching terms for S, S' are found by $\delta_{S,S'}$ anywhere in the output of the coproduct, then the Merge operator $\mathfrak{M}_{S,S'}$ outputs the original workspace unchanged. This modification produces a map of sets that is no longer a linear map of vector spaces (as a formal sum $F+F'$ with $\delta_{S,S'}(F)=0$ and $\delta_{S,S'}(F')\neq 0$ would go to $\mathfrak{M}_{S,S'}(F')$, not to $F+\mathfrak{M}_{S,S'}(F')$). In terms of computational implementations, it is generally an advantage to be able to formulate operations in terms of linear algebra. Thus, we will adopt here the form (1.3.7), with a minor deviation from the formulation of [40].

We can equivalently write (1.3.7) using the other forms of the coproduct Δ^ω of (1.2.8) with $\omega = c,\rho$. Namely, let $\Pi_{\rho,d}$, $\Pi_{c,\rho}$ and $\Pi_{c,d}$ denote the following linear projections:

- The projection (with the notation as in (1.2.13))
$$\Pi_{d,\rho} : \mathcal{V}(\tilde{\mathfrak{F}}^{\leq 2}_{SO_0}) \to \mathcal{V}(\mathfrak{F}_{SO_0}),$$
which maps an "at most binary" tree to the unique maximal binary tree in $\mathfrak{T}_{SO_0}$ obtained by edge contraction (extended componentwise to forests)
- The projection
$$\Pi_{\rho,c} : \mathcal{V}(\mathfrak{F}_{SO_0}) \to \mathcal{V}(\tilde{\mathfrak{F}}^{\leq 2}_{SO_0}),$$
which deletes any edge terminating in a leaf with a label of the form $\not{T}$ (see Definition 1.2.4)
- The projection $\Pi_{d,c} = \Pi_{d,\rho}\circ\Pi_{\rho,c}$, as follows:
$$\Pi_{d,c} : \mathcal{V}(\mathfrak{F}_{SO_0}) \to \mathcal{V}(\mathfrak{F}_{SO_0}).$$

We then have
$$\Delta^d = (\mathrm{id}\otimes\Pi_{d,\rho})\circ\Delta^\rho = (\mathrm{id}\otimes\Pi_{d,c})\circ\Delta^c.$$

We can then equivalently write (1.3.7) as
$$\begin{aligned}\mathfrak{M}_{S,S'} &= \sqcup\circ(\mathcal{B}\otimes\mathrm{id})\circ(\mathrm{id}\otimes\Pi_{d,\rho})\circ\delta_{S,S'}\circ\Delta^\rho \\ &= \sqcup\circ(\mathcal{B}\otimes\mathrm{id})\circ(\mathrm{id}\otimes\Pi_{d,c})\circ\delta_{S,S'}\circ\Delta^c.\end{aligned}\tag{1.3.10}$$

Note that here, we are using the fact that $\delta_{S,S'}$ is acting on the left-channel of the coproduct, while the coproducts Δ^ω differ only in the right-channel. Thus, the projections $\Pi_{d,\rho}$ and $\Pi_{d,c}$ commute with $\delta_{S,S'}$. The vanishing of $\delta_{S,S'}(\Delta(F))$ does not depend on which Δ^ω is used.

Lemma 1.3.6. *The expression* (1.3.7) *agrees with the description of the action of Merge on workspaces in [28], [29], and* Elements *[40] in all cases where* $\delta_{S,S'}(\Delta(F))\neq 0$.

Proof. We need to check that indeed, when matches for S, S' exist, the Merge operator $\mathfrak{M}_{S,S'}$ searches for copies of the syntactic terms S and S' among the accessible terms of a given workspace F, extracts those accessible terms on which to perform the Merge operation, and then finally cancels copies from the workspace, producing the resulting new workspace.

In (1.3.7), we see this description as in [40] by following, step by step, the chain of compositions in (1.3.7):

1. The first operation, the coproduct Δ, produces (in the left-channel) all possible collections $F_{\underline{v}} = T_{v_1} \sqcup \cdots \sqcup T_{v_n}$ of accessible terms T_{v_i} of the workspace. In the right-channel, it keeps track of the corresponding remaining terms $F/F_{\underline{v}}$.
2. The second operation, $\delta_{S,S'}$, searches for a matching term for S and S'. If none is found, then the chain of compositions ends. If a term with $F_{\underline{v}} = S \sqcup S'$ exists, then this term is selected and the next step receives an input of the form
$$(S \sqcup S') \otimes F/F_{\underline{v}},$$
as in (1.3.3) or as in (1.3.4).
3. The third operation, $\mathcal{B} \otimes \mathrm{id}$, takes the term $S \sqcup S'$ in the left-channel and replaces it by $\mathfrak{M}(S, S')$, while leaving the right-channel unchanged.
4. Finally, the product $\sqcup$ reassembles the new workspace by merging the two channels of the coproduct into a single one; hence, the resulting new workspace looks like
$$\mathfrak{M}(S, S') \sqcup F/F_{\underline{v}}.$$
In this new workspace, the two accessible terms S, S' have been extracted and merged into a new structure $\mathfrak{M}(S, S')$, while in the remaining part of the original workspace $F/F_{\underline{v}}$, the deeper copies of S and S' have been cancelled.

This matches the description of the action of Merge given in [28] and [29], modulo the small modification discussed in Remark 1.3.5. Workspaces that contain matching accessible terms are replaced by a new workspace given by merging the matching terms and by cancelling the deeper copies, in the form

$$\sum_{v,w:T_v=S,T_w=S'} \mathfrak{M}(T_v, T_w) \sqcup (T/T_v) \sqcup (T'/T_w) \sqcup \hat{F}, \tag{1.3.11}$$

where we include the sum over possibilities in case of multiple possible matches in the same workspace, and $\hat{F}$ is the rest of the workspace that is not involved in the operation. We describe these various cases in more detail in section 1.4.

Remark 1.3.7. In the second step, only the terms of $\Delta(F)$ that have $F_{\underline{v}} = T_{v_1} \sqcup T_{v_2}$ with two components can be selected by $\delta_{S,S'}$, while all other terms are mapped to zero at this step. This makes it appear as if a large part of the coproduct is superfluous. This is not the case, as we will discuss in section 1.16, as they are used by another operations required in Minimalism, (namely, the FormSet operation).

1.3.4 Examples

We can now compare again the mathematical formulation given here with the same action described in *Elements* [40]. Let us consider the example of equations (15), (16), and (17) in *Elements* [40]. There, one considers a workspace of the form WS = [eaten, {the, apple}], which we write in the mathematical notation as the forest $F = T_1 \sqcup T_2$, with T_1 as the tree with a single vertex labeled by the lexical item α =eat(en), and

$$T_2 = \overset{\frown}{\beta \quad \gamma} \quad \text{with } \beta = \text{the} \ \ \gamma = \text{apple}\,.$$

Then equation (16) of *Elements* [40] tells us that we want to use the Merge operator

$$\mathfrak{M}_{S,S'} \quad \text{with } S = \alpha \ \text{ and } \ S' = T_2\,.$$

According to our description of the action of Merge on workspaces,

$$\mathfrak{M}_{S,S'} = \sqcup \circ (\mathcal{B} \otimes \mathrm{id}) \circ \delta_{S,S'} \circ \Delta\,,$$

and the first step in applying $\mathfrak{M}_{S,S'}$ to F is to compute the coproduct $\Delta(F)$. This is computed as $\Delta(F) = \Delta(T_1)\Delta(T_2)$, using (1.2.8). We obtain

$$\Delta(F) = F \otimes 1 + 1 \otimes F + \alpha \otimes T_2 + T_2 \otimes \alpha$$

$$+\alpha \sqcup \beta \otimes \gamma + \alpha \sqcup \gamma \otimes \beta + \beta \sqcup \gamma \otimes \alpha + \alpha \sqcup \beta \sqcup \gamma \otimes 1\,,$$

where the last two lines are the terms with simultaneous extraction of more than one accessible term. Equivalently,

$$\Delta(\text{eaten} \sqcup \overset{\frown}{\text{the} \quad \text{apple}}) = \text{eaten} \sqcup \overset{\frown}{\text{the} \quad \text{apple}} \otimes 1 + 1 \otimes \text{eaten} \sqcup \overset{\frown}{\text{the} \quad \text{apple}}$$

$$+\text{eaten} \otimes \overset{\frown}{\text{the} \quad \text{apple}} + \overset{\frown}{\text{the} \quad \text{apple}} \otimes \text{eaten}$$

$$+ \text{eaten} \sqcup \text{the} \otimes \text{apple} + \text{eaten} \sqcup \text{apple} \otimes \text{the} + \text{the} \sqcup \text{apple} \otimes \text{eaten}$$

$$\text{eaten} \sqcup \text{the} \sqcup \text{apple} \otimes 1\,.$$

The second step in the action of $\mathfrak{M}_{S,S'}$ is to apply the operator $\delta_{S,S'}$ to the output of the coproduct to check for matching terms. The only term of the coproduct that extracts two accessible terms matching S and S' is $F \otimes 1 = \alpha \sqcup T_2 \otimes 1$. In all the other summands, the terms extracted do not match the chosen pair S, S', so $\mathfrak{M}_{S,S'}$ cannot act. These terms are eliminated by $\delta_{S,S'}$, just as the linguistic theory requires.

In the cases with matching terms, $\delta_{S,S'}$ acts as in (1.3.3) and finds the matching term

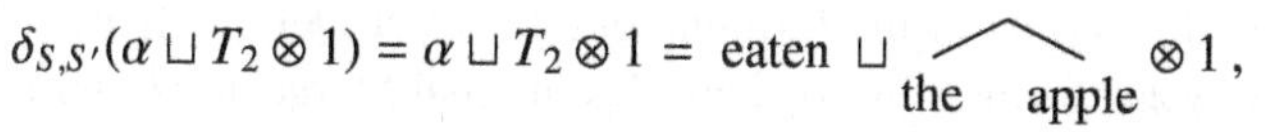

The last term here is just 1 since in this case, both S, S' match components of F rather than accessible terms inside components, and there are no other remaining, unused components. Applying $(\mathcal{B} \otimes \mathrm{id})$ to this result yields

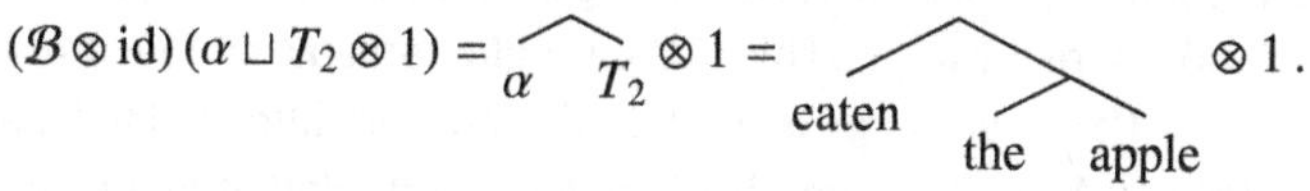

In the last step, applying $\sqcup$ multiplies the two entries of the tensor product, which gives

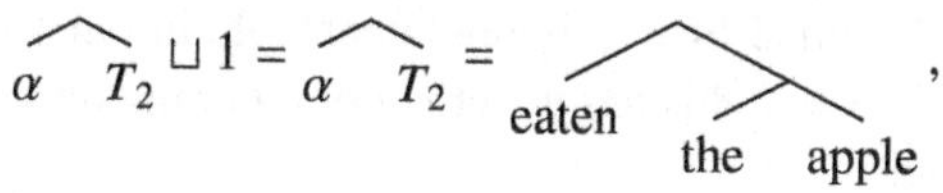

since 1 is the unit of the product $\sqcup$, so we have obtained the workspace of (17) of *Elements* [40]:

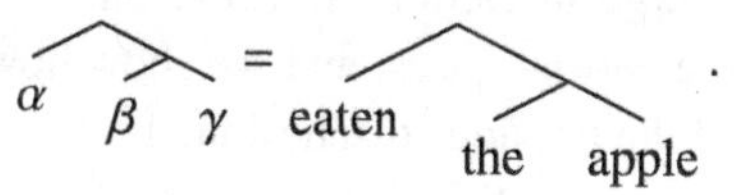

Remark 1.3.8. Note that in the action of External Merge, the operation $\delta_{S,S'}$ singles out a term in the coproduct that belongs to its *primitive part* (1.2.9); namely, the part where each component T_a contributes only through $T_a \otimes 1 + 1 \otimes T_a$, and not through the *extraction* of accessible terms. This is because External Merge applies to components of the workspace rather than to accessible terms contained *inside* components. The more interesting, non-primitive part of the coproduct is needed to have a linguistic model that accounts for Internal Merge (hence movement).

Remark 1.3.9. With our formulation of the action of $\mathfrak{M}_{S,S'}$ we also keep track of some amount of extra information, which is counting how many of the possibilities for extraction of accessible terms yield a match, but this additional information about counting possibilities is not carried over in the next step of the derivation (in accordance with the Markovian principle, as stated in section 3.3.3 of [40]); see section 1.9 in this chapter. Different possible matching terms yield new workspaces, to which one can then apply the next Merge action, individually.

1.4 Forms of Merge

One of the drawbacks of the formulation (1.3.7) of Definition 1.3.4 is that it allows additional forms of Merge, besides Internal Merge and External Merge, that are not considered desirable in certain linguistic accounts, such as Sideward Merge. We discuss here how a simple modification of Definition 1.3.4 that incorporates a formulation of Minimal Search suffices to eliminate these cases and to retain only the presumably linguistically desirable cases of External Merge and Internal Merge. This is the usual argument provided in linguistics, where only External Merge and Internal Merge are intended to be retained based on Minimal Search, except that here we reformulate it in a way that fits our algebraic setting (see section 1.5). First, we review in this section how the different cases of Merge are incorporated in (1.3.7), and then in section 1.5, we describe how Minimal Search is implementable in our Hopf algebra setting. Finally, we show that this has the effect of retaining only External Merge and Internal Merge.

1.4.1 Different Forms of Merge

In the description of Merge in Definition 1.3.4, one can distinguish several cases. We recall here the various cases, and we show how they are realized in the formulation given in the previous discussion.

Two syntactic objects, $\alpha, \beta \in \mathcal{SO} = \mathfrak{T}_{\mathcal{SO}_0}$, can occur in a workspaces $F \in \mathfrak{F}_{\mathcal{SO}_0}$ either as "members" (i.e., as connected components of the forest F), or as "accessible terms" of members, in the terminology of [40]. We write $T \in F$ to indicate that a certain syntactic object $T \in \mathcal{SO}$, seen as a tree, is a connected component of the forest F. We write $T \in Acc(T')$ to indicate that T occurs as an accessible term T'_v of a syntactic object $T' \in F$.

Thus, we have the following three possibilities:

1. $\alpha = T_i$ and $\beta = T_j$, with $T_i, T_j \in F$ and $i \neq j$
2. $\alpha = T_i \in F$ and $\beta \in Acc(T_j)$, for some $T_j \in F$, with two subcases:
   (a) $i = j$
   (b) $i \neq j$
3. $\alpha \in Acc(T_i)$ and $\beta \in Acc(T_j)$, for some $T_i, T_j \in F$, with two subcases:

(a) $i = j$

(b) $i \neq j$

1.4.2 External Merge

Case 1 describes External Merge: for a workspace $F = \sqcup_a T_a$, the Merge operation $\mathfrak{M}_{T_i,T_j}$ replaces the pair T_i, T_j of elements of F with a new syntactic object given by the tree $T_{ij} = \{T_i, T_j\} = \mathfrak{M}(T_i, T_j)$, and produces the new workspace

$$F' = T_{ij} \sqcup \bigsqcup_{a \neq i,j} T_a \,,$$

where the two components T_i, T_j of F have been removed and replaced by the new tree $T_{ij} = \{T_i, T_j\}$. External Merge decreases by 1 the number of syntactic objects, and increases by 2 the number of accessible terms by adding T_i and T_j to the set $Acc(T_i) \cup Acc(T_j)$. In the case of External Merge, the statement of the following lemma is immediately evident.

Lemma 1.4.1. *External Merge is achieved by the operators $\mathfrak{M}_{T_i,T_j}$ of Definition 1.3.4 when the syntactic objects (trees) T_i, T_j match two connected components of the workspace $F = \sqcup_a T_a$. The operator δ_{T_i,T_j} in this case selects the term of the coproduct of the form*

$$(T_i \sqcup T_j) \otimes \hat{F} \,.$$

with $\hat{F} = \sqcup_{a \neq, i,j} T_a$. The operator $\mathcal{B} \otimes \mathrm{id}$ then maps this term to

$$\mathcal{B} \otimes \mathrm{id} : (T_i \sqcup T_j) \otimes \hat{F} \mapsto \mathfrak{M}(T_i, T_j) \otimes \hat{F} \,.$$

Finally, applying the product $\sqcup$ gives

$$\mathfrak{M}_{T_i,T_j}(F) = \mathfrak{M}(T_i, T_j) \sqcup \hat{F} \,;$$

that is, the two components T_i, T_j are merged while the rest of the workspace $\hat{F}$ remains unchanged.

1.4.3 Internal Merge

Case 2(a) describes Internal Merge: in this case, the new workplace F' contains a new component of the form $\mathfrak{M}(\beta, T_i/\beta)$, where $\beta \in Acc(T_i)$ is an accessible term of T_i and T_i is a component of the given workspace F. The quotient T_i/β indicates that the deeper copy of β as an accessible term of T_i is no longer an accessible term of $\mathfrak{M}(\beta, T_i/\beta)$ in F', as β already occurs as an accessible term at a *higher* level in the new syntactic object $\mathfrak{M}(\beta, T_i/\beta)$ formed by Merge. In this case, the realization of Internal Merge by the operators $\mathfrak{M}_{S,S'}$ of Definition 1.3.4 is more interesting, as it involves the composition of two such operators and the role of the multiplicative unit 1 of the Merge magma, given by the trivial tree.

Proposition 1.4.2. *Internal Merge is realized by the operators introduced in Definition 1.3.4 as composition:*

$$\mathfrak{M}_{T/\beta,\beta} \circ \mathfrak{M}_{\beta,1}, \tag{1.4.1}$$

where 1 *is the unit of the Merge magma and the tree* β *is an accessible term of a connected component of* F *isomorphic to* T.

Proof. The operator $\delta_{\beta,1}$ in $\mathfrak{M}_{\beta,1}$, acting on the workspace F, will select a term $\beta \otimes T/\beta$ in the coproduct $\Delta(F)$, where β occurs as an accessible term of a component T of F, using the identification $\beta \otimes T/\beta = (\beta \sqcup 1) \otimes T/\beta$, since 1 is the product unit. Then $\mathcal{B} \otimes \mathrm{id}$ acts as the identity on $(\beta \sqcup 1) \otimes T/\beta$ since $\mathcal{B}(\beta \sqcup 1) = \mathfrak{M}(\beta, 1) = \beta$ and the product then maps this to $\beta \sqcup T/\beta$, producing two components $\beta = \mathfrak{M}(\beta, 1)$ and T/β. The operator $\mathfrak{M}_{T/\beta,\beta}$ can then be applied to $\beta \sqcup T/\beta$, as an External Merge, where $\delta_{\beta,T/\beta}$ selects the term $\beta \sqcup T/\beta \otimes 1$ of the coproduct, $\mathcal{B} \otimes \mathrm{id}$ maps it to $\mathfrak{M}(\beta, T/\beta) \otimes 1$ and the product maps this to $\mathfrak{M}(\beta, T/\beta) \sqcup 1 = \mathfrak{M}(\beta, T/\beta)$, which is the Internal Merge. □

Remark 1.4.3. Note that in (1.4.1), Internal Merge *appears* to involve the repeated application of two External Merge operations, one of them involving the magma unit, which has the effect of extracting an accessible term and adding it to the new workspace, together with the cancellation of its deeper copy. At first glance, this appears to be a return to older formulations of Minimalism, in which Internal Merge was considered a composite operation. These were later discarded on the grounds that External Merge and Internal Merge should both be the same simple operation. While our formulation gives the impression that Internal Merge is more complex than External Merge, in fact this is not the case. The formulation (1.4.1) is *not* equivalent to such older formulations: we will see by examining Minimal Yield, as well as by counting the size and number of accessible terms, that the operation $\mathfrak{M}_{\beta,1}$ in fact can occur only in the combination $\mathfrak{M}_{T/\beta,\beta} \circ \mathfrak{M}_{\beta,1}$ (i.e., as Internal Merge), *not on its own.* Hence Internal Merge cannot be further decomposed; see also section 1.4.3.1.

1.4.3.1 Is Internal Merge composite?

If we look more closely at the form (1.4.1) of Internal Merge, we see that the role of the first operation $\mathfrak{M}_{\beta,1}$ is to move the term T/β from the right to the left channel of the coproduct, where $\mathcal{B}$ can then combine it with β to form the merge $\mathfrak{M}(\beta, T/\beta)$, which is the desired outcome of Internal Merge.

Thus, the main reason for the apparent composite nature of Internal Merge is the fact that in (1.3.7), the grafting operation $\mathcal{B}$ acts only on the left channel of the coproduct. We explain next why this is both mathematically and linguistically desirable.

Suppose that instead, we were to define the Merge operation $\mathfrak{M}_{S,S'}$ by accessing pairs of terms on the two sides of the coproduct. If applied to a workspace consisting of a single syntactic object T, then this would be no problem, and we could simply define an operator $\tilde{\delta}_{S,S'}$ that searches for S in the left channel and for S' in the right channel (and vice versa for symmetry), selecting a coproduct term of the form $\beta \otimes T/\beta$, with $S \simeq \beta$ and $S' \simeq T/\beta$. One could then apply

$$\mathfrak{M} : \mathfrak{T}_{SO_0} \otimes \mathfrak{T}_{SO_0} \to \mathfrak{T}_{SO_0}$$

to this term and get $\mathfrak{M}(\beta, T/\beta)$ as desired. This looks like a simpler description of Merge than what we used in (1.3.7), and Internal Merge is clearly not a composite operation in this description.

The problem with this description arises when the workspace F has several components, with several syntactic objects as members; for simplicity, say $F = T_1 \sqcup T_2 \sqcup T_3$. Suppose that we proceed as proposed here, and we want to perform $\mathfrak{M}_{S,S'}$ on this workspace. In the case of External Merge we already run into difficulty. Suppose that $S \simeq T_1$ and $S' \simeq T_2$. Then $\tilde{\delta}_{S,S'}$ can find S in the left channel and S' in the right channel in the terms $T_1 \otimes (T_2 \sqcup T_3)$ and $(T_1 \sqcup T_3) \otimes T_2$, or S' in the left channel and S in the right channel in $T_2 \otimes (T_1 \sqcup T_3)$ and $(T_3 \sqcup T_2) \otimes T_1$. (These are some of the terms of $\Delta(F) = \sqcup_{i=1}^{3} \Delta(T_i)$ that come from multiplying with $\sqcup$ the primitive parts (1.2.9) of each $\Delta(T_1)$.) Thus, there are four terms selected by $\tilde{\delta}_{S,S'}$. Now, however, we cannot apply

$$\mathfrak{M} : \mathfrak{T}_{SO_0} \otimes \mathfrak{T}_{SO_0} \to \mathfrak{T}_{SO_0}$$

to these terms, because in each of them, one or the other side has *two* components and therefore is in $\mathfrak{F}_{SO_0}$ not in $\mathfrak{T}_{SO_0}$. The only way to make $\mathfrak{M}$ act on inputs in $\mathfrak{F}_{SO_0}$ is to *mark* which component $\mathfrak{M}$ should act on. Such a marking operation would be difficult to formalize, and it is also undesirable linguistically: having to keep track of specific components by marking them is a case of what is usually referred to in linguistics as a *coindexing* operation—and that would violate the *No Tampering Condition (NTC)*, according to which Merge does not in any way modify the combined objects.

If one instead makes $\mathfrak{M}$ always derive its input entirely from the left channel of the coproduct as the $\mathcal{B} \otimes \mathrm{id}$ operator, there is *no coindexing* involved and NTC is preserved. This choice, however, requires the possibility of moving a term from the right to the left channel of the coproduct. This can again be done easily without any need for coindexing, using the operation $\mathfrak{M}_{\beta,1}$ as previously described.

Note that this does *not* mean that Internal Merge is really a composite of Merge operations, as $\mathfrak{M}_{\beta,1}$ is not in itself a Merge operation. This is so because, since 1 is the unit element, nothing is merged. We will discuss in more detail the properties of $\mathfrak{M}_{\beta,1}$ and show that it does not in fact exist on its own, so the composite nature of (1.4.1) is purely illusory.

In Chapter 2, we will compare these forms of Merge with older versions of Minimalism, such as Edward P. Stabler's Computational Minimalism. We will see that, while here the same mechanism (1.3.7) accounts for both Internal Merge and External Merge, in older versions these two operations exhibit very different algebraic structures, and need to be introduced as two separate mechanisms.

1.4.4 Internal Merge: An Example

The description of Internal Merge given in Proposition 1.4.2 may at first seem at odds with the usual description of IM in the literature (as given for instance in §3.3.2 of *Elements* [40]). However, we can check via an explicit example that our description is indeed the same as in [40].

Consider then the case of equations (18), (19), and (20) of *Elements* [40]. Here, we start with a workspace of the form $WS = [\{\text{was}, \{\text{eaten}, \{\text{the}, \text{apple}\}\}\}]$ that we write as

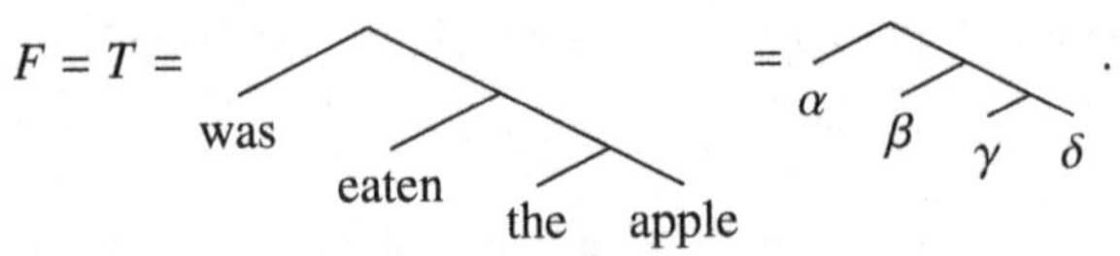

We consider here, according to (19) of *Elements* [40], the action of Merge of the form:

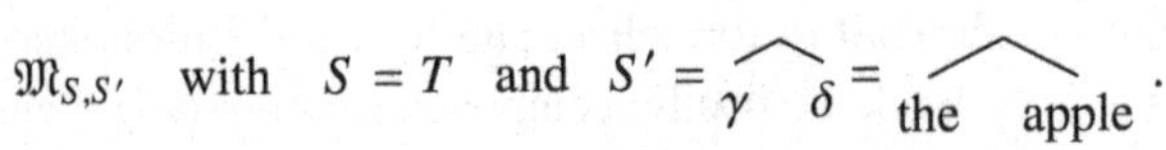

We write

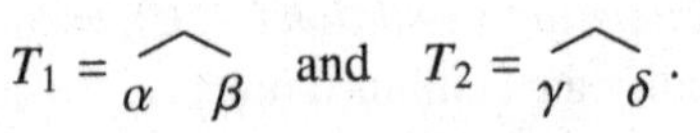

According to Proposition 1.4.2, we first act using $\mathfrak{M}_{S',1}$. The corresponding $\delta_{S',1}$ finds a match in the term of the coproduct $\Delta(T)$ of the form

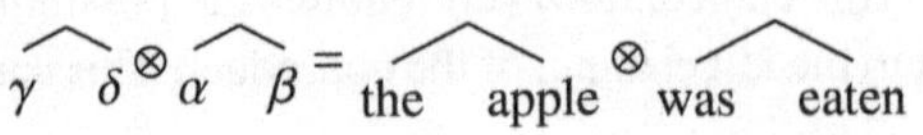

where

$$T_1 = \widehat{\alpha\ \ \beta} = T/^d\, \widehat{\gamma\ \ \delta} = T/^d T_2 \,. \tag{1.4.2}$$

The latter can be also seen as in (1.3.10) as

$$\widehat{\text{was}\ \ \text{eaten}} = T/^d T_2 = \Pi_{d,c}(T/^c T_2) = \Pi_{d,c}\left(\right).$$

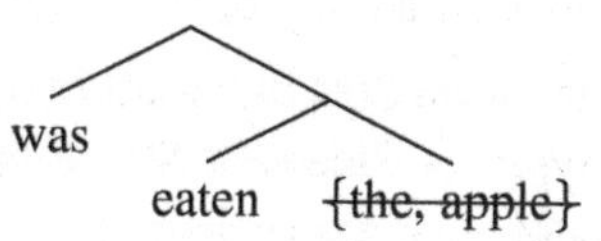

We can identify

$$\widehat{\gamma\ \ \delta} \otimes \widehat{\alpha\ \ \beta} = \widehat{\gamma\ \ \delta} \sqcup 1 \otimes \widehat{\alpha\ \ \beta}\,.$$

Since $\mathcal{B}(T \sqcup 1) = T$, we then obtain as a result of $\mathfrak{M}_{S',1}$ the forest

$$\widehat{\gamma\ \ \delta} \sqcup \widehat{\alpha\ \ \beta}\,.$$

Then, following Proposition 1.4.2, we apply $\mathfrak{M}_{T/S',S'}$ to this forest. Here, $\delta_{T/S',S'}$ finds a match in the term

$$\widehat{\gamma\ \ \delta} \sqcup \widehat{\alpha\ \ \beta} \otimes 1$$

of the coproduct, yielding under $\sqcup \circ (\mathcal{B} \otimes \mathrm{id})$ the following new workspace:

$$\widehat{\widehat{\gamma\ \ \delta}\ \ \widehat{\alpha\ \ \beta}} = \quad .$$

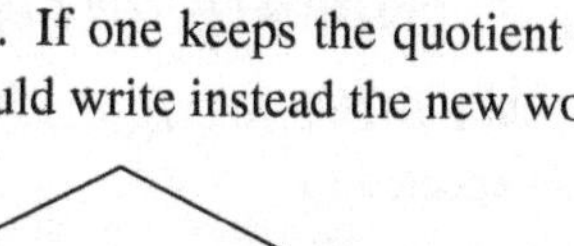

Note here that this differs from the way the resulting workspace is written in equation (20) of *Elements* [40], as the deeper *copy* of T_2 has been cancelled when taking the quotient (1.4.2). If one keeps the quotient notation explicit, using the quotient $T/^c T_2$, one could write instead the new workspace as

$$\widehat{T_2\ \ T/^c T_2} = \quad ,$$

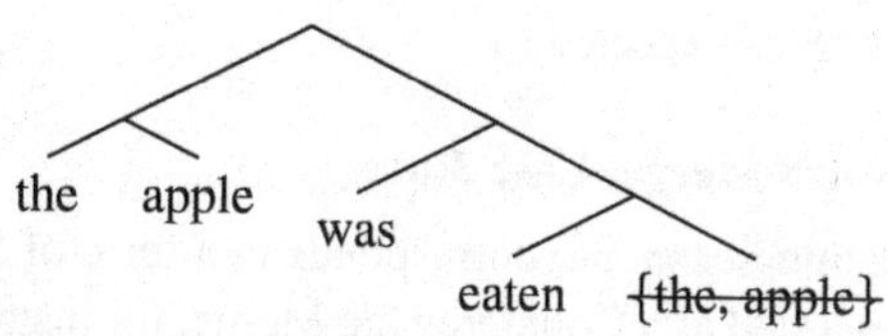

which would more directly be our analog of (20) of *Elements* [40], where the quotient notation $T/^c T_2$ reminds us that this T_2 is a *copy*, and it is the *deeper copy* that gets deleted.

1.4.5 Sideward Merge, Cases 2(b) and 3(b)

Case 2(b) corresponds to a case of Sideward Merge. Here, one obtains in the new workspace F' a component of the form $\mathfrak{M}(T_i,\beta)$ and a component of the form T_j/β. Similarly, case 3(b) also represents a case of Sideward Merge, where in the resulting workspace F', one has new components: $\mathfrak{M}(\alpha,\beta)$, as well as T_i/α and T_j/β.

In the form 2(b) Sideward Merge was originally proposed in [97], and these two types of Sideward Merge operations have been proposed as models for various constructions in linguistics. However, Sideward Merge is criticized as a proposal by Chomsky in [29] and [31].

These cases of Sideward Merge also occur in the formulation of Merge of Definition 1.3.4, as Lemma 1.4.4 clearly shows.

Lemma 1.4.4. *The two cases of Sideward Merge 2(b) and 3(b) are realized by the Merge operators of* (1.3.7) *with $\mathfrak{M}_{T_i,\beta}$, with T_i occurring as a component of F and β as an accessible term of a different component T_j of F, and $\mathfrak{M}_{\alpha,\beta}$, with $\alpha \in Acc(T_i)$ and $\beta \in Acc(T_j)$, for two components $i \neq j$ of F.*

Proof. In the Sideward Merge 2(b) the operator $\delta_{T_i,\beta}$ picks the term of the coproduct $\Delta(F)$, for $F = \sqcup_a T_a$ of the form

$$(T_i \sqcup \beta) \otimes (T_j/\beta \sqcup \hat{F}),$$

with $\hat{F} = \sqcup_{a \neq i,j} T_a$. Then $\mathcal{B} \otimes \mathrm{id}$ acts on this term, producing $\mathfrak{M}(T_i,\beta) \otimes (T_j/\beta \sqcup \hat{F})$, and applying the product $\sqcup$ to this, we obtain $\mathfrak{M}(T_i,\beta) \sqcup T_j/\beta \sqcup \hat{F}$ as expected. Case 3(b) of the Sideward Merge is analogous, with $\delta_{\alpha,\beta}$ selecting the term

$$(\alpha \sqcup \beta) \otimes (T_i/\alpha \sqcup T_j/\beta \sqcup \hat{F}),$$

which is mapped to $\mathfrak{M}(\alpha,\beta) \otimes (T_i/\alpha \sqcup T_j/\beta \sqcup \hat{F})$ by $\mathcal{B} \otimes \mathrm{id}$ and then to $\mathfrak{M}(\alpha,\beta) \sqcup T_i/\alpha \sqcup T_j/\beta \sqcup \hat{F}$ as expected. □

1.4.6 Sideward Merge, Case 3(a)

The last remaining case (3a) corresponds to a form of Sideward Merge that is sometimes regarded as a Countercyclic Merge, for instance in [67]. In general, in linguistics one refers of *countercyclic* constructions to indicate operations where structure formation/modification on a tree is performed away from the root vertex. Case 3(a) can be seen as a special case of this situation since it results in modifying the component T of the workspace away from its root. On the other hand, countercyclic movement usually refers to "growing" a tree away from its root; namely, movements that violate the Extension Condition, which states that Merge should only grow trees *at the root* (we will discuss countercyclic movement in section 1.7). With this definition, the extraction of two substructures T_v and T_w from a tree T and their merging into a new component $\mathfrak{M}(T_v, T_w)$ of the workspace does not, in itself, constitute countercyclic movement (it does not violate the Extension Condition), so it is preferable to regard this case as a further form of Sideward Merge. For example, case 3(a) was considered in [193] as a form of Sideward Merge, where both T_v and T_w simultaneously perform sideward movement, and was proposed in [193] as a model for cleft sentences with multiple phrases in the focus position in Japanese. This form of Merge is also criticized in [29] and [31], and arguments for ruling out these extensions of Merge are discussed in [67] and [102].

In case 3(a), the new workspace F' contains a new component $\mathfrak{M}(\alpha,\beta)$ and a modified component $T_i/(\alpha,\beta)$, where we write $T_i/(\alpha,\beta)$ for the cancellation from the accessible terms of the deeper copies of α and β inside T_i. We will also write this as $T_i/(\alpha \sqcup \beta)$.

This type of Merge can also be obtained through our formulation (1.3.7). In this case, as with Internal Merge, one uses a composition of operators $\mathfrak{M}_{S,S'}$.

Lemma 1.4.5. *Case 3(a) of Merge is realized as $\mathfrak{M}_{\alpha,\beta}$, where matching terms in $F = \sqcup_i T_i$ are found as disjoint accessible terms $\alpha \simeq T_v$, $\beta \simeq T_w$ of the same component T_a of the workspace, corresponding to an admissible cut on two edges, and to a term of the coproduct of the form $T_v \sqcup T_w \otimes (T_a/(T_v \sqcup T_w) \sqcup \hat{F})$, with $\hat{F} = \sqcup_{i \neq a} T_i$.*

Proof. Indeed, in this case the operator $\delta_{\alpha,\beta}$ selects a term of the form $T_v \sqcup T_w \otimes (T/(T_v \sqcup T_w) \sqcup \hat{F})$ in the coproduct, with $\alpha \simeq T_v, \beta \simeq T_w$, which $\mathcal{B}\otimes$id then maps to $\mathfrak{M}(T_v, T_w) \otimes (T/(T_v \sqcup T_w) \sqcup \hat{F})$ and $\sqcup$ maps to $\mathfrak{M}(T_v, T_w) \sqcup T/(T_v \sqcup T_w) \sqcup \hat{F}$. □

This brief discussion shows that, in addition to Internal Merge and External Merge, our construction allows three extensions of Merge, which take the forms of 2(b), 3(a), and 3(b) that correspond to forms of Sideward Merge (and possibly Countercyclic Merge), which are considered undesirable in certain linguistic accounts such as *Elements*. We show in section 1.5 how one can introduce a simple weight (cost function) on the Merge operation described in (1.3.7) that will retain only Internal Merge and External Merge. This will correspond to implementing Minimal Search to eliminate the unwanted extensions of Merge.

Remark 1.4.6. Countercyclic Merge more generally refers to other constructions on trees performed at non-root vertices (especially insertions of structures at internal vertices, violating the Extension Condition) that are *not* attainable in our setting through the Hopf algebra coproduct and the grafting operations. This is because the grafting $\mathcal{B} \otimes \mathrm{id}$, acting on the image of the Hopf coproduct, always involves the root vertex. In contrast, all countercyclic constructions involving insertions at internal non-root vertices are a priori not directly implementable as extensions of Merge in our formalism. In section 1.7, we discuss more precisely how countercyclic insertion should be understood in Hopf algebra theory. All forms of Countercyclic Merge are considered controversial as linguistic models. Further results on Sideward Merge and Countercyclic Merge, in this mathematical setting, will be presented in [142].

1.5 Minimal Search

To avoid misunderstandings regarding the purpose of the construction in the coming sections 1.5.1 and 1.5.3, it is worth stressing that our goal here is simply to implement the usual mechanism by which, in linguistics, only Internal Merge and External Merge are generally retained, eliminating other extensions of Merge, namely the mechanism known as *Minimal Search.* Where our presentation differs from the usual formulation in linguistics is that we provide a somewhat different-looking, but in fact equivalent, description of Minimal Search.

The reason why we introduce a reformulation of Minimal Search is the following. We are arguing here that *all* the key properties of Merge follow directly from underlying Hopf-algebraic properties. In particular, to show that this is the case, we need to reformulate all the necessary aspects of the linguistic description of the key Merge operation in such algebraic terms, including Minimal Search. This requires describing the "minimality" property of Minimal Search in terms of a minimization procedure that can be made sense of entirely in terms of the associated algebraic structure. We argue in sections 1.5.1 and 1.5.3 that this can indeed be done, with minimality expressed in the form of extraction of a leading-order term, with respect to a suitable grading (cost function) assigned to the terms of the coproduct.

1.5.1 The Minimality of Minimal Search

In the formulation of Merge in Chomsky's accounts in [28] and [29], the search for matching copies of S, S' in the workspace components and accessible terms to serve for application of $\mathfrak{M}_{S,S'}$ is performed according to a "Minimal Search" principle, where accessible terms in the higher levels of trees are preferentially searched, before those occurring in the deeper levels.

This should be compared with the discussion of Minimal Search presented in section 3.3.2 of *Elements* [40], which analyzes this situation in terms of the following question: Given the resources available in the workspace, what does Search locate, and why? In the description given in *Elements* [40], Search first locates a connected component (member) of the workspace and then searches for another term to merge with it, in the form of either another component (External Merge) or an accessible term of the component first selected (Internal Merge). All other forms of Merge are expected to be less cost-effective than this form of search, which is therefore called "Minimal Search."

To clarify why we want to provide a mathematical formulation for this description of Minimal Search, and to explain its "minimality," consider the following problem. Consider a workspace $F = T_1 \sqcup T_2$, where T_1 and T_2 are the trees shown in Figure 1.4. Suppose then that we are acting on this workspace with the Merge operator $\mathfrak{M}_{S,S'}$ where $S = T_1$ and the other syntactic objects $S' \in \mathcal{SO}$ is the tree β in the figure, which has a match in the workspace F both as an accessible term of T_1 and as an accessible term of T_2. We need to justify why Minimal Search as described here would more easily (at a lower computational cost) find the occurrence of S inside T_1, even if it is much deeper into the tree and farther from the root, rather than the occurrence in T_2, even though that is located very near the root of a smaller tree, and apparently easier to find in a search. This question will be answered using a cost function, which we define by assigning *weights* to the terms occurring in the coproduct, which will, in a natural way, assign a lower cost to performing the Merge $\mathfrak{M}(T_1, \beta)$ using the occurrence of β inside T_1 than the one inside T_2.

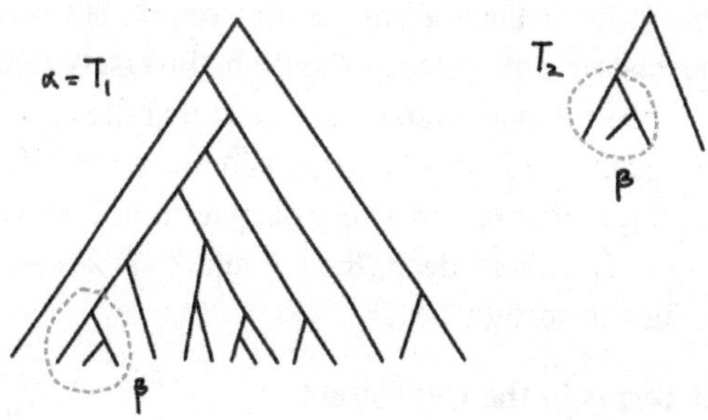

Figure 1.4
Minimal Search for $\mathfrak{M}_{S,S'}$ with $S = T_1$ and $S' = \beta$, should assign a lower cost to finding the copy of β inside T_1 than the one inside T_2.

What we will do here is to quantify, in our description of the action of Merge, in what exact sense the Minimal Search described in section 3.3.2 of *Elements* [40] is indeed minimal, demonstrating that External Merge and Internal Merge are the most efficient (leading-order terms in our formulation) possibilities, while other forms of Merge are not. Since in our description of the action of Merge, the possible material available for search is given by the terms of the coproduct of the Hopf algebra, a quantitative measure of cost/efficiency of different choices should be a way of assigning a weight to the terms occurring in the coproduct that reflects a measure of how computationally hard it is to locate that term and extract it from the object being decomposed by the coproduct, and perform the corresponding cancellation of the deeper copy. What we mean by saying that our choice of cost function (of weights) should be "natural" is that it should be defined in a way that does not introduce any arbitrary new data and only uses the form of the coproduct that we already have in the formalism for the Merge operator. This means that such a weight/cost function is not an arbitrary choice and is already implicitly built into (1.3.7).

The coproduct extracts the entire list of accessible terms (simultaneously implementing the cancellation of deeper copies). It is then natural that a weight measuring the cost of the extraction of a particular accessible term will depend on the distance from the root vertex of that accessible term $T_v \subset T$ from the root of T. This is not all, though, in the sense that while this will assign a cost to the extraction of T_v, we also need to assign a cost to simultaneously perform the corresponding quotient operation T/T_v in the right-channel of the coproduct. The cost of this operation will also naturally depend on the location of $T_v \subset T$ in terms of its distance from the root vertex, but it is clear that the dependence of the cost on this distance should be inversely related in the two cases. Indeed, if on the one hand, one can expect that the cost of extracting a term T_v that is very deep into T should be high compared to extracting near the root of T, the cost of quotienting out a very deep term in T should be very low (since heuristically if T_v is very deep, then T and T/T_v do not differ much). We formalize this idea in section 1.5.2

1.5.2 Weighted Terms in the Coproduct

We want to assign a cost to the operation $\mathfrak{M}(\alpha,\beta)$, where α,β is material obtained from a given workspace F. We will let this cost function depend on a parameter $\epsilon > 0$. This is such that the *minimal cost* can be reinterpreted as the identification of a *leading term*, when one takes a limit on the parameter $\epsilon \to 0$.

We assign a cost to the structures obtained from the given workspace as terms in the coproduct in the following way:

1. *Assess the weight of extracted accessible terms.* It is reasonable to assume that the cost increases with the depth at which the accessible term is located (the depth of where the admissible cut has to be performed), so we assign to $T_v \subset T$ a weight of ϵ^{d_v}, with $d_v = \mathrm{dist}(v, v_T)$, where v_T is the root of T.
2. *Assess the weight of quotient terms.* It is also reasonable to assign a cost to performing the operation $T \mapsto T/T_v$. The larger d_v is, the more similar T and T/T_v are (i.e., the farther down the tree one needs to search to notice the difference). Thus, it is reasonable to assign to the quotient T/T_v a weight of ϵ^{-d_v}.
3. *Assess the weight of multiple extractions/quotients.* If a collection of disjoint accessible terms $F_{\underline{v}} = T_{v_1} \sqcup \cdots \sqcup T_{v_n}$ is simultaneously extracted from the workspace F, the weight associated with the extraction is taken to be the product of the weights, $\epsilon^{d_{\underline{v}}}$, with $d_{\underline{v}} = d_{v_1} + \cdots + d_{v_n}$. Correspondingly, in the quotient term $F/F_{\underline{v}}$, every quotient of a component of $F_{\underline{v}}$ produces a weight of $\epsilon^{-d_{v_i}}$. Thus, a term of the form $F_{\underline{v}} \otimes F/F_{\underline{v}}$ in the coproduct $\Delta(F)$ has weight $\epsilon^{d_{\underline{v}}}$ in the left channel and weight $\epsilon^{-d_{\underline{v}}}$ in the right channel.
4. *Assess the cost of grafting.* We can then assign a cost to the operation of grafting together two trees to a common root. The Merge $\mathfrak{M}(T, 1) = T$ should be a no-cost operation, regardless of the weight of T, since nothing is merged in this case; therefore, we assign to $\mathfrak{M}(T, 1) = T$ a cost $c(\mathfrak{M}(T, 1)) = 0$. In the nontrivial case, we assign to $\mathfrak{M}(T, T')$ a cost that is obtained from the weights of the two merged components, so if T has weight ϵ^d and T' has weight $\epsilon^{d'}$, then we set $c(\mathfrak{M}(T, T')) = d + d'$ and weight $\epsilon^d \cdot \epsilon^{d'} = \epsilon^{c(\mathfrak{M}(T,T'))}$.
5. *Assess the cost/weight of derivations.* Consider a derivation (in the linguistic sense) consisting of a finite sequence of Merge operations $\varphi = \mathfrak{M}_{S_n,S'_n} \circ \cdots \circ \mathfrak{M}_{S_1,S'_1}$. Then the total cost is the sum of the costs of the individual Merge operations, $c(\varphi) = \sum_i c(\mathfrak{M}_{S_i,S'_i})$ and the weight is the product of the weights, $\epsilon^{c(\phi)} = \prod_i \epsilon^{c(\mathfrak{M}_{S_i,S'_i})}$.

Taking the parameter in the range $\epsilon > 1$ gives a larger weight (a multiplicative form of cost) to an extraction of an accessible term T_v with larger d_v (farther from the root) and lower weight (cost) to a deeper quotient (with larger d_v), as discussed in section 1.5.1. In the range where $\epsilon < 1$, terms T_v with smaller d_v dominate in the limit where $\epsilon \to 0$, and terms with larger d_v are subdominant and suppressed.

1.5.3 Minimal Search: Internal and External Merge

Given this accounting, to evaluate the resulting cost of a Merge operation, we keep track of these weights and the cost function by making them explicit in (1.3.7) by taking

$$\mathfrak{M}^{\epsilon}_{S,S'}(F) = \sqcup \circ (\mathcal{B}^{\epsilon} \otimes \mathrm{id}) \circ \delta_{S,S'} \circ \Delta\,, \tag{1.5.1}$$

where the weighted grafting operator $\mathcal{B}^{\epsilon}$ is defined as

$$\mathcal{B}^{\epsilon}(\alpha \sqcup \beta) = \epsilon^{c(\mathfrak{M}(\alpha,\beta))}\, \mathcal{B}(\alpha \sqcup \beta)\,. \tag{1.5.2}$$

Proposition 1.5.1. *Consider the weights/costs as before and $\mathfrak{M}^{\epsilon}_{S,S'}$ as in* (1.5.1).

- *The only zero-cost Merge operations are Internal Merge and External Merge. All other forms of Sideward Merge (i.e., 2(b), 3(a), and 3(b)) have higher costs.*
- *For $\epsilon < 1$, Internal Merge and External Merge are the leading-order terms in any derivation.*
- *In the limit $\epsilon \to 0$, only derivations in which all the Merge operations are Internal Merge and External Merge remain.*

Proof. External Merge applied to a workspace $F = \sqcup_a T_a$ merges two components of the workspace, $\mathfrak{M}(T_i, T_j)$, with cost $c(\mathfrak{M}(T_i, T_j)) = 0$, since the distance to the root is zero for both. Thus, (1.5.1) produces

$$\epsilon^{c(\mathfrak{M}(T_i,T_j))}\mathfrak{M}(T_i, T_j) \sqcup \hat{F} = \mathfrak{M}(T_i, T_j) \sqcup \hat{F}\,,$$

with $\hat{F} = \sqcup_{a\neq i,j} T_a$. Thus, External Merge has cost 0 and weight 1.

Internal Merge is obtained as $\mathfrak{M}^{\epsilon}_{T_v,T_i/T_v} \circ \mathfrak{M}^{\epsilon}_{T_v,1} = \mathfrak{M}^{\epsilon}_{T_v,T_i/T_v} \circ \mathfrak{M}_{T_v,1}$, which gives

$$\epsilon^{c(\mathfrak{M}(T_v,T_i/T_v))}\, \mathfrak{M}(T_v, T_i/T_v) \sqcup \hat{F}$$

for $\hat{F} = \sqcup_{a\neq i} T_a$. We have $c(\mathfrak{M}(T_v, T_i/T_v)) = c(T_v) + c(T_i/T_v) = d_v - d_v = 0$. So Internal Merge also has cost 0 with weight 1.

In the Sideward Merge of type 2(b) we have $\mathfrak{M}^{\epsilon}(T_v, T')$, where T' is a component of the workspace F and T_v is an accessible term $T_v \subset T$ of a different component $T \neq T'$. This Merge operation has cost $c(\mathfrak{M}^{\epsilon}(T_v, T')) = c(T_v) + c(T') = c(T_v) = d_v > 0$, so in this case, we obtain $\epsilon^{d_v}\, \mathfrak{M}^{\epsilon}(T_v, T') \sqcup T/T_v \sqcup \hat{F}$.

In case 3(a), case we have $\mathfrak{M}^{\epsilon}(T_v, T_w)$, with T_v, T_w disjoint accessible terms of the same component T of the workspace F. We have $c(\mathfrak{M}^{\epsilon}(T_v, T_w)) = d_v + d_w > 0$, and we obtain $\epsilon^{d_v+d_w}\, \mathfrak{M}^{\epsilon}(T_v, T_w) \sqcup T/(T_v \sqcup T_w) \sqcup \hat{F}$.

In case 3(b), we have $\mathfrak{M}^{\epsilon}(T_v, T_w)$ with $T_v \subset T$, $T_w \subset T'$ accessible terms of components $T \neq T'$ of the workspace F. This has $c(\mathfrak{M}^{\epsilon}(T_v, T_w)) = d_v + d_w > 0$, so we obtain $\epsilon^{d_v+d_w} \mathfrak{M}^{\epsilon}(T_v, T_w) \sqcup T/T_v \sqcup T'/T_w \sqcup \hat{F}$.

Clearly, for $\epsilon < 1$, the dominant terms are those involving Internal Merge and External Merge as the others are scaled by some $\epsilon^d < 1$ factor. In particular, in the limit $\epsilon \to 0$, any chain $\varphi = \mathfrak{M}_{S_n,S'_n} \circ \cdots \circ \mathfrak{M}_{S_1,S'_1}$ that contains at least one term $\mathfrak{M}_{S_i,S'_i}$ that is a Sideward Merge of type 2(b), 3(a), or 3(b) will have weight $\epsilon^{c(\varphi)} \to 0$, so the only derivations that remain are those involving only External Merge and Internal Merge. □

In particular, in the example in Figure 1.4, $\mathfrak{M}^{\epsilon}(T_1, \beta)$ with the occurrence $\beta \subset T_1$ (Internal Merge) has cost zero (weight 1), while, with the occurrence $\beta \subset T_2$ (Sideward Merge of type 2(b)) has cost $d_v > 0$ and weight ϵ^{d_v}, so the first option is more cost effective than the second, despite the fact that $\beta \subset T_1$ is deeper in the tree than $\beta \subset T_2$.

The cost function described here then implements the minimality of Minimal Search as expected by linguistic theory; see section 3.3.2 of [40].

1.6 Other Linguistic Properties

We verify in this section that the action of Merge on workspaces defined as in (1.3.7) satisfies other desired linguistic properties of the linguistic theory in *Elements* [40]. We have seen in section 1.4 that (1.3.7) accounts for the usual types of Merge: External Merge and Internal Merge, and for three forms of Sideward Merge. We showed in section 1.5 that our model also provides a natural choice of a cost function that implements Minimal Search, eliminating the linguistically undesirable extended forms.

We now show that several properties that are imposed empirically on Merge are in fact naturally built into the mathematical formulation. In particular, we discuss the requirement that Merge does not decrease the total size of workspaces in the course of a derivation; rather, it increases it at most by 1 at each derivational step, as required in *Elements*, [40]. We show that these empirical constraints on workspace size are indeed satisfied by the dominant (Minimal Search) part $\mathfrak{M}^{\epsilon}_{S,S'}|_{\epsilon=0}$ obtained as in Proposition 1.5.1, which recovers Internal/External Merge, while violations occur when one also includes Sideward Merge, confirming what is known from linguistic theory.

Moreover, in section 1.8, we will also show that the cancellation of copies of accessible terms in the resulting workspaces is dictated not only by "economy principles" like these, as usually stated in the linguistics literature, but also more directly by algebraic constraints, dictated by the coassociativity property of the coproduct Δ^c.

1.6.1 Minimal Yield and Complexity

We now analyze the effect of the action of Merge on workspaces in terms of the size of the workspace and the number of accessible terms.

Our discussion here should be directly compared with the discussion on Merge and the non-expansion of the workspaces in section 4 of *Elements* [40]. The results obtained in this subsection, on the effect of different forms of Merge on various measures of workspace size, quantify what is described in section 4 of *Elements* [40] as "Resource Restriction" and "Minimal Yield" (namely, the principle that Merge yields the fewest possible new terms accessible for further computation). These restrictions are intended to be associated with notions of limitations on computational resources; for instance, so-called Third Factor' principles; see *Elements* [40] for further discussion.

Recall from Definition 1.2.2 that one can evaluate the size of the workspace in terms of one of the following measures:

- The number b_0 of connected components of the workspace $F = \sqcup_{a\in\mathcal{I}} T_a$; that is, $b_0(F) = \#\mathcal{I}$: the number of syntactic objects T_a (members of the workspace)
- The number α of accessible terms of the components:

$$\alpha(F) = \#\mathrm{Acc}(F) = \sum_a \#\mathrm{Acc}(T_a)$$

- The number σ of accessible terms of the workspace (components and accessible terms of each component):

$$\sigma(F) = \#\mathrm{Acc}'(F) = b_0(F) + \#\mathrm{Acc}(F) = b_0(F) + \sum_{a\in\mathcal{I}} \#\mathrm{Acc}(T_a) \quad (1.6.1)$$

The notation b_0 for the number of connected components is the usual Betti number notation in topology.

The assessment of the effect of Merge on the number of accessible terms depends on the cancellation of the deeper copy, as we will see more precisely next; that is, on the use of the coproduct Δ^c or Δ^d. This bookkeeping gives rise to slightly different counting of the effect on the size of the workspace. We will see that it is in some way more natural, in view of the formulation of Minimal Yield given in Definition 1.6.1, to use the form Δ^c of the coproduct to evaluate this counting. However, we also explain how one can use a different counting, directly based on the grading of the Hopf algebra, that has two main advantages over the measures b_0, α, σ explained previously: it is not sensitive to the difference between the forms of quotient used in the different coproducts Δ^c, Δ^d, Δ^ρ and does eliminate in a very transparent and straightforward way all the unwanted forms (2(b), 3(b), and 3(a)) of Sideward Merge discussed in Section 1.4.1.

In *Elements* [40], Resource Restriction in its instantiation as the Minimal Yield principle is described as the empirical constraint that Merge yields the fewest possible new terms accessible to further operations. We formulate this principle here as the following requirement.

Definition 1.6.1. A transformation of workspaces $\Phi : \mathfrak{F}_{SO_0} \to \mathfrak{F}_{SO_0}$ satisfies the Minimal Yield principle if the following conditions hold:

$$\begin{array}{ll} b_0(\Phi(F)) \leq b_0(F) & \text{(no divergence)} \\ \alpha(\Phi(F)) \geq \alpha(F) & \text{(no information loss)} \\ \sigma(\Phi(F)) = \sigma(F) + 1 & \text{(minimality of yield)} \end{array} \tag{1.6.2}$$

The requirement $b_0(\Phi(F)) \leq b_0(F)$ is meant to ensure the *convergence* of derivations, while $\alpha(\Phi(F)) \geq \alpha(F)$ is meant to avoid syntactic *information loss*, and $\sigma(\Phi(F)) = \sigma(F)+1$ is a constraint that balances these two conditions (since $\sigma = b_0 + \alpha$) and encodes the "minimality" of Minimal Yield. (One can consider a weaker form by only requiring the two bounds on b_0 and α.)

We also present a different Resource Restriction requirement—an alternative to the constraints expressed in Definition 1.6.1. What we want to describe here is the idea that Merge recursively builds hierarchical structures of increasing complexity. Consequently, a natural constraint on Merge operations arises by requiring that the complexity of the syntactic objects present in the workspace is *nondecreasing* when Merge is applied. Namely, the components of the workspace either remain untouched by Merge or contribute to the building of a new structure that should have greater (or at least the same) complexity. A very simple measure of the level of complexity of the syntactic objects in a workspace is their degree as elements in the Hopf algebra. The Hopf algebra is graded by the number of leaves of the trees/forests. We can see how this provides an estimate of the complexity of the structure, as a binary rooted tree T with $n = \#L(T)$ leaves has a depth of at least $\log_2(n)$, and the depth (the longest path from the root to a leaf) is the most basic measure of the complexity of a binary rooted tree.

In the free, nonassociative, commutative magma $\mathcal{SO}$ of syntactic objects, this measures the number of iterations of the Merge magma multiplication $\mathfrak{M}$ needed to generate that structure. The lower bound on the depth given in terms of the Hopf algebra grading behaves better than the depth itself, as it is more stable. We therefore consider the constraint, which we are going to call the "No Complexity Loss" principle, stating that this lower bound on complexity (measured by the Hopf algebra) is nondecreasing over components of the workspace.

Definition 1.6.2. Consider a transformation of workspaces $\Phi : \mathfrak{F}_{SO_0} \to \mathfrak{F}_{SO_0}$. For $F \in \mathfrak{F}_{SO_0}$, let $\pi_0(F)$ denote the set of connected components of F; for $F = \sqcup_{a \in \mathcal{I}} T_a$, one has $\pi_0(F) \simeq \mathcal{I}$. The induced map $\Phi_0 : \pi_0(F) \to \pi_0(\Phi(F))$ maps a component $a \in \pi_0(F)$ to the component $\Phi_0(a)$ of $\pi_0(\Phi(F))$ that contains the image of the root vertex of the component T_a of F. We say that $\Phi : \mathfrak{F}_{SO_0} \to \mathfrak{F}_{SO_0}$ satisfies the No Complexity Loss principle if, for all $a \in \pi_0(F)$,

$$\deg(\Phi_0(a)) \geq \deg(a), \tag{1.6.3}$$

where $\deg(a)$ is the degree of the component T_a as an element of the Hopf algebra; that is, $\deg(a) = \#L(T_a)$. We define the associated "degree loss" function as

$$DL(\Phi, F) = \min_{a \in \pi_0(F)} (\deg(\Phi_0(a)) - \deg(a)), \tag{1.6.4}$$

with (1.6.3) equivalent to $DL(\Phi, F) \geq 0$.

Observe that the overall degree $\deg(F) = \#L(F)$ of the workspace $F \in \mathfrak{F}_{SO_0}$ is a conserved quantity throughout all the forms of Merge described by (1.3.7), as none of the lexical items originally in the workspace is ever removed. However, what (1.6.3) is measuring is how the degrees $\deg(T_a) = \#L(T_a)$ of the individual components of the workspace change under the effect of the various forms of Merge, by tracking which structures combine and are altered by the transformation, so it is a much more refined counting than just the total degree $\deg(F)$. The No Complexity Loss principle expresses the idea that Merge is a structure formation operation and should always generate structures with increasing, or at least nondecreasing, minimal complexity.

We show that this Minimal Yield property of Definition 1.6.1 holds for the dominant term (i.e., the term that remains when $\epsilon = 0$) of the Merge action of Proposition 1.5.1 (namely, Internal/External Merge), while it fails for the forms 3(b) and 3(a) of Sideward Merge of Section 1.4.1. This recovers an observation already known from linguistics. We also analyze how the measures b_0, α, σ of the size of workspaces described in Definition 1.2.2 transform under Merge, and we give further supportive evidence, based on this counting, of the fact that Internal Merge is *not* a "composite" operation, despite the appearance (see section 1.4.3.1). We also show, however, that the Sideward Merge of type 2(b) cannot be distinguished from Internal Merge, solely on the basis of the size measures of Definition 1.2.2, so it is not eliminated by Minimal Yield requirements as in Definition 1.6.1. On the other hand, we show that our No

Complexity Loss of Definition 1.6.2 suffices to easily rule out all the forms (2(b), 3(b), and 3(a)) of Sideward Merge, leaving only Internal Merge and External Merge.

1.6.2 Cases of Merge and Size Counting

We look now at the change in each of the measures b_0, α, and σ produced by the action of Merge $F \mapsto F' = \mathfrak{M}_{S,S'}(F)$. We first discuss the cases of External/Internal Merge. We then discuss the other forms of Merge that are eliminated by Minimal Search, according to Proposition 1.5.1.

Lemma 1.6.3. *For a workspace F, the counting α of accessible terms satisfies*

$$\alpha(\mathfrak{M}(T_v, T_w)) = \alpha(T_v) + \alpha(T_w) + 2 \tag{1.6.5}$$

$$\sigma(\mathfrak{M}(T_v, T_w)) = \sigma(T_v) + \sigma(T_w) + 1 \tag{1.6.6}$$

for any $v, w \in V(F)$, and for an accessible term $T_v \subset T$,

$$\alpha(T) = \alpha(T_v) + \alpha(T/^d T_v) + 2 \tag{1.6.7}$$

$$\alpha(T) = \alpha(T_v) + \alpha(T/^c T_v) + 1\,. \tag{1.6.8}$$

Correspondingly, we also have

$$\sigma(T) = \sigma(T_v) + \sigma(T/^d T_v) + 1\,. \tag{1.6.9}$$

$$\sigma(T) = \sigma(T_v) + \sigma(T/^c T_v)\,. \tag{1.6.10}$$

Proof. In the first case, the merge $\mathfrak{M}(T_v, T_w)$ contains all the accessible terms of T_v and T_w. Moreover, the root vertices of T_v and T_w also correspond to two accessible terms of $\mathfrak{M}(T_v, T_w)$ given by the full components T_v and T_w, respectively. This gives (1.6.5), while the total number of vertices increases by 1 (the new root vertex), which gives (1.6.6). In the case of the quotient $T/^d T_v$, according to Definition 1.2.5, $T/^d T_v$ is obtained by removing T_v and contracting the edge above the root of T_v and the other edge adjacent to it at the vertex above the root of T_v. Thus, all the vertices of T_v and one additional vertex of T are removed to form $T/^d T_v$. Thus, the total number of vertices satisfies (1.6.9). When counting accessible terms, we then obtain (1.6.7). Similarly, in the case of the quotient $T/^c T_v$, where the subtree T_v is contracted to its root vertex, all the accessible terms (non-root vertices) of T_v are removed from T to form $T/^c T_v$, leaving $\alpha(T) - \alpha(T_v)$. Moreover, the leaf of $T/^c T_v$ corresponding to the root vertex of T_v, with label $\not{T_v}$, is no longer counted as an accessible term of $T/^c T_v$ (cancellation of the deeper copy), so we have a resulting counting as in (1.6.8). □

Note that, in the case of $\sigma(T/^c T_v)$, this is not exactly the total number of vertices of $T/^c T_v$, as it does not count the leaf labeled by $\not{T_v}$ among the accessible terms in $\mathrm{Acc}'(T/^c T_v)$. It is in fact the total number of vertices of $\rho_C(T)$, with C the elementary admissible cut with $\pi_C(T) = T_v$.

Proposition 1.6.4. *Under External Merge and Internal Merge, the effect on the counting functions of Definition 1.2.2 is given by the following table, where we display the difference between the values of each counting function after and before the application of Merge.*

Type of Merge	*Coproduct*	b_0	α	σ
External	Δ^c *and* Δ^d	-1	$+2$	$+1$
Internal	Δ^c	0	$+1$	$+1$
Internal	Δ^d	0	0	0

Thus, the counting with Δ^c satisfies all the constraints of Definition 1.6.1, while the counting with Δ^d satisfies the weaker form only.

Proof. In the case of External Merge, for $F = \sqcup_a T_a$, we have $F' = \mathfrak{M}_{S,S'}(F)$, given by

$$F' = \mathfrak{M}(T_i, T_j) \sqcup \hat{F}^{(i,j)},$$

with $T_i \simeq S$, $T_j \simeq S'$, where S, S' are assumed to be nontrivial syntactic objects (i.e., not equal to 1), and where $\hat{F}^{(i,j)} = F \smallsetminus (T_i \sqcup T_j)$ is the union of the remaining components. Thus, the number of connected components (of syntactic objects in the workspace) decreases by 1, and we obtain $b_0(F') = b_0(F) - 1$. The number of accessible terms $\mathrm{Acc}(F)$ increases by 2 by (1.6.5), and the total number of vertices (accessible terms $\mathrm{Acc}'(F)$ increases by 1 by (1.6.6). The counting for External Merge is independent of the choice of the coproduct, as Δ^c and Δ^d have the same primitive part where no quotient term T/T_v is involved. The special case $\mathfrak{M}_{S,1}$, where $S' = 1$ is the trivial object (empty tree), is discussed in Remark 1.6.7.

In the case of Internal Merge, the number of connected components (the number of syntactic objects in the workspace) remains unchanged, as Internal Merge operates on a single tree, so $b_0(F') = b_0(F)$. The number of accessible terms is obtained by combining (1.6.5) and (1.6.7) or (1.6.8). In the case of the coproduct Δ^d, we obtain

$$\alpha(\mathfrak{M}(T_v, T/^d T_v)) = \alpha(T_v) + \alpha(T/^d T_v) + 2 = \alpha(T),$$

so the number of accessible terms is conserved under Internal Merge, while with respect to the coproduct Δ^c we obtain

$$\alpha(\mathfrak{M}(T_v, T/^c T_v)) = \alpha(T_v) + \alpha(T/^c T_v) + 2 = \alpha(T) + 1 ,$$

so the number of accessible terms increases by 1. The number σ of $\mathrm{Acc}'(F)$ is also preserved in the case of Δ^d and increases by 1 in the case of Δ^c. □

Remark 1.6.5. The number of connected components decreases by 1 under External Merge while remaining unchanged under Internal Merge, so the number of components of the workspace decreases overall during the course of a derivation, as expected, leading to the desired "convergence."

Remark 1.6.6. We have shown that with External Merge, the number of syntactic objects decreases by 1 at each derivation step and the number of accessible terms increases by 2, while under Internal Merge with Δ^d, both the number of syntactic objects and the number of accessible terms remain the same. This fact is consistent with the counting described in [67]. Note that if one takes the quotient $T/^c T_v$ of the coproduct Δ^c instead, then the number of accessible terms (counted by α or by σ) in Internal Merge would increase by 1: this is the counting considered by Riny Huijbregts. With this choice of quotient and counting, both Internal Merge and External Merge would increase the number σ of accessible terms $\mathrm{Acc}'(F)$ by exactly 1. This makes using the counting with Δ^c more natural in terms of what is empirically expected of Merge. This difference in counting with Δ^c and Δ^d only reflects different measures of size (thus, instead of σ, we may call σ_d and σ_c, and similarly for α), not a difference in the Merge operation itself, since (1.3.10) is the same as (1.3.7).

Remark 1.6.7. In the special case of a Merge operation with the form $\mathfrak{M}_{S,1}$, where $S' = 1$ is the trivial syntactic object (empty tree), if S is matched by a component tree T_i of the workspace forest $F = \sqcup_a T_a$, then $\mathfrak{M}(T_i, 1) = T_i$. Thus, $F' = \mathfrak{M}_{S,1}(F) = F$ and the operation is just the identity. In the case of interest for us, where S is matched by a subtree T_{i,v_i} of a component T_i of F, then

$$\mathfrak{M}_{S,1}(F) = \mathfrak{M}(T_{i,v_i}, 1) \sqcup T_i/T_{i,v_i} \sqcup \hat{F} = T_{i,v_i} \sqcup T_i/T_{i,v_i} \sqcup \hat{F} ,$$

for $\hat{F} = \sqcup_{a \neq i} T_a$. In this case, the number of connected components grows by 1, as the original component T_i is separated into two parts. We also have

$$\sigma(T_{i,v_i} \sqcup T_i/^c T_{i,v_i}) = \sigma(T_{i,v_i}) + \sigma(T_i/^c T_{i,v_i}) = \sigma(T_i) ,$$

where the leaf with label $\not{T}_{\overline{i,v_i}}$ is not included in $\mathrm{Acc}'(T_i/^c T_{i,v_i})$, so the root vertex of T_{i,v_i} is counted only once, while

$$\sigma(T_{i,v_i} \sqcup T_i/^d T_{i,v_i}) = \sigma(T_{i,v_i}) + \sigma(T_i/^d T_{i,v_i}) = \sigma(T_i) - 1 ,$$

because of the edge contraction and resulting identification of two vertices in forming the quotient $T_i/^d T_{i,v_i}$. The counting α of accessible terms then satisfies

$$\alpha(T_{i,v_i} \sqcup T_i/^c T_{i,v_i}) = \alpha(T_i) - 1$$

$$\alpha(T_{i,v_i} \sqcup T_i/^d T_{i,v_i}) = \alpha(T_i) - 2 .$$

Thus, we have the following table:

Merge	Coproduct	b_0	α	σ
$\mathfrak{M}_{S,1}$	Δ^c	$+1$	-1	0
$\mathfrak{M}_{S,1}$	Δ^d	$+1$	-2	-1

Thus, $\mathfrak{M}_{S,1}$ violates all the constraints of Minimal Yield of Definition 1.6.1, and this implies that a Merge of the form $\mathfrak{M}_{S,1}$ can occur *only* in compositions $\mathfrak{M}_{\beta,T/\beta} \circ \mathfrak{M}_{\beta,1}$ that give Internal Merge as in Proposition 1.4.2, but *not* alone, since otherwise we would violate such constraints. This is consistent with the fact that the weight in ϵ in Proposition 1.5.1 also excludes the occurrence of a Merge operation of type $\mathfrak{M}_{S,1}$ by itself rather than in a composition that forms an Internal Merge.

In light of Remark 1.6.7, we can revisit the discussion on the apparent "composite" nature of Internal Merge in section 1.4.3.1. The operation $\mathfrak{M}_{S,1}$ used in writing Internal Merge in the form $\mathfrak{M}_{S,T/S} \circ \mathfrak{M}_{S,1}$ does not stand on its own as a merge operation, because it violates the expected properties of nonincreasing number of components and nondecreasing number of accessible terms. However, these bad properties disappear when $\mathfrak{M}_{S,1}$ occurs in the combination $\mathfrak{M}_{S,T/S} \circ \mathfrak{M}_{S,1}$ which satisfies both requirements. One can think of an analogy, where $\mathfrak{M}_{S,1}$ in the decomposition $\mathfrak{M}_{S,T/S} \circ \mathfrak{M}_{S,1}$ plays a role similar to the "virtual particles" in physics terminology, which appear in the computations of interactions but do not exist independently, as they have the wrong physical parameters since they do not obey the correct energy-momentum relation (this behavior is called "off mass shell" in physics). The decomposition of interactions into exchanges of virtual particles that happens in quantum field theory does not correspond to an actual physical decomposition, as the virtual particles are not detectable inputs/outputs. The situation with the apparent "composite" nature of Internal Merge is somewhat similar, in the sense that the "virtual" merge operation $\mathfrak{M}_{S,1}$ appears only as part of a computation but does not have an independent existence as a form of Merge.

The remaining cases of the Merge operation (1.3.7), which are subdominant in $\epsilon \to 0$ in (1.5.1) (and hence eliminated by Minimal Search) have a different behavior with respect to the counting functions of Definition 1.2.2.

In the following list of cases, we assume that the admissible cuts extracting the accessible terms do not happen at the edges immediately below the root of a component of the workspace. This special case can be handled separately and can be easily seen to also violate, in these extended forms of Merge, the empirical constraints Minimal Yield, so we do not discuss it explicitly.

We carry out here all the size computations with respect to the coproduct Δ^c, as that gives a more transparent interpretation of the Minimal Yield condition, in view of what we have seen in Proposition 1.6.4.

Proposition 1.6.8. *In the cases of the Sideward/Countercyclic forms of Merge (1.3.7) of the types 2(b), 3(a), and 3(b), we have the following change in the counting functions of Definition 1.2.2:*

Merge	Coproduct	b_0	α	σ
3(b)	Δ^c	+1	0	+1
3(b)	Δ^d	+1	−2	−1
2(b)	Δ^c	0	+1	+1
2(b)	Δ^d	0	0	0
3(a)	Δ^c	+1	0	+1
3(a)	Δ^d	+1	−2	−1

Thus, all but type 2(b) are ruled out by the Minimal Yield principle, in the strong form (for Δ^c) or in the weak form (for Δ^d).

Proof. In the case of Sideward Merge, case 3(b) of Section 1.4.1, we have, for $F = \sqcup_i T_i$,

$$F' = \mathfrak{M}(T_{a,v_a}, T_{b,w_b}) \sqcup T_a/T_{a,v_a} \sqcup T_b/T_{b,w_b} \sqcup \hat{F}^{(a,b)},$$

with $\hat{F}^{(a,b)} = \sqcup_{i \neq a,b} T_i$. Thus, the number of connected components increases by 1, because of the new component $\mathfrak{M}(T_{a,v_a}, T_{b,w_b})$. By (1.6.5), (1.6.7), and (1.6.8), the number of accessible terms satisfies

$$\alpha(\mathfrak{M}(T_{a,v_a}, T_{b,w_b})) = \alpha(T_{a,v_a}) + \alpha(T_{b,w_b}) + 2,$$

so we have

$$\alpha(F') = \alpha(T_{a,v_a}) + \alpha(T_{b,w_b}) + 2 + \alpha(T_a/^d T_{a,v_a}) + \alpha(T_b/^d T_{b,w_b}) + \alpha(\hat{F}^{(a,b)})$$

$$= \alpha(T_a) + \alpha(T_b) - 2 + \alpha(\hat{F}^{(a,b)}) = \alpha(F) - 2$$

in the case of Δ^d, and also, in the case of Δ^c,

$$\alpha(F') = \alpha(T_{a,v_a}) + \alpha(T_{b,w_b}) + 2 + \alpha(T_a/^c T_{a,v_a}) + \alpha(T_b/^c T_{b,w_b}) + \alpha(\hat{F}^{(a,b)})$$

$$= \alpha(T_a) + \alpha(T_b) + \alpha(\hat{F}^{(a,b)}) = \alpha(F) .$$

Thus, the number of accessible terms in Acc(F) is preserved in the case of Δ^c and decreases by 2 in the case of Δ^d. Thus, $\sigma(F') = \sigma(F) - 1$ in the case of Δ^d and $\sigma(F') = \sigma(F) + 1$ for Δ^c.

In case 2(b) of Section 1.4.1, we have

$$F' = \mathfrak{M}(T_a, T_{b,w_b}) \sqcup T_b/T_{b,w_b} \sqcup \hat{F}^{(a,b)} ,$$

where the number of components remains the same, $b_0(F') = b_0(F)$, while in the counting of accessible terms, we have, in the case of Δ^d,

$$\alpha(F') = \alpha(\mathfrak{M}(T_a, T_{b,w_b})) + \alpha(T_b/^d T_{b,w_b}) + \alpha(\hat{F}^{(a,b)})$$

$$= \alpha(T_a) + \alpha(T_{b,w_b}) + 2 + \alpha(T_b/^d T_{b,w_b}) + \alpha(\hat{F}^{(a,b)})$$

$$= \alpha(T_a) + \alpha(T_b) + \alpha(\hat{F}^{(a,b)}) = \alpha(F) ,$$

so $\sigma(F)$ is also preserved, while for the case of Δ^c,

$$\alpha(F') = \alpha(T_a) + \alpha(T_{b,w_b}) + 2 + \alpha(T_b/^c T_{b,w_b}) + \alpha(\hat{F}^{(a,b)})$$

$$= \alpha(T_a) + \alpha(T_b) + 1 + \alpha(\hat{F}^{(a,b)}) = \alpha(F) + 1 .$$

Finally for case 3(a) of Section 1.4.1, we have

$$F' = \mathfrak{M}(T_{a,v_a}, T_{a,w_a}) \sqcup T_a/(T_{a,v_a} \sqcup T_{a,w_a}) \sqcup \hat{F}^{(a)} ,$$

with $\hat{F}^{(a)} = \sqcup_{i \neq a} T_i$. The number of connected components increases by 1, as the new component $\mathfrak{M}(T_{a,v_a}, T_{a,w_a})$ is created. The counting of admissible terms gives, for Δ^d,

$$\alpha(F') = \alpha(\mathfrak{M}(T_{a,v_a}, T_{a,w_a})) + \alpha(T_a/^d(T_{a,v_a} \sqcup T_{a,w_a})) + \alpha(\hat{F}^{(a)})$$

$$= \alpha(T_{a,v_a}) + \alpha(T_{a,w_a}) + 2 + \alpha(T_a/^d(T_{a,v_a} \sqcup T_{a,w_a})) + \alpha(\hat{F}^{(a)}) .$$

We are now extracting from the tree T_a a forest $F_{\underline{v}} = T_{a,v_a} \sqcup T_{a,w_a} = \pi_C(T_a)$ of accessible terms, corresponding to an admissible cut C on two edges, and the counting of accessible terms follows the analog of (1.6.7). In $T_a/^d\pi_C(T_a)$, two edges are contracted, and hence the counting of (non-root) vertices decreases by 2, while the root vertex of each of the two components of $\pi_C(T_a)$ is not counted as an accessible term of $\pi_C(T_a)$, while being counted as an accessible term of T_a. This gives an overall discrepancy of 4 in the counting between $\alpha(\pi_C(T_a)) + \alpha(T_a/^d\pi_C(T_a))$ and $\alpha(T_a)$. Thus, we get

$$\alpha(F') = \alpha(T_a) - 2 + \alpha(\hat{F}^{(a)}) = \alpha(F) - 2 .$$

From the relation

$$\alpha(\pi_C(T_a)) + \alpha(T_a/^d\pi_C(T_a)) + 4 = \alpha(T_a) ,$$

we obtain

$$\sigma(\pi_C(T_a)) + \sigma(T_a/^d\pi_C(T_a))) + 2 =$$
$$(2 + \alpha(\pi_C(T_a)) + (1 + \alpha(T_a/^d\pi_C(T_a))) + 2 = 1 + \alpha(T_a) = \sigma(T_a) .$$

We then have

$$\sigma(F') = \sigma(\mathfrak{M}(T_{a,v_a}, T_{a,w_a}) + \sigma(T_a/^d\pi_C(T_a))) + \sigma(\hat{F}^{(a)}) =$$
$$\sigma(\pi_C(T_a)) + 1 + \sigma(T_a/^d\pi_C(T_a))) + \sigma(\hat{F}^{(a)})$$
$$= \sigma(T_a) - 1 + \sigma(T_a/^d\pi_C(T_a))) + \sigma(\hat{F}^{(a)}) = \sigma(F) - 1 .$$

In the case of Δ^c, we similarly have

$$\alpha(\pi_C(T_a)) + \alpha(T_a/^c\pi_C(T_a)) + 2 = \alpha(T_a),$$

as the two root vertices of $\pi_C(T_a)$ are not counted as accessible terms in either component. This gives $\alpha(F') = \alpha(F)$ and $\sigma(F') = \sigma(F) + 1$. Note that for case 3(a) we do not need to consider explicitly cases with $T_{a,v_a} \subset T_{a,w_a}$ or $T_{a,w_a} \subset T_{a,v_a}$, as those do not correspond to admissible cuts of the tree T_a. □

Remark 1.6.9. The Sideward Merge of type 2(b) cannot be distinguished solely in terms of its effect on the sizes b_0, α, and σ from Internal Merge. Indeed, it is clear that the transformations of the workspaces

$$T \sqcup T' \mapsto \mathfrak{M}(T_v, T') \sqcup T/T_v$$
$$T \sqcup T' \mapsto \mathfrak{M}(T_v, T/T_v) \sqcup T'$$

have the same effect on the number of connected components b_0, the number of accessible terms α, and the size of $\mathrm{Acc}'(F)$, as measured by σ. This can be seen directly in the table from Proposition 1.6.8. We show in section 1.6.3 that they are instead easily distinguished by the No Complexity Loss constraint of Definition 1.6.2. A further analysis of the different forms of Sideward Merge, their minimality violations, and direct applications to linguistics questions will be presented in [142].

1.6.3 No Complexity Loss Constraint

We now check that the No Complexity Loss constraint of Definition 1.6.2 is satisfied only by Internal Merge and External Merge and is violated by all other forms (2(b), 3(a), and 3(b)) of Sideward Merge of Section 1.4.1. It is also violated by $\mathfrak{M}_{S,1}$, so the same argument of Remark 1.6.7 (that $\mathfrak{M}_{S,1}$ cannot exist in isolation and that Internal Merge is therefore not composite) is obtained from No Complexity Loss.

Proposition 1.6.10. *Only Internal Merge and External Merge satisfy the No Complexity Loss constraint of Definition 1.6.2. All other forms of Merge (2(b), 3(a), and 3(b)) as well as $\mathfrak{M}_{S,1}$, would violate this principle. In particular, No Complexity Loss distinguishes Internal Merge and Sideward Merge of type 2(b), eliminating the latter.*

Proof. For External Merge, the two components T_i, T_j that are merged to the new component $\mathfrak{M}(T_i, T_j)$ have $\deg(\mathfrak{M}(T_i, T_j)) = \deg(T_i) + \deg(T_j)$, which is greater than or equal to both $\deg(T_i)$ and $\deg(T_j)$. All the remaining components of the workspace not used by Merge maintain the same degree. For Internal Merge, similarly, $\deg(T_v, T/T_v) = \deg(T)$. Sideward Merge of type 2(b) produces modified components of the form $\mathfrak{M}(T_v, T') \sqcup T/T_v$, where the component T' of the original workspace contributes to a new component of increased degree $\deg(\mathfrak{M}(T_v, T')) > \deg(T') + \deg(T_v)$. However, the root of the component T is mapped (as in Definition 1.6.2) to the root of the component T/T_v in the new workspace F', with $\deg(T/T_v) < \deg(T)$; thus, it violates the No Complexity Loss constraint. Similarly, in case 3(b), which produces components $\mathfrak{M}(T_v, T_w) \sqcup T/T_v \sqcup T'/T_w$, the root vertices of the old components T and T' both map to components T/T_v and T'/T_w of lower degree. Case 3(a) is analogous since it produces $\mathfrak{M}(T_v, T_w) \sqcup T/(T_v \sqcup T_w)$ and the root vertex of the original component T is mapped to a component of lower degree, so it does not participate in the formation of a structure of nondecreasing complexity. The case of $\mathfrak{M}_{T_v,1}$ is analogous, as it produces $T_v \sqcup T/T_v$ with $\deg(T/T_v) < \deg(T)$. □

We have seen that the arguments of Minimal Search as cost minimization for all the operations involved in the definition of Merge (extraction of accessible terms, cancellation of deeper copies, grafting), as well as Minimal Yield constraints on size countings for workspaces and analogous complexity arguments, select Internal Merge and External Merge as the primary operations and discard other proposed extensions like Sideward Merge.

There is a further argument that one can make, however, which identifies a sense in which External Merge is more complex than Internal Merge. It was observed in [37] that External Merge is "unboundedly more complex" than Internal Merge. This is not evident if one thinks in terms of the search for matching terms inside a fixed workspace, as we have argued for Minimal Search. However, we can also consider the following setting. Suppose that we fix one of the two syntactic objects S, S' of the Merge operation $\mathfrak{M}_{S,S'}$ as defined in (1.3.7), and we consider an operation of the form

$$\mathfrak{M}^S(T) := \mathfrak{M}(S,T),$$

$$T \mapsto \sqcup \circ (\mathfrak{M}^S \otimes \mathrm{id}) \circ \delta_{S'} \circ \Delta(T),$$

where $\delta_{S'}(T \otimes T') = S' \otimes T'$ if $T \simeq S'$ and zero otherwise. (We will be discussing a similar form of the Merge operation in Lemma 1.12.1, later in this chapter.) This means that we block one of the terms of Merge and look at the resulting operation $\mathfrak{M}(S,\cdot)$ as acting on the set of all possible syntactic objects. It is then clear why External Merge is "unboundedly more complex" than Internal Merge, as the possible arguments of $\mathfrak{M}(S,\cdot)$ that produce Internal Merge are a finite set (the accessible terms of the chosen S), while the set of possible arguments that produce External Merge countably infinite, consisting of arbitrary syntactic objects.

1.7 Countercyclicity and Extensions of Merge?

We argued in the previous sections of this chapter that the Merge operation described in (1.3.7) includes, a priori, possible extensions of Merge beyond External Merge and Internal Merge, resulting in three forms of Sideward Merge (one of which is sometimes also regarded as a form of Countercyclic Merge), given by the cases 2(b) and 3(b), and case 3(a) of section 1.4.1, respectively. We have shown that these additional forms of Merge are subdominant terms for the Minimal Search procedure, and also cause violations of optimality constraints on the effect on the size of the workspace and on complexity. Thus, they can be excluded on the basis of these failures of optimality. (For a more detailed discussion of the effect of optimality violations, see [142].)

In this section, we discuss briefly what our model can say about other proposed extensions of Merge that have been considered in the literature, such as Countercyclic and Late Merge, involving operations that grow a tree structure at internal vertices away from the root (see Figure 1.5). These operations violate the Extension Condition, which states that Merge should only grow trees at the root, while insertions as illustrated in Figure 1.5 happen by inserting a new vertex in one of the edges and attaching a new subtree to this vertex, introducing a growth of the tree away from the root vertex.

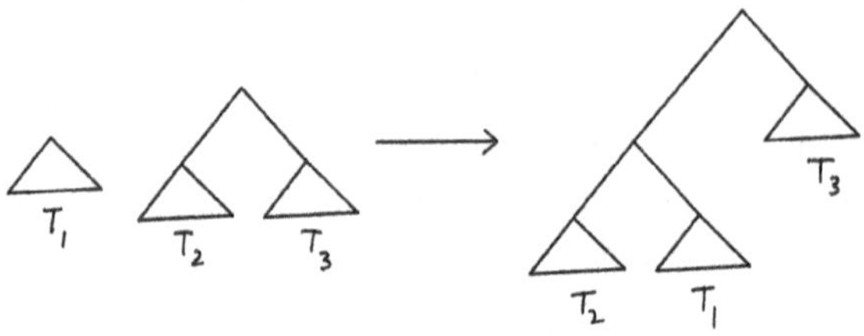

Figure 1.5
Some proposed extensions of Merge (like Countercyclic Merge and Late Merge) involve the grafting of trees at internal vertices, away from the root vertex.

We do not discuss here in detail such proposed extensions of Merge. Their viability as linguistic models is criticized in [29] and elsewhere. We only provide a general argument for how these proposed extensions of Merge can be analyzed in terms of the algebraic structure underlying our description of the Merge operation of (1.3.7), in particular the Hopf algebra of workspaces described in sections 1.2 and 1.2.1, and why this view strongly indicates that constructions accounted for by such extensions should in fact be directly obtainable just in terms of the formulation of Internal/External Merge of (1.3.7) (possibly with the use of a form of Sideward Merge that minimally violates optimality, as discussed in [142]).

We will show that insertions at internal vertices obey their own algebraic structure (namely, a Lie algebra), and this has a well-known explicit relation to the Hopf algebra structure on the set of workspaces that we have been discussing in the previous sections. This relation is part of a general duality structure for Hopf algebras. This duality relation in particular implies that those instances of Merge extensions involving countercyclic movement and insertion at lower levels of the trees will in fact produce structures that already exist in what is obtainable in our formulation of the free symmetric Merge, and therefore should be otherwise obtainable from External/Internal Merge alone (possibly with some cases of Sideward Merge, see [142]).

To see this, we first describe the algebraic structure underlying insertion operations of the type illustrated in Figure 1.5.

1.7.1 Premise: Lie Algebras and Hopf Algebras

We first recall here another important algebraic structure: Lie algebras. In particular, we discuss their relation to the notion of Hopf algebra that we have already encountered. We discuss in section 1.7.2 how this structure relates to the insertion operation illustrated in Figure 1.5.

A Lie algebra is a vector space $\mathcal{L}$ over a field $\mathbb{K}$ endowed with a bilinear operation $[\cdot,\cdot] : \mathcal{L} \otimes \mathcal{L} \to \mathcal{L}$ (Lie bracket) satisfying $[L, L] = 0$ (hence $[L_1, L_2] = -[L_2, L_1]$ by bilinearity) and the Jacobi identity

$$[L_1, [L_2, L_3]] + [L_2, [L_3, L_1]] + [L_3, [L_1, L_2]] = 0 \qquad (1.7.1)$$

for all L, L_1, L_2, L_3 in $\mathcal{L}$.

A *right pre-Lie structure* on a vector space $\mathcal{L}$ is a bilinear map

$$\triangleleft : \mathcal{L} \otimes \mathcal{L} \to \mathcal{L}$$

satisfying the identity

$$(L_1 \triangleleft L_2) \triangleleft L_3 - L_1 \triangleleft (L_2 \triangleleft L_3) = (L_1 \triangleleft L_3) \triangleleft L_2 - L_1 \triangleleft (L_3 \triangleleft L_2) \quad (1.7.2)$$

for all $L_1, L_2, L_3 \in \mathcal{L}$. A right pre-Lie structure determines a Lie algebra structure by setting

$$[L_1, L_2] := L_1 \triangleleft L_2 - L_2 \triangleleft L_1 \,. \qquad (1.7.3)$$

The pre-Lie identity (1.7.2) ensures that the Jacobi identity (1.7.1) holds for the bracket of (1.7.3). A left pre-Lie structure is defined analogously and determines a Lie algebra in the same way.

There is a close relation between Lie algebras and Hopf algebras. Namely, given a Lie algebra $\mathcal{L}$, one can form an associative algebra $U(\mathcal{L})$, called the *universal enveloping algebra* of the Lie algebra $\mathcal{L}$. This is obtained as the quotient $U(\mathcal{L}) = \mathcal{T}(\mathcal{L})/\mathcal{I}(\mathcal{L})$ of the tensor algebra $\mathcal{T}(\mathcal{L})$ by the ideal generated by the relation

$$L_1 \otimes L_2 - L_2 \otimes L_1 = [L_1, L_2] \,. \qquad (1.7.4)$$

More explicitly, this means that elements of $U(\mathcal{L})$ are polynomials (in noncommuting variables) in the elements of $\mathcal{L}$, with the commutation relation (1.7.4) given by the Lie bracket of $\mathcal{L}$. The universal enveloping algebra is in fact also a Hopf algebra, with the cocommutative coproduct determined by the property that all the elements $L \in \mathcal{L}$ are primitive, $\Delta_{U(\mathcal{L})}(L) = L \otimes 1 + 1 \otimes L$. Given a Hopf algebra $\mathcal{H}$, there is a *dual* Hopf algebra $\mathcal{H}^\vee$ structure on the underlying dual vector space, where product and coproduct exchange roles in the sense that the dual product is induced by the original coproduct and vice versa. The unit and counit similarly exchange roles. The dual $U(\mathcal{L})^\vee$ is a commutative Hopf algebra since $U(\mathcal{L})$ is cocommutative. A well-known result in the theory of Hopf algebras, the Milnor–Moore theorem ([153], Theorem 5.18) shows that a commutative, graded connected Hopf algebra $\mathcal{H}$ (see Chapter 4 for details) is isomorphic to $\mathcal{H} = U(\mathcal{L})^\vee$ for a Lie algebra $\mathcal{L}$. This Lie algebra also has a characterization as follows.

Let $\mathcal{H}$ be a combinatorial Hopf algebra over a field $\mathbb{K}$, in the sense recalled in section 1.2.1. Consider the vector space of linear functionals $\phi : \mathcal{H} \to \mathbb{K}$. In particular, consider those linear functionals that satisfy the relation

$$L(xy) = L(x)\epsilon(y) + \epsilon(x)L(y)\,, \tag{1.7.5}$$

where $\epsilon : \mathcal{H} \to \mathbb{K}$ is the counit of the Hopf algebra. We define a product operation $L_1 \star L_2$ on these functionals, defined by the relation

$$(L_1 \star L_2)(x) := (L_1 \otimes L_2)(\Delta(x))\,, \tag{1.7.6}$$

and we set

$$[L_1, L_2] := L_1 \star L_2 - L_2 \star L_1\,. \tag{1.7.7}$$

Note that the product (1.7.6) on linear functionals on $\mathcal{H}$ is dual to the coproduct of $\mathcal{H}$. Similarly, there is a coproduct dual to the product of $\mathcal{H}$, and (1.7.5) corresponds to the condition that L is a primitive element with respect to this dual coproduct, with the counit ϵ playing the role of unit in the dual.

We have not discussed this yet, but the form of the counit $\epsilon : \mathcal{H} \to \mathcal{K}$ in a graded, connected, commutative Hopf algebra $\mathcal{H}$, is simply given by an isomorphism $\epsilon : \mathcal{H}_0 \to \mathbb{K}$ (with inverse given by the unit) and with $\mathrm{Ker}(\epsilon) = \oplus_{k\geq 1}\mathcal{H}_k$; namely, every $x \in \mathcal{H}$ not of degree zero (i.e., $\deg(x) \geq 1$) is mapped to zero by the counit. Moreover, a combinatorial Hopf algebra $\mathcal{H}$, with respect to the product operation, is a polynomial algebra in a set B of generators (of $\mathcal{H}$ as an algebra); namely, as a vector space, $\mathcal{H}$ is the linear span of monomials of the form $\prod_i b_i$ with $b_i \in B_{\ell_i}$ of degree $\deg(b_i) = \ell_i$. These two facts combine to show that condition (1.7.5) on the linear functional $L : \mathcal{H} \to \mathbb{K}$ implies that L is supported on the basis elements (of $\mathcal{H}$ as a vector space) that are single elements $b \in B$, rather than products $\prod_i b_i$ of several such elements.

The coassociativity of the coproduct Δ of $\mathcal{H}$ ensures the associativity of the dual product of (1.7.6), and this in turn implies that the bracket defined by (1.7.7) satisfies the Jacobi identity (1.7.1). Thus, the linear functionals satisfying (1.7.5), endowed with the bracket (1.7.7), form a Lie algebra, which we denote by $\mathcal{L}$, associated with the Hopf algebra $\mathcal{H}$, which is the Lie algebra of primitive elements in the dual Hopf algebra $\mathcal{H}^\vee$, as described previously.

An equivalent, though more technical, description of the Lie algebra associated as before to a graded, connected, commutative Hopf algebra $\mathcal{H}$ is obtained in the following way. Such commutative Hopf algebras can be described equivalently in terms of an *affine group scheme* G. This is defined in the following way: for any commutative algebra $\mathcal{R}$, the set of morphisms (of commutative algebras) $G(\mathcal{R}) = \mathrm{Hom}(\mathcal{H}, \mathcal{R})$ is not just a set; it also has the structure of a *group*, with multiplication operation

$$(\phi_1 \star \phi_2)(x) = (\phi_1 \otimes \phi_2)(\Delta(x))\,.$$

In other words, the group operation on morphisms in $\mathrm{Hom}(\mathcal{H}, \mathcal{R})$ is induced by the coproduct operation on $\mathcal{H}$. The inverses in the group are similarly determined by the antipode of the Hopf algebra $\mathcal{H}$. In particular, G can be regarded as a Lie group whose Lie algebra agrees with the Lie algebra of primitive elements of the dual Hopf algebra. We will not discuss here the viewpoint based on this notion of affine group scheme, but we mention it briefly because we will encounter a variation on this idea in chapter 3, and especially in section 3.5.

1.7.2 Insertion Lie Algebra

We now show that Lie algebras are the natural algebraic structure that accounts for insertion operations at the internal vertices of a tree, of the type involved in the forms of Countercyclic Merge and Late Merge. More precisely, the insertions determine a pre-Lie structure.

Definition 1.7.1. Consider two nonplanar binary rooted trees $T_1, T_2 \in \mathfrak{T}_{SO_0}$, where T_1 has a nonempty set of edges. For a given edge $e \in E(T_1)$, we define an insertion operation where

$$T_1 \triangleleft_e T_2$$

denotes the binary rooted tree obtained by splitting the edges e with the insertion of a new (valence 2) vertex v, attaching to v a new edge e' and attaching to the other end of e' the root of T_2. We then set

$$T_1 \triangleleft T_2 = \sum_{e \in E(T_1)} T_1 \triangleleft_e T_2\,, \tag{1.7.8}$$

where the sum provides the list of all the possible ways in which the tree T_2 can be inserted in the tree T_1 with the operation described by $T_1 \triangleleft_e T_2$.

The insertion $T_1 \triangleleft_e T_2$ is exactly the type of insertion operation illustrated in Figure 1.5, used in the construction of Countercyclic Merge and Late Merge.

Lemma 1.7.2. *The insertion operation $T_1 \triangleleft T_2$ of* (1.7.8), *on the subspace of $\mathcal{V}(\mathfrak{T}_{SO_0})$ spanned by trees with a nonempty set of edges, satisfies the right* *pre-Lie identity* (1.7.2), *hence determining on this subspace the structure of Lie algebra with the Lie bracket*

$$[T_1, T_2]_R = T_1 \lhd_e T_2 - T_2 \lhd_e T_1 \,. \tag{1.7.9}$$

We can similarly define a left pre-Lie structure $T_1 \rhd T_2$, *given by the sum of all possible insertions of* T_1 *in* T_2 *and bracket* $[T_1, T_2]_L = T_1 \rhd_e T_2 - T_2 \rhd_e T_1$ *with* $[T_1, T_2]_L = -[T_1, T_2]_R$.

Proof. Consider the term $(T_1 \lhd T_2) \lhd T_3$. Here, one first performs all the possible insertions of T_2 at the edges of T_1, and then the insertions of T_3 at the edges of the resulting tree. There are three different possibilities of where the second insertion happens:

1. At an edge of T_2
2. At an edge of T_1 different from the edge used for the insertion of T_2
3. At one of the three new edges produced by the insertion of T_2 into T_1 (the two split parts e_1, e_2 of e and the new e' from the new vertex to the root of T_2)

In the first case, insertions of T_3 at edges of T_2 can be performed before the insertion at the edge of T_1, so they are the same terms that give $T_1 \lhd (T_2 \lhd T_3)$. Thus, the difference $(T_1 \lhd T_2) \lhd T_3 - T_1 \lhd (T_2 \lhd T_3)$ consists of the sum of terms in cases (2) and (3). Now consider similarly the expression $(T_1 \lhd T_3) \lhd T_2 - T_1 \lhd (T_3 \lhd T_2)$. After a similar cancellation of terms, one is left with two cases: the insertion of T_2 at an edge of T_1 different from the one used for the insertion of T_3, and the insertion of T_1 at one of the three new edges produced by the insertion of T_2 in T_1. The case of insertions at different edges matches the same case in the previous list, as the two insertions in this case can be done in either order without affecting the result. The terms obtained in case (3) match, as shown in Figure 1.6. □

1.7.3 Insertion Lie Algebra of Workspaces

Let $\mathcal{H}$ denote the Hopf algebra of workspaces, as described in sections 1.2 and 1.2.1, with the coproduct (1.2.8) of the form Δ^c. The relation between the Lie algebra $\mathcal{L}$ of primitive elements in the dual Hopf algebra and the insertion Lie algebra described in section 1.7.2 is given as follows (see also the discussion in section 4.3 of [19]).

Lemma 1.7.3. *The insertion Lie algebra of Lemma 1.7.2 is the Lie algebra of primitive elements in the dual Hopf algebra of the Hopf algebra of workspaces.*

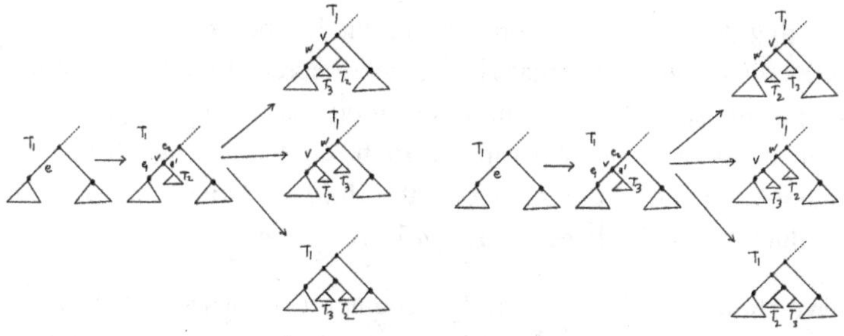

Figure 1.6
Insertion of T_3 (or T_2) at one of the edges produced by the insertion of T_2 (or T_3) in T_1.

Proof. We take as basis elements for the dual $\mathcal{H}^\vee$ (as a vector space) the delta functions δ_F, for $F \in \mathfrak{F}_{SO_0}$, with $\delta_F(F') = 1$ if $F = F'$ and zero otherwise, extended by linearity to $\mathcal{V}(\mathfrak{F}_{SO_0})$. Since the coproduct of $\mathcal{H}^\vee$ is dual to the product of $\mathcal{H}$, we have that the primitive elements among the δ_F are given by the δ_T for $T \in \mathfrak{T}_{SO_0}$, since

$$\Delta_{\mathcal{H}^\vee}(\delta_T)(x \otimes y) = \delta_T(xy) = (\delta_T \otimes \epsilon + \epsilon \otimes \delta_T)(x \otimes y),$$

where ϵ is the counit of $\mathcal{H}$ (unit of $\mathcal{H}^\vee$). The product of $\mathcal{H}^\vee$ is dual to the coproduct

$$(\delta_{T_1} \star \delta_{T_2})(x) = (\delta_{T_1} \otimes \delta_{T_2})\Delta(x) = \sum_v \delta_{T_1}(T_v)\delta_{T_2}(T/T_v).$$

Thus, the product can be written in the form

$$\delta_{T_1} \star \delta_{T_2} = \sum_T c^T_{T_1,T_2}\, \delta_T ,$$

where

$$c^T_{T_1,T_2} = \#\{v \in V(T) \,|\, T_v \simeq T_1 \text{ and } T/T_v \simeq T_2\}$$

counts the number of matches for the syntactic objects T_1 and T_2 as, respectively, an accessible term of T and its complementary piece. The Lie algebra of primitive elements of $\mathcal{H}^\vee$ then has as its Lie bracket

$$\begin{aligned}
[\delta_{T_1}, \delta_{T_2}] &= \delta_{T_1} \star \delta_{T_2} - \delta_{T_2} \star \delta_{T_1} \\
&= \sum_T c^T_{T_1,T_2}\, \delta_T - \sum_T c^{T'}_{T_2,T_1}\, \delta_{T'} \\
&= \sum_{e \in E(T_2)} \delta_{T_1 \rhd_e T_2} - \sum_{e' \in E(T_1)} \delta_{T_2 \rhd_{e'} T_1} = \delta_{T_1 \rhd T_2 - T_2 \rhd T_1} ,
\end{aligned}$$

so we see that the insertion Lie bracket and the Lie bracket of the Lie algebra of primitive elements of the dual Hopf algebra agree. Finally, notice that the restriction to the span of trees with a nonempty set of edges in Lemma 1.7.2 for the Lie algebra structure corresponds to the quotient by the ideal generated by the elements $1 - \alpha$ with $\alpha \in \mathcal{SO}_0$ in the Hopf algebra of workspaces with the coproduct Δ^c, as discussed in section 1.2.1. □

We have shown here that the kind of countercyclic insertions described by the operations $T \lhd_e T'$ and $T \rhd_{e'} T'$ are naturally accounted for by the "dual Lie algebra" (meaning the Lie algebra of primitive elements in the dual Hopf algebra) of the Hopf algebra of workspaces. This identification suggests that this type of insertions do *not* represent new operations and genuine extensions of Merge, but it should in fact be equivalently describable in terms of the original generative process involving only Internal Merge and External Merge (with possibly a role for some forms of Sideward Merge, as discussed in [142]). Indeed, the identification in Lemma 1.7.3 between $\delta_{T_1} \star \delta_{T_2} - \delta_{T_2} \star \delta_{T_1}$ and the dual of $T_1 \rhd T_2 - T_2 \rhd T_1$ shows that the insertion $T_1 \rhd T_2$ can be seen as a way of inverting the extraction of accessible terms performed by the coproduct, resulting in a structure $T = T_1 \rhd T_2$ that is *already accounted for* in the products of the Merge operation on workspaces. The argument given here is incomplete, as it needs to also take into consideration how the structure $T = T_1 \rhd T_2 \in \mathcal{SO}$ behaves with respect to the head, phases, and labeling algorithm, which we have not yet discussed. We will return to discuss these additional data in sections 1.13.3, 1.14, and 1.15 of this chapter, while here, we discuss briefly a simple example to illustrate the point.

The kind of linguistic phenomena for which Late Merge is proposed as an explanation typically involve syntactic objects that occur in a certain position as accessible terms of a larger structure, but that behave as if absent from that position, thus suggesting the idea that they are "late-merged" by the kind of insertion described here, into the rest of the structure already formed. An example is given by a sentence like

[These pictures of John$_i$]$_j$ seemed to him$_i$ [$-_j$ to be very good].

This sentence is an instance of "raising of subject" (a case of A-movement). The apparent problem here is that *John* and *him* are coindexed, and the coindexing of these terms, when the first occurs in the base position $-_j$, would result in a violation of Condition C of Binding Theory, which requires that R-expressions (referring expressions) cannot be bound by a coindexed syntactic object. This is interpreted in the Late Merge proposal as an indication that *of John* is late-merged into its position by an insertion at the edge above the substructure *These pictures*. However, this is not a problem within the Merge formulation of syntax in SMT, as the subject raising in this sentence is a *single phase* and the binding conditions B and C do not apply. So this does not require the introduction of a different form of Merge, and indeed the structure obtained via this insertion is already realizable with the original form of External Merge and Internal Merge, in the way that the mathematical argument described here also suggests. For a discussion of specific linguistic cases where some form of Sideward Merge may also be involved, see [142].

1.8 Cancellation of Copies

As we have already discussed in sections 1.2 and 1.2.1, in the form (1.3.7) of the action of Merge on workspaces, the cancellation of the deeper copies of the accessible terms used by Merge is implemented by the coproduct Δ of (1.2.8) through the quotient terms $T/^{c}T_v$ (and the corresponding $T/^{d}T_v$ with the cancellation actually performed, as will be the case in Externalization).

Cancellation of copies is usually posited as an "economy principle" in linguistics, and it is usually assumed that cancellation always happens in the deeper copies. In our algebraic model, we can say something more, and perhaps go a bit deeper than this.

A first observation is that, in the formalism that we are using, the fact that cancellation is implemented in the deeper copy is directly built into the structure of the coproduct, and it does not have to be included as an additional requirement since in the terms $T_v \otimes T/^{c}T_v$ the copy of T_v on the left-hand-side (the one that contributed to Merge) has a lower depth than the copy inside T, which is cancelled on the right-hand-side.

A second observation is that cancellation of copies is necessary to have a "good," well-behaved, coassociative coproduct Δ^c. Indeed, one needs to quotient out the copy of T_v inside T on the right-hand-side of the coproduct for coassociativity to work. We have seen the role of the quotients $T/^{c}T_v$ (and more generally $T/^{c}F_{\underline{v}}$) in the coassociativity property of Δ^c in Lemma 1.2.10. One can see that a coproduct of the form $T \mapsto \sum_v T_v \otimes T$ without the cancellation would no longer have the coassociativity property since we would have $(\Delta \otimes \mathrm{id}) \circ \Delta(T) = \sum_{v,w} T_{v,w} \otimes T_v \otimes T$, with $T_{v,w} \subset T_v$ accessible terms of T_v, but we would have $(\mathrm{id} \otimes \Delta) \circ \Delta(T) = \sum_{u,v} T_v \otimes T_u \otimes T$, with $T_u \subset T$ accessible terms of T. Those terms where $T_u \supset T_v$ are accounted for in $(\mathrm{id} \otimes \Delta) \circ \Delta(T)$, but other terms are not, so coassociativity will fail (for all forms Δ^c, Δ^ρ, Δ^d of the coproduct). Thus, the cancellation of the deeper copies is required for intrinsic algebraic reasons.

Note, moreover, that although we refer to the quotient term T/T_v as "cancellation" of a copy of T_v, nothing is really cancelled for two reasons:

- A copy of T_v is still present in the left-channel of the term $T_v \otimes T/T_v$ of the coproduct.
- In the case of the coproducts Δ^c, a trace of the cancellation of the deeper copy also remains in the term $T/^cT_v$ in the right-channel of the coproduct in the form of a label $\cancel{T_v}$. This trace is removed in the projection $\Pi_{d,c}(T/^cT_v) = T/^dT_v$, which accounts for the form that goes to Externalization, while $\cancel{T_v}$ remains for interpretation at the interface with semantics.

The basic structure of the coproduct separates out trees (in all possible ways) into a subforest $F_{\underline{v}}$ and a quotient $T/F_{\underline{v}}$. One can simply then read a subtree T_v as the "creation of a copy" and the quotient tree as the corresponding "cancellation of the original (deeper) copy" when Internal Merge is applied. The different forms of quotients $T/^cT_v$ and $T/^dT_v$ which we discussed in sections 1.2 and 1.2.1, account for the description of $\cancel{T_v}$ given in *Elements* [40]—namely, that "the element is present in the syntactic object (for Conceptual-Intensional interpretation), but not pronounced at the Sensory–Motor interface". Thus, $T/^cT_v$, with the explicit remaining leaf with label $\cancel{T_v}$, represents the form that the Conceptual-Intensional interface receives for interpretation, while its projection $T/^dT_v = \Pi_{d,c}(T/^cT_v)$ is what is received by the Sensory–Motor (SM) interface. These two channels of interface are not unrelated, as we will be discussing extensively in chapter 3, and the form that interpolates between these two and accounts for the interaction of these two channels is provided by the third quotient $T/^\rho T_v$ discussed in sections 1.2 and 1.2.1, which is related to these by two projections $T/^dT_v = \Pi_{d,\rho}(T/^\rho T_v) = \Pi_{\rho,c}(T/^cT_v)$.

Also, observe that there are distinct roles in the model that we are discussing here for *copies* versus *repetitions*, a point covered in some detail in *Elements* [40]. Repetitions are accounted for in this setting because we define the workspace as a forest (a disjoint union of trees; i.e., of syntactic objects). This allows the presence of repetitions since a forest is not a set but a multiset of trees. Copies, on the other hand, are created only during the application of the coproduct and ultimately play a role only in the operation of Internal Merge. (We will discuss a different type of operation that forms copies in section 1.16, in terms of restriction to diagonals: this will be a different algebraic structure than the copies and cancellations discussed here, coming from the coproduct and involved in Internal Merge.)

As noted, it is worth comparing the discussion here about copies and repetitions with sections 3.2 and 3.4 of *Elements* [40]. It is important to keep in mind that in the formulation that we are using here, copies and repetitions are *always kept separate*.

The workspaces $F \in \mathfrak{F}_{SO_0}$ are forests that can have several connected components isomorphic to the same tree, the same syntactic object $T \in \mathfrak{T}_{SO_0}$. These are just *repetitions*, never copies. We can view it this way: the disjoint union of sets can be characterized by a universal property in the category of sets. One can also obtain a realization of the disjoint union of a family $\{A_i\}$ of sets as the set of pairs (x, i) with $x \in A_i$. This means, for example, that in our setting, the forest $F = T \sqcup T$ can be seen as $F = T \times \{0\} \cup T \times \{1\}$. This way of writing disjoint unions (which is cumbersome and thus is not typically followed) does have an advantage: it shows clearly the two *repetitions* of the same syntactic object T, where both $T \times \{i\} \simeq T$ are isomorphic to T but are not related by the identity (are not copies in the linguistics sense). There are isomorphisms that implement the repetition relation through permutations σ of the set of indices $T \times \{i\} \to T \times \{\sigma(i)\}$. For example, in the case of $F = T \sqcup T$, the repetition relation is implemented by the isomorphism $T \times \{0\} \to T \times \{1\}$.

On the other hand, *copies* in our formulation are created only by the coproduct $\Delta : \mathcal{H} \to \mathcal{H} \otimes \mathcal{H}$, when accessible terms T_v are extracted from a syntactic object T in the workspace and deposited in the left-channel of the coproduct output, while simultaneously deleted in the output in the other channel.

The key algebraic property here is that the coproduct outputs in $\mathcal{H} \otimes \mathcal{H}$ and *not* in $\mathcal{H}$, so copies are kept distinct from repetitions, and the Boolean valued function *FormCopy* of section 3.4 of *Elements* [40] is determined here only by the distinction between $x \otimes 1 \in \mathcal{H} \otimes \mathcal{H}$ and $x \in \mathcal{H}$. As already mentioned, when the coproduct outputs an accessible term (a copy) in the left $\mathcal{H}$ factor in the tensor product $\mathcal{H} \otimes \mathcal{H}$, it simultaneously also outputs a term T/T_v in the right $\mathcal{H}$ factor. This means that we simultaneously have the creation of one copy T_v and the cancellation of the deeper copy (resulting in T/T_v). Since these two happen simultaneously on the two parallel channels of the coproduct output, we never actually *see* any copy. The two factors of the coproduct output are reassembled in the Merge action so the result is a new workspace (containing possible repetitions but no copies).

Even though with $T/^c T_v$ one keeps a *trace* of the deeper copy for interpretation at the Conceptual-Intensional interface, this does not appear as a copy in the new workspace as it occurs only as *label* $\cancel{T_v}$ in the new leaf of the quotient $T/^c T_v$. This is a key point in the algebraic properties discussed here—namely, when iterating the coproduct (as in the coassociativity argument), one is no longer extracting from the cancelled deeper copy $\cancel{T_v}$ (otherwise the coassociativity would be violated): the fact that the cancelled $\cancel{T_v}$ is now a label in $T/^c T_v$

means that it is indecomposable for the coproduct, unlike the extracted copy T_v in the left-channel, which continues to be decomposable to further applications of the coproduct.

One may worry here that, when labels of the form $\not{T}_{\bar{v}}$ are introduced by the coproduct Δ^c, the set of labels of leaves will become an infinite set, rather than the finite set $\mathcal{SO}_0$. However, for the purposes of interpretation at the CI interface, it suffices to retain in $\not{T}_{\bar{v}}$ certain syntactic features, since the specific structure T_v remains present in the remaining copy. Thus, one can still assume that the resulting label is an element of the finite set $\mathcal{SO}_0$.

There is another important point to clarify here, and we can do this explicitly by considering the example discussed in section 3.4 of [40], specifically the examples in equation (23) and in equation (26) of [40]. In (23) one considers a workspace of the form

$$WS = [\{\text{many, people}\}, \{\text{praised}, \{\text{many, people}\}\}],$$

or in our notation, $F = T_1 \sqcup T_2$, with

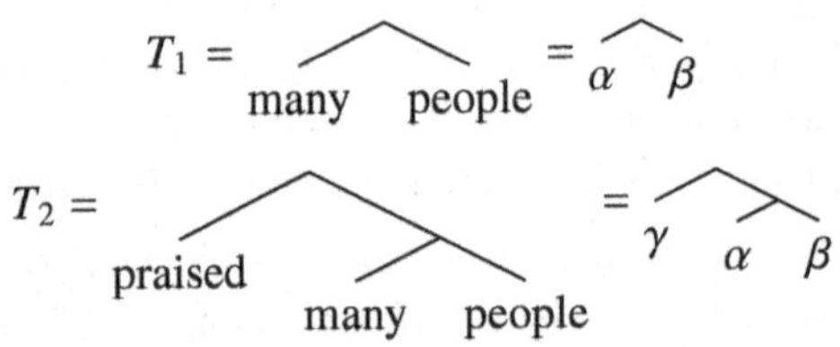

In fact, we can regard the syntactic object T_2 as already being the result of an External Merge operation:

$$T_2 = \mathfrak{M}(\gamma, T) \quad \text{with} \quad T =$$ 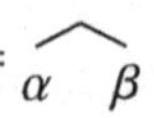

Thus, we can start by considering the workspace

$$WS' = [\{\text{many, people}\}, \text{praised}, \{\text{many, people}\}],$$

or $F' = T_1 \sqcup \gamma \sqcup T$ in our notation, from which the previous WS was obtained by one application of External Merge. In this workspace F', two of the connected components, T_1 and T, are both *isomorphic* to the same element of $\mathfrak{T}_{\mathcal{SO}_0}$, namely

$$T_1 \cong T \cong$$ 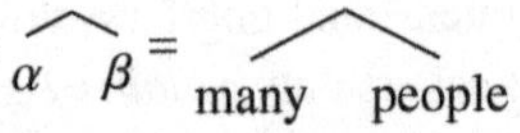

When we apply the External Merge $\mathfrak{M}_{S,S'}$ with $S = \gamma$ and $S' =$ 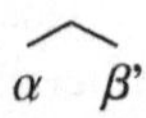, the operator $\delta_{S,S'}$ in our description of the action of Merge searches for *isomorphic* copies of S and S' among the accessible terms of the workspace. In the case of F', it finds *two* matches for S'. These matches are *distinct choices*, each of them isomorphic to S'.

Now consider instead the example discussed in equation (26) of *Elements* [40]. In this case, we have a workspace consisting of a single component of the form

$$WS = [\{\text{were}, \{\text{praised}, \{\text{many}, \text{people}\}\}\}]$$

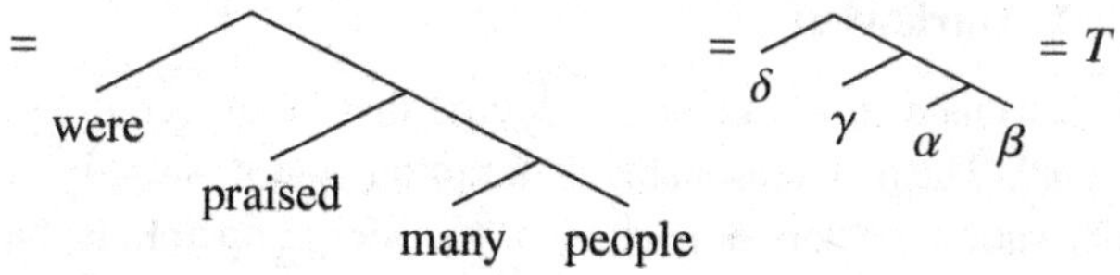

and an application of Internal Merge that yields the resulting workspace

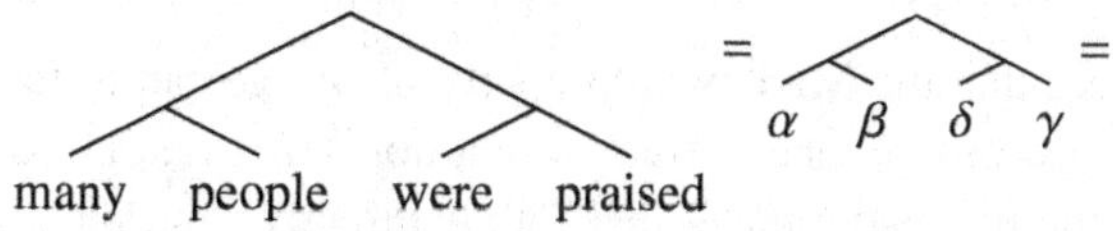

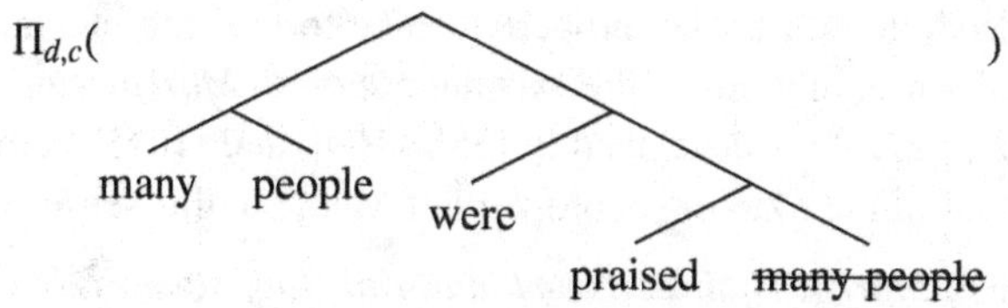

in the same way that we discussed in the example of Internal Merge in section 1.4. In particular, in our description of the action of Merge, this resulting workspace comes from the term of the Hopf algebra coproduct of the form

$$\widehat{\alpha\ \beta} \otimes T/^d\, \widehat{\alpha\ \beta},$$

with the quotient

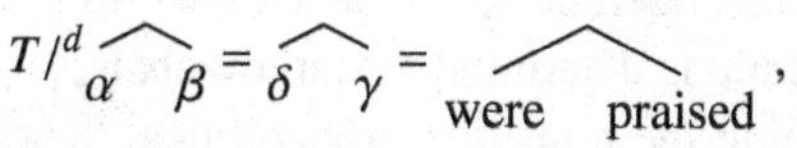

or in terms of the coproduct Δ^c,

$$\widehat{\alpha\ \beta} \otimes T/^c\, \widehat{\alpha\ \beta},$$

with the quotient

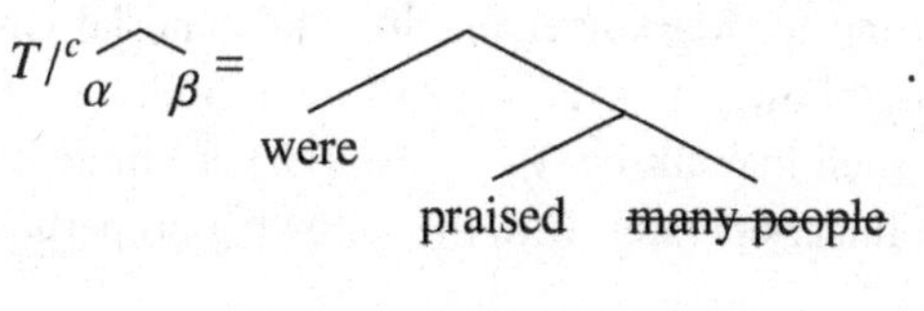

In this case, the tree $\overset{\wedge}{\alpha \quad \beta}$ that occurs in the left channel of the coproduct output as the extracted accessible term and the same tree that occurs in the right channel of the coproduct output as the cancelled term in the quotient are *not* isomorphic objects—they are *the same. This is indeed a conceptual key to the distinction between repetitions and copies in linguistics.* It is exactly the same as the important distinction between *isomorphism* and *identity* in mathematics.

1.9 Merge Is Markovian

It is usually assumed in the theory of Merge that "the operations of syntax are Markovian". This is a reasonable assumption, and it is used in the course of Merge derivations, where at each step, the Merge operations have access to only the current state of the workspace. This is discussed, for instance, in *Elements* text [40] in sections 3.3.1, 3.3.3, and 3.4.

We show here that the Markovian property of Merge can in fact be *proved* and does not need to be taken as an assumption. It is one more instance of properties of the linguistic model that follow directly from the algebraic formalism. Indeed, the Markovian property can be formulated in a more precise mathematical sense, in terms of the existing notion of *Markov chains on combinatorial Hopf algebras* developed in [55], [160], and [161]. In this section, we give a proof of the Markovian property of Merge in this sense.

The notion of *Markov chains on combinatorial Hopf algebras* of [55], [160], and [161] arises from analyzing the properties of the composition $\sqcup \circ \Delta$ of the coproduct followed by the product, as well as the iterated forms (given associativity and coassociativity)

$$\sqcup^a \circ \Delta^a : \mathcal{H} \to \mathcal{H}^{\otimes a} \to \mathcal{H}$$

that were considered in section 4.5 of [121] as a form of what is known in mathematics as "Adams operations". It was shown in [55] that, up to a total rescaling, these operations determine a Markov chain, when one views their matrix form as defining the transition probabilities. It was further shown in [160] and [161] that one can consider more general Markov chains in combinatorial Hopf algebras by modifying the compositions of (possibly iterated) coproducts and products using projections onto some of the graded subspaces. These projections are known as *descent operators* and were studied in [163].

The general setting for Markov chains in combinatorial Hopf algebras can be summarized as follows. Let $\mathcal{V} = \oplus_\ell \mathcal{V}_\ell$ be a graded vector space with a basis $\mathcal{B} = \cup_\ell \mathcal{B}_\ell$, such that all the $\mathcal{V}_\ell = \text{span}(\mathcal{B}_\ell)$ are finite dimensional. Let $\mathcal{K} : \mathcal{V} \to \mathcal{V}$ be a linear operator, with the following properties:

- The operator $\mathcal{K} = \oplus_\ell \mathcal{K}_\ell$ preserves the graded subspaces, $\mathcal{K}_\ell : \mathcal{V}_\ell \to \mathcal{V}_\ell$
- The matrix $K_{\mathcal{B}_\ell}$ representing the operator $\mathcal{K}$ in the basis $\mathcal{B}_\ell$ of $\mathcal{V}_\ell$ has non-negative entries
$$K_{\mathcal{B}_\ell}(x,y) \geq 0 \;\; \forall x, y \in \mathcal{B}_\ell$$
- For every input $x \in \mathcal{B}_\ell$ there is a $y \in \mathcal{B}_\ell$ such that $K_{\mathcal{B}_\ell}(x,y) > 0$.

Then there is an associated Markov chain whose set of states is $\mathcal{B}_\ell$, and with transition matrix
$$\tilde{K}_{\mathcal{B}_\ell}(x,y) = c(x)^{-1} K_{\mathcal{B}_\ell}(x,y)\,,$$
with a "local rescaling" $c(x) = \sum_y K_{\mathcal{B}_\ell}(x,y) > 0$, such that $\tilde{K}_{\mathcal{B}_\ell}$ is a *stochastic matrix*; namely,
$$\tilde{K}_{\mathcal{B}_\ell}(x,y) \geq 0 \;\forall x,y \in \mathcal{B}_\ell \;\; \text{and} \;\; \sum_y \tilde{K}_{\mathcal{B}_\ell}(x,y) = 1 \;\forall x \in \mathcal{B}_\ell\,.$$

Moreover, if $K_{\mathcal{B}_\ell}$ satisfies the following additional condition:

- there is an eigenfunction $\sum_y K_{\mathcal{B}_\ell}(x,y)\eta(y) = \eta(x)$ with $\eta(x) > 0$ for all $x \in \mathcal{B}_\ell$,

then
$$\hat{K}_{\mathcal{B}_\ell}(x,y) = \frac{\eta(y)}{\eta(x)} K_{\mathcal{B}_\ell}(x,y)$$
is the transition matrix of a Markov chain; namely, it is a stochastic matrix:
$$\hat{K}_{\mathcal{B}_\ell}(x,y) \geq 0 \;\forall x,y \in \mathcal{B}_\ell \;\; \text{and} \;\; \sum_y \hat{K}_{\mathcal{B}_\ell}(x,y) = 1 \;\forall x \in \mathcal{B}_\ell\,.$$

Markov chains in combinatorial Hopf algebras are data as stated previously, where the vector space $\mathcal{V}$ is also a graded connected commutative Hopf algebra (or bialgebra) $\mathcal{H} = (\mathcal{V}, \sqcup, \Delta)$, the linear basis $\mathcal{B}_\ell$ consists of monomials $F = \sqcup_a T_a$, with $\sum_a \deg(T_a) = \ell$, and the T_a in the set of generators of $\mathcal{H}$ as a commutative algebra.

Definition 1.9.1. A *weak* Hopf algebra Markov chain, on a graded connected commutative Hopf algebra $\mathcal{H}$ with linear basis $\mathcal{B}$, is a linear operator $\mathcal{K}$ of the form
$$\mathcal{K} = \sqcup \circ Q \circ \Delta \tag{1.9.1}$$
for a linear operator $Q : \mathcal{H} \otimes \mathcal{H} \to \mathcal{H} \otimes \mathcal{H}$ that preserves the grading $\mathcal{K} = \oplus_\ell \mathcal{K}_\ell$. The associated matrix K_ℓ has the property that, for all $x \in \mathcal{B}_\ell$, there is a $y \in \mathcal{B}_\ell$ with $K_\ell(x,y) > 0$. A *strong* Hopf algebra Markov chain is an operator as discussed previously, such that there is a global normalization factor $\rho = \rho(K_\ell)$ such that $\mathcal{K}_{\ell,\rho} = \rho^{-1}\mathcal{K}_\ell$ has an eigenfunction $\sum_y K_{\ell,\rho}(x,y)\eta(y) = \eta(x)$ with $\eta(x) > 0$ for all $x \in \mathcal{B}_\ell$.

The cases considered in [55], [160], and [161] are strong Hopf algebra Markov chains in the sense of Definition 1.9.1 after a "global rescaling" of the operator. The eigenfunction η provides a rescaling of the basis $\mathcal{B}$ by $x \mapsto \eta(x)^{-1}x$ with the property that a global rescaling $\mathcal{K}_\rho$ of the operator is a Markov chain in the rescaled basis. We show that the action of Merge on workspaces determines both weak and a strong Hopf algebra Markov chains.

Proposition 1.9.2. *The Merge operation of* (1.3.7) *defines a weak Hopf algebra Markov chain.*

Proof. First, observe that the operation $\sqcup \circ \Delta$, in the case of the Hopf algebra of workspaces, has the following effect:

$$\sqcup \circ \Delta(F) = 2F + \sum_{\underline{v}} F_{\underline{v}} \sqcup F/F_{\underline{v}} \,. \tag{1.9.2}$$

The first term, $2F$, comes from the primitive part $F \otimes 1 + 1 \otimes F$ of the coproduct after application of the product $\sqcup$. Notice in particular that the sum in (1.9.2) contains all possible terms of the form

$$T_v \sqcup T/T_v \sqcup \hat{F}\,, \quad \text{for } F = T \sqcup \hat{F}\,,$$

for accessible terms T_v used by Internal Merge.

We consider the Merge operators of (1.3.7), taking the form

$$\mathfrak{M}_{S,S'} = \sqcup \circ (\mathcal{B} \otimes \mathrm{id}) \circ \delta_{S,S'} \circ \Delta\,, \tag{1.9.3}$$

and we assemble them into the single transformation that performs all the possible Merge operations on a given workspace. Namely, we take the sum $\mathcal{K} := \sum_{S,S'} \mathfrak{M}_{S,S'}$. While this is written formally as an infinite sum, whenever it applies to a workspace F, it reduces to a finite sum $\mathcal{K}(F) = \sum_{S,S'} \mathfrak{M}_{S,S'}(K)$, as all but finitely many of the terms in the sum map F to zero. The terms that remain are exactly those corresponding to all the terms in $\Delta(F)$ of the form $F' \otimes F''$, where $F' = S \sqcup S'$ has two components. We write $\Pi_{(2)}$ for the projection onto the span of the subspace of $\mathcal{H} \otimes \mathcal{H}$ generated by basis elements of the form $S \sqcup S' \otimes F''$. We have

$$\mathcal{K} = \sqcup \circ (\mathcal{B} \otimes \mathrm{id}) \circ \Pi_{(2)} \circ \Delta\,. \tag{1.9.4}$$

To also include all possible applications of Internal Merge (see sections 1.4.3 and 1.4.3.1), we consider the composition $\mathcal{K} \circ \Xi$ with the operator

$$\Xi := \sqcup \circ \Pi_{(1)} \circ \Delta\,, \tag{1.9.5}$$

where $\Pi_{(1)}$ is the projection onto the subspace of $\mathcal{H} \otimes \mathcal{H}$ spanned by basis elements of the form $T \otimes F'$, with $T \in \mathfrak{T}_{SO_0}$ and $F' \in \mathfrak{F}_{SO_0}$.

Consider, on $\mathcal{V}(\mathfrak{F}_{SO_0}) = \oplus_\ell \mathcal{V}(\mathfrak{F}_{SO_0,\ell})$, the grading by number of leaves of $F \in \mathfrak{F}_{SO_0}$. The operators $\mathcal{K} = \oplus_\ell \mathcal{K}_\ell$ and $\Xi = \oplus_\ell \Xi_\ell$ preserve the graded subspaces,

$$\mathcal{K}_\ell : \mathcal{V}(\mathfrak{F}_{SO_0,\ell}) \to \mathcal{V}(\mathfrak{F}_{SO_0,\ell}) \quad \text{and} \quad \Xi_\ell : \mathcal{V}(\mathfrak{F}_{SO_0,\ell}) \to \mathcal{V}(\mathfrak{F}_{SO_0,\ell}).$$

We write $K_\ell(F, F')$ for the coefficient of F' in $\mathcal{K}_\ell(F)$. Similarly, we write $K\Xi_\ell(F, F')$ for the coefficient of F' in $\mathcal{K}_\ell(\Xi_\ell(F))$.

The subspace $\mathcal{V}(\mathfrak{F}_{SO_0,1})$ is spanned by the workspaces consisting of a single lexical item or syntactic feature in $\mathcal{SO}_0$ (a tree consisting of a single leaf node). Such workspaces cannot be used for structure formation, as Merge can only act trivially. Thus, we consider only $\mathcal{K}_\ell$ and Ξ_ℓ for $\ell \geq 2$.

Given any $F \in \mathfrak{F}_{SO_0,\ell}$ with $\ell \geq 2$, we can find $F' \in \mathfrak{F}_{SO_0,\ell}$ such that $K_\ell(F, F') > 0$, since given two leaves $\alpha, \beta \in L(F)$, the coproduct $\Delta(F)$ will contain a term of the form $\alpha \sqcup \beta \otimes \hat{F}$, with $\hat{F}$ possibly 1; hence, $F' = \mathfrak{M}(\alpha, \beta) \sqcup \hat{F}$ has $K_\ell(F, F') > 0$. The case of $K\Xi_\ell(F, F')$ is similar. Thus, given $F \in \mathfrak{F}_{SO_0,\ell}$ with $\ell \geq 2$, we can define

$$c_{\mathcal{K},\ell}(F) := \sum_{F' \in \mathfrak{F}_{SO_0,\ell}} K_\ell(F, F') \quad \text{and} \quad c_{\mathcal{K}\Xi,\ell}(F) := \sum_{F' \in \mathfrak{F}_{SO_0,\ell}} K\Xi_\ell(F, F').$$

We have $c_{\mathcal{K},\ell}(F) > 0$ and $c_{\mathcal{K}\Xi,\ell}(F) > 0$, so we can assign to $\mathcal{K}_\ell$ and $\mathcal{K}_\ell \circ \Xi_\ell$ the transition matrices

$$\tilde{K}_\ell(F, F') := c_{\mathcal{K},\ell}(F)^{-1} K_\ell(F, F') \quad \text{and} \quad \tilde{K\Xi}_\ell(F, F') := c_{\mathcal{K},\ell}(F)^{-1} K\Xi_\ell(F, F'). \tag{1.9.6}$$

These are nonnegative and satisfy

$$\sum_{F' \in \mathfrak{F}_{SO_0,\ell}} \tilde{K}_\ell(F, F') = 1 \quad \text{and} \quad \sum_{F' \in \mathfrak{F}_{SO_0,\ell}} \tilde{K\Xi}_\ell(F, F') = 1;$$

hence, they define the transition matrices of a Markov chain. This satisfies the properties of a *weak Hopf algebra Markov chain* as in Definition 1.9.1. □

Remark 1.9.3. The weak Hopf algebra Markov chain property, which relies on the fact that every basis element F has $K\Xi_\ell(F, F') > 0$ for some F', would also be satisfied in the limit $\epsilon \to 0$ of the Minimal Search weight function, which selects only External Merge and Internal Merge. However, it would not be satisfied just by External Merge, as $F = T$ with a single component would have no F' obtainable from it by External Merge alone.

We now consider the more interesting *strong* Hopf algebra Markov chain property of Definition 1.9.1, which ensures that after a rescaling of the basis elements, a global rescaling of the transformation is a Markov chain in the sense that its representing matrix in the rescaled basis is stochastic and defines the transition probabilities of the Markov chain.

Proposition 1.9.4. *Any multiset Ω of lexical items and syntactic features in $\mathcal{SO}_0$, with $\ell = \#\Omega > 2$, determines an invariant subspace $\mathcal{V}_\Omega$ of $\mathcal{V}(\mathfrak{F}_{\mathcal{SO}_0,\ell})$ for the operators $\mathcal{K}_\ell$ and $\mathcal{K}\Xi_\ell$. The restriction to such a subspace $\mathcal{V}_\Omega$ defines a strong Hopf algebra Markov chain.*

Proof. First, observe that one can obtain an invariant subspace $\mathcal{W}_\Omega$ (slightly larger than the one in which we're interested) from $\Omega = \alpha_1 \sqcup \cdots \sqcup \alpha_\ell$, with $\alpha_i \in \mathcal{SO}_0$, by taking the span of all the possible $F \in \mathfrak{F}_{\mathcal{SO}_0}$ with leaves set $L(F) = \Omega$. We take the subspace $\mathcal{V}_\Omega \subset \mathcal{W}_\Omega$ spanned by the basis elements F of $\mathcal{W}_\Omega$ that have a nonempty set of edges. We denote this spanning set as $\mathcal{B}_{\Omega,\ell} \subset \mathcal{B}_\ell$. We discuss the case of $\mathcal{K}_\ell$. The argument for $\mathcal{K}\Xi_\ell$ is similar. We associate to the restriction $\mathcal{K}_{\Omega,\ell}$ of $\mathcal{K}_\ell$ to $\mathcal{V}_\Omega$ (and its matrix representation $K_{\mathcal{B}_{\Omega,\ell}}$) a directed graph $G_{\Omega,\mathcal{K}_\ell}$ with vertices given by the elements of $\mathcal{B}_{\Omega,\ell}$, and with a directed edge from F to F' iff $K_{\mathcal{B}_{\Omega,\ell}}(F, F') > 0$. The restriction $\mathcal{K}_{\Omega,\ell}$ is *irreducible* iff the graph $G_{\Omega,\mathcal{K}_\ell}$ is strongly connected; namely, given any two vertices F, F', there is a directed path in the graph that connects them. If we know that this is the case, then the Perron–Frobenius theorem for the irreducible $\mathcal{K}_{\Omega,\ell} : \mathcal{V}_\Omega \to \mathcal{V}_\Omega$, with all entries $K_{\mathcal{B}_\ell}(F, F') \geq 0$, ensures the existence of a Perron–Frobenius eigenfunction $\eta_{\Omega,\ell}$, with

$$\sum_{F' \in \mathcal{B}_{\Omega,\ell}} K_{\mathcal{B}_\ell}(F, F')\, \eta_{\Omega,\ell}(F') = \lambda_{\Omega,\ell}\, \eta_{\Omega,\ell}(F),$$

with $\eta_{\Omega,\ell}(F) > 0$ for all $F \in \mathcal{B}_{\Omega,\ell}$ and with the Perron–Frobenius eigenvalue $\lambda_{\Omega,\ell} > 0$. Let $\tilde{\mathcal{K}}_{\Omega,\ell} = \lambda_{\Omega,\ell}^{-1} \mathcal{K}_{\Omega,\ell}$ be the global rescaling of the operator and $\tilde{K}_{\mathcal{B}_\ell}(F, F')$ be the corresponding matrix elements in the basis $\mathcal{B}_{\Omega,\ell}$. This satisfies

$$\sum_{F' \in \mathcal{B}_{\Omega,\ell}} \tilde{K}_{\mathcal{B}_\ell}(F, F')\, \eta_{\Omega,\ell}(F') = \eta_{\Omega,\ell}(F).$$

We then have the strong Hopf algebra Markov chain property: after rescaling the basis to

$$\hat{\mathcal{B}}_{\Omega,\ell} := \{\eta(F)^{-1} F \mid F \in \mathcal{B}_{\Omega,\ell}\},$$

the matrix of $\tilde{\mathcal{K}}_{\Omega,\ell} : \mathcal{V}_\Omega \to \mathcal{V}_\Omega$ in the rescaled basis is stochastic and gives the transition probabilities of a Markov chain. Thus, we need to show that the irreducibility condition holds—namely, that given two $F, F' \in \mathcal{B}_{\Omega,\ell}$, there is a path of vertices $F_0 = F, \dots, F_k = F'$, with $K_{\mathcal{B}_\ell}(F_i, F_{i+1}) > 0$. It suffices to show this for two trees $T, T' \in \mathcal{B}_{\Omega,\ell}$, as the argument for forests then follows similarly. These two trees have the same set of leaves $L(T) = L(T') = \Omega$. Each of them is a syntactic object obtained by repeated applications of the magma operation $\mathfrak{M}$, starting with elements in Ω. If we want to obtain T from $F_0 = T'$ using combinations of operations $\mathfrak{M}_{S,S'}$ we can start by identifying (following a bottom-up construction of T) pairs of substructures of T' (consisting of single leaves or larger) that are joined in T. Given two such accessible terms $T_v, T_w \subset T'$, such that $\mathfrak{M}(T_v, T_w)$ is an accessible term of T, we take $F_1 = \mathfrak{M}(T_v, T_w) \sqcup T'/(T_v \sqcup T_w)$. Note that we are using the fact that the Merge operations of (1.3.7) include this kind of Sideward Merge, not just External/Internal Merge. By repeatedly applying this kind of operation, we arrive at a workspace $F_r = \mathfrak{M}(T_{v_1}, T_{w_1}) \sqcup \cdots \sqcup \mathfrak{M}(T_{v_r}, T_{w_r})$, where the T_{v_i} and T_{w_i} are accessible terms of the previous quotient term of T' and each $\mathfrak{M}(T_{v_i}, T_{w_i})$ is an accessible term of T. Repeated applications of External Merge then assemble these terms as they occur in T, producing successive $F_{r+1}, \dots, F_k = T$, where at each step, $K_{\mathcal{B}_\ell}(F_i, F_{i+1}) > 0$. □

Remark 1.9.5. As we have already discussed in section 1.4, the Merge operations of (1.3.7) include the forms of Sideward Merge described in sections 1.4.5 and 1.4.6. We know that these operations are subdominant with respect to the leading terms External Merge and Internal Merge, when implementing Minimal Search as in section 1.5 in terms of weights accounting for a natural cost function as in section 1.5.2. We have seen in section 1.5.3 that in the limit $\epsilon \to 0$, only External Merge and Internal Merge are retained. We have also seen in section 1.6, that Sideward Merge violates some optimality constraints that Internal Merge and External Merge satisfy. However, what we see here is that Sideward Merge plays a structural role in ensuring a strong Markovian property that makes the Merge operations of (1.3.7) a Hopf algebra Markov chain in the strong sense of [55], [160], and [161]. With only External/Internal Merge, this property would be lost. Thus, one should view the selection of External/Internal Merge by Minimal Search as the leading behavior for small $\epsilon > 0$, with the subdominant terms of Sideward Merge ensuring the strong Markovian property, while External/Internal Merge primarily perform structure formation and movement. A more detailed discussion of the role of Sideward Merge can be found in [142].

1.10 The Core Computational Structure of Merge

The description of Merge and its action on workspaces that we have presented up to now follows closely the formulation given in the recent linguistic literature, as we have noted several times. We discuss here a further simplification of the structure of Merge, which extracts its core computational structure as presented in Chomsky's work, e.g., [32], [36].

Let $\mathfrak{T}$ be the set of binary rooted trees without planar structure (and without labeling of the leaves), and $\mathcal{V}(\mathfrak{T})$ be the free $\mathbb{Z}$-module (or the $\mathbb{Q}$-vector space) spanned by the set $\mathfrak{T}$. Lemma 1.10.1 presents the analog of Definition 1.1.1 and Remark 1.1.2.

Lemma 1.10.1. *The set $\mathfrak{T}$ is identified with the free, nonassociative, commutative magma whose elements are the balanced bracketed expressions in a single variable x, with the binary Merge operation $(\alpha,\beta) \mapsto \mathfrak{M}(\alpha,\beta) = \{\alpha,\beta\}$. Correspondingly, $\mathcal{V}(\mathfrak{T})$ is identified with the free, commutative, nonassociative algebra generated by a single variable x.*

Proof. We can identify the binary rooted trees without planar structure with the balanced bracketed expressions in a single variable x. For example, we identify

$$\{\{x\{xx\}\}x\} \longleftrightarrow$$

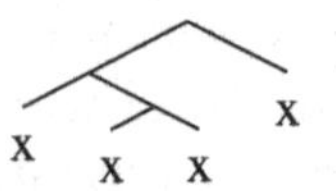

.

The Merge operation $\mathfrak{M}(\alpha,\beta) = \{\alpha,\beta\}$ takes two such bracketed expressions α and β and forms a new one of the form $\{\alpha,\beta\}$, which corresponds to attaching the roots of the two binary trees to a common root,

$$\mathfrak{M}(T,T') = T \wedge T' :=$$

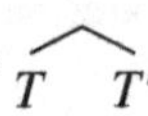

.

□

Equivalently, $\mathcal{V}(\mathfrak{T})$ is the free algebra over the *quadratic operad* freely generated by the single commutative binary operation $\mathfrak{M}$ (see [96]).

The generative process for the set $\mathfrak{T}$ via the Merge operation can be equivalently described as a recursive procedure encoded in the form of a fixed-point equation.

Proposition 1.10.2. *Let $\mathcal{V}(\mathfrak{T}) = \oplus_\ell \mathcal{V}(\mathfrak{T})_\ell$ with the grading by length (number of leaves) as before, with $\mathfrak{M} : \mathcal{V}(\mathfrak{T})_\ell \times \mathcal{V}(\mathfrak{T})_{\ell'} \to \mathcal{V}(\mathfrak{T})_{\ell+\ell'}$, where $\mathfrak{M}$ is extended by linearity in each variable. Consider formal infinite sums X =*

$\sum_{\ell \geq 1} X_\ell$, with $X_\ell \in \mathcal{V}(\mathfrak{T})_\ell$, and the recursive equation

$$X = \mathfrak{M}(X, X)\,. \tag{1.10.1}$$

Then the generative process for $\mathfrak{T}$ via the Merge operation is equivalent to the recursive construction of a solution of (1.10.1) *with the initial condition $X_1 = x$.*

Proof. We have $\mathfrak{M}(\sum_\ell X_\ell, \sum_{\ell'} X_{\ell'}) := \sum_{\ell,\ell'} \mathfrak{M}(X_\ell, X_{\ell'})$. In particular, the term of degree n in $\mathfrak{M}(X, X)$ is given by

$$\mathfrak{M}(X, X)_n = \sum_{j=1}^{n-1} \mathfrak{M}(X_j, X_{n-j})\,;$$

therefore, the fixed-point equation (1.10.1) reduces to the recursive relation

$$X_n = \sum_{j=1}^{n-1} \mathfrak{M}(X_j, X_{n-j})\,.$$

Starting with $X_1 = x$, the recursion produces $X_2 = \{xx\}$, $X_3 = \{x\{xx\}\} + \{\{xx\}x\} = 2\{x\{xx\}\}$, $X_4 = 2\{x\{x\{xx\}\}\} + \{\{xx\}\{xx\}\}$, and so on. These first terms X_n list all the possible nonplanar binary rooted trees with n leaves, with multiplicities that account for the various planar structures. Given a nonplanar binary rooted tree T with n leaves, we can always write it as $T = \mathfrak{M}(T', T'')$, where T', T'' are the two binary rooted trees with roots at the two internal vertices of T connected to the root of T, with $j = \#L(T')$ and $n - j = \#L(T'')$ for some $j \in \{1, \ldots, n-1\}$. Since the Merge product is commutative, it does not matter in which order we list T' and T''. Thus, each $T \in \mathfrak{T}_n$ can be mapped uniquely to an *unordered* pair $\{T', T''\}$, and conversely, any pair of trees T', T'' with numbers of leaves ℓ' and ℓ'', respectively, determines uniquely a tree $\mathfrak{M}(T', T'')$ with $\ell' + \ell''$ leaves. Thus, inductively, if each X_j for $1 \leq j < n$ consists of a list (i.e., a formal sum) of all the possible nonplanar binary rooted trees with j leaves, then X_n also consists of a sum of all the possible nonplanar binary rooted trees with n leaves. One sees similarly that the integer coefficients in the sum count different planar structures (planar embeddings). □

As we discuss further in section 1.17, this demonstrates that the generative process for the core computational structure of Merge in the Minimalist Model of syntax is in fact the most fundamental basic case of the Dyson-Schwinger equations in physics. We provide a quick summary here of what we will discuss in more detail in section 1.17.

In general, the Dyson-Schwinger (DS) equation implements in perturbative quantum field theory the construction of solutions for the equations of motion.

It is a way of encoding the variational principle of least action for the equations of motion in classical physics in a form that is suitable for quantum fields, via a recursive method of solution that can be performed order by order in the perturbative expansion. There are two main conceptual aspects to single out here. One is the fact that the construction of solutions of DS equations becomes a combinatorial problem in terms of Feynman graphs and their associated trees, expressible as a solution to a fixed-point equation, of which (1.10.1) is the most fundamental example. The general version of such a *combinatorial Dyson–Schwinger equation* takes the form of a (possible n-ary) Merge operation, given by the grafting operator $\mathcal{B}$ of Definition 1.3.2, and a polynomial fixed-point equation in a Hopf algebra, which takes the general form $X = \mathcal{B}(P(X))$, for a polynomial P, and a variable $X = \sum_\ell X_\ell$ in a completion of a Hopf algebra of rooted trees. This equation is then solved recursively, as in the fundamental case of (1.10.1).

The other aspect is the usual requirement for classical solutions of the equations of motion: the action functional is stationary under infinitesimal variations. This is transformed in the case of quantum fields into corresponding equations for the quantum correlation functions. In the formulation of perturbative quantum field theory in terms of Hopf algebras, these in turn arise from the combinatorial solution, which is entirely determined in terms of the underlying Hopf-algebraic structure, together with the evaluation of a renormalized Feynman rule, to obtain the physical solution from the combinatorial one.

As we discuss further in section 1.17, the first observation identifies the generative process of syntactic objects through Merge with the basic case of the structure of generative processes of fundamental physics. The second observation suggests that the optimality that the core computational structure of Merge ought to satisfy has the same conceptual nature as the least action principle of physics, when the latter manifests itself in combinatorial form. In this way, Minimalism's fundamental call for the imposition of similar "least effort" principles (sometimes called "Third Factor" principles) shares a deep commonality, it appears, with physics.

The procedure described in this section, which encodes the generative process of free symmetric Merge into a formal series $X = \sum_{\ell \geq 1} X_\ell$ satisfying the fixed-point equation $\mathfrak{M}(X, X) = X$, is somewhat reminiscent of the more familiar Chomsky–Schützenberger results of [39], where the generative process of an unambiguous context-free language is encoded through a power series $L(x) = \sum_k a_k x^k$ with coefficients a_k counting the number of words of length k in the language, satisfying a polynomial relation $P(x, L(x)) = 0$ for some polynomial P with rational coefficients. These two settings share a common underlying idea of encoding a generative grammar into a series that is subject to a specific type of polynomial equation. The main differences between the case discussed here and the Chomsky–Schützenberger setting of [39] are these:

- The result of [39] requires unambiguous context-free grammars, while here we work with the generative process of free symmetric Merge rather than with the theory of formal languages
- In [39] the important information captured by the infinite series $L(x)$ is in the coefficients a_k, while in the case of Merge, it is in the terms X_ℓ, of degree ℓ in the Hopf algebra grading (and the coefficients only give multiplicities that detect the different counting of planar and nonplanar trees)
- In the Merge case, the recursive solution of the polynomial (quadratic) fixed-point equation determines all the X_ℓ as formal sums of the objects generated by Merge (the syntactic objects), while in [39], the polynomial equation satisfied by $L(x)$ is not meant to be solved for $L(x)$: it simply establishes the fact that $L(x)$ lies in the special class of power series that are algebraic over $\mathbb{Q}(x)$.

1.11 Constraints on Merge: The *n*-arity Question

An important question regarding the Merge operation is whether the same generative power would be achievable with a similar operation that is n-ary, for some $n \geq 3$, rather than binary. This is discussed in section 3.3.1 of *Elements* [40], where it is argued that principles of simplicity and least effort favor a binary over an n-ary Merge. We show here that there are also *intrinsic computational* properties that rule out the n-ary case for $n \geq 3$.

Riny Huijbregts presented in [101] strong empirical linguistic evidence for why a ternary Merge would be inadequate, in the sense that such a ternary operation would produce both *undergeneration* and *overgeneration* with respect to binary Merge, in the following sense:

- *Undergeneration* refers to syntactic constructions that can be derived through the binary Merge but would not be generated by a ternary Merge.
- *Overgeneration* refers to the generation of ungrammatical sentences that would be generated by a ternary Merge, but not by a binary Merge.

While the undergeneration problem could in principle be bypassed by hypothesizing the simultaneous presence of a binary and a ternary Merge (but a principle of least effort would then rule out this if just a single binary Merge suffices), the overgeneration problem cannot be similarly handled.

Our goal in this section is to explain why any n-ary Merge operation, for any $n \geq 3$, would necessarily lead to both undergeneration and overgeneration (with respect to the binary Merge) as a simple consequence of the algebraic structure described in the previous sections.

In particular, within this formulation, one can see that undergeneration and overgeneration have two somewhat different origins:

- Undergeneration depends only on the magma of syntactic objects.
- Overgeneration involves the action of Merge on workspaces.

1.11.1 The n-ary Merge Magma

Here, we assume the existence of a hypothetical n-ary Merge, for some $n \geq 3$, and we discuss how the structure of the magma of syntactic objects changes with respect to the binary case. We assume the same initial set $\mathcal{SO}_0$ of lexical terms and syntactic features.

Definition 1.11.1. *An n-magma consists of a set X, together with an n-ary operation*

$$\mathfrak{M}_n : \underbrace{X \times \cdots \times X}_{n\text{-times}} \to X\,, \quad (x_1, \ldots, x_n) \mapsto \mathfrak{M}_n(x_1, \ldots, x_n)\,.$$

We say that $(X, \mathfrak{M}_n)$ *is an n-magma over a set Y if all elements of X are obtained by iterated applications of* $\mathfrak{M}_n$ *starting with n-tuples of elements in Y.*

We write $\{x_1, \ldots, x_n\} := \mathfrak{M}_n(x_1, \ldots, x_n)$ to refer to the element of X that is obtained by applying $\mathfrak{M}_n$ to the n-tuple $(x_1, \ldots, x_n)$. In particular, the set X consists of a set X_1 consisting of all elements of the form $\{y_1, \ldots, y_n\} := \mathfrak{M}_n(y_1, \ldots, y_n)$, with all the $y_i \in Y$, a set X_{2n-1} consisting of all elements of the form

$$\mathfrak{M}_n(y_1, \ldots, y_{i-1}, \mathfrak{M}_n(a_{i,1}, \ldots, a_{i,n}), y_{i+1}, \ldots, y_n) =$$
$$\{y_1, \ldots, y_{i-1}, \{a_{i,1}, \ldots, a_{i,n}\}, y_{i+1}, \ldots, y_n\}$$

for $i = 1, \ldots, n$ and with all the $y_i, a_{i,j} \in Y$, a set X_{3n-2} consisting of all elements of the form

$$\{y_1, \ldots, y_{i-1}, \{a_{i,1}, \ldots, a_{i,n}\}, y_{i+1}, \ldots, y_{j-1}, \{b_{j,1}, \ldots, b_{j,n}\}, y_{j+1}, \ldots y_n\} \text{ and}$$

$$\{y_1, \ldots, y_{i-1}, \{a_{i,1}, \ldots, a_{i,j-1}, \{b_{j,1}, \ldots, b_{j,n}\}, a_{i,j+1}, a_{i,n}\}, y_{i+1}, \ldots, y_n\}\,,$$

with $i \neq j$, $i, j = 1, \ldots, n$, and all the $y_i, a_{i,k}, b_{j,k} \in Y$, and so on. Thus, we have

$$X = \bigsqcup_{k \geq 1} X_{k(n-1)+1}\,. \tag{1.11.1}$$

We refer to the subset $X_{k(n-1)+1}$ as the set of elements of length $k(n-1)+1$ in the n-magma.

The n-magma is associative if all the elements of length $k(n-1)+1$ are identified (i.e., if bracketing is irrelevant). This is not the case for n-ary Merge, so the resulting magma is nonassociative. It is commutative if elements $\{x_1, \dots, x_n\}$ with entries that differ by a permutation in the symmetric group S_n are identified (i.e., if every set within brackets is unordered). This is the case for n-ary Merge, so the resulting n-magma is commutative. Since there are no other relations (except commutativity), we obtain the free, commutative, nonassocoative n-magma.

We then have the following description of the set of syntactic objects produced by a hypothetical n-ary Merge $\mathfrak{M}_n$, given in Definition 1.11.2.

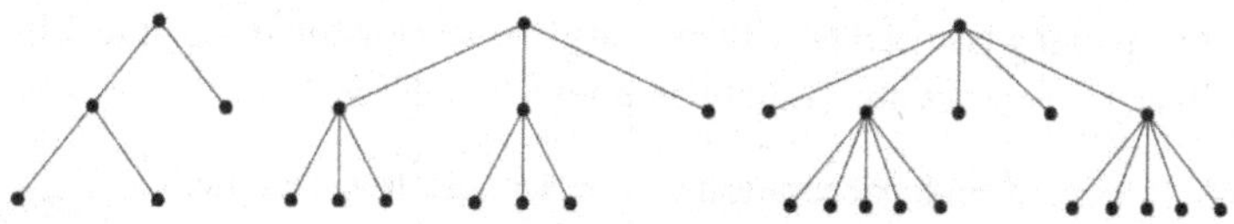

Figure 1.7
Examples of syntactic objects produced by a binary, ternary, or 5-ary Merge.

Definition 1.11.2. *The set $\mathcal{SO}^{(n)}$ of n-ary syntactic objects is the free, nonassociative, commutative n-magma on the set $\mathcal{SO}_0$,*

$$\mathcal{SO}^{(n)} = \text{Magma}^{(n)}_{na,c}(\mathcal{SO}_0, \mathfrak{M}_n)\,, \tag{1.11.2}$$

with

$$\mathcal{SO}^{(n)} = \bigsqcup_{k\geq 1} \mathcal{SO}^{(n)}_{k(n-1)+1}\,. \tag{1.11.3}$$

Remark 1.11.3. We can identify the elements of $\mathcal{SO}^{(n)}$ with rooted n-ary trees (see Figure 1.7):

$$\mathcal{SO}^{(n)} \simeq \mathfrak{T}^{(n)}_{\mathcal{SO}_0}\,; \tag{1.11.4}$$

namely, trees where all the non-leaf vertices have n descendants, without a planar structure, and with leaves labeled by elements of the set $\mathcal{SO}_0$. (Note that, as in the binary case, what we call here "n-ary trees" are *full* n-ary trees.)

The set $\mathcal{SO}^{(n)}_{k(n-1)+1}$ is the set of rooted n-ary trees (with no assigned planarity) with $k(n-1)+1$ leaves, and therefore with k non-leaf vertices. The number k of non-leaf vertices is the number of applications of $\mathfrak{M}_n$ in the process of generating elements of $\mathcal{SO}^{(n)}$ where each non-leaf vertex is the graphical representation of a Merge operation.

1.11.2 Undergeneration

Given the structure (1.11.3) of the set of n-ary syntactic objects, we can show that there are two different forms of undergeneration (with respect to the binary Merge), and that both of them inevitably occur for any n-ary Merge with $n \geq 3$. The two forms of undergeneration correspond, respectively, to the following two phenomena:

- Certain lengths (number of leaves, length of resulting sentence) are not achievable through an n-ary Merge construction for any given $n \geq 3$ (only binary Merge can achieve all lengths)
- Certain syntactic parsing ambiguities are not accounted for, in cases where only one parsing is possible with an n-ary Merge construction while different inequivalent parsings are available with binary Merge.

The first form of undergeneration can be seen as described next.

Lemma 1.11.4. *Only strings of elements of $\mathcal{SO}_0$ of length $k(n-1)+1$, for some $k \geq 1$, can be achieved through an n-ary Merge. In particular, only the binary Merge can achieve all lengths.*

Proof. The number of leaves of an n-ary tree with k non-leaf vertices is $k(n-1)+1$. Thus, the only possible strings of elements of $\mathcal{SO}_0$ that can be obtained through k successive applications of an n-ary Merge $\mathfrak{M}_n$ are of length $k(n-1)+1$, as in the decomposition (1.11.3) of the set of n-ary syntactic objects. Only in the case $n = 2$ does the set $\{k(n-1)+1\}_{k\geq 1}$ contain all the positive integers greater than or equal to 2. □

Known empirical linguistic examples of this kind of undergeneration include, for instance, the fact that sentences like "*It rains*" are in $\mathcal{SO}_2$ while we have $\mathcal{SO}_2^{(3)} = \emptyset$, and hence they are not realizable by a ternary Merge $\mathfrak{M}_3$.

The second form of undergeneration can be seen through counting and comparing the sizes of the sets $\mathcal{SO}_{k(n-1)+1}$ and $\mathcal{SO}^{(n)}_{k(n-1)+1}$, for $n \geq 3$. The counting formula for rooted trees are simpler in the case of trees with an assigned planar structure, rather than for abstract trees with no assigned planarity. Thus, we count the resulting trees after the Externalization step that introduces planar structures.

Let $\mathfrak{T}^{pl}_{\mathcal{SO}_0} = \sqcup_\ell \mathfrak{T}^{pl}_{\mathcal{SO}_0,\ell}$ and $\mathfrak{T}^{(n),pl}_{\mathcal{SO}_0} = \sqcup_k \mathfrak{T}^{(n),pl}_{\mathcal{SO}_0,k(n-1)+1}$ denote, respectively, the sets of binary and of n-ary rooted trees with a choice of planar embedding.

Lemma 1.11.5. *For any given $n \geq 3$, and for $\ell = k(n-1)+1$ with any $k \geq 2$, we have*

$$\#\mathfrak{T}^{pl}_{SO_0,\ell} > \#\mathfrak{T}^{(n),pl}_{SO_0,\ell}.$$

Proof. The number of planar, binary, rooted trees with $\ell = r+1$ leaves (hence r non-leaf vertices) is given by the Catalan number

$$C_r = \frac{1}{r+1}\binom{2r}{r}.$$

Thus, for $\ell = k(n-1)+1$, we have the counting

$$C_{k(n-1)} = \frac{1}{(n-1)k+1}\binom{2k(n-1)}{k}.$$

The number of planar, n-ary, rooted trees with $(n-1)k+1$ leaves (hence k non-leaf vertices) is correspondingly given by the Fuss–Catalan number

$$C_k^{(n)} = \frac{1}{(n-1)k+1}\binom{nk}{k}.$$

The different assignments of labels at the leaves contribute in both cases a factor $S^{(n-1)k+1}$, where $S := \#SO_0$. When we compare the counting, we see that

$$S^{(n-1)k+1}(C_{k(n-1)} - C_k^{(n)}) = \frac{S^{(n-1)k+1}}{(n-1)k+1}\left(\binom{2k(n-1)}{k} - \binom{nk}{k}\right) > 0 \quad (1.11.5)$$

since $2k(n-1) > nk$ for $n \geq 3$. □

Thus, at the level of planar trees, counting detects an *undergeneration* that is present at all levels $k \geq 1$ of the construction of the sets of syntactic objects. This phenomenon shows that there are always strings of elements of SO_0 of length $k(n-1)+1$ that have ambiguous parsings when realized in terms of binary Merge, while the same ambiguity cannot be accounted for with an n-ary Merge.

As a simple example of this type of undergeneration, the two parsings of the ambiguous sentence "I saw someone with a telescope" depend on the difference between the two binary trees

which would disappear entirely if the terms $\alpha, \beta, \gamma, \delta$ are assembled through a 4-ary Merge to form the tree

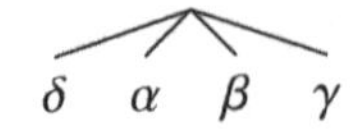

where the ambiguity would no longer be detectable, as only a single 4-ary parsing is possible.

1.11.3 The Structure of a Hypothetical n-ary Merge

Given the set $\mathcal{SO}^{(n)}$ of syntactic objects associated with a hypothetical n-ary Merge, obtained as in (1.11.3), we can consider the same type of action of Merge on workspaces that we have introduced previously for a binary Merge. We will see in section 1.11.4 that when the same structure is implemented through an n-ary Merge with $n \geq 3$, it *inevitably* leads to overgeneration.

As in the binary case, we introduce the set of workspaces as finite collections (multisets) of syntactic objects, which in the n-ary case are elements of the set $\mathcal{SO}^{(n)} \simeq \mathfrak{T}^{(n)}_{\mathcal{SO}_0}$ of nonplanar, n-ary, rooted trees. We again consider the vector space $\mathcal{V}(\mathfrak{F}^{(n)}_{\mathcal{SO}_0})$, where $\mathfrak{F}^{(n)}_{\mathcal{SO}_0}$ is the set of finite forests with connected components in $\mathfrak{T}^{(n)}_{\mathcal{SO}_0}$. To write the extraction of accessible terms and the cancellation of copies in the form of a coproduct, as well as the Merge operation on accessible terms, we consider the relevant algebraic structure on $\mathcal{V}(\mathfrak{F}^{(n)}_{\mathcal{SO}_0})$; namely, the product given by the disjoint union $\sqcup$ and the coproduct as in (1.2.8).

Remark 1.11.6. In defining a coproduct of the form (1.2.8) for an n-ary tree, one no longer has the option of taking a quotient of the form $T/^d T_v$ because after removal of the subtree T_v, contractions of edges in the resulting tree $T \smallsetminus T_v$ will produce vertices with either fewer or more than n descendants. Unlike with binary trees, there is no unique maximal n-ary tree obtainable from the deletion of the subtree T_v. To have a quotient T/T_v, which is itself an n-ary tree, one can use the quotient construction $T/^c T_v$ obtained by contracting T_v to its root vertex, which will carry a label $\not{T}_v$, as in the binary case. We can also still consider a quotient of the form $T/^\rho T_v$, using admissible cuts of the n-ary trees, where the extracted $\pi_C(T)$ is an n-ary tree and the remaining term $\rho_C(T) = T/^\rho \pi_C(T)$ is an "at most n-ary" tree, which can have vertices of lower valence.

We consider the coproduct on $\mathcal{V}(\mathfrak{F}^{(n)}_{\mathcal{SO}_0})$, given by

$$\Delta^c(T) = \sum_{\underline{v}} F_{\underline{v}} \otimes T/^c F_{\underline{v}},$$

where $F_{\underline{v}} = \pi_C(T)$ is the forest given by the collection of accessible terms obtained as the result of an admissible cut on T, and

$$\Delta^c(F) = \sqcup_a \Delta^c(T_a) \quad \text{for} \quad F = \sqcup_a T_a .$$

We can assume that the form of the action of Merge on workspaces will be of the same form as in the binary case of (1.3.7). Thus, we can write the desired form for the n-ary Merge action on workspaces as follows.

Given a collection $S = (S_i)_{i=1}^n$ of n-ary syntactic objects $S_i \in \mathfrak{T}^{(n)}_{\mathcal{SO}_0}$, we also define an operator

$$\delta_{S_1,\dots,S_n} : \mathcal{V}(\mathfrak{F}^{(n)}_{\mathcal{SO}_0}) \otimes \mathcal{V}(\mathfrak{F}^{(n)}_{\mathcal{SO}_0}) \to \mathcal{V}(\mathfrak{F}^{(n)}_{\mathcal{SO}_0}) \otimes \mathcal{V}(\mathfrak{F}^{(n)}_{\mathcal{SO}_0})$$

in the same way as the $\delta_{S,S'}$ defined in the binary case. We set

$$\delta_{S_1,\dots,S_n} = \gamma_{S_1,\dots,S_n} \otimes \mathrm{id} ,$$

where

$$\gamma_{S_1,\dots,S_n} : \mathcal{V}(\mathfrak{F}^{(n)}_{\mathcal{SO}_0}) \to \mathcal{V}(\mathfrak{F}^{(n)}_{\mathcal{SO}_0})$$

$$\gamma_{S_1,\dots,S_n}(F) = \begin{cases} F & F = S_1 \sqcup \cdots \sqcup S_n \\ 0 & \text{otherwise.} \end{cases}$$

Thus, the operator $\delta_{S_1,\dots,S_n}$ selects terms $F_{\underline{v}} \otimes T/^c F_{\underline{v}}$ of the coproduct, where

$$F_{\underline{v}} = T_{v_1} \sqcup \cdots \sqcup T_{v_n} \simeq S_1 \sqcup \cdots \sqcup S_n .$$

which means that up to a permutation σ of indices $T_{\sigma(i)} \simeq S_i$. We then obtain the action of the hypothetical n-ary Merge on workspaces in the form described in Definition 1.11.7.

Definition 1.11.7. *The action of n-ary Merge on workspaces consists of a collection of operators*

$$\{\mathfrak{M}_{S_1,\dots,S_n}\}_{S_i \in \mathfrak{T}^{(n)}_{\mathcal{SO}_0}}, \quad \mathfrak{M}_{S_1,\dots,S_n} : \mathcal{V}(\mathfrak{F}^{(n)}_{\mathcal{SO}_0}) \to \mathcal{V}(\mathfrak{F}^{(n)}_{\mathcal{SO}_0}),$$

parameterized by n-tuples $(S_i)_{i=1}^n$ *of n-ary syntactic objects. These operators act on* $\mathcal{V}(\mathfrak{F}^{(n)}_{\mathcal{SO}_0})$ *by*

$$\mathfrak{M}_{S_1,\dots,S_n} = \sqcup \circ (\mathcal{B} \otimes \mathrm{id}) \circ \delta_{S_1,\dots,S_n} \circ \Delta^c , \qquad (1.11.6)$$

with the same grafting operation $\mathcal{B}$ *as in Definition 1.3.2.*

Note that the n-ary analog of Lemma 1.3.3 also holds, so (1.11.6) is obtained analogously.

This action of Merge on workspaces has the same structure as in the binary case; namely, for each of the n inputs of the n-ary Merge $\mathfrak{M}_n$ a search is made over the workspace by extracting accessible terms and comparing them with the corresponding n-ary syntactic object S_i. Nonmatching terms are left unchanged in the new workspace, while the n-ary Merge operation is applied to n-tuples of matching terms among the extracted accessible terms for each Merge input. The new workspace then has these Merge outputs, along with the terms coming from the quotient part of the coproducts, where cancellation of the deeper copies of the accessible terms used by Merge is performed.

The analog of (1.4.1) for Internal Merge would take the form

$$\mathfrak{M}_{S_1,\dots,S_n,T/(S_1\sqcup\cdots\sqcup S_n)} = \mathfrak{M}_n \circ_i \mathfrak{M}_{S_i,1,\dots,1} \tag{1.11.7}$$

in a hypothetical n-ary case (whereas in the binary case, the $\mathfrak{M}_{S_i,1,\dots,1}$ do not exist as independent Merge forms and only take place in composition).

One can envision other possible generalizations of a binary Merge action on workspaces to the n-ary case, using a coproduct with higher arity instead of Δ. We will not discuss them here, since (1.11.6) is the simplest direct generalization of (1.3.7), and it suffices to show the inevitability of overgeneration (that would occur for the same reasons in other such generalizations as well).

1.11.4 Overgeneration

We can now view the *overgeneration* phenomenon as a different type of comparison between the sets $\mathcal{SO}$ and $\mathcal{SO}^{(n)}$, with respect to the undergeneration discussed previously. Unlike undergeneration, overgeneration depends not only on the structure of the set $\mathcal{SO}^{(n)}$ of syntactic objects, but also on the action of on workspaces as described previously.

Indeed, consider the following empirical linguistic example of overgeneration by a hypothetical ternary Merge. We assume a workspace given by a ternary forest of the form

$$F = \{\alpha,\beta,\gamma\} \sqcup \delta \sqcup \eta ,$$

with $\alpha,\beta,\gamma,\delta,\eta \in \mathcal{SO}^{(3)} \simeq \mathfrak{T}^{(3)}_{\mathcal{SO}_0}$. The action of a hypothetical ternary Merge on workspaces described by (1.11.6) with $n = 3$ and with $S = (S_1, S_2, S_3)$ given by $S_1 = \alpha$, $S_2 = \beta$, and $S_3 = \{\alpha,\beta,\gamma\}$ gives the internal Merge

$$\mathfrak{M}_{S_1,S_2,S_3}(F) = \{\alpha,\beta,\{\alpha,\beta,\gamma\}\} \sqcup \delta \sqcup \eta .$$

The same action with $S_1 = \delta$, $S_2 = \eta$, and $S_3 = \{\alpha,\beta,\gamma\}$ gives the external Merge

$$\mathfrak{M}_{S_1,S_2,S_3}(F) = \{\delta,\eta,\{\alpha,\beta,\gamma\}\} .$$

These ternary Merge operations are responsible for generating ungrammatical sentences such as[6] *peanuts monkeys children will throw* (as opposed to *children will throw monkeys peanuts*), resulting from

$$\{\text{peanuts}, \text{monkeys}, \{\text{children}, \text{will}, \{\text{throw}, \text{monkeys}, \text{peanuts}\}\}\} \,. \qquad (1.11.8)$$

In the example in (1.11.8), one sees that α and β are accessible terms of $\{\alpha,\beta,\gamma\}$, and hence with a ternary Merge, one can form $\{\alpha,\beta,\{\alpha,\beta,\gamma\}\}$. On the other hand, in the case of binary Merge, $\{\alpha,\beta\}$ is *not* an accessible term of $\{\{\alpha,\gamma\},\beta\}$, and hence the analogous Merge construction is not possible with binary Merge. Thus, ternary Merge overgenerates with respect to binary Merge by allowing expressions that are not allowed by binary Merge (and non grammatical, as in the example).

This example indicates that the illustrated overgeneration phenomenon is caused by the existence of accessible terms for an n-ary Merge that are not part of the set of accessible terms for the binary Merge. We can explain that more in detail in the following way, to show that this is in fact a general phenomenon and not a peculiarity of the example described here.

In the case of an arbitrary hypothetical n-ary Merge with $n \geq 3$, the overgeneration phenomenon is caused by a simple fact.

Lemma 1.11.8. *Consider a set L of $\ell = k(n-1)+1$ items in SO_0, where we assume that $k \geq 3$. Let $T^{(n)}$ be an n-ary tree in $\mathfrak{T}^{(n)}_{SO_0}$ with leaves set $L(T^{(n)}) = L$. There is always a collection of n disjoint accessible terms S_i of $T^{(n)}$, with the property that the set $L' \subset L$ given by $L' = L(S_1) \sqcup \cdots \sqcup L(S_n)$ cannot be the set of leaves $L' = L(T_1) \sqcup L(T_2)$ of two accessible terms T_i of a binary tree $T^{(2)}$ with the same set $L = L(T^{(2)})$ of leaves.*

Proof. It is enough to exhibit the existence of one such example. For larger values of k, examples with larger accessible terms are possible, with a similar construction. Since $n \geq 3$ and $k \geq 3$, we have $k \geq 3 > (2n-1)/(n-1)$ and $2n < k(n-1)+1$, so we can choose $2n$ points among the $k(n-1)+1$ leaves in $L(T^{(n)}) = L$. We chose them as follows: first choose a planar embedding of the tree $T^{(n)}$, which in turn fixes a linear ordering of the set $L(T^{(n)})$ of leaves, $\ell_1, \dots, \ell_{k(n-1)+1}$. We can position the leaves in this planar embedding in such a way that they all lie on the x-axis in this order. Choose as accessible terms of $T^{(n)}$ the first n odd-numbered leaves $\ell_1, \ell_3, \dots, \ell_{2n-1}$. Now suppose that there is a binary tree $T^{(2)}$ with the same set $L = L(T^{(2)})$ of leaves and two disjoint,

[6] This example was communicated to us by Riny Huijbregts. For a more detailed discussion of this and other examples, see [101].

nonempty, accessible terms T_1, T_2 of $T^{(2)}$ such that $L(T_1) \sqcup L(T_2) = L' = L(S_1) \sqcup \cdots \sqcup L(S_n)$. Consider the first three leaves ℓ_1, ℓ_2, ℓ_3 in the set L with the ordering obtained from the planar embedding of $T^{(n)}$. We know that ℓ_1 and ℓ_3 are in L'. Take the paths from ℓ_1 to the root of $T^{(2)}$ and from ℓ_2 to the root of $T^{(2)}$. Let v be the internal (non-leaf) vertex of $T^{(2)}$ where the paths first meet (and then continue as the same path from there to the root). If this vertex is the root, then the subtree of $T^{(2)}$ containing ℓ_1 and ℓ_3 is all of $T^{(2)}$, and hence we cannot have two disjoint, nonempty, accessible terms T_1, T_2 of $T^{(2)}$ with $L(T_1) \sqcup L(T_2) = L'$. Thus, v is not the root of $T^{(2)}$. If ℓ_1 and ℓ_3 belong to the same accessible term, for instance T_1, then the accessible term $T_v \subset T_1$. Now look at the leaf ℓ_2. It does not belong to L', so in particular it should not be in $L(T_1)$. But the path from ℓ_2 to the root has to meet either the path from ℓ_1 to v or the path from ℓ_3 to v (by the Jordan curve theorem in the plane, since it starts inside the region bounded by these two paths and the x-axis and it ends at the root that is outside this region). This implies that ℓ_2 in fact is a leaf of T_v, and hence a leaf of T_2, giving rise to a contradiction. If ℓ_1 and ℓ_3 do not belong to the same accessible term T_i, we can repeat the same argument with two leaves ℓ_{2k-1}, ℓ_{2j-1} that belong to the same accessible term. There will be at least one such pair because there are $n \geq 3$ leaves in L' by construction, so at least two must be in the same component of $T_1 \sqcup T_2$. □

An immediate consequence of this fact is that an n-ary Merge always overgenerates with respect to a binary Merge. We see that considering constructions obtained via Internal Merge.

Corollary 1.11.9. *For any n-ary syntactic object $T^{(n)} \in \mathcal{SO}_n$, with the set of leaves $L = L(T^{(n)})$, there is a choice $S_1, \ldots, S_{n-1}$ of disjoint, accessible terms of $T^{(n)}$ such that the subset $L' = L(S_1) \sqcup \cdots \sqcup L(S_{n-1}) \subset L$ cannot be realized as the set of leaves $L' = L(T_v)$ of an accessible term T_v of a binary tree $T^{(2)}$ with $L(T^{(2)}) = L$. Therefore, the ordered set of leaves of any planar embedding of the n-ary Internal Merge*

$$\mathfrak{M}_n(S_1, \ldots, S_n, T^{(n)}/(S_1 \sqcup \cdots \sqcup S_n))$$

cannot be realized as the ordered set of leaves of a planar embedding of a binary external Merge

$$\mathfrak{M}(T_v, T^{(2)}/T_v)$$

with $L(T^{(2)}) = L$ and $L(T_v) = L(S_1) \sqcup \cdots \sqcup L(S_{n-1})$.

1.12 A Model of Externalization

In this section, we discuss an algebraic model of Externalization. This is the process that interfaces the core computational mechanism of syntax, based on free symmetric Merge, with the Sensory–Motor (SM) system that externalizes language in the form of a time-ordered string of words, in the form of speech, sign, writing. The key feature of this step is that the syntactic objects produced by free symmetric Merge, which we have identified with abstract (nonplanar) binary rooted trees, acquire a planar embedding, which induces a linear ordering of the leaves of the tree (namely, a particular ordering of the words in a sentence). While at the level of the core computational system of syntax, only structural hierarchical relations exist, at the level of Externalization, proximity relations in the ordering also become relevant. Manifestly, Externalization is language dependent, in the sense that different word-order constraints are followed in different languages, depending on corresponding syntactic parameters.

Given our mathematical model of free symmetric Merge, building a compatible mathematical model of Externalization revolves around formalizing the assignment of planar structures to abstract, binary, rooted trees. As we will discuss in this section, there are some important algebraic properties related to this operation of planarization that fix the structure that Externalization can take. In particular, the algebra accounts for two main properties of Externalization, which we will discuss later in more detail:

1. Merge can act either *before* or *after* Externalization, *but not both* simultaneously and compatibly. (In the setting of *Elements* [40], Merge acts before Externalization.)
2. Externalization is language dependent, with constraints determined by syntactic parameters.

We discuss the general structure in section 1.12.1 before getting into more technical details about this topic.

1.12.1 Externalization: A Preliminary Discussion

Our mathematical model of Externalization is summarized by a diagram of the following form:

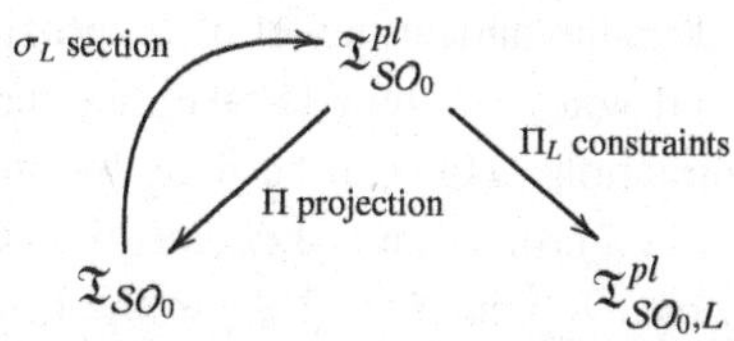

(1.12.1)

We will give further details about this diagram in the rest of this section and in section 1.13.1. We outline here just the main ideas.

When one considers the difference between the abstract (nonplanar embedded) binary rooted trees produced by free symmetric Merge and the planar binary rooted trees that one should obtain through the process of Externalization, this difference can be described in algebraic terms by two different magma structures. On the one hand, we have the commutative, nonassociative, free magma of syntactic objects

$$\mathcal{SO} = \text{Magma}_{na,c}(\mathcal{SO}_0, \mathfrak{M}) = \mathfrak{T}_{\mathcal{SO}_0},$$

which we have already discussed, which is the generative process of the free symmetric Merge. On the other hand, for planar trees, we have a similar algebraic structure:

$$\mathcal{SO}^{nc} = \text{Magma}_{na,nc}(\mathcal{SO}_0, \mathfrak{M}^{nc}) = \mathfrak{T}^{pl}_{\mathcal{SO}_0}.$$

This is the free, *noncommutative*, nonassociative magma over the same set $\mathcal{SO}_0$ of lexical items, with a binary operation *asymmetric* Merge:

$$\mathfrak{M}^{nc}(T_1^\pi, T_2^\pi) = \overset{\frown}{T_1^\pi \quad T_2^\pi} \neq \overset{\frown}{T_2^\pi \quad T_1^\pi} = \mathfrak{M}^{nc}(T_2^\pi, T_1^\pi),$$

where we use the notation T^π for the trees here as a reminder that they are endowed with the choice π of some planar embedding. We write $\mathfrak{T}^{pl}_{\mathcal{SO}_0}$ for the set of *planar*, binary, rooted trees with leaves labeled by elements of the set $\mathcal{SO}_0$. We will return to discuss these magma structures in more detail in section 1.13.1.

As we will show more explicitly in Lemma 1.13.1, there is a natural projection map $\Pi : \mathfrak{T}^{pl}_{\mathcal{SO}_0} = \mathcal{SO}^{nc} \twoheadrightarrow \mathcal{SO} = \mathfrak{T}_{\mathcal{SO}_0}$, which just forgets the planar embedding of the trees, $\Pi : T^\pi \mapsto T$. This projection map has very good properties: it is a morphism of magmas, it is defined everywhere, and it is canonical in the sense that it does not depend on any auxiliary choice to be defined. In Lemma 1.13.1, we will also give an explicit proof of why a map with the same properties going in the opposite direction, from $\mathfrak{T}_{\mathcal{SO}_0}$ to $\mathfrak{T}^{pl}_{\mathcal{SO}_0}$, cannot exist.

The problem with this projection map, however, is that it goes in the *wrong direction* with respect to Externalization. So the first question to answer in order to develop a good mathematical model of Externalization is how to get around this problem, or if you wish, how to take that one-way street $\mathfrak{T}^{pl}_{\mathcal{SO}_0} \to \mathfrak{T}_{\mathcal{SO}_0}$ in the opposite direction. This can be done by taking a *section* of the projection Π. A section of a projection is a *choice* of an element in each fiber (preimage set) $\Pi^{-1}(T)$ over a point $T \in \mathfrak{T}_{\mathcal{SO}_0}$ in the target space of the projection. If one writes $\sigma(T) \in \Pi^{-1}(T)$ for the chosen element of the fiber, one obtains a map $\sigma : \mathfrak{T}_{SO_0} \to \mathfrak{T}^{pl}_{SO_0}$ with the property that the composition $\Pi \circ \sigma = \text{id}$ is the identity on $\mathfrak{T}_{SO_0}$. This relation $\Pi \circ \sigma = \text{id}$ is the defining property of a *section* of the projection Π.

So one can take the one-way street $\Pi : \mathfrak{T}^{pl}_{SO_0} \twoheadrightarrow \mathfrak{T}_{SO_0}$ in the opposite direction using a section σ of the projection, but one pays a price for doing this, which is the loss of some of the "nice" properties of the map Π. In particular, any section σ of the projection Π will have two properties that determine some important aspects of Externalization:

- Unlike the projection Π, its section σ is *not* a morphism of magmas
- Constructing a section of the projection is *noncanonical*, (meaning that it is nonunique and it depends on choices)

These two properties capture two important linguistic properties of Externalization that have sometimes been raised:

- Merge can *only* act either *before* Externalization (as free symmetric Merge of SMT) or *after* the assignment of planar structures (as asymmetric Merge in older forms of Minimalism that we will discuss in chapter 2), but it cannot simultaneously and compatibly act in both ways.
- Externalization is language dependent: different languages, with different syntactic parameters, have Externalization that is realized by different sections of the same projection Π.

To emphasize the second property, we write the section as

$$\sigma_L : \mathfrak{T}_{SO_0} \to \mathfrak{T}^{pl}_{SO_0},$$

where the subscript L stands for the dependence of this section on a particular language L.

This explains the left-hand-side of the diagram (1.12.1), where one uses a section σ_L to "climb" the projection Π in the opposite way, assigning a planar structures to trees in a language-dependent way. Here, the language dependence is determined by the fact that the choice of planar structure has to be compatible with syntactic parameters dictating constraints on word order.

It remains to explain what the right-hand-side of the diagram (1.12.1) represents. When we look at the image $\sigma_L(\mathfrak{T}_{SO_0}) \subset \mathfrak{T}^{pl}_{SO_0}$, we find for each syntactic object $T \in \mathcal{SO}$ produced by free symmetric Merge a corresponding planar tree (hence a corresponding linearly ordered sequence of the lexical items attached to the leaves) constructed in such a way that it does not violate any of the word-order constraints of the given language L. However, among all the planar trees contained in $\sigma_L(\mathfrak{T}_{SO_0})$, there will still be many that may violate some other syntactic parameters of the language L that are not specifically about word order. Note that identifying explicitly which syntactic parameters disentangle from word-order constraints is not a simple task. For example, for a simple case of a syntactic parameter that appears unrelated to word order, consider ProDrop, which allows the subjects of sentences to remain unexpressed. While in principle this property does not prescribe word-order constraints, relations between ProDrop and word-order structures may occur; see [61]. It is a difficult general question to identify relations between syntactic parameters. Nonetheless, as a first approximation, let us assume that one can isolate a subset of syntactic parameters that suffices to determine all the word-order constraints, so that any word-order relation implicitly present in other parameters would be already determined by this subset, and any remaining parameter would therefore only impose further constraints that do not affect word order.

So one still needs to eliminate those trees in $\sigma_L(\mathfrak{T}_{SO_0})$ that violate other constraints imposed by syntactic parameters that are not just about word order. In addition, as we will be discussing shortly, one needs to eliminate constructions that cannot be labeled: these are additional constraints that come from additional data of head, complement, and Phase Theory. Eliminating these unwanted elements, or equivalently imposing all the constraints coming from syntactic parameters, means that we are taking a quotient of $\sigma_L(\mathfrak{T}_{SO_0})$ (the operation that eliminates an unwanted part of the structure) and that quotient is the other arrow on the right-hand-side of the diagram (1.12.1), that lands in a language-dependent collection $\mathfrak{T}^{pl}_{SO_0,L}$ of planar trees, which are the actual results of Externalization.

We will expand on this basic idea, in a more detailed form, in the rest of this section. A more extensive discussion of how to model Externalization within the mathematical formalism introduced here will appear in the forthcoming work [141].

1.12.2 Externalization as Correspondence

Our description of the action of Merge on workspaces suggests a possible way of thinking about the process of Externalization, which accounts for the fact that the core computational structure of free symmetric Merge needs to be followed by a procedure (called an "Externalization procedure") that models interaction with the SM system; see [7]. It is in this Externalization process that additional constraints are imposed, such as the presence of a linear ordering on sentences (in the form of planar embeddings of binary rooted trees), as well as constraints coming from any other universal grammar principles, called "First Factor" principles in *Elements* [40]. One also needs to account for the observed syntactic diversity across different human languages (syntactic parameters); for instance, see [62].

We can look first at the step of Externalization that introduces planar structures, and hence a linear ordering on the leaves of the trees (i.e., an ordering on the resulting sentence). At first, it may seem, intuitively, that introducing a linear ordering is a way of imposing a constraint and therefore should give rise to some kind of quotient map. In fact, the quotient map runs in the opposite direction, as the map that identifies the abstract (nonplanar) tree behind all its different planar embeddings. It can also be described as the quotient that maps noncommuting variables (where order matters) to corresponding commuting variables (where it does not).

The part of the Externalization process that fixes a planar structures consists of the choice of a section of this projection morphism, as we discussed previously. In particular, this section depends on choices. Indeed, this is not surprising, as this simply says that the choice of planar embeddings cannot be universal and is in fact language-dependent: it involves specific word-order constraints, as is familiar in linguistic theory. Thus, the construction of this section of the projection is the first instance where one can see the role of syntactic parameters in fixing the visible *externalized* language, in this case specifically in the form of word-order parameters.

We denote by $\mathfrak{T}_{SO_0}$ and $\mathfrak{F}_{SO_0}$ as before, the sets of binary rooted trees (respectively, forests) with leaf labels in SO_0, and we denote by $\mathfrak{T}^{pl}_{SO_0}$ and $\mathfrak{F}^{pl}_{SO_0}$ the corresponding sets of *planar* binary rooted trees (respectively, forests) with leaf labels in SO_0.

As we will be discussing more extensively in section 1.13.1, the free, nonassociative, commutative magma of syntactic objects, or equivalently of abstract (nonplanar) binary rooted trees, has a counterpart for *planar* trees. These are equivalently described as *ordered* words in the alphabet SO_0 with matching parentheses, such as

$(\alpha,(\beta,\gamma)) =$ 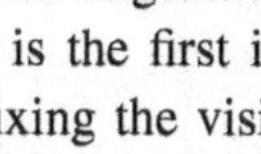 $\neq$ 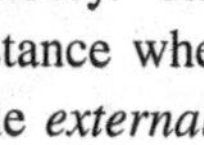 $= ((\beta,\gamma),\alpha)$,

where now the planar embedding of the tree matters, so the two cases $(\alpha,(\beta,\gamma))$ and $((\beta,\gamma),\alpha)$ are now different objects. The set $\mathfrak{T}^{pl}_{SO_0}$ also has a magma structure

$$\mathfrak{T}^{pl}_{SO_0} = SO^{nc} = \text{Magma}(SO_0, \mathfrak{M}^{nc}), \qquad (1.12.2)$$

the *free, nonassociative, noncommutative magma* of *planar* syntactic objects generated by the set SO_0.

Equivalently, we can describe the magma SO^{nc} through its Malcev representation. This is defined in the following way: consider a new variable c and use it to mark the opening parenthesis in the strings describing the objects of SO^{nc}. The position of the closing parenthesis is determined, so for example, instead of $(\alpha, ((\beta, \gamma), \delta)))$, one can just unambiguously write $c\alpha c^2\beta\gamma\delta$; see [95]. The magma operation $\mathfrak{M}^{nc}$ in the Malcev representation takes the form

$$\mathfrak{M}^{nc}(\alpha, \beta) = c\,\alpha\,\beta\,.$$

When we identify, as before, the set of ordered words in SO_0 with matched parentheses and the set of binary rooted trees with a choice of planar embedding, $\mathfrak{T}^{pl}_{SO_0} = SO^{nc}$, we see that in terms of the Malcev representation, the variable c marks the opening parenthesis that corresponds to an internal vertex of the planar tree in $\mathfrak{T}^{pl}_{SO_0}$.

As we will discuss in section 1.13.1, there is a surjective morphism of magmas,

$$\Pi : \mathfrak{T}^{pl}_{SO_0} \to \mathfrak{T}_{SO_0}\,,$$

which identifies all the planar trees that have the same underlying abstract tree.

An assignment of a planar structure can then be seen as a section σ_L of the projection Π:

$$\mathfrak{T}^{pl}_{SO_0} \underset{\Pi}{\overset{\sigma_L}{\rightleftarrows}} \mathfrak{T}_{SO_0}\,; \tag{1.12.3}$$

namely, a map satisfying $\Pi \circ \sigma_L = \mathrm{id}$, where the section is dependent on a particular language L and exists as a map of sets, but not as a morphism of magmas (this will be explained in more detail in section 1.13.1). These properties express the fact that the assignment of a linear ordering of sentences is not directly generated by Merge itself, but rather requires an additional mechanism (which is part of the Externalization procedure), and cannot be implemented in a universal language-independent way; see the discussion in section 1.12.7.

As we have seen in the previous sections, the computational mechanism described by the action of the free symmetric Merge on workspaces encodes the fundamental computational structure of syntax, which is independent of the variation of syntactic structures across languages. Where this variation actually occurs is only in the Externalization process that we are describing here. At the level of the syntactic objects, given by the trees in $\mathfrak{T}^{pl}_{SO_0}$, and of the workspaces, given by the forests in $\mathfrak{F}^{pl}_{SO_0}$, the Externalization that corresponds to a particular language L introduces quotient maps

$$\Pi_L : \mathfrak{T}^{pl}_{SO_0} \twoheadrightarrow \mathfrak{T}^{pl,L}_{SO_0}$$
$$\Pi_L : \mathfrak{F}^{pl}_{SO_0} \twoheadrightarrow \mathfrak{F}^{pl,L}_{SO_0}, \tag{1.12.4}$$

where $\mathfrak{T}^{pl,L}_{SO_0}$ and $\mathfrak{F}^{pl,L}_{SO_0}$ are the set of planar binary rooted trees (respectively, forests) with leaf labels in $\mathcal{SO}_0$, that are *possible* syntactic trees for the given language L. This quotient map very significantly reduces the combinatorial explosion of Merge, as only a small fraction of all the possible binary rooted trees generated by the Merge magma are realizable as syntactic trees of a specific given language. We discuss in section 1.12.7 the role of syntactic parameters in determining the quotient map (1.12.4). We will also discuss in sections 1.13.3 and 1.14 the role of head and phases in distinguishing possible objects. A further discussion of the modeling of syntactic parameters will appear in [141].

As we will discuss more in section 1.12.7, we see here two distinct roles for syntactic parameters in the model of Externalization that we propose: they influence the choice of the section σ_L through constraints on word order, and they contribute to the quotient map Π_L, which significantly cuts down the combinatorial explosion of Merge.

Thus we obtain in this way the two-step process described in the diagram (1.12.1). This type of two-step procedure, which we use for going from the set $\mathfrak{T}_{SO_0}$ to the set $\mathfrak{T}^{pl,L}_{SO_0}$, passing through the set $\mathfrak{T}^{pl}_{SO_0}$ (and similarly for forests) is an example of the general notion of *correspondence*, which is used in mathematics as a useful generalization of the notion of *function*.

1.12.3 Correspondences

We recall here the main idea behind the mathematical notion of *correspondence* and how it generalizes the concept of function and mapping, and we discuss a more category-theoretic way of interpreting correspondences. We refer the reader to the background material covered in chapter 4 for a review of the notions of category and 2-category that we mention here.

The notion of correspondence is a natural generalization of the concept of function or map, and it has played a crucial role in contemporary mathematics. It is generally understood that correspondences provide a better notion of morphisms than functions, such as by including the possibility of multivalued functions, as well as more general relations. In the case of a category of geometric spaces (or the underlying category of sets), one typically replaces the usual notion of a function $f : X \to Y$ with correspondences that take the form

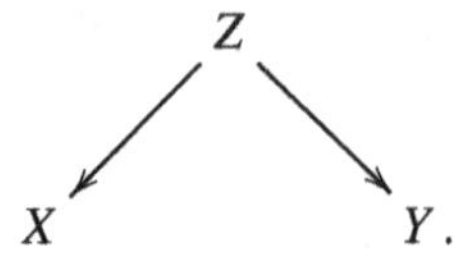

(1.12.5)

The case of a function is recovered as the special case where $Z = G(f) \subset X \times Y$ is the graph of the function $G(f) = \{(x, y) \mid y = f(x)\}$, with the two projection maps to X and Y. Correspondences, however, are more general than functions. Given a correspondence Z, one can transfer structures (e.g., vector bundles, spaces of functions, etc.) from X to Y by pulling them back to Z and then pushing them forward to Y via the two maps of the correspondence.

Thus, in this setting, given a category C that has such pullbacks, one can view correspondences as 1-morphisms in a 2-category of *spans* in C. (For more on the notion of 2-category, see Definition 4.1.4 in the review material in chapter 4.) This is the 2-category Spans(C) that has the following:

- Objects given by the objects of C
- 1-morphisms given by C-diagrams of the form (1.12.5), with the composition given by the pullback

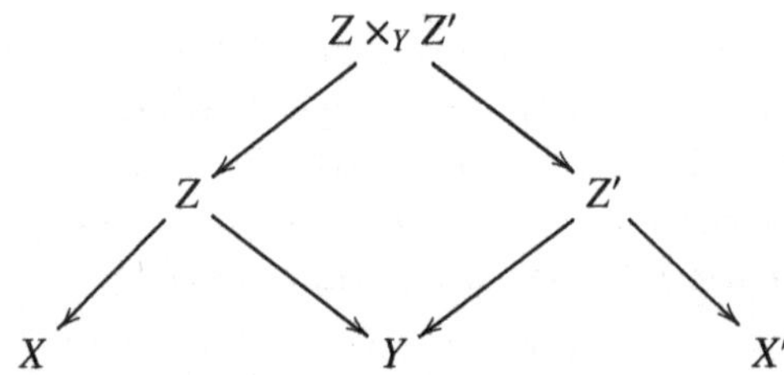

- 2-morphisms between spans $X \leftarrow Z_1 \to Y$ and $X \leftarrow Z_2 \to Y$ are morphisms $Z_1 \to Z_2$ in C that give a commutative diagram

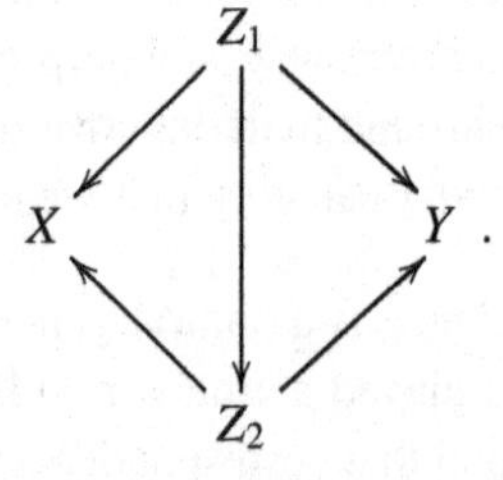

Since correspondences are usually described in this way as *spans* in the case of geometric spaces, they are usually described dually as *cospans* in the case of algebras—namely, as diagrams of the following form:

$$\mathcal{A} \longrightarrow \mathcal{E} \longleftarrow \mathcal{A}'.$$

This construction further extends to the typical case where correspondences of algebras are defined as bimodules. However, one can also consider the case of correspondences (or co-correspondences) given by spans of algebras

$$\mathcal{A} \longleftarrow \mathcal{E} \longrightarrow \mathcal{A}'.$$

(and co-spans of spaces), with the composition

$$(\mathcal{A}' \stackrel{g}{\leftarrow} \mathcal{E}' \to \mathcal{A}') \circ (\mathcal{A} \leftarrow \mathcal{E} \stackrel{f}{\to} \mathcal{A}'')$$

given by the pullback; that is, the restricted direct sum

$$\mathcal{E} \oplus_{\mathcal{A}'} \mathcal{E}' = \{(e, e') \mid f(e) = g(e')\}.$$

This is the kind of correspondences that we see in the description of the Externalization of Merge as outlined here. To formulate it in terms of this formalism, we can pass from the maps Π, σ_L, Π_L of sets and magmas to corresponding maps of vector spaces, algebras, and modules. We will develop this viewpoint in the rest of this section.

Since this description is more technical and will not be directly needed in the following discussion, readers can simply refer to the description of Externalization given previously in terms of the diagram of maps (1.12.1), which is all that we will need in the following parts of the book, and skip the rest of this discussion in sections 1.12.4, 1.12.5, 1.12.2, and 1.12.6, jumping directly to section 1.12.7.

1.12.4 Magma and Nonassociative Algebra

The action of Merge on workspaces described in (1.3.7) and Definition 1.3.4 also can be interpreted as a representation of a nonassociative algebra in the following way. (As elsewhere in this book, all vector spaces and algebras are taken over the field $\mathbb{Q}$.)

First, note that the magma structure on $\mathcal{SO} = \mathfrak{T}_{\mathcal{SO}_0}$ of (1.1.2) gives to the vector space $\mathcal{V}(\mathfrak{T}_{\mathcal{SO}_0})$ the structure of a nonassociative, commutative algebra, see [94] and [95], where the binary Merge operation $\mathfrak{M}$ gives the product operation.

Note that the coproduct (1.2.8) does not induce a bialgebra structure on $\mathcal{V}(\mathfrak{T}_{SO_0})$ with respect to this product because it does not satisfy the required compatibility requirement:

$$\Delta \circ \mathfrak{M} \neq (\mathfrak{M} \otimes \mathfrak{M}) \circ \tau \circ (\Delta \otimes \Delta) .$$

This is unlike the compatibility of $\sqcup$ and Δ on $\mathcal{V}(\mathfrak{F}_{SO_0})$ in Lemma 1.2.12. The reason is that $\Delta(\mathfrak{M}(T, T'))$ has only terms of the form

$$\mathfrak{M}(T_v, T') \otimes T/T_v, \quad \mathfrak{M}(T, T'_w) \otimes T'/T'_w, \quad T_v \otimes \mathfrak{M}(T/T_v, T'), \quad T'_w \otimes \mathfrak{M}(T, T'/T'_w),$$

while the right-hand-side applied to $T \otimes T'$ also has all terms of the form

$$\mathfrak{M}(T_v, T'_w) \otimes \mathfrak{M}(T/T_v, T'/T'_w) .$$

However, a modified form of the coproduct (1.2.8) *does* give $\mathcal{V}(\mathfrak{T}_{SO_0})$ the structure of a nonassociative, commutative, co-commutative, co-associative Hopf algebra (see [94] and [95]), with

$$\Delta(T) = \sum_{L \subset L(T)} T|_L \otimes T|_{L^c} , \tag{1.12.6}$$

where, for a subset $L \subset L(T)$ (with $L^c = L(T) \setminus L$), we write $T|_L$ to denote the binary rooted tree obtained by removing all the leaves in L^c and then performing the edge contractions needed to obtain a binary tree. The difference between this coproduct and (1.2.8) lies in the fact that the coproduct of (1.12.6) would correspond to a notion of accessible terms that includes all possible subsets of the set of leaves of the corresponding trees, not just those of the form $L = L(T_v)$. This structure, however, would not correspond to the linguistic properties of Merge because, for example, it would produce cancellations not only of the deeper copies, but also of the copies extracted by Internal Merge. Thus, this is not the right Hopf algebra structure to consider in our setting.

However, here we only need to consider the nonassociative commutative algebra structure $\mathcal{A}_{na,c} = (\mathcal{V}(\mathfrak{T}_{SO_0}), \mathfrak{M})$, without the comultiplication. The remark given here above about the coproduct is only included for completeness.

1.12.5 Merge Representation

We describe here the range of the Merge operations $\mathfrak{M}_{S,S'}$ acting on on $\mathcal{V}(\mathfrak{T}_{SO_0})$ in terms of a representation of a nonassociative algebra.

The notion of representation and module over a nonassociative algebra is much weaker than its associative counterpart. If $\mathcal{A}$ is a *nonassociative* algebra and $\mathcal{V}$ is a vector space, an $\mathcal{A}$-module structure on $\mathcal{V}$ is simply given by a *linear* map:

$$\rho : \mathcal{A} \to \mathrm{End}(\mathcal{V}) .$$

This map is not a morphism of algebras when $\mathcal{A}$ is nonassociative. We can equivalently view ρ as a linear map $\rho : \mathcal{A} \times \mathcal{V} \to \mathcal{V}$. We say that a vector space $\mathcal{V}$ is a module over a nonassociative algebra $\mathcal{A}$ if it is endowed with a representation of $\mathcal{A}$ on $\mathcal{V}$ in the sense described here.

Lemma 1.12.1. *The vector space $\mathcal{V}(\mathfrak{F}_{SO_0})$ is a module over the algebra $\mathcal{A}_{na,c}$ through the representation given by the maps*

$$\rho(T)(F) = \sqcup \circ (\mathfrak{M}^T \otimes 1) \circ \Delta(F) = \sqcup_a(\mathfrak{M}(T, T_{a,v}) \sqcup T_a/T_{a,v}), \qquad (1.12.7)$$

where $F = \sqcup_a T_a$ and $\mathfrak{M}^T(T_a) := \mathfrak{M}(T, T_a)$.

It then suffices to show that the representation (1.12.7) is enough to determine the Merge operators $\mathfrak{M}_{S,S'}$ as described in (1.3.7) in Definition 1.3.4. In other words, we want to show that every new workspace obtained through the action of Merge described in (1.3.7) can also be obtained through (1.12.1). The following observation indeed shows that the same terms that can be obtained via Internal Merge and External Merge can be obtained through the procedure described by (1.12.7) in the representation of Lemma 1.12.1. Note that this does not mean that (1.12.7) is the same as (1.3.7): it simply means that when we consider all the $\rho(T)$, their images will recover the images of the operators $\mathfrak{M}_{S,S'}$ of (1.3.7):

$$\bigcup_T \rho(T)(\mathcal{V}(\mathfrak{F}_{SO_0})) = \bigcup_{S,S'} \mathfrak{M}_{S,S'}(\mathcal{V}(\mathfrak{F}_{SO_0})) ,$$

with the union here meant as the common span as vector spaces.

Lemma 1.12.2. *The representation* (1.12.7) *suffices to determined the Merge operations* (1.3.7) *on workspaces in $\mathcal{V}(\mathfrak{F}_{SO_0})$.*

Proof. First, observe that in (1.12.1), the operator $\mathfrak{M}^T \otimes 1$ selects the terms of $\Delta(F)$ of the form $T_v \otimes F/T_v = T_v \otimes (T_i/T_v \sqcup \hat{F})$, for $F = \sqcup_a T_a$ and $\hat{F} = \sqcup_{a \neq i} T_a$. It then produces from each such term a new forest of the form $\mathfrak{M}(T, T_v) \sqcup T_i/T_v \sqcup \hat{F}$. Each term can also be obtained from (1.3.7) by applying $\mathfrak{M}_{S,S'}$ with $S \simeq T$ and $S' \simeq T_v$ to a workshop F' of the form $F' = T \sqcup F$, (1.3.7), with $\mathfrak{M}_{S,S'}$ selecting the term $T \sqcup T_v \otimes T_i/T_v \sqcup \hat{F}$ in $\Delta(F')$. Note that the form of Merge obtained can be one of the cases 2(b), 3(a), 3(b) of Sideward Merge: we do not worry about it here, as we know that they will be subdominant terms by Minimal Search, and it is simpler here to work with the full expression (1.3.7), without worrying about the various subcases. Conversely, a workspace of the form

$$\mathfrak{M}_{S,S'}(F) = \mathfrak{M}(S,S') \sqcup T_i/S \sqcup T_j/S' \sqcup \hat{F}\,,$$

with $\hat{F} = \sqcup_{a \neq i,j} T_a$, is contained in the range $\rho(S)(T_j \sqcup \hat{F})$. □

In the case of planar binary rooted trees and the free, nonassociative, noncommutative magma $\mathcal{SO}^{nc} = \mathfrak{T}^{pl}_{\mathcal{SO}_0}$ described here, one similarly obtains a nonassociative and noncommutative algebra:

$$\mathcal{A}_{na,nc} = (\mathcal{V}(\mathfrak{T}^{pl}_{\mathcal{SO}_0}), \mathfrak{M}^{nc})\,.$$

Lemma 1.12.3. *The morphism of magmas* $\Pi : \mathfrak{T}^{pl}_{\mathcal{SO}_0} \to \mathfrak{T}_{\mathcal{SO}_0}$ *and a section (map of sets)* $\sigma_L : \mathfrak{T}_{\mathcal{SO}_0} \to \mathfrak{T}^{pl}_{\mathcal{SO}_0}$ *with* $\Pi \circ \sigma_L = \mathrm{id}$ *on* $\mathfrak{T}_{\mathcal{SO}_0}$ *induces a morphism of algebras:*

$$\Pi : \mathcal{A}_{na,nc} = (\mathfrak{T}^{pl}_{\mathcal{SO}_0}, \mathfrak{M}^{nc}) \to \mathcal{A}_{na,c} = (\mathfrak{T}_{\mathcal{SO}_0}, \mathfrak{M})\,, \tag{1.12.8}$$

and a linear map of vector spaces:

$$\sigma_L : \mathcal{V}(\mathfrak{T}_{\mathcal{SO}_0}) \to \mathcal{V}(\mathfrak{T}^{pl}_{\mathcal{SO}_0})\,, \tag{1.12.9}$$

satisfying $\Pi \circ \sigma_L = \mathrm{id}$ *on* $\mathcal{V}(\mathfrak{T}_{\mathcal{SO}_0})$.

Proof. Consider the projection map $\Pi : \mathfrak{T}^{pl}_{\mathcal{SO}_0} \to \mathfrak{T}_{\mathcal{SO}_0}$, which assigns to a planar-embedded tree an underlying abstract tree, while dropping the planar embedding; that is, lumping together all the different planar embeddings of the same abstract tree. This map induces, via extension by linearity, a map of vector spaces $\Pi : \mathcal{V}(\mathfrak{T}^{pl}_{\mathcal{SO}_0}) \twoheadrightarrow \mathcal{V}(\mathfrak{T}_{\mathcal{SO}_0})$.

The algebra $\mathcal{A}_{na,nc}$ is the free, nonassociative, noncommutative algebra generated by the set $\mathcal{SO}_0$ with a nonassociative, noncommutative product, which we denote as $\mathfrak{M}^{nc}$. Unlike the nonassociative, commutative Merge product $\mathfrak{M}$ of $\mathcal{A}_{na,c}$, we have in general $\mathfrak{M}^{nc}(\alpha,\beta) \neq \mathfrak{M}^{nc}(\beta,\alpha)$; hence, we can identify $\mathcal{A}_{na,nc}$ with the algebra associated to the nonassociative, noncommutative magma $\mathcal{SO}^{nc} = \mathrm{Magma}(\mathcal{SO}_0, \mathfrak{M}^{nc})$. The quotient map is exactly Π, which identifies different planar embeddings of the same underlying abstract tree, and hence the linear map Π described here is in fact also a morphism of algebras:

$$\Pi : \mathcal{A}_{na,nc} \to \mathcal{A}_{na,c}\ ;$$

namely, the morphism that kills the commutators and has as its kernel the ideal generated by the elements $\mathfrak{M}^{nc}(T,T') - \mathfrak{M}^{nc}(T',T)$. Elements in this ideal are by construction formal differences between pairs of trees that differ in planar embeddings at one or more of the internal vertices since every application of $\mathfrak{M}^{nc}$ corresponds to an internal vertex of the resulting planar tree.

The first step of Externalization, described previously as the construction of a section σ_L of the projection Π (as a map of sets, not a morphism of magmas), induces a section (that we still call σ_L) of the projection $\Pi : \mathcal{A}_{na,nc} \to \mathcal{A}_{na,c}$. Such a section cannot be a morphism of algebras, as that would not map a commutative to a noncommutative algebra (see section 1.13.1 for why $\mathcal{A}_{na,nc}$ does not have a commutative subalgebra that $\mathcal{A}_{na,c}$ can map to). So the section σ_L determines a map $\sigma_L : \mathcal{V}(\mathfrak{T}_{SO_0}) \to \mathcal{V}(\mathfrak{T}^{pl}_{SO_0})$ that is just a linear map of vector spaces, with $\Pi \circ \sigma_L = \mathrm{id}$ on $\mathcal{V}(\mathfrak{T}_{SO_0})$. □

Proposition 1.12.4. *There is a corresponding quotient map on workspaces:*

$$\mathcal{V}(\mathfrak{F}^{pl}_{SO_0}) \twoheadrightarrow \mathcal{V}(\mathfrak{F}_{SO_0}) .$$

The representation (1.12.7) *extends to a representation:*

$$\rho^{pl} : \mathcal{A}_{na,nc} \to \mathrm{End}(\mathcal{V}(\mathfrak{F}^{pl}_{SO_0}))$$

so the following diagram commutes:

$$\begin{array}{ccc} \mathcal{A}_{na,nc} \otimes \mathcal{V}(\mathfrak{F}^{pl}_{SO_0}) & \xrightarrow{\rho^{pl}} & \mathcal{V}(\mathfrak{F}^{pl}_{SO_0}) \\ \downarrow{\scriptstyle \Pi\otimes\Pi} & & \downarrow{\scriptstyle \Pi} \\ \mathcal{A}_{na,c} \otimes \mathcal{V}(\mathfrak{F}_{SO_0}) & \xrightarrow{\rho} & \mathcal{V}(\mathfrak{F}_{SO_0}) . \end{array} \qquad (1.12.10)$$

Proof. As with Lemma 1.12.3, in $\mathfrak{F}^{pl}_{SO_0}$ forests are now planarly embedded, and hence the components T_a form an ordered set, which we describe by writing $F = \sqcup^{nc}_a T_a$, where $\sqcup^{nc}$ means that the order of the T_a matters; namely, $\sqcup^{nc}$ is the union as planarly embedded trees, in a sequential order compatible with an ordering of the union of their leaves. By defining ρ^{nc} as

$$\rho^{nc}(T)(F) = \sqcup^{nc} \circ (\mathfrak{M}^{T,nc} \otimes 1) \circ \Delta(F) = \sqcup^{nc}_a (\mathfrak{M}^{nc}(T, T_{a,v}) \sqcup T_a/T_{a,v}) ,$$

with $F = \sqcup^{nc}_a T_a$ and $\mathfrak{M}^{T,nc}(T_a) := \mathfrak{M}^{nc}(T, T_a)$, one obtains compatibility, as expressed by the commutativity of the diagram in the statement. □

1.12.6 Externalization Correspondence and Algebras

To formulate the quotient map Π_L of (1.12.4) at the level of the algebra $\mathcal{A}_{na,nc}$ and its action ρ^{pl} on the space $\mathcal{V}(\mathfrak{F}^{pl}_{SO_0})$ of workspaces with planar structure, we need to use the notion of *partial algebra*, which is a vector space endowed with a partially defined bilinear multiplication. Other widely used mathematical examples of partial algebras include the span of paths in a directed graph with the composition product.

Lemma 1.12.5. *The projection map of vector spaces* $\Pi_L : \mathcal{V}(\mathfrak{T}^{pl}_{SO_0}) \twoheadrightarrow \mathcal{V}(\mathfrak{T}^{pl,L}_{SO_0})$ *induced by the quotient map of* (1.12.4) *determines a nonassociative, noncommutative, partial algebra* $\mathcal{A}_{na,nc,L}$*, with an induced action* $\rho^{pl,L}$ *of* $\mathcal{A}_{na,nc,L}$ *on* $\mathcal{V}(\mathfrak{F}^{pl,L}_{SO_0})$*, with a commutative diagram:*

$$\begin{array}{ccc} \mathcal{A}_{na,nc,L} \otimes \mathcal{V}(\mathfrak{F}^{pl,L}_{SO_0}) & \xrightarrow{\rho^{pl,L}} & \mathcal{V}(\mathfrak{F}^{pl,L}_{SO_0}) \\ \Pi_L \big\uparrow & & \Pi_L \big\uparrow \\ \mathcal{A}_{na,nc} \otimes \mathcal{V}(\mathfrak{F}^{pl}_{SO_0}) & \xrightarrow{\rho^{pl}} & \mathcal{V}(\mathfrak{F}^{pl}_{SO_0}) . \end{array} \tag{1.12.11}$$

Proof. As a vector space, $\mathcal{V}(\mathfrak{T}^{pl,L}_{SO_0})$ is spanned by those trees in $\mathfrak{T}^{pl}_{SO_0}$ that are realizable as syntactic trees of the given language L, and as such it can be viewed either as a quotient space, under the projection Π determined by (1.12.4), or as a subspace of $\mathcal{V}(\mathfrak{T}^{pl}_{SO_0})$. This subspace $\mathcal{V}(\mathfrak{T}^{pl,L}_{SO_0})$ is a priori not a subalgebra with respect to the Merge product $\mathfrak{M}^{nc}$. However, it is a partial algebra, where the induced Merge $\mathfrak{M}^{nc,L}$ acts as $\mathfrak{M}^{nc}$ on the domain given by the set of pairs $T, T' \in \mathcal{V}(\mathfrak{T}^{pl,L}_{SO_0})$ with the property that $\mathfrak{M}^{nc}(T, T') \in \mathcal{V}(\mathfrak{T}^{pl,L}_{SO_0})$. This gives a nonassociative, noncommutative, partial algebra

$$\mathcal{A}_{na,nc,L} = (\mathcal{V}(\mathfrak{T}^{pl,L}_{SO_0}), \mathfrak{M}^{nc,L}) .$$

The vector space $\mathcal{V}(\mathfrak{F}^{pl,L}_{SO_0})$ can similarly be regarded either as a quotient of $\mathcal{V}(\mathfrak{F}^{pl}_{SO_0})$ under the quotient map Π_L or as a subspace. We can consider on $\mathcal{V}(\mathfrak{F}^{pl,L}_{SO_0})$ a coproduct induced by the coproduct Δ of $\mathcal{V}(\mathfrak{F}^{pl}_{SO_0})$, determined by setting

$$\Delta_L(T) = \sum_{v \in V_{int}(T): T_v, T/T_v \in \mathfrak{T}^{pl,L}} T_v \otimes (T/T_v) .$$

The induced action of $\mathcal{A}_{na,nc,L}$ on $\mathcal{V}(\mathfrak{F}^{pl,L}_{SO_0})$ is given by

$$\rho^{pl,L}(T)(F) = \Pi_L \circ (\mathfrak{M}^{T,nc} \otimes 1) \circ \Delta_L(F) ,$$

and satisfies by construction the stated compatibility. □

We encounter here a first instance of a phenomenon that we will discuss more extensively in chapter 2—namely, the fact that, if one tries to have Merge act after Externalization, then one necessarily has to deal everywhere with partially defined operations. As we will see in chapter 2, this indeed reflects the situation with earlier models of Minimalism (such as Stabler's Computational Minimalism). The partially defined nature of the Merge operation in this setting causes a compounding problem of domain checking when iterations of Merge are considered in the course of derivation, significantly increasing the computational complexity.

We can now see that the combination of the procedure described by the section σ_L of the projection Π for introducing a linear ordering of sentences, and the quotient procedure that eliminates trees that are not realizable as syntactic trees of a specific language describes the Externalization process in the form of a correspondence, in the sense outlined in section 1.12.3.

Remark 1.12.6. *Externalization is a correspondence given by the span of algebras (or partial algebras) and associated modules:*

$$
\begin{array}{ccccc}
 & & \mathcal{A}_{na,nc,L}\otimes\mathcal{V}(\mathfrak{F}^{pl,L}_{SO_0}) & \xrightarrow{\rho^{pl,L}} & \mathcal{V}(\mathfrak{F}^{pl,L}_{SO_0}) \\
 & \nearrow^{\Pi_L\otimes\Pi_L} & & \nearrow^{\Pi_L} & \\
\mathcal{A}_{na,nc}\otimes\mathcal{V}(\mathfrak{F}^{pl}_{SO_0}) & \xrightarrow{\rho^{pl}} & \mathcal{V}(\mathfrak{F}^{pl}_{SO_0}) & & \\
\downarrow{\scriptstyle \Pi\otimes\Pi} & & \downarrow{\scriptstyle \Pi} & & \\
\mathcal{A}_{na,c}\otimes\mathcal{V}(\mathfrak{F}_{SO_0}) & \xrightarrow{\rho} & \mathcal{V}(\mathfrak{F}_{SO_0}). & &
\end{array} \tag{1.12.12}
$$

1.12.7 The Role of Syntactic Parameters

In the Minimalist Model, where the core structure of syntax is described by the Merge operation of binary set formation, *syntactic parameters*, which account for syntactic variation across languages, become part of Externalization. The notion of syntactic parameters was originally introduced in the context of the earlier Principles and Parameters model, in [21] and [38]. A recent extensive study of syntactic parameters can be found in [176]. Here, we explore some the formal consequences of our model with respect to parameters. A more extensive discussion of syntactic parameters within this same mathematical model will be presented in the forthcoming work [141].

For simplicity, we can assume that syntactic parameters are binary variables. This may not account for phenomena such as some kind of entailment relations between parameters, observed for example in [127], but it is still accurate to a large extent. We can describe the set of syntactic parameters as a subset $\mathcal{P} \subset \mathbb{F}_2^N$, where N is a large number of binary variables that record various syntactic features of languages, and the locus $\mathcal{P} \subset \mathbb{F}_2^N$ accounts for the set of "possible languages" (see [154]), which can be viewed as regions of possible values of parameters that are realizable by actual human languages.

It is sometimes argued (see, for instance, [154]) that the class of possible languages is a more suitable object of study than the set of syntactic parameters. On the other hand, from a mathematical perspective, the use of binary variables describing syntactic parameters provides an ambient geometric space where the set of possible or impossible languages may be describable in algebraic and geometric terms, as discussed for instance in [158] and [187]. This approach has the advantage of being able to adopt several mathematical tools and techniques that are available in those contexts. So we restrict our discussion here, by simply adopting the framework of syntactic parameters.

The set $\mathcal{P}$ incorporates all the possible relations between parameters. One knows that a significant number of relations is expected: the presence of relations can be detected in various ways, such as through the geometric and topological data analysis techniques applied to databases of syntactic features; see, for instance, [73], [158], [172], and [187]. The exact nature of these relations is not known, but one can hypothesize that $\mathcal{P}$ may be realizable as an algebraic set (or algebraic variety) over $\mathbb{F}_2$, embedded in the affine space $\mathbb{F}_2^N$. Regardless of any specific assumption on the geometry of the set $\mathcal{P}$, we have that a language L determines a corresponding point $\pi_L \in \mathcal{P}$, which is a vector $\pi_L \in \mathbb{F}_2^N$ that lists as entries the binary values of the N syntactic parameters for that particular language.

In the description of Externalization proposed in section 1.12.2 and 1.12.6, one expects that syntactic parameters will be involved in determining both the section σ_L of (1.12.3) and the projection Π_L of (1.12.4).

Since the first part of Externalization, which corresponds to the section σ_L of (1.12.3), only depends on syntactic parameters that govern word order, while the projection Π_L of (1.12.4) depends on all other parameters, we can single out a subset of $M < N$ parameters that affect word order. We denote by $q : \mathbb{F}_2^N \to \mathbb{F}_2^M$ the corresponding projection map that only keeps the word-order parameters, and we denote by $\bar{\mathcal{P}} = q(\mathcal{P})$ the image under this projection of the locus of parameters, with $\bar{\pi} = q(\pi)$, for $\pi \in \mathcal{P}$. The parameters $q(\pi_L)_i$, $i = 1, \dots, M$ of a point $q(\pi_L)$ in this space $\bar{\mathcal{P}} \subset \mathbb{F}_2^M$ cut out a subset of $\mathfrak{T}^{pl}_{SO_0}$ that consists of those planar structures for trees in $\mathfrak{T}_{SO_0}$ that are compatible with the word-order properties of the given language L. These define the range of the section σ_L, and similarly for workspaces $\mathfrak{F}^{pl}_{SO_0}$.

On the other hand, given the set of all planar binary trees and forests in $\mathfrak{T}^{pl}_{SO_0}$ and $\mathfrak{F}^{pl}_{SO_0}$, respectively, the syntactic parameters specified by the point $\pi_L \in \mathcal{P}$ have the effect of selecting which syntactic trees are realizable in the given language L, thus determining the sets $\mathfrak{T}^{pl,L}_{SO_0}$ and $\mathfrak{F}^{pl,L}_{SO_0}$. We can give the following *geometric* description of this procedure, which has the advantage that it allows the possible use of tools from algebraic geometry in order to model more closely the Externalization process. We give here some examples of questions that can be naturally formulated in this mathematical framework.

As discussed, the vector spaces $\mathcal{V}(\mathfrak{T}^{pl}_{SO_0})$ and $\mathcal{V}(\mathfrak{F}^{pl}_{SO_0})$ are graded by the number of leaves (sentence length):

$$\mathcal{V}(\mathfrak{T}^{pl}_{SO_0}) = \oplus_\ell \mathcal{V}(\mathfrak{T}^{pl}_{SO_0})_\ell \quad \text{and} \quad \mathcal{V}(\mathfrak{F}^{pl}_{SO_0}) = \oplus_\ell \mathcal{V}(\mathfrak{F}^{pl}_{SO_0})_\ell , \tag{1.12.13}$$

with finite dimensional graded pieces.

Proposition 1.12.7. *Let $\mathcal{L}$ denote the set of languages $L \in \mathcal{L}$ and $\mathrm{Gr}(d,n)$ denote the Grassmannian, the parameterizing space for d-dimensional linear subspaces in an n-dimensional vector space. The identification of the spaces $\mathcal{V}(\mathfrak{T}^{pl,L}_{SO_0})$ and $\mathcal{V}(\mathfrak{F}^{pl,L}_{SO_0})$ by specifying the syntactic parameters $\pi_L \in \mathbb{F}_2^N$ for a language L is described by a collection of maps:*

$$E_{i,\ell} : \mathcal{P} \to \mathrm{Gr}(d_{\pi_i,\ell}, d_\ell) , \tag{1.12.14}$$

where for $\pi = (\pi_i)_{i=1}^N \in \mathcal{P}$, the image $E_{i,\ell}(\pi) \subset \mathcal{V}(\mathfrak{T}^{pl}_{SO_0})_\ell$ is the subspace spanned by the trees that are compatible with the constraints imposed by the value of the ith syntactic parameter π_i, with $d_\ell = \dim \mathcal{V}(\mathfrak{T}^{pl}_{SO_0})_\ell$. Thus, a point $\pi \in \mathcal{P}$ determines a subspace $E_\ell(\pi) = \cap_i E_{i,\ell}(\pi)$, and similarly for $\mathcal{V}(\mathfrak{F}^{pl}_{SO_0})$. The assignment $\pi : \mathcal{L} \to \mathcal{P}$ of syntactic parameters to languages $L \mapsto \pi_L$ in turn determines $\mathcal{V}(\mathfrak{T}^{pl,L}_{SO_0})$ as

$$\mathcal{V}(\mathfrak{T}^{pl,L}_{SO_0}) = \oplus_\ell E_{i,\ell}(\pi_L) , \tag{1.12.15}$$

and similarly for $\mathcal{V}(\mathfrak{F}^{pl,L}_{SO_0})$.

Proof. As in Lemma 1.12.5, we can view $\mathcal{V}(\mathfrak{T}^{pl,L}_{SO_0})$ and $\mathcal{V}(\mathfrak{F}^{pl,L}_{SO_0})$ as subspaces (rather than quotient spaces) of $\mathcal{V}(\mathfrak{T}^{pl}_{SO_0})$ and $\mathcal{V}(\mathfrak{F}^{pl}_{SO_0})$, respectively. Each syntactic parameter π_i of a point $\pi = (\pi_i)_{i=1}^N \in \mathcal{P} \subset \mathbb{F}_2^N$ determines a subspace $\mathcal{V}(\mathfrak{T}^{pl}_{SO_0})_{\pi_i} \subset \mathcal{V}(\mathfrak{T}^{pl}_{SO_0})$ ($\mathcal{V}(\mathfrak{F}^{pl}_{SO_0})_{\pi_i} \subset \mathcal{V}(\mathfrak{F}^{pl}_{SO_0})$, respectively), such that, for $\pi = \pi_L$ for some language $L \in \mathcal{L}$,

$$\bigcap_{i=1}^N \mathcal{V}(\mathfrak{T}^{pl}_{SO_0})_{\pi_{L,i}} = \mathcal{V}(\mathfrak{T}^{pl,L}_{SO_0}) , \tag{1.12.16}$$

and similarly for $\mathcal{V}(\mathfrak{F}^{pl}_{SO_0})_{\pi_{L,i}}$. Given the graded structure (1.12.13), we can consider the procedure (1.12.16) of cutting out the subspaces $\mathcal{V}(\mathfrak{T}^{pl,L}_{SO_0})$ and $\mathcal{V}(\mathfrak{F}^{pl,L}_{SO_0})$ step by step, by degrees. For a given $\ell \in \mathbb{N}$, there are integers $c_{\pi_i,\ell}$,

$i = 1, \dots, N$ that specify the codimensions of the subspaces $\mathcal{V}(\mathfrak{T}^{pl}_{SO_0})_{\pi_i,\ell}$ inside $\mathcal{V}(\mathfrak{T}^{pl}_{SO_0})_\ell$. If $d_\ell = \dim \mathcal{V}(\mathfrak{T}^{pl}_{SO_0})_\ell$ with $d_{\pi_i,\ell} = d_\ell - c_{\pi_i,\ell}$ as the dimensions, we then have maps

$$E_{i,\ell} : \mathcal{P} \to \mathrm{Gr}(d_{\pi_i,\ell}, d_\ell)$$

to the Grassmannian of $d_{\pi_i,\ell}$-dimensional subspaces inside the d_ℓ-dimensional space $\mathcal{V}(\mathfrak{T}^{pl}_{SO_0})_\ell$, such that

$$E_{i,\ell}(\pi_L) = \mathcal{V}(\mathfrak{T}^{pl}_{SO_0})_{\pi_{L,i}} \in \mathrm{Gr}(d_{\pi_{L,i},\ell}, d_\ell),$$

and similarly for $\mathcal{V}(\mathfrak{F}^{pl}_{SO_0})_\ell$. Similarly, if $d_{L,\ell} = \dim \mathcal{V}(\mathfrak{T}^{pl,L}_{SO_0})_\ell$ is the dimension of the resulting intersection (which need not be transversal due to relations between syntactic parameters), the vector $\pi_L \in \mathbb{F}_2^N$ of parameters for the language L determines a map as follows:

$$E_\ell \circ \pi : \mathcal{L} \to \bigcup_d \mathrm{Gr}(d, d_\ell) \quad E_\ell(\pi_L) = \mathcal{V}(\mathfrak{T}^{pl,L}_{SO_0})_\ell \in \mathrm{Gr}(d_{L,\ell}, d_\ell),$$

with $\mathcal{V}(\mathfrak{T}^{pl,L}_{SO_0}) = \oplus_\ell E_\ell(\pi_L)$. □

This viewpoint suggests natural geometric questions. For instance, when comparing the syntax of languages $L, L' \in \mathcal{L}$, one can consider the resulting comparison between the systems of subspaces $\mathcal{V}(\mathfrak{T}^{pl,L}_{SO_0}) = \oplus_\ell E_\ell(\pi_L)$ and $\mathcal{V}(\mathfrak{T}^{pl,L'}_{SO_0}) = \oplus_\ell E_\ell(\pi_{L'})$. Syntactic proximity can be viewed in terms of the geometric position of these subspaces. For example, mathematically, a special case of pairs E, E' of infinite dimensional subspaces inside an infinite dimensional space $\mathcal{V}$ is given by the Fredholm pairs, where the intersection $E \cap E'$ is finite dimensional and the span of the union $E \cup E'$ has finite codimension. These would represent the situation of maximal differentiation. Moreover, there are models of semantics based on the geometry of Grassmannians [133], and one can consider in this context the possibility of algebro-geometric models of a syntax–semantics interface.

We will not discuss these questions further in this book, but we will describe a more geometrically explicit relation between Externalization and the syntax–semantics interface in section 3.4.4 in chapter 3.

1.13 Externalization and Planarization

In the new formulation of Minimalism, as we discussed previously, Merge occurs in the free symmetric form described by the free, commutative, nonassociative magma $SO = \mathrm{Magma}_{na,c}(SO_0, \mathfrak{M})$ of (1.1.2), which constructively defines syntactic objects, which are binary rooted trees with no assignment of planar structure.

In this formulation of Minimalism, the assignment of planar structure to trees happens *after* the action of Merge has taken place, in a further process of Externalization.

This is in contrast with older versions of Minimalism (including the case of Stabler's formulation, which we will discuss in chapter 2), where Merge is applied directly to planar trees.

There are suggestions, such as Richard Kayne's Linear Correspondence Axiom (LCA) (see [109] and [110]), which proposes the replacement of Externalization with a more implicit (and unique) choice of planar embeddings for trees. Regardless of the tenability of this proposal on linguistic grounds, we discuss in this section some difficulties with its implementation, that arise at the formal algebraic level.

First, a comment about terminology: in the linguistics literature, it is customary to use the term *linearization* to refer to the choice of a linear ordering for the leaves of a binary rooted tree. Since this creates a terminology conflict with the more common mathematical use of the word "linearization," and the choice of a linear ordering of the leaves is one induced by the choice of a planar embedding of the tree, we will adopt here the terminology *planarization* (of trees) instead of *linearization* (of the set of leaves). Thus, we will refer to the LCA proposal as a "planarization" rather than as a "linearization algorithm," as usually described. We trust that this will not be a source of confusion for readers.

1.13.1 Commutative and Noncommutative Magmas

The first important observation to note is that consistently, the Merge operation can take place either *before* (as in New Minimalism) or *after* the assignment of planar structure to trees (as in Old Minimalism), but *not* both at the same time. This is the same observation that we already made for the Externalization procedure. What we mean by this is the following simple mathematical observation.

Just as we consider the free, commutative, nonassociative magma

$$\mathrm{Magma}_{na,c}(\mathcal{SO}_0, \mathfrak{M})$$

of New Minimalism, we can similarly consider, as in section 1.12, the free, noncommutative, nonassociative magma over the same set $\mathcal{SO}_0$:

$$\mathcal{SO}^{nc} := \mathrm{Magma}_{na,nc}(\mathcal{SO}_0, \mathfrak{M}^{nc})$$

As we have discussed in the context of Externalization in section 1.12, the elements of the free, noncommutative, nonassociative magma $\mathcal{SO}^{nc}$ are the planar binary rooted trees with leaves labeled by the set $\mathcal{SO}_0$. We write the elements of $\mathcal{SO}^{nc}$ as T^π, where T is an abstract (nonplanarly embedded) binary rooted tree and π is a planar embedding of T. The noncommutative, nonassociative magma operation is given by

$$\mathfrak{M}^{nc}(T_1^{\pi_1}, T_2^{\pi_2}) = \widehat{T_1^{\pi_1} \quad T_2^{\pi_2}} =: T^\pi ,$$

where now the trees $T_1^{\pi_1}$ and $T_2^{\pi_2}$ are planar and the tree T^π is assigned the planar embedding π, where $T_1^{\pi_1}$ is to the left of $T_2^{\pi_2}$.

In particular, in the free, noncommutative, nonassociative magma $\mathcal{SO}^{nc}$, we have $\mathfrak{M}^{nc}(T_1^{\pi_1}, T_2^{\pi_2}) \neq \mathfrak{M}^{nc}(T_2^{\pi_2}, T_1^{\pi_1})$, unlike the case of the free, commutative, nonassociative magma $\mathfrak{M}$ of $\mathcal{SO}$. We now explain in more detail the following fact, which we already mentioned in section 1.12.

Lemma 1.13.1. *There is a morphism of magmas $\mathcal{SO}^{nc} \to \mathcal{SO}$ that simply "forgets" the planar structure of trees: $T^\pi \mapsto T$. On the other hand, there is* no possible *morphism of magmas that goes in the opposite direction, from $\mathcal{SO}$ to $\mathcal{SO}^{nc}$.*

Proof. The forgetful morphism $\mathcal{SO}^{nc} \to \mathcal{SO}$ is well defined as a morphism of magmas since the two planar trees $\mathfrak{M}^{nc}(T_1^{\pi_1}, T_2^{\pi_2})$ and $\mathfrak{M}^{nc}(T_2^{\pi_2}, T_1^{\pi_1})$ have the same underlying abstract (nonplanar) tree $\mathfrak{M}(T_1, T_2)$.

On the other hand, if a morphism of magmas $\mathcal{SO}$ to $\mathcal{SO}^{nc}$ existed, then its image would necessarily be a commutative submagma of $\mathcal{SO}^{nc}$, but $\mathcal{SO}^{nc}$ does not contain any nontrivial, commutative submagma. This can be seen easily as follows: if a tree T^π is contained in a commutative submagma of $\mathcal{SO}^{nc}$, then $\mathfrak{M}^{nc}(T^\pi, T^\pi)$ is as well, but $\mathfrak{M}^{nc}(\mathfrak{M}^{nc}(T^\pi, T^\pi), T^\pi)) \neq \mathfrak{M}^{nc}(T^\pi, \mathfrak{M}^{nc}(T^\pi, T^\pi))$, contradicting the fact that the submagma is commutative. □

This fact has the immediate consequence that we already stated—namely, that if a free symmetric Merge takes place *before* any assignment of planar structure to trees, then it cannot also consistently apply *after* a choice of planarization. In other words, if σ is any choice of a section of the projection $\mathcal{SO}^{nc} \to \mathcal{SO}$ (i.e., an assignment of planarization), then one *cannot have* compatible Merge operations satisfying $\mathfrak{M}^{nc}(\sigma(T_1), \sigma(T_2)) = \sigma(\mathfrak{M}(T_1, T_2))$. Merge can act on abstract trees, as in New Minimalism, or on planar trees, as in Old Minimalism, but these two views are mutually exclusive.

1.13.2 Planarization versus Externalization

Proposals such as Kayne's LCA planarization of trees, as in [109] and [110] (see also [118] and chapter 7 of [98] for a short summary), suggest the replacement of Externalization with a different way of constructing planarization. This relies on the idea that the abstract trees are endowed with additional data, related to heads, maximal projections, and c-command (constituent command) relations, that permit a *canonical* choice of planar structure. Next, we discuss two different mathematical difficulties inherent in this proposal.

First, let us assume that indeed, these additional data on the abstract syntactic tree suffice to endow it with a unique canonical choice of linearization. (We will discuss the difficulties with this assumptions shortly.) In this case, we would have a map, which we call σ^{LCA}, which assigns to an abstract, binary, rooted tree T a corresponding, uniquely defined, nonplanar tree $\sigma^{LCA}(T) = T^{\pi_{LCA}}$, with π_{LCA} denoting the planar embedding constructed by the LCA algorithm.

By our previous observations on morphisms of magmas in Lemma 1.13.1, we will necessarily have in general that

$$\mathfrak{M}^{nc}(\sigma^{LCA}(T_1), \sigma^{LCA}(T_2)) \neq \sigma^{LCA}(\mathfrak{M}(T_1, T_2)),$$

so σ^{LCA} *cannot be compatible* with an asymmetric Merge of planar trees.

The only way to make this compatible with Merge would be to define Merge on the image of σ^{LCA}, not as the asymmetric Merge of planar trees but as

$$\mathfrak{M}^{LCA}(\sigma^{LCA}(T_1), \sigma^{LCA}(T_2)) := \sigma^{LCA}(\mathfrak{M}(T_1, T_2)),$$

with $\mathfrak{M}^{LCA}$ only defined on the image of σ^{LCA} and not on all of $\mathcal{SO}^{nc}$ like $\mathfrak{M}^{nc}$. This, however, simply creates an isomorphic copy of the commutative magma $\mathcal{SO} = \text{Magma}_{na,c}(\mathcal{SO}_0, \mathfrak{M})$ (by a choice of a particular representative in each equivalence class of the projection $\mathcal{SO}^{nc} \to \mathcal{SO}$). This would imply that application of the planarization σ^{LCA} has no effect, in terms of the properties of Merge, with respect to working directly with free symmetric Merge on abstract, nonplanar trees.

As in Externalization, the alternative is to not require any compatibility between σ^{LCA} and Merge: in other words, *even if* the LCA replaces Externalization, Merge still only occurs in the form of free symmetric Merge, at the level of the abstract, nonplanar trees, not after planarization.

We now look more specifically at the proposals for how planarization σ^{LCA} should be obtained, in order to highlight a different kind of difficulty.

Definition 1.13.2. In an abstract, binary, rooted tree T, two vertices, v_1 and v_2, are sisters if there is a vertex v of T above (i.e., closer to the root) and connected to both v_1 and v_2. A vertex v of T dominates another vertex w if v is on the unique path from the root of T to w. A vertex v in T c-commands another vertex w if neither dominates the other and the lowest vertex that dominates v also dominates w. A vertex v asymmetrically c-commands a vertex w if v c-commands w, and v and w are not sisters. If in the tree T, every subtree T_v has a well-defined head, a maximal projection is a subtree T_v of T that is not strictly contained in any larger T_w with the same head.

The asymmetric c-command relation defines a partial ordering of the leaves of T. To extend this partial ordering relation, instead of directly using the asymmetric c-command relation to define the order structure, one uses maximal projections. Namely, one requires that a leaf ℓ precedes another leaf ℓ' iff either ℓ asymmetrically c-commands ℓ' or a maximal projection dominating ℓ c-commands ℓ'.

Even with this extension using maximal projections, there are issues in making this a total ordering, as discussed for instance in chapter 7 of [98] and in [118].

In the case of the LCA, the problem arises from the fact that a result of Merge need not have the heads of the two merged trees in an asymmetric c-command relation. This implies that even with the introduction of heads and maximal projections, the assumption that all trees have a head-marked leaf cannot always be satisfied; hence, one cannot obtain a total ordering of the leaves (a unique choice of planar embedding). Partial corrections to this problem in the LCA are suggested using movement, by introducing "null heads" in the structure, or by morphological reanalysis that hides certain items from LCA (see [98], pp. 230–231). In any case, the fundamental difficulty in constructing a planarization algorithm based on heads and maximal projections can be described using a formalization of the notion of head as in Definition 1.13.3.

Definition 1.13.3. We define a *head function* on an abstract, binary, rooted tree T as a function $h_T : V^o(T) \to L(T)$ from the set $V^o(T)$ of non-leaf vertices of T to the set $L(T)$ of leaves of T, with the property that if $T_v \subseteq T_w$ and $h_T(w) \in L(T_v) \subseteq L(T_w)$, then $h_T(w) = h_T(v)$. We write $h(T)$ for the value of h_T at the root of T.

This general definition is designed to abstract the properties of the head in the syntactic sense. We will discuss this further in section 1.13.3.

Lemma 1.13.4. *There are exactly $2^{\#V^o(T)}$ possible head functions on an abstract, binary, rooted tree T, with $V^o(T)$ the set of non-leaf vertices of T.*

Proof. Consider pairs (T, h_T) and $(T', h_{T'})$ of trees with given head functions. There are exactly two choices of a head function on the Merge $\mathfrak{M}(T, T')$, corresponding to whether $h(T)$ or $h(T')$ is equal to $h(\mathfrak{M}(T, T'))$. Since the trees T, T' are not planar and $\mathfrak{M}$ is the symmetric Merge, there is no consistent way of making one rather than the other choice for $h_{\mathfrak{M}(T,T')}$ at each application of $\mathfrak{M}$. This implies that on a given binary rooted tree T, there are $2^{\#V^o(T)}$ possible head functions. We can think of any such choice as the assignment, at each vertex $v \in V^o(T)$, of a marking to either one or the other of the two edges exiting v in the direction away from the root (or a choice of one black and one white coloring on those two edges). □

Lemma 1.13.5. *The set of all possible head functions h_T on an abstract, binary, rooted tree T can be identified with the set of all possible choices of a planar embedding of T.*

Proof. By thinking of h_T as an assignment of a marking to one of the two edges below each vertex, the head function h_T determines a planar embedding of T by putting under each vertex the marked edge to the left. Thus, the problem of constructing planar embeddings of abstract binary rooted trees can be transformed into the problem of constructing head functions. □

This planar embedding associated to a head function can be seen as the "harmonic head-initial" embedding, in the linguistic terminology. An analogous choice can be made by assigning to a head function the planar embedding that always puts the marked edge to the right. This would be the "harmonic head-final" choice. We make here the harmonic head-initial choice whenever we refer to the planar embedding determined by a head function.

The LCA algorithm aims at obtaining a special assignment $T \mapsto h_T$ of head functions h_T to abstract binary rooted trees T, which is somehow determined uniquely by the properties of the labeling set $\mathcal{SO}_0$ of the leaves of the trees. Let us denote by $\lambda(\ell)$ the label in $\mathcal{SO}_0$ assigned to the leaf $\ell \in L(T)$.

This is where the main difficulty arises. For instance, if the labeling set $\mathcal{SO}_0$ happens to be a totally ordered set (which is not a realistic linguistic assumption), as long at two trees (T, h_T) and $(T', h_{T'})$ have head functions with labels $\lambda(h(T)) \neq \lambda(h(T'))$, there is always a preferred choice of head function on $\mathfrak{M}(T, T')$, which is the one in which the two subtrees T and T' are ordered according to the ordering of the labels $\lambda(h(T))$ and $\lambda(h(T'))$ in $\mathcal{SO}_0$. However, this excludes the case where the leaves $h(T)$ and $h(T')$ may have the same label in $\mathcal{SO}_0$. So even under the unrealistically strong assumption that labels are taken from a totally ordered set, a planarization algorithm based on the construction of a head function cannot be defined on all the syntactic objects in $\mathcal{SO}$.

At the level of the underlying algebraic structure, the issue with planarization is therefore twofold. It does not provide an alternative to Merge acting on the nonplanar trees, for the reasons mentioned previously, regarding maps of magmas. At the same time, a choice of planar embedding that is independent of syntactic parameters and is based only on heads of trees would require a canonical construction of head functions from properties of the labeling set $\mathcal{SO}_0$, but this cannot be done consistently on the entire set of syntactic objects $\mathcal{SO}$ produced by a free symmetric Merge $\mathfrak{M}$.

1.13.3 Abstract Head Functions

In a syntactic tree, as is familiar, in general the syntactic category of the head determines the category of the phrase (verb for Verb Phrase, etc.; here we use the traditional terminology of "Verb Phrase (VP)" even though this is actually described by a set in the notation of [40] or as an abstract binary rooted tree in our notation). Moreover, the syntactic head determines the type of objects described; hence, it can be regarded as part of the mechanism that interfaces syntax with semantics.

In Definition 1.13.3, we showed that one can define an abstract *head function* on binary rooted trees T (with no assigned planar structure). We refine the definition in the following way.

Definition 1.13.6. A *head function* is a function h defined on a subdomain $\mathrm{Dom}(h) \subset \mathfrak{T}_{\mathcal{SO}_0}$, which assigns to a $T \in \mathrm{Dom}(h)$ a map $h : T \mapsto h_T$ where h_T is a function

$$h_T : V^o(T) \to L(T) \tag{1.13.1}$$

from the set $V^o(T)$ of non-leaf vertices of T to the set $L(T)$ of leaves of T, as in Definition 1.13.3: if $T_v \subseteq T_w$ and $h_T(w) \in L(T_v) \subseteq L(T_w)$, then $h_T(w) = h_T(v)$. We write $h(T)$ for the value of h_T at the root of T.

This notion summarizes the main properties of the syntactic head, although of course, one can have many more abstract head functions that do not correspond to the actual syntactic head.

To see this, note that our notion of head function of Definition 1.13.6 can be directly derived from the formulation of the notion of head and projection given by Chomsky in section 4 of [23], The equivalence of these formulations follows immediately by observing that in section 4 of [23] the syntactic head is characterized by the following inductive properties:

1. For $T = \mathfrak{M}(\alpha,\beta)$, with α and $\beta \in \mathcal{SO}_0$, the head $h(T)$ should be one or the other of the two items α,β. The item that becomes the head $h(T)$ is said to *project.*
2. In further projections, the head is obtained as the "head from which they ultimately project, restricting the term head to terminal elements", in the terminology of [23].
3. Under Merge operations $T = \mathfrak{M}(T_1,T_2)$, one of the two syntactic objects $T_1,T_2 \in \mathcal{SO}$ projects, and its head becomes the head $h(T)$. The label of the structure T formed by Merge is the head of the constituent that projects.

Lemma 1.13.7. *The three properties listed here are equivalent to the properties of heads in Definition 1.13.6.*

Proof. First, observe that the three properties from section 4 of [23] listed here determine a function $h_T : V^o(T) \to L(T)$ from the set $V^o(T)$ of non-leaf vertices of T to the set $L(T)$ of leaves of T. The function is defined by "following the head" determined by the three listed properties. In other words, the root vertex of the tree carries a label, which by the listed requirements is obtained as "the head from which it ultimately projects", which is assumed to be "a terminal element". This means that we are assigning to the root vertex a label $h(T)$ that is one of the items in $\mathcal{SO}_0$ attached to the leaves $L(T)$. Similarly, for any other internal vertex v of T, one can view the subtree (accessible term) T_v as the Merge of two subtrees $T_v = \mathfrak{M}(T_{v_1},T_{v_2})$ where T_{v_i} are the two subtrees with roots at the vertices below v. The same listed properties then ensures that we are mapping v to a leaf $\ell(v) \in L(T_v)$, which agrees with either the head of T_{v_1} or the head of T_{v_2}. Moreover, this also ensures that the property of Definition 1.13.6 is satisfied by the function $h_T : V^o(T) \to L(T)$ obtained in this way. Indeed, consider $T_v \subseteq T_w$. If the function determined by the three properties given here satisfies $h_T(w) \in L(T_v)$, then it means that it is T_v that projects, according to the definition of [23]; hence, $h_T(w) = h_T(v)$. This shows that the definition of "head" in section 4 of [23] implies the one given in Definition 1.13.6.

Conversely, suppose that we have an abstract head function as in Definition 1.13.6. We can see that it has to satisfy the three properties of section 4 of [23] in the following way. The first property is immediate from the fact that $h_T : V^o(T) \to L(T)$ is a function, which means that if we consider any subtree of T consisting of two leaves with a common vertex above them (i.e., $T_v = \mathfrak{M}(\alpha,\beta)$), then $h_T(v)$ has to be either α or β. To see that the second and third properties also hold, consider first the full tree T. Since this is a binary rooted tree, it is uniquely describable in the form $T = \mathfrak{M}(T_1, T_1)$ for two other binary rooted trees T_1, T_2. Since the function h_T takes values in the set $L(T) = L(T_1) \sqcup L(T_2)$, the head $h(T)$ is in either $L(T_1)$ or in $L(T_2)$. Suppose that it is in $L(T_1)$. The other case is analogous. Then, by Definition 1.13.6, we have $h(T) = h(T_1)$, where we write $h(T_v) := h_T(v)$. Continuing in the same way for each successive nodes, with the corresponding unique decompositions $T_v = \mathfrak{M}(T_{v,1}, T_{v,2})$, we obtain for each internal vertex a path to a leaf, which follows the head, and provides the "head from which it ultimately projects" as desired in the second property, while at each step, the third property holds. □

Remark 1.13.8. There are two important remarks to make regarding the two equivalent formulations of Definition 1.13.6 and Lemma 1.13.7. As we discussed in section 1.13.2, a consistent definition (compatible with the Merge operation) of a head function h does not extend to the entire $\mathcal{SO}$ but rather is defined on a domain $\mathrm{Dom}(h) \subset \mathcal{SO}$, so the identification between the descriptions of Definition 1.13.6 and of section 4 of [23] also holds for such a domain. Moreover, as we also discussed in section 1.13.2, on a given $T \in \mathcal{SO}$, there are $2^{\#V^o(T)}$ choices of a head function (which are in bijective correspondence with the choices of a planar structure for T). This is why we are saying that, for a given T, there are more abstract head functions than just the one that corresponds to the syntactic head (when the latter is well defined). This does not matter, as for most of the arguments we are using that involve head functions, the formal property of Definition 1.13.6 is the only characterization required. In terms of explicit linguistics examples, one can think of the usual syntactic head as presented in [23].

As shown in section 1.13, it follows directly from this definition that assigning a head function h_T to a tree T is equivalent to assigning a planar embedding π_{h_T} (every head function determines a planar embedding and conversely).

Thus, we can equivalently think of an assignment

$$h : T \mapsto h_T \tag{1.13.2}$$

of a head function to every tree $T \in \mathfrak{T}_{\mathcal{SO}_0}$ as a function

$$h : \mathrm{Dom}(h) \subset \mathfrak{T}_{\mathcal{SO}_0} \to \Sigma^*[\mathcal{SO}_0] \tag{1.13.3}$$

to the set of all finite ordered sequences, of arbitrary length, in the alphabet $\mathcal{SO}_0$, given by

$$h(T) = L(T^{\pi_{h_T}}),$$

where $T^{\pi_{h_T}}$ is the planar embedding of T determined by the head function, and $L(T^{\pi_{h_T}})$ is its ordered set of leaves. Since it is equivalent to describe $h(T)$ as the ordered set $L(T^{\pi_{h_T}})$ or as a single leaf (the head) in $L(T)$, we will switch between these two descriptions without changing the notation. However, one should keep in mind here that the map (1.13.3) does not realize all possible orderings: not all the $n!$ linear orderings of a set of n leaves correspond to orderings induced by planar embeddings of the tree (i.e., not all ordering are compatible with the tree structure): only 2^{n-1} of them are. We will discuss this difference in more detail in section 3.4.

We have shown in section 1.13.2 that one does not have a well-defined head function on the entire $\mathfrak{T}_{\mathcal{SO}_0}$; hence, we write h as defined on a domain $\mathrm{Dom}(h) \subset \mathfrak{T}_{\mathcal{SO}_0}$. The obstacle to the extension of a head function to the entire set $\mathfrak{T}_{\mathcal{SO}_0}$ derives from the well-known issue of *exocentric* constructions (e.g., in the traditional division of sentences into Subjects and Predicates); namely, cases of syntactic objects $T \in \mathcal{SO} = \mathfrak{T}_{\mathcal{SO}_0}$ that are obtained as the result of External Merge $T = \mathfrak{M}(T', T'')$, where even if a head function is well defined on T' and T'', there is no good way of comparing $h(T')$ and $h(T'')$ to decide which one should become the head of $T = \mathfrak{M}(T', T'')$. Abstract heads are thus partially defined functions.

It is interesting to observe here that this fact makes abstract heads amenable to treatment according to the renormalization model used in the theory of computation, where the source of "meaningless infinities" arises from what lies outside of the domain where a function is computable (see [129] and [130]). We will indeed use this approach in Chapter 3 to construct a very simple illustrative model of our proposed view of the syntax–semantics interface; see section 3.2.

1.14 Phase Theory

The notion of an abstract head function, which we discussed in section 1.13.3, also allows us to formulate, within our general formalism, the notion of *phases*, which we discuss in this section, as well as the main aspects of the labeling algorithm for syntactic objects as introduced in [26] and [27]. The latter replaces, under "Bare Phrase Structure" in the Minimalist model, the usual labels, such as NP and VP. We will discuss labeling in section 1.15. We first focus on the notion of phases.

Phase Theory is designed to reduce the computational complexity of the generative process based on the action of Merge on workspaces, by identifying substructures that, once generated, are no longer modified by further computation. Phase Theory aims at identifying, in an optimal way, syntactic objects with this property. This is motivated by the general principle of economy of computation (also referred to as "Third Factor Principle"). We follow here the definition of phases and interior and edge of phases given by Chomsky in [26], which we present in a form compatible with our terminology and notation.

If a syntactic object T has a syntactic head h_T, then this determines the syntactic category of T (its "projection"), and it also determines a *complement*–namely, all the elements that the head $h(T)$ *must* combine with. One says in this case that the complement is *selected by the head.* In addition to the head and the complement, a phrase can contain other elements, *modifiers*, that are not selected by the head.

The *interior* and *edge* of phases are described in [26] in the following way:

- "If H is a phase head with complement Z, then Z is the interior of the phase."
- "The edge is H along with anything merged to $\{H, Z\}$."
- "It is the interior that is subject to no further modification. Elements of the edge–H and a sister of $\{H, Z\}$ (and a sister of the resulting SO, etc.)–can be modified in the next higher phase."

We describe here how this formulation fits in our mathematical setting.

In the original formulation, as in [25], phases can be identified in terms of heads and the notion of maximal projection, which we recalled in Definition 1.13.2. The use of maximal projection is superseded by the current formulation of the labeling algorithm as in [26] and [27]. However, it is useful here, within our formulation, to maintain a notion equivalent to that of maximal projection, which we formulate geometrically as follows.

Lemma 1.14.1. *Let $h_T : V^o(T) \to L(T)$ be a head function in the sense of Definition 1.13.3. Then h_T determines a partition of the vertices of T into sets $V(T) = \sqcup_\ell V_\ell$, indexed by the leaves $\ell \in L(T)$, where all vertices $v \in V_\ell$ have the property that $h_T(v) = \ell$. The vertices of V_ℓ form a path γ_ℓ that extends from the leaf ℓ to the vertex v_ℓ of the maximal projection.*

Proof. For any given $v \in V^o(T)$, let $\gamma_{h_T(v),v}$ denote the unique path in T from v to the leaf $h_T(v)$. Let $w_0, \ldots, w_{n-1}$ be the vertices on this path with $w_0 = h_T(v)$ and $w_{n-1} = v$. By the property of the head function, all accessible terms T_{w_i}, for vertices w_i on this path, have the same head, $h_T(w_i) = h_T(v)$. The path extends with this property up to a maximal length corresponding to the

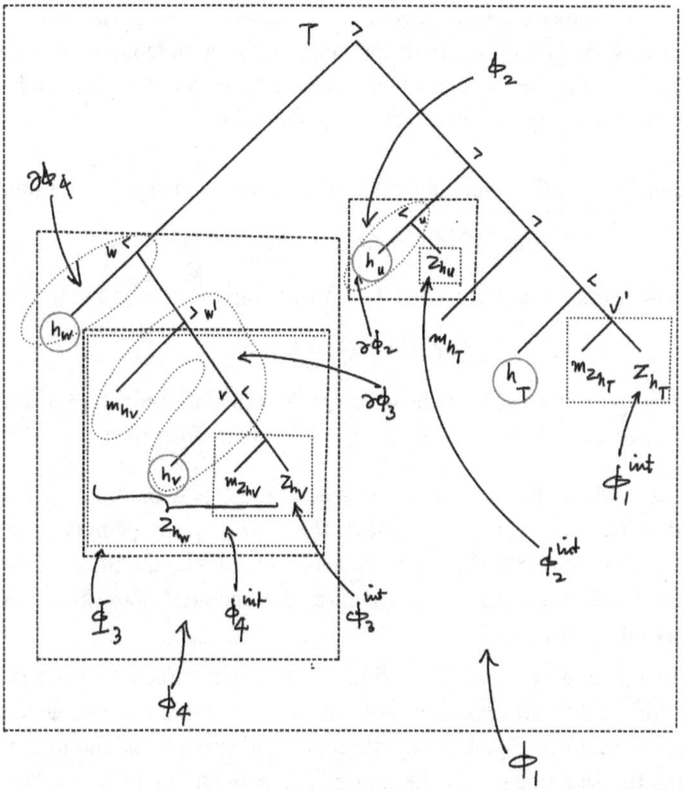

Figure 1.8
Decomposition of a syntactic object $T \in \mathrm{Dom}(h)$ into phases Φ_k, with interiors Φ_k° and edges $\partial\Phi_k$, with the paths γ_ℓ obtained by following the $>$ and $<$ marks at vertices.

maximal projection of the head. Given any leaf $\ell \in L(T)$, therefore, one has a path γ_ℓ in T that connects ℓ to the maximal projection v_ℓ of ℓ, such that, for all vertices $w \in \gamma_\ell$, the head is $h_T(w) = \ell$, with n_ℓ the length of the path. Every vertex $v \in V(T)$ is on one (and exactly one) of the paths γ_ℓ, which therefore define a partition of the set of vertices. On the other hand, not all edges of T lie on one of the paths γ_ℓ. □

We say that a path γ_ℓ as in Lemma 1.14.1 is nontrivial if it contains at least one internal (non-leaf) vertex of T, and it is trivial when $\gamma_\ell = \{\ell\}$.

The head function h_T assigns a head $h_T(v)$ to all the substructures given by the accessible terms $T_v \subset T$. For Phase Theory, one needs, in addition to the head $h_{T(v)}$, the identification of the *complement of the head*. We can extend the definition of an abstract head function to a notion of *complemented* abstract head function that identifies both the head and a complement. (In the case where the head function is the actual syntactic head, the complement is given by all the elements that the head *must* combine with.)

Definition 1.14.2. A complemented abstract head function $h_{T,Z}$ is a function

$$h_{T,Z} : V^o(T) \to L(T) \times (\mathrm{Acc}(T) \cup \{1\})$$

from the set of non-leaf vertices of T, with 1 being the empty tree $\emptyset$,

$$h_{T,Z}(v) = (h_T(v), Z_v)\,,$$

where $v \mapsto h_T(v)$ is an abstract head function as in Definition 1.13.3, and Z_v is a (possibly empty) $Z_v \subset T_{s_{h_T(v)}}$, with $s_{h_T(v)}$ the sister vertex of $h_T(v)$ in T.

The cases where $Z_v = \emptyset$ are those where $T_{s_{h_T(v)}}$ is a modifier of the head $h_{T(v)}$ rather than the head's complement. If $Z_v \subset T_{s_{h_T(v)}}$ is nonempty, then $T_{s_{h_T(v)}}$ consists of the complement Z_v of the head combined with modifiers of Z_v. (The modifiers of the complement are structures that the head does not necessarily have to combine with.)

Note that here, what we call "modifiers" are just constructed by External and Internal Merge, like all structures: we simply refer in this way to constructions (like the phase Φ_3 in Figure 1.8), where the new structure added to the phase maintains the same head. In the figure, this happens going from T_v to $T_{w'}$, which still has $h_T(w') = h_T(v)$; we then say that the added structure m_{h_v} is a "modifier of the head". Thus, we use this term and the notation m_{h_v} simply to describe what is Merged to the structured formed by the head and the complement, with the same head, as in the description of phases recalled from [26].

We illustrate the algorithm for identifying phases in a syntactic object $T \in \mathrm{Dom}(h) \subset \mathfrak{T}_{SO_0}$, with a head function h_T, by first showing how it works in the example of Figure 1.8.

- The head function h_T determines a choice of one the two edges below each non-leaf vertex: the choice is indicated in the figure by the marks > and < at the vertices. These in turn determine the paths γ_ℓ of Lemma 1.14.1.
- The leaves $\ell \in L(T)$ such that γ_ℓ contains interior (non-leaf) vertices are circled in Figure 1.8 and have labels h_T, h_w, h_v, h_u.
- For each of these leaves, there is a phase, Φ_k, $k = 1, \ldots, 4$, as shown in the figure.
- The phase Φ_2 has head h_u and interior $\Phi_2^\circ = Z_{h_u}$; the edge $\partial\Phi_2$ consists of the vertices h_u and u.

- The phase Φ_3 has head h_v and complement Z_{h_v} with modifier $m_{Z_{h_u}}$. The accessible term $T_{u'} = \{m_{Z_{h_u}}, Z_{h_v}\} = \mathfrak{M}(m_{Z_{h_u}}, Z_{h_v})$ is the interior Φ_3°, and the edge $\partial\Phi_3$ consists of the head h_v, the modifier m_{h_v} of the head, and the vertices v and w'.
- The phase Φ_4 has head h_w and interior Φ_4° given by $T_{w'}$; the edge $\partial\Phi_4$ is given by h_w and w.
- In the case of the phase Φ_1 with head h_T, the phase interior Φ_1° is given by $T_{v'}$ (so $T_{v'}$ is *inaccessible* for further computation), while T_w and its head h_w, T_u and its head h_u, the head h_T, and its modifier m_{h_T} are all in the edge $\partial\Phi_1$ so they remain accessible for further computation.
- The phase structure of the syntactic object T identifies the interior $T_{v'}$ of h_T, the interior Z_{h_u} of h_u, and the interior $T_{w'}$ of h_w, as well as all of their accessible terms, as unavailable for further computation. (This includes h_v and its complement and modifiers.)

Thus, we obtain a general strategy for identifying phases of a syntactic object $T \in \mathrm{Dom}(h) \subset \mathfrak{T}_{\mathcal{SO}_0}$ with a head function h_T that extends to a complemented head function $h_{T,Z}$, as in Definition 1.14.2.

Definition 1.14.3. Construction of the *Phase algorithm:*

1. Consider the partition of the vertices of T into the paths γ_ℓ of Lemma 1.14.1.
2. The set $L_\Phi(T)$ of phases of T is given by

$$L_\Phi(T) = \{\ell \in L(T) \,|\, \#V(\gamma_\ell) > 1\}, \qquad (1.14.1)$$

   the set of $\ell \in L(T)$ such that γ_ℓ contains interior (non-leaf) vertices. Let v_ℓ be the vertex at the end of the path γ_ℓ (the maximal projection). We write Φ_ℓ for the phase associated to $\ell \in L_\Phi(T)$:

$$\Phi_\ell = \{T_v \in \mathrm{Acc}'(T) \,|\, T_v \subseteq T_{v_\ell}\}. \qquad (1.14.2)$$

3. For each $\ell \in L_\Phi(T)$, let v be the mother vertex above ℓ on the path γ_ℓ, and let s_ℓ be the sister vertex of ℓ under v. If the complement $Z_\ell \subset T_{s_\ell}$, determined by the complemented head $h_{T,Z}$, is nonempty, the interior of the phase Φ_ℓ is given by the accessible term T_{s_ℓ} and all of its accessible terms:

$$\Phi_\ell^\circ := \{T_v \in \mathrm{Acc}(T) \,|\, T_v \subseteq T_{s_\ell}\}. \qquad (1.14.3)$$

   If the complement $Z_\ell \subset T_{s_\ell}$ is empty, then we set $\Phi_\ell^\circ = \emptyset$.
4. For $Z_\ell \neq \emptyset$, the edge $\partial\Phi_\ell$ of the phase Φ_ℓ is given by

$$\partial\Phi_\ell := \{T_v \in \mathrm{Acc}'(T) \,|\, T_w \subseteq T_{v_\ell} \text{ and } T_w \nsubseteq T_{s_\ell}\}, \qquad (1.14.4)$$

all the accessible terms of T_{v_ℓ} that are not in the interior of the phase. For $Z_\ell = \emptyset$, we set $\partial\Phi_\ell = \Phi_\ell$.

There is a partial ordering on the set $L_\Phi(T)$ of phases induced by inclusion: $\ell < \ell'$ if $\Phi_\ell \subset \Phi_{\ell'}$. We say in this case that Φ_ℓ is a *lower phase* and $\Phi_{\ell'}$ a *higher phase*. The highest phase of T is the phase Φ_{h_T}, which corresponds to the path γ_{h_T} that contains the root vertex of T. We also define the set of *inaccessible terms* of the phase Φ_ℓ as

$$\Upsilon_\ell := \left\{ T_v \in \mathrm{Acc}(T) \,|\, T_v \in \bigcup_{\ell' < \ell} \Phi^\circ_{\ell'} \right\} ; \tag{1.14.5}$$

namely, all the terms that are in the interior of any of the lower phases. The set given by the complement $\Phi_\ell \smallsetminus \Upsilon_\ell$ is the set of terms *available for computation* in the phase Φ_ℓ.

Remark 1.14.4. In the previous description, we are counting, as in [26], the head as part of the edge of the phase. On the other hand, there are reasons in the current theory of Merge for excluding head movement, which would indicate that the head itself should also not remain accessible. A more restrictive Phase Theory would then also include the heads of the lower phases in the set Υ_ℓ.

As we have discussed, phases are a way of identifying terms that are or are not available for further computation in the process of structure formation, given by the edge and the interior, respectively. We formulate this condition in more detail to explain the meaning of the sets Υ_ℓ defined in (1.14.5).

Consider again the example of Figure 1.8. The structure T shown in the figure has been built in steps by Merge starting from the lexical items at the leaves. Suppose that the substructure T_u is the first one built. The head h_u is merged to its complement Z_u. This forms phase Φ_2. So this structure T_u has a single phase. At this point, there are two ways in which computation can continue. Either some other structure in the workspace is merged with External Merge to this structure T_u, in which case the vertex u in the edge $\partial\Phi_2$ is involved or some accessible term of T_u is used by Internal Merge to modify T_u and obtain a new structure, $T' = \mathfrak{M}(T_{u,a}, T/T_{u,a})$. This action of Internal Merge has the effect of taking an accessible term $T_{u,a} \subset T_u$ that is *in the interior* of the phase $\Phi^\circ_2 = Z_u$ and *move it to the edge of the phase* (of the resulting object T'). Note that accessible terms $T_{u,a} \in \Phi^\circ_2$ are available for this kind of Internal Merge computation, which acts within phase Φ_2, before any additional structure is merged to it by External Merge: an accessible term can be moved to the edge of the phase by such Internal Merge operation, where it remains available for further computation. Similarly, when phase Φ_3 is formed, with $T_{w'}$ being a component of the resulting workspace, the accessible terms $T_{w',a} \subset \Phi_3^\circ$ are accessible to Internal Merge, which can move them to the edge $\partial\Phi_3$ of the resulting object. After h_w is merged via External Merge to $T_{w'}$, forming the new structure T_w and the phase Φ_4, the accessible terms in $\Phi_4^\circ \smallsetminus \Phi_3^\circ$ are available to Internal Merge acting within phase Φ_4, but the terms in Φ_3° (or what remains in Φ_3° after Internal Merge in the previous phase Φ_3 has moved some terms to the edge $\partial\Phi_3$) are now *inaccessible* to Internal Merge acting on Φ_4. Similarly, once the entire structure T is formed, and we are considering the highest phase Φ_1, Internal Merge can access terms in Φ_1° and move them to the edge $\partial\Phi_1$, but it cannot access the terms in Φ_2°, Φ_3°, and Φ_4°, as these are all unavailable for computation. If T is then used to build a new phase, by External Merge with some other objects in the workspace, the terms in Φ_1° will become inaccessible to Internal Merge in this new phase, unless they have been moved by Internal Merge in Φ_1 to the edge of the phase.

This means that the statement that accessible terms in the interior of the phase are unavailable for further computation means, in a more expanded form, the following two statements:

- At the stage of derivation of T where Φ_ℓ is the highest phase, the accessible terms in Φ_ℓ° are accessible to Internal Merge, which moves them to the edge $\partial\Phi_\ell$ of the resulting syntactic object, where they remain accessible for further computation in the higher phases.
- When a new higher phase $\Phi_{\ell'} \supset \Phi_\ell$ is formed by External Merge of Φ_ℓ with other objects in the workspace, all the accessible terms that are in the interiors of the lower phases Φ_ℓ° and $\Phi_{\ell''}^\circ$ for all $\ell'' < \ell$ are inaccessible to Internal Merge acting in the phase $\Phi_{\ell'}$.

We see then that the set Υ_ℓ defined in (1.14.5) describes the set of terms that are *unavailable for further computation in Phase* Φ_ℓ (before any further structure is added to Φ_ℓ).

Now we can turn into a mathematical statement the description of terms that are or are not available for computation in different phases. Saying that certain accessible terms of T are *unavailable for further computation* is equivalent to saying that those terms are *not extracted by the coproduct*, as the Hopf algebra coproduct is exactly what selects accessible terms and makes them available for computation. Thus, the way to interpret Phase Theory in our setting is as a restriction of the form of the coproduct of the Hopf algebra of workspaces. Indeed, one can see in very simple examples that the coproducts Δ^ω of (1.2.8) produce a large number of terms available for the Merge action, which has the problem of a combinatorial explosion in the size of the computation. Thus, we can view the main idea of Phase Theory as an algorithm designed to significantly reduce the number of terms that can be extracted by the coproduct. This means modifying the coproduct as discussed in Definition 1.14.5.

Definition 1.14.5. Consider the subspace $\mathcal{V}^h(\mathfrak{F}_{SO})) \subset \mathcal{V}(\mathfrak{F}_{SO})$ spanned by forests $F = \sqcup_a T_a$ with all the components $T_a \in \mathrm{Dom}(h) \subset \mathfrak{T}_{SO}$. The *phase coproduct* $\Delta^c_\Phi : \mathcal{V}^h(\mathfrak{F}_{SO})) \to \mathcal{V}^h(\mathfrak{F}_{SO})) \otimes \mathcal{V}^h(\mathfrak{F}_{SO}))$ is given by

$$\Delta^c_\Phi(T) = \sum_{\underline{v} \in \Phi_{h_T} \smallsetminus \Upsilon_{h_T}} F_{\underline{v}} \otimes T/^\omega F_{\underline{v}} \,, \tag{1.14.6}$$

for $T \in \mathrm{Dom}(h)$, where we write $\underline{v} \in \Phi_{h_T} \smallsetminus \Upsilon_{h_T}$ for $\underline{v} = (v_1, \dots, v_n)$ to mean that $T_{v_i} \in \Phi_{h_T} \smallsetminus \Upsilon_{h_T}$ for all $i = 1, \dots, n$. We set $\Delta^c_\Phi(F) = \sqcup_a \Delta^c_\Phi(T_a)$ for $F = \sqcup_a T_a$.

Lemma 1.14.6. *The coproduct* $\Delta^c_\Phi : \mathcal{V}^h(\mathfrak{F}_{SO})) \to \mathcal{V}^h(\mathfrak{F}_{SO})) \otimes \mathcal{V}^h(\mathfrak{F}_{SO}))$ *is well defined and coassociative.*

Proof. We need to check that, if $T \in \mathrm{Dom}(h)$, then both the term $F_{\underline{v}} = T_{v_1} \sqcup \cdots \sqcup T_{v_n}$ in the left-channel of the coproduct and the term $T/F_{\underline{v}}$ in the right-channel are also in $\mathrm{Dom}(h)$. Each component T_{v_i} of $F_{\underline{v}}$ is an accessible term of T, and hence h_T induced a head function $h_{T_{v_i}} = h_T|_{T_{v_i}}$. Since we are using the coproduct Δ^ω with $\omega = c$, the quotient term $T/^c F_{\underline{v}}$ has each of the components T_{v_i} in T replaced by a single leaf vertex ℓ_{v_i} with label $\overline{T_{v_i}}$. Thus, the head function h_T induces a head function on $T/^c F_{\underline{v}}$ where, for $w \in V(T/^c F_{\underline{v}})$,

$$h_{T/^c F_{\underline{v}}}(w) = \begin{cases} \ell_{v_i} & \text{if } h_T(w) \in L(T_{v_i}) \\ h_T(w) & \text{otherwise.} \end{cases}$$

This shows that Δ^c_Φ indeed maps $\mathcal{V}^h(\mathfrak{F}_{SO}))$ to $\mathcal{V}^h(\mathfrak{F}_{SO})) \otimes \mathcal{V}^h(\mathfrak{F}_{SO}))$. The coassociativity follows from the same argument given for Δ^c in Lemma 1.2.10. We need to check that this is still the case if the extracted terms T_{v_i} are only in $\Phi_{h_T} \smallsetminus \Upsilon_{h_T}$. It suffices to observe that the terms in the coproduct $\Delta^c_\Phi(T)$ can be bijectively mapped to the terms in the coproduct $\Delta^c(T/^c\pi_C(T))$, where $\pi_C(T)$ is the admissible cut that cuts each edge above each vertex s_ℓ. The bijection consists of replacing the labels $\overline{T_{s_\ell}}$ at the new leaves of $T/^c\pi_C(T)$ with the restored T_{s_ℓ}. Thus, the coassociativity identity

$$(\Delta^c \otimes \mathrm{id}) \circ \Delta^c(T/^c\pi_C(T)) = (\mathrm{id} \otimes \Delta^c) \circ \Delta^c(T/^c\pi_C(T))$$

for $T/^c\pi_C(T)$ gives the coassiciativity identity

$$(\Delta^c_\Phi \otimes \mathrm{id}) \circ \Delta^c_\Phi(T) = (\mathrm{id} \otimes \Delta^c_\Phi) \circ \Delta^c_\Phi(T) \,.$$

□

1.15 Labeling Algorithm

The linguistic model of Minimalism adopts what is referred to as *bare phrase structure*, and avoids the conventional notions from X-bar theory and older formulations of "labels" for phrases such as VP, NP, and S. To replace these, Chomsky introduced in [26] and [27] a notion of labeling and a labeling algorithm. This is based on the idea that in Bare Phrase Structure (as formulated in [23]), no new specifications are introduced beyond what is in the lexical items at the leaves. We discuss here how this is formulated in our notation and terminology.

We discussed in sections 1.13.2 and 1.13.3 the fact that a head function is defined only on a domain $\mathrm{Dom}(h) \subset \mathcal{SO}$, not on the entire set of syntactic objects. In particular, the domain of definition $\mathrm{Dom}(h)$ is in general *not* a submagma of the magma $\mathcal{SO}$ of syntactic objects; namely, there may be pairs of syntactic objects $T_1, T_2 \in \mathrm{Dom}(h)$, on which h determines a head function, h_{T_1}, h_{T_2} respectively, but $T = \mathfrak{M}(T_1, T_2) \notin \mathrm{Dom}(h)$. This is indeed the case when the abstract head function h is the actual syntactic head.

Despite the fact that head functions do not extend to the entire $\mathcal{SO}$ and they are not always compatible with the magma operation $\mathfrak{M}$ of $\mathcal{SO}$, it is possible to obtain a good *labeling algorithm* for the objects in $\mathrm{Dom}(h)$, as well as for *some* objects in $\mathcal{SO}$ that take the form $T = \mathfrak{M}(T_1, T_2)$, with $T_1, T_2 \in \mathrm{Dom}(h)$, which are not themselves in $\mathrm{Dom}(h)$.

A labeling algorithm is designed to make these objects interpretable at the syntax–semantics interface, and it can be described as an assignment of labels at the non-leaf vertices of the tree, determined by the lexical items and syntactic features assigned to the leaves, and the head function. The labeling algorithm we discuss here is the one originally derived by Chomsky in [26] and [27]. A somewhat different form of the labeling algorithm is presented by Rizzi [174]. We will follow here the formulation of [26] and [27]. As we already noted (see section 1.1.3), we follow here the tree notation for syntactic objects, because it is the common use in mathematics, rather than the set notation adopted in [26] and [27]; hence, the labeling algorithm will be described here in terms of an *assignment of labels to the internal (non-leaf) vertices of the tree*, This is equivalent to the set notation formulation through identifications of the form

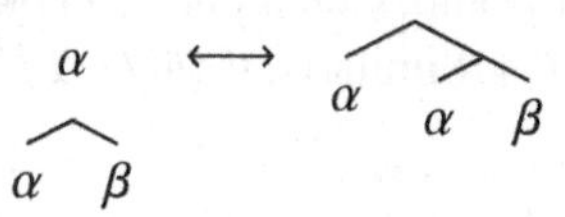

$\longleftrightarrow \{\alpha, \{\alpha, \beta\}\}\,.$

The discussion in section 1.14, and in particular Lemma 1.14.1, immediately suggests a labeling algorithm for all $T \in \mathrm{Dom}(h) \subset \mathfrak{T}_{\mathcal{SO}_0}$, where labeling of internal vertices is performed by the head function h_T. The labeling algorithm, however, also can perform labeling on certain syntactic objects $T \in \mathfrak{T}_{SO_0}$ that do not belong to Dom(h), provided that the head function h satisfies the condition given in Definition 1.15.1 with respect to the action of Internal Merge.

Definition 1.15.1. We say that a head function h is *raising* if the following holds:

- For any $T \in \mathrm{Dom}(h)$ and any accessible term $T_v \subset T$ with
$$h(T) = h(T/^d T_v)\,,$$
the Internal Merge satisfies
$$\mathfrak{M}(T_v, T/^c T_v) \in \mathrm{Dom}(h)\,,$$
with $h(\mathfrak{M}(T_v, T/^c T_v)) = h(T/^d T_v)$.
- For any $T \in \mathfrak{T}_{SO_0}$ and any accessible term $T_v \subset T$ with Internal Merge
$$\mathfrak{M}(T_v, T/^c T_v) \in \mathrm{Dom}(h)$$
and $T/^d T_v \in \mathrm{Dom}(h)$, we have
$$h(\mathfrak{M}(T_v, T/^c T_v)) = h(T/^d T_v)\,.$$

The notion defined here of *raising head function* can be related to Moro's dynamical asymmetry, [155]. It reflects the expected linguistic properties: for instance, in a syntactic object of the form
$$T = \mathfrak{M}(T_1, T_2)$$
obtained by External Merge of the objects T_i, $i = 1, 2$, assume that one of the T_i (which we can take to be T_1, since the other case is analogous) raises through Internal Merge, which in our setting corresponds to having Internal Merge $\mathfrak{M}(T_1, T/^c T_1) \in \mathrm{Dom}(h)$. Then the structure

$T/^c T_1 =$ 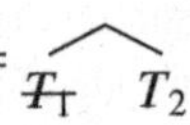

has the head $h(T/^c T_1) = h(T_2) = h(T/^d T_2)$. Thus, if T_1 and T_2 are in Dom(h) and one knows that $\mathfrak{M}(T_1, T/^c T_1) \in \mathrm{Dom}(h)$, then one can conclude that $T \in \mathrm{Dom}(h)$, with $h(T) = h(T_2) = h(T/^c T_2)$. Similarly, if $\mathfrak{M}(T_2, T/^c T_2) \in \mathrm{Dom}(h)$, then $T \in \mathrm{Dom}(h)$, with $h(T) = h(T_1)$.

We can then describe the labeling algorithm as in Definition 1.15.2.

Definition 1.15.2. Let h be a raising head function, as in Definition 1.15.1.

- If $T \in \mathrm{Dom}(h) \subset \mathfrak{T}_{SO_0}$, we label every internal (non-leaf) vertex $v \in V^o(T)$ by the corresponding head $h_T(v) \in L(T)$.
- For $T_1, T_2 \in \mathrm{Dom}(h)$, with External Merge $T = \mathfrak{M}(T_1, T_2)$, if the Internal Merge $\mathfrak{M}(T_1, T/^c T_1) \in \mathrm{Dom}(h)$, then we label the root vertex of T with $h(T_2)$, while all other vertices are labeled by the head function on T_1 or T_2. The case where $\mathfrak{M}(T_2, T/^c T_2) \in \mathrm{Dom}(h)$ is analogous, with the label $h(T_1)$ at the root vertex of T.
- If T has an accessible term T_v with $T/^d T_v \in \mathrm{Dom}(h)$, and the Internal Merge satisfies
$$\mathfrak{M}(T_v, T/^c T_v) \in \mathrm{Dom}(h),$$
then we label the root of T by $h(T/^d T_v)$.
- If none of these conditions is satisfied but the heads $h(T_1)$ and $h(T_2)$ in $L(T)$ carry labels in $\mathcal{SO}_0$ that share the same syntactic feature, then the root vertex of T can be labelled by that syntactic feature. Note that in this last case, $T \notin \mathrm{Dom}(h)$, as this labeling does not define a head function on T. So the labeling can be extended to some cases where $T \notin \mathrm{Dom}(h)$.

This algorithm can in principle still leave some remaining objects $T \notin \mathrm{Dom}(h)$ that cannot be labelled by the algorithm in the last case. These cases where the labeling algorithm fails should be regarded as part of the objects that will be eliminated during the Externalization procedure by the quotient map Π_L, and they will be similarly eliminated also at the syntax–semantics interface as nonparsable objects. (We will discuss these interface channels further in chapter 3.)

1.16 Form Set

We discuss here another operation that plays a role in the current version of Minimalism in generative linguistics: the FormSet operation. It was introduced by Chomsky [31], and further discussed in [34] and [70].

When we introduced the action of Merge on workspaces, written in the form of (1.3.7):
$$\mathfrak{M}_{S,S'}(F) = \sqcup \circ (\mathcal{B} \otimes \mathrm{id}) \circ \delta_{S,S'} \circ \Delta(F),$$
the operator $\delta_{S,S'}$ selects terms of the form $S \sqcup S' \otimes F'$ in the coproduct
$$\Delta(F) = 1 \otimes F + F \otimes 1 + \sum_{\underline{v}} F_{\underline{v}} \otimes F/F_{\underline{v}}.$$
The grafting operator $\mathcal{B}$ then acts on $S \sqcup S'$ to produce $\mathfrak{M}(S, S') = \mathcal{B}(S \sqcup S')$. If we allow all possible choices of the syntactic objects S, S', we can simply

write

$$\mathfrak{M} = \sqcup \circ (\mathcal{B} \otimes \mathrm{id}) \circ \Pi_{(2)} \circ \Delta \,,$$

as in (1.9.4), where $\Pi_{(2)} = \gamma_{(2)} \otimes \mathrm{id}$ is the projection, where $\gamma_{(2)}(F) = F$ if $F = T_1 \sqcup T_2$ has two components, and zero otherwise. This just says that we restrict the operator $\mathcal{B}$ to grafting only two components. Note that, as we saw in Proposition 1.4.2, while External Merge uses only terms of the coproduct with two components in the left-channel, Internal Merge also requires the use of terms with one component, through the $\mathfrak{M}_{S,1}$ operation; see (1.9.5).

All the remaining terms with $k \geq 3$ components (the part of the coproduct that we referred to as $\Delta_{(k)}$ in (1.2.10)) are not used by Merge; rather, they are necessarily present for structural reasons (i.e., the coassociativity of the coproduct). The fact that these terms are necessary for algebraic reasons suggests that they should play a role and should encode another part of the structure of the theory of Minimalism. We show here that they do exactly that, as they are responsible for the FormSet operation. Our first observation here is that the primitive part of the coproduct on workspaces performs all the possible partitions of the workspace.

Recall that, as we have seen (see (1.1.1) and the surrounding discussion), the grafting operator $\mathcal{B}$ is originally defined as acting on any arbitrary forest $F = \sqcup_{a=1}^{n} T_a$ by

$\mathcal{B}(F) =$  .

Here again, the tree is not endowed with any choice of planar embedding, so the order of the T_i does not matter. It is then clear that there is another family of operations that one can consider on workspaces, which is given by

$$\sqcup \circ (\mathcal{B} \otimes \mathrm{id}) \circ \Pi_{(k)} \circ \Delta \qquad \text{for } k \geq 3 \,,$$

where

$$\Pi_{(k)} = \gamma_{(k)} \otimes \mathrm{id} \quad \text{with} \quad \gamma_{(k)}(F) = \begin{cases} F & \#\mathcal{I} = k \\ 0 & \text{otherwise} \end{cases} \quad \text{for } F = \sqcup_{a \in \mathcal{I}} T_a \,. \tag{1.16.1}$$

These operations apply the grafting operator $\mathcal{B}$ only to terms of the coproduct where the left channel $F = T_1 \sqcup \cdots \sqcup T_k$ has exactly k components.

We know a priori that operations of this form have to be available by algebraic consistency: once the Merge operation is available, then these other operations are also defined, as they follow essentially the same structure. The question, then, is what their linguistic meaning is. As we will discuss next, this operation is exactly the FormSet operation introduced in Chomsky's [31] for a suitable form of the coproduct Δ. This is described in Definition 1.16.1.

Definition 1.16.1. As in Lemma 1.2.11, let $\tilde{\mathfrak{F}}_{SO_0}$ denote the set of all forests (not necessarily binary) with leaf labels in the set SO_0. Then FormSet is a family of operators

$$\mathcal{FS}^{(k)} : \mathcal{V}(\tilde{\mathfrak{F}}_{SO_0}) \to \mathcal{V}(\tilde{\mathfrak{F}}_{SO_0}) \quad \text{with } k \geq 3$$

of the form

$$\mathcal{FS}^{(k)} = \sqcup \circ (\mathcal{B} \otimes \mathrm{id}) \circ \Pi_{(k)} \circ \Delta_P , \tag{1.16.2}$$

where $\Pi_{(k)}$ is as in (1.16.1) and Δ_P is the coproduct

$$\Delta_P : \mathcal{V}(\tilde{\mathfrak{F}}_{SO_0}) \to \mathcal{V}(\tilde{\mathfrak{F}}_{SO_0}) \otimes \mathcal{V}(\tilde{\mathfrak{F}}_{SO_0}) ,$$

which is uniquely identified by the following properties:

1. $\Delta_P(T) = 1 \otimes T + T \otimes 1$ for all $T \in \mathfrak{T}_{SO_0}$
2. $\Delta_P(F) = \sqcup_a \Delta_P(T_a)$ for all $F = \sqcup_a T_a$

The range of $\mathcal{FS}^{(k)}$ in $\mathcal{V}(\tilde{\mathfrak{F}}_{SO_0})$ is contained in a subspace $\mathcal{V}(\tilde{\mathfrak{F}}^R_{SO_0})$ with

$$\tilde{\mathfrak{F}}^R_{SO_0} = \{F \in \tilde{\mathfrak{F}}_{SO_0} \,|\, \pi_{C_R}(F) \in \mathfrak{F}_{SO_0}\} , \tag{1.16.3}$$

where π_{C_R} is the admissible cut that cuts the edges connected to the root of each component of F. Namely, $\tilde{\mathfrak{F}}^R_{SO_0}$ consists of forests where each component is either a binary tree or a tree that may have higher valence only at the root vertex and is binary everywhere below the root. We refer to the set $\tilde{\mathfrak{F}}^R_{SO_0}$ of (1.16.3) as the set of *extended workspaces*.

Note that the extended workspaces defined here would become more complicated if we allow iterations of the FormSet operation, and simultaneously Merge operations forming the syntactic objects. A more precise description of the resulting mathematical structure will be analyzed elsewhere.

Remark 1.16.2. In (1.16.2), we use the coproduct Δ_P on $\mathcal{V}(\tilde{\mathfrak{F}}_{SO_0})$ in which all trees are primitive elements (the coproduct $\Delta_P(T)$ is the primitive part $P(T)$ as in (1.2.9) of the coproduct Δ of (1.2.8)). As we discussed in the introduction, this is the simplest possible way of introducing a decomposition operation on workspaces. In the case of the construction of Merge, we opted for the more complicated form (1.2.8) of the coproduct in order to make the extraction of accessible terms available for Internal Merge. However, in the case of FormSet, the analogous extraction operation does not seem to be required on empirical grounds, so the simplified form of the coproduct suffices. It remains an open question whether there are linguistic phenomena that require an extension of the FormSet operation $\mathcal{FS}^{(k)}$ to the nonprimitive terms of the coproduct Δ. These would be an analog for the FormSet operation of the Sideward forms of Merge, which we discussed in section 1.4.

Remark 1.16.3. Note that in [31], [34], and [70], the operation $\mathcal{FS}^{(k)}$ would be written in set notation as

$$\mathcal{B} : T_1 \sqcup \cdots \sqcup T_k \mapsto \{T_1, \ldots, T_k\},$$

while we write it here in tree notation as

$$\mathcal{B} : T_1 \sqcup \cdots \sqcup T_k \mapsto \begin{array}{c} \diagup \; \diagup \; \diagdown \; \diagdown \\ T_1 \quad T_2 \quad \cdots \quad T_n \end{array},$$

which is normally used in mathematical physics for the grafting operator $\mathcal{B}$, but these notations are completely equivalent (as already discussed in section 1.1.3).

In particular, this means that there is a natural interpretation for the objects of the set $\tilde{\mathfrak{F}}^R_{SO_0}$.

Remark 1.16.4. We should interpret forests $F \in \tilde{\mathfrak{F}}^R_{SO_0}$ as a workspace $\pi_{C_R}(F) = \sqcup_{a \in \mathcal{I}} T_a$ in our usual set $\mathfrak{F}_{SO_0}$ of workspaces, where a certain subcollection of components $F' = \sqcup_{b \in \mathcal{J} \subset \mathcal{I}} T_a$ has a common structure, which we represent by writing $F = \mathcal{B}(F') \sqcup \hat{F} \in \tilde{\mathfrak{F}}^R_{SO_0}$ with $\hat{F} = \sqcup_{a \notin \mathcal{J}} T_a$.

Two main types of applications of this FormSet operation are as follows:

- *Unbounded unstructured sequences*: these include examples of the following form (see [31] and [70]):

$$\mathcal{B}(\text{John} \sqcup \text{Bill} \sqcup \text{my friends} \sqcup \text{the actor who won the Oscar}),$$

$$\mathcal{B}(\text{ran} \sqcup \text{danced} \sqcup \text{took a vacation})$$

- *Assignment of diagonals* (a combination of FormSet and FormCopy, in the terminology of [31]): this will be discussed in section 3.8.2, and it accounts for structure building processes that lead to examples of the following form (see [70]):

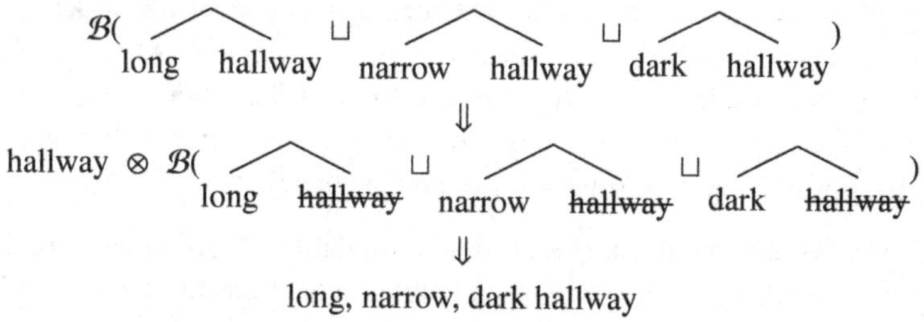

1.16.1 FormSet and Workspaces

To further elaborate on the observation of Remark 1.16.4, when we incorporate in our model the FormSet operation, we view workspaces not just as collections of syntactic objects (as forests $F \in \mathfrak{F}_{SO_0}$), but we introduce the possibility of grouping together a subcollection of the components of F. Thus, instead of working with the set $\mathfrak{F}_{SO_0}$ of workspaces, we work with the set $\tilde{\mathfrak{F}}^R_{SO_0}$ of extended workspaces, where grouping together subcollections of components is taken into account.

If we make this modification to the set of workspaces, to account for the structures created by FormSet, we need to ensure that the action of Merge on workspaces, with both Internal Merge and External Merge, as we constructed it based on the operation in (1.3.7), continues to make sense for the extended workspaces. Since Merge requires the form (1.2.8) of the coproduct, instead of the simpler form Δ_P used by FormSet, we need to ensure that the original coproduct (1.2.8) and the associated Hopf algebra structure can be formulated for extended workspaces without any significant changes.

As in the case of workspaces, we form the linear span $\mathcal{V}(\tilde{\mathfrak{F}}^R_{SO_0})$. The same product given by the disjoint union $\sqcup$ gives $\mathcal{V}(\tilde{\mathfrak{F}}^R_{SO_0})$ the structure of associative and commutative algebra, but the coproduct structure needs to be adjusted.

Lemma 1.16.5. *The coproducts* Δ^ω *of* (1.2.8) *extend to coproducts on extended workspaces by setting*

$$\Delta^\omega(F) = F \otimes 1 + 1 \otimes F + \sum_C \pi_C(F) \otimes F/^\omega \pi_C(F), \tag{1.16.4}$$

where the sum is over those admissible cuts C that do not cut any of the edges adjacent to a root vertex of valence $k \geq 3$.

The same arguments that we used in Lemmas 1.2.10, 1.2.11, and 1.2.12 to describe the properties of the coproducts Δ^ω of (1.2.8) also show that the same properties remain valid for the coproduct of (1.16.4). The restriction on the choice of admissible cuts ensures that the grafting edges introduced by the operator $\mathcal{B}$ in the FormSet action are not regarded as part of the structure that Merge acts on. In other words, this restriction means that, while accessible terms that are inside the components collected by FormSet into a common structure remain accessible to Merge for further computation, the structure itself that FormSet has constructed is not undone by Merge.

We should also point out that, in our formulation, the FormSet operation does not include the operation that assembles collections of syntactic objects into workspaces. So, in this way, it differs slightly from the formulation given in Chomsky's [31]. Indeed, for us, the formation of workspaces is just the product structure $\sqcup$ of the Hopf algebra, while FormSet is an operation of structure formation that acts on workspaces. The reason why it is reasonable to keep these two operations distinct lies in the fact that the workspace is just a "transitory" structure that defines the locus where the steps of a derivation that build the final resulting sentence structure are taking place. On the other hand, the grouping together of components via FormSet creates a structure that remains in the ultimate form of the resulting sentence (as in the examples discussed previously and in the examples reported in [31] and [70]). Our formalism automatically keeps these two roles distinct.

1.16.2 FormSet: Not an *n*-ary Merge

We have discussed in section 1.11 the properties of a hypothetical n-ary Merge and ruled it out on account that it both undergenerates and overgenerates with respect to the binary Merge. Since both the FormSet operation $\mathcal{FS}^{(k)}$ and a k-ary Merge $\mathfrak{M}_k$ involve the grafting $\mathcal{B}$ of a collection of k trees to the same root, it may at first appear that the FormSet operation $\mathcal{FS}^{(k)}$ is itself a form of k-ary Merge. We show here that this is *not* the case, since $\mathcal{FS}^{(k)}$ and $\mathfrak{M}_k$ satisfy very different algebraic properties, so they are not directly comparable operations.

The main significant difference between $\mathcal{FS}^{(k)}$ and $\mathfrak{M}_k$ is in the kind of objects that are grafted together (the kind of structures that are formed):

- The k-ary Merge $\mathfrak{M}_k$ grafts together objects $T_1, \dots, T_k \in \mathcal{SO}^{(k)} = \mathfrak{T}^{(k)}_{\mathcal{SO}_0}$, where all the T_i are *k-ary* rooted trees.
- The FormSet operation $\mathcal{FS}^{(k)}$ grafts together objects $T_1, \dots, T_k \in \mathcal{SO} = \mathfrak{T}_{\mathcal{SO}_0}$, which are *binary* rooted trees (namely, ordinary syntactic objects generated by the binary Merge).

Another difference is in the fact that their action on workspaces has a different structure, as follows:

- The k-ary Merge maps k-ary workspaces (forests in $\mathfrak{F}^{(k)}_{SO_0}$, where all the components are k-ary trees) to new k-ary workspaces.
- The FormSet operation $\mathcal{FS}^{(k)}$ maps *binary workspaces* to *binary extended workspaces*, where the extended structure (the grouping of components) is k-ary but all the components involved and their accessible terms are binary.

Finally, another significant difference is the way that the grafting edges behave with respect to the Merge action:

- The structure building of a k-ary Merge can be disassembled by the extraction of accessible terms (as needed by Internal Merge).
- The structure building performed by the FormSet operation $\mathcal{FS}^{(k)}$ is not undone by Merge: the built structure acts as a single object for the purpose of Merge, and the edges that realize the k-ary grouping are not available for cutting in the extraction of accessible terms (individual components of the grouping are not extracted once the grouping is formed).

We can leave it as a question whether there are empirical reasons for this last property of the FormSet operations $\mathcal{FS}^{(k)}$ to be relaxed. We will also discuss elsewhere the structure under iterations of the FormSet operation.

In the rest of the book, for simplicity, we will use only our original setting with the workspaces $F \in \mathfrak{F}_{SO_0}$, without keeping track of extended workspaces and FormSet. For most of what we will be discussing, this extension is not needed, so we will use the more essential and minimal structure. Only in section 3.8.2 will we need to return to this extended setting, where the grouping together of workspace components by FormSet plays a role.

1.17 Merge and Fundamental Combinatorial Recursions in Physics

In classical physics, a *least action principle* governs the solutions of equations of motion of physical systems, in the form of minimization (or more generally stationarity) of the action functional (namely, a minimization with respect to energy). The equations of motion are then expressed as the Euler–Lagrange equations, which describe the stationarity of the action functional under infinitesimal variation. In quantum physics, and more precisely in quantum field theory, the classical equations of motion become equations in the quantum correlation functions of the fields (see section 9.6 of [166]). More precisely, the Euler–Lagrange equations are satisfied by the Green functions of the quantum field theory, up to terms that reflect the noncommutativity of field operators. The resulting quantum equations of motion are known in physics as Dyson–Schwinger equations, as mentioned erarlier. They represent the optimization process of the least action principle, implemented at the quantum level.

Quantum physics, in the form of perturbative quantum field theory, is governed by a combinatorial generative process that determines the terms of the perturbative expansion. The combinatorial objects involved are the Feynman graphs of the theory, and the generative process can be described either by formal languages (in the form of graph grammars, see [143]) or in a more efficient way in terms of Hopf algebras (the Connes–Kreimer Hopf algebras of Feynman graphs and of rooted trees; see [45] and [46]). We provide a more detailed discussion of these viewpoints in section 4.6 of chapter 4.

These two descriptions of the generative process that produces the Feynman graphs of quantum field theory can be compared to what happens with older formulations of the Minimalist Model in generative linguistics, where one can give both a formal languages description (see [199]) and a description in terms of (Internal/External) Merge operators, where the latter is computationally significantly more efficient (see [6]).

We discuss in chapter 2 (based on [139]) how to compare older versions of the Minimalist Model to the new version of [28] and [29] that we analyzed in this chapter, at the level of the Hopf algebra structure, and how one sees in those terms the advantage of the more recent formulation.

Here, the main point that we want to stress is that, in the setting of quantum physics, the best description of the generative process of the hierarchy of Feynman graphs organized by the increasing loop number in the asymptotic expansion, is also determined by a Hopf algebra. There are two main advantages of this algebraic formalism in physics:

- The algebraic structure governs the construction of the quantum solutions of the equations of motion, through the DS equations, so solutions can be constructed through a combinatorial recursive procedure.
- The Hopf algebra formalism also transparently explains the renormalization process in physics (namely the elimination of infinities, that is, the consistent extraction of finite (meaningful) values from divergent Feynman integrals).

We will discuss in more detail here the role of the algebraic formalism in quantum field theory in the recursive construction of solutions to the DS equations, as this is the aspect that is more closely related to the properties of Merge that we discussed in the previous sections. We will return to discuss the relevance of the algebraic formulation of renormalization in chapter 3 where we introduce our algebraic model of syntax–semantics interface.

1.17.1 The Recursive Construction of Dyson–Schwinger Equations

In the physics of renormalization in quantum field theory, the generative process for the hierarchical structure of Feynman graphs, is described equivalently by the Connes–Kreimer Hopf algebra of Feynman graphs [45], or by a Hopf algebra of planar, rooted trees (which are not necessarily binary), where the tree structure describes the way in which subgraphs are nested inside Feynman graphs (see [66], [115]). When formulated in terms of the Hopf algebra of trees, one can obtain a recursive construction of the solutions of the equations of motion of the quantum system, the DS equations, in terms of the combinatorics of trees; see [5], [65], and [203].

This happens in the following way, as we outlined briefly in sections 1.3 and 1.10. The Hopf algebra $\mathcal{H}$ of planar, rooted trees and forests has a product given by the disjoint union and coproduct given by the coproduct $\Delta = \Delta^\rho$, which we already discussed:

$$\Delta(T) = \sum_C \pi_C(T) \otimes \rho_C(T),$$

where the left-hand-side $\pi_C(T)$ of the coproduct is a forest obtained by cutting subtrees of T using an admissible cut and the right-hand-side $\rho_C(T)$ is the tree that remains attached to the root when the cut is performed. Note that this is one of the forms of the coproduct that we also used in (1.2.8).

One defines the operator $\mathcal{B} : \mathcal{H} \to \mathcal{H}$, as in Definition 1.3.2. Namely, $\mathcal{B}$ acts on a forest $T_1 \sqcup \cdots \sqcup T_m$ by creating a new rooted tree T, in which all the roots $v_{r_1}, \ldots, v_{r_m}$ of the trees $T_1, \ldots, T_m$ are attached to a single new root vertex:

$$\mathcal{B}(T_1 \sqcup \cdots \sqcup T_m) = T =$$

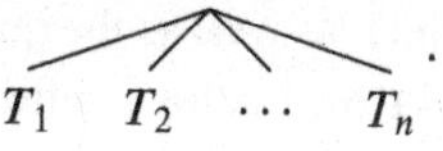

As we observed in sections 1.3 and 1.10, this has the exact structure of a Merge operator (though not necessarily binary, as it can take an arbitrary number of input trees) and of the FormSet operations. The operator $\mathcal{B}$ satisfies the identity

$$\Delta(\mathcal{B}(X)) = \mathcal{B}(X) \otimes 1 + (Id \otimes \mathcal{B}) \circ \Delta(X) \qquad (1.17.1)$$

for all $X \in \mathcal{H}$. This identity is the Hochschild 1-cocycle condition (see [45], [5], and [66]). A further discussion of the relevance of the cocycle condition in the context of Merge and linguistics will be presented in [142].

A *combinatorial Dyson–Schwinger equation* then takes the form of a fixed-point equation:

$$X = \mathcal{B}(P(X)), \qquad (1.17.2)$$

where $X = \sum_{k\geq 1} x_k$ is a formal series of elements $x_k \in \mathcal{H}_k$ in the graded pieces of the Hopf algebra; and $P(t) = \sum_{k\geq 0} a_k t^k$ with $a_0 = 1$ is a formal power series (or polynomial). The simplest and most fundamental such equation is the case where P is quadratic, $P(X) = X^2$, which is the form that we have encountered in (1.10.1), and it governs the generative process of the core structure of Merge, discussed in section 1.10. The equation (1.17.2) has a unique solution $X = \sum_{k\geq 1} x_k$ ([5] and [66]) that can be written in the recursive form

$$x_{n+1} = \sum_{k=1}^{n} \sum_{j_1+\cdots+j_k=n} a_k \mathcal{B}(x_{j_1} \cdots x_{j_k}), \tag{1.17.3}$$

with an initial step $x_1 = \mathcal{B}(1)$. It is shown in [5] and [66] that the cocycle property (1.17.1) of the operator $\mathcal{B}$ is required to ensure that the coordinates x_n of the solution of a DS equation determine a Hopf subalgebra, although the construction of the solution (1.17.3) itself does not require the cocycle condition (1.17.1). In the case of the linguistic Merge, the basic combinatorial structure is the same, with the recursion (1.17.3) corresponding to the core generative process of Merge, as described in section 1.10.

For a short overview of how these kinds of combinatorial DS equations recover the physical equations of motion in quantum field theory, see [201]. For a more detailed treatment of combinatorial DS equations, see [203]. A discussion of the use of DS equations in the context of the theory of computation is given in [50], following the *Renormalization and Computation* approach developed by Yuri I. Manin in [129] and [130].

As we mentioned at the beginning of this section, the usual classical Euler–Lagrange equations of motion express an optimality process governed by a least action principle, as well as the quantum equations of motion given by the recursively solved DS equations, reflect this form of optimization in the quantum setting. In this sense, the core computational structure of syntax defined by the Merge operator can also be seen as being optimal and most fundamental, as this reflects the structure of the physical DS equation (for the appropriate Hopf algebra) and for the most basic quadratic form of the recursion.

One can ask whether there are any other characterizations of syntactic Merge that involve optimality. Optimization is usually done with respect to some real-valued cost functional (energy/action in the case of physical systems). There are also other ways of thinking about optimization that do not require an evaluation via a function with values that are real numbers. For example, it is possible to formulate optimization processes in a purely categorical framework (e.g., see [136]), and an optimality property for the syntactic Merge may similarly take some more abstract categorical form. On the other hand, it is also possible to consider minimization conditions with respect to other types of "action functionals" that replace energy in the case of computational systems. For example, it is argued by Manin [131] that complexity provides a suitable replacement for energy in the context of the theory of computation. We leave these open questions to future investigation.

2 Minimalism Old and New: A Hopf Algebra Comparison

2.1 Introduction

The *Minimalist Program* of generative linguistics, introduced by Chomsky in the 1990s [22], underwent a significant simplifying reformulation in more recent work ([26], [28], [29], and [30]); see also [7], [8], and [114]. In this chapter, we will examine this transition in light of the algebraic model presented in chapter 1.

To recap, in chapter 1, we showed that the newest formulation of Merge has a very natural mathematical description in terms of magmas and Hopf algebras. This mathematical formulation lets one derive several desirable linguistic properties of Merge directly, as consequences of the mathematical setting. We also showed that this mathematical formulation of Merge has the same structure as the mathematical theory underlying fundamental interactions in physics, such as the renormalization process of quantum field theory and the recursive solution of equations of motion via combinatorial Dyson-Schwinger (DS) equations. An analogous recursive generative process of hierarchies of graphs plays a crucial role in both cases, as Feynman graphs in the physical case and as syntactic trees in the linguistic case.

In this chapter, we demonstrate how this same mathematical formalism based on Hopf algebras can be used to compare the new formulation of Merge with older forms of the Minimalist Model. For instance, advantages of the new formulation have been analyzed in linguistic terms in [28], [29], and [33]. What we argue here is that one can also see some advantages of this New Minimalism in terms of the underlying mathematical structure, demonstrating how to use the algebraic structure that we have introduced.

More precisely, we first consider Edward P. Stabler's formalization of the earlier version of Minimalism, known as *Computational Minimalism*, [192]; that is, before Chomsky's reformulation of Minimalism in [26] and [28], among others. We choose this account for comparison because it is one of the most clearly formulated versions of the earlier ideas. Further, among the older versions of Minimalism, it is one that tends to be more widely known to mathematicians (as well as to theoretical computer scientists) through its relation to formal languages. Indeed, mathematicians familiar with the theory of formal languages are usually aware of the fact that Stabler's Computational Minimalism is describable in terms of a class of *minimalist grammars* (MGs), which are equivalent to the *multiple context-free grammars*, a class of strictly context-sensitive formal languages that include all the context-free and regular languages, as well as other classes such as the tree-adjoining grammars (TAGs); see [199]. The MGs can also be characterized in terms of linear context-free rewrite systems (LCFRSs), [152] and the so-called *mildly context-sensitive languages*.

However, as shown by Berwick [6], this equivalence between Stabler's Computational Minimalism and multiple context-free grammars hides an important difference in terms of a *succinctness gap*. Namely, Computational Minimalism is exponentially more succinct than otherwise equivalent multiple context-free grammars (see [6] and [192]). As shown in [6], a similar gap exists between transformational generative grammar and generalized phrase structure grammar. Another problem with thinking in terms of formal languages is that they are designed to describe strings (ordered sets) produced as ordered sequences of transitions in an automaton that computes the language. This time-ordered description of languages hides its more intrinsic and fundamental description in terms of hierarchical structures (binary rooted trees without an assigned planar embedding, in mathematical terms). Indeed, the current formulation of Minimalism not only provides a more efficient encoding of the generative process of syntax, but it also proposes a model where the core computational structure of syntax is entirely based on hierarchical structures rather than on linear order. The latter (equivalently, planar embeddings of trees) is superimposed to this core computational structure, in a later Externalization phase, as we discussed in chapter 1.

The goal of the present chapter is to elaborate on the difference between this older' version of Minimalism and its most recent version. We also want to stress the point that, while formal languages have long been considered the mathematical theory of choice for application to generative linguistics, it is in fact neither the only one nor the most appropriate. As argued in the previous chapter, the algebraic formalism of Hopf algebras can be a suitable mathematical tool for theoretical linguistics and a useful way to move beyond the traditional thinking in terms of formal languages. Thus, in this chapter, we will use this Hopf algebraic viewpoint to make comparisons between older forms of Minimalism like Stabler's and the current form based on free symmetric Merge, as well as to demonstrate the utility of the our algebraic perspective.

Consequently, we analyze Stabler's Computational Minimalism from the point of view of Hopf algebra structures, which we showed in the previous chapter to be an appropriate algebraic language for the formulation of Merge in Minimalism. We show here that in the older setting, Internal Merge and External Merge correspond to very different types of mathematical structures. External Merge is expressible in terms of the notion of "operated algebra," while Internal Merge can be described in terms of right-ideal coideals in a Hopf algebra and corresponding quotient right-module coalgebras. While these are both interesting mathematical structures, the very different forms of Internal Merge and External Merge make it difficult to reconcile them as two forms of a single, underlying, basic Merge operation. This is unsatisfactory because linguistically, one expects Internal Merge and External Merge to be manifestations of the same fundamental computational principle. We also show that the need for keeping track of labels and projection in the Old Minimalism creates a problem with several algebraic operations being partially defined, as well as the checking of domains along a sequence of derivations. This problem is resolved in the New Minimalism, where this is separated from the generative process of Merge and included only in the subsequent Externalization phase.

As we will see in more detail in the rest of this chapter, the main sources of difficulties in an algebraic formulation of Old Minimalism are summarized as follows:

- Working only with trees instead of workspaces (forests) complicates the algebraic structure: the natural composition operation (product), which in New Minimalism just combines syntactic objects into workspaces, here requires grafting trees together into another tree, which is a more involved operation of less evident interpretation. One of the consequences is that the decomposition operation (coproduct) that is compatible with this product is then less transparently related to the extraction of substructures needed by Internal Merge.
- Applying the structure formation of Merge directly to planar trees (rather than before planarization) further complicates the product structure, making it necessarily noncommutative.
- Incorporating labeling within the Merge operation (instead of a labeling algorithm taking place after the Merge action, as in New Minimalism) makes Merge only partially defined, and the compounding problem of domain check-

ing over iterations of Merge significantly increases the computational complexity.

- There is no unifying description of the operations of External Merge and Internal Merge, which present different algebraic properties.

We then show that the mathematical formulation of the approach introduced in the previous chapter completely bypasses this problem by directly presenting a unified framework for both Internal Merge and External Merge.

2.2 Old Minimalism and Loday–Ronco Hopf Algebra

In this section, we first recall some mathematical structures, in particular the Loday–Ronco Hopf algebra of *planar* binary rooted trees, and their associated explicit forms of product and coproduct. (Readers can consult the additional material in section 4.2 of chapter 4, where we recall the basic definition and properties of Hopf algebras.) In section 2.2.2, we introduce the specific case of planar binary rooted trees, with the product and coproduct described in section 2.2.4.

As mentioned previously, the fundamental combinatorial objects involved in the older formulations of Minimalism differ, from the algebraic perspective, from the more recent formulation in two important ways:

- Considering *planar* trees, namely rooted trees endowed with a particular choice of a *planar embedding*, instead of abstract (nonplanar) trees
- Working only with *trees* (syntactic objects) instead of *forests* (action of Merge on workspaces)

Fixing the embedding is equivalent to fixing an ordering of the leaves of the tree, compatibly with the tree structure, which corresponds linguistically to considering a linear order on sentences. While this is assumed in older versions of Minimalism, the newer version eliminates the assignment of planar structures at the level of the fundamental computational mechanism of Merge, relegating the choice of linear ordering to a later Externalization procedure that interfaces the fundamental computational mechanism of Merge (where ordering is not assumed) to the Sensory–Motor (SM) system, where ordering is imposed via the Externalization of language into speech, sign, or writing. The difference between imposing a priori planar embeddings on trees or not, together with the difference between working only with trees (syntactic objects) rather than with forests (workspaces), lead to significantly different mathematical formulations of the resulting Hopf algebras and of the Merge operations.

Another very significant difference between older versions of Minimalism and the newer formulation is the calculus of labels of syntactic features associated with trees, as well as how such labels determine and are determined by the transformations implemented by Internal Merge and External Merge. We introduce labeling in section 2.2.6. For a discussion of *labels and projection* in the various versions of Minimalism, see [26]. This should be compared with our discussion of the labeling algorithm in New Minimalism in section 1.15.

We demonstrate in section 2.2.7 how in older formulations of Minimalism, the conditions on planarity and labels impose conditions on the applicability of Merge operations, for both External Merge and Internal Merge. This makes the operations only partially defined on specific domains, reflecting conditions on matching/related labels and projection. We discuss the specific case of External Merge in section 2.2.8 and Internal Merge in section 2.2.9.

In sections 2.2.10 and 2.2.11, we describe the mathematical structure underlying the formulation of Internal Merge in the older versions of Minimalism. In section 2.2.10, we show that the issue of labels and domains requires a modification of the Loday–Ronco coproduct, while in section 2.2.11, we show that a similar modification is required for the product and the problem of heads and projection results in the fact that the domain of Internal Merge is a right ideal (but not a left ideal) with respect to the product and a coalgebra with respect to the modified coproduct.

Moreover, in section 2.2.12, we show that the presence of domains that make Merge partially defined, due to labeling conditions, creates an additional mathematical complication related to the iterated application of Merge, viewed as the composition of partially defined operations.

We observe in section 2.2.14 that a significant difference with respect to the Hopf algebra formulation of New Minimalism described in chapter 1 arises in comparison with the analogous structure in fundamental physics. In the case of Old Minimalism discussed here, Internal Merge has an intrinsically asymmetric structure due to the constraints coming from labeling and projection, and from the imposition of working with trees with planar structures. Because of this, instead of finding Hopf ideals as in the setting of theoretical physics, one can only obtain right-ideal coideals, which correspond to a much weaker form of *generalized quotients* in Hopf algebra theory.

We then discuss in more detail the mathematical structure of the old formulation of External Merge in sections 2.2.15 and 2.2.16. We show that these exhibit a very different mathematical structure with respect to Internal Merge. It can be described in terms of the notion of "operated algebra," where again one needs to extend the notion to a partially defined version because of the conditions on domains coming from the problem of labels and projection.

In section 2.5, we discuss comparison between New Minimalism and other linguistic models, including TAGs, as well as some tensor models.

2.2.1 Preliminaries on Hopf Algebras of Rooted Trees

We begin here by reviewing the main constructions of Hopf algebras of rooted trees, focusing on the *Loday–Ronco Hopf algebra*, which is based on *planar binary rooted trees*. This is the appropriate machinery for formalizing the original version of Minimalism within our algebraic context.

Associative algebras can be regarded as the natural algebraic structure that describes linear strings of letters in a given alphabet with the operation of concatenation of words. On the other hand, in linguistics, the emphasis is put preferably on the syntactic trees that provide sentence *structure* rather than on the strings of words produced by a given grammar. This is considered one of the fundamental aspects of human language. When dealing with trees instead of strings as the main objects, the appropriate algebraic formalism is no longer associative algebras, but Hopf algebras and more generally operads. These are mathematical structures essentially designed for the parsing of hierarchical compositional rules, by encoding (in a coproduct operation) all the possible "correct parsings"; that is, all available decompositions of an element into its possible building blocks.

As we have already seen, heuristically, the main properties that a Hopf algebra $\mathcal{H}$ describes can be summarized as follows. As we will see in the specific example of trees here, as a vector space, $\mathcal{H}$ consists of formal linear combinations of a specific class of objects (planar binary rooted trees, Feynman graphs, etc.). Operations will be defined on these generators and extended by linearity. The multiplication structure is a combination operation and the coproduct structure is a decomposition operation that lists all the possible different decompositions (parsings). The antipode is like a group inverse, and it establishes compatibility between multiplication and comultiplication, unit and counit.

We already encountered Hopf algebras in our formulation of free symmetric Merge and the action on workspaces in the previous chapter. We will now discuss a different Hopf algebra, one that is associated with the older formulations of Minimalism.

In the supplementary material in chapter 4, we recall in section 4.2 the general definition of Hopf algebra, and we discuss basic facts that we will be using in this chapter. As in the case of the Hopf algebra of workspaces that we used in chapter 1, the Hopf algebra that we will discuss here is graded connected, and hence the antipode can be inductively constructed (as discussed in section 4.2). Thus, in the following text, we will focus only on the bialgebra structure and not discuss the antipode map explicitly.

2.2.2 Binary Rooted Trees and Admissible Cuts

We recall in section 4.3 some general facts about binary rooted trees. Here, we consider the case of *planar* binary rooted trees, as the older formulations of Minimalism assume this kind of assignment of planar structures (an ordering of the terminal leaves on trees).

We use here the same notion of *admissible cut* on rooted trees that we introduced in Definition 1.2.6. The definition here is the same, but it applies to *planar* trees rather than to abstract (nonplanar) trees; hence both the extracted forest $\pi_C(T)$ and the remaining tree $\rho_C(T)$ are regarded as planar, with the planar embedding induced by that of T.

Definition 2.2.1. An *admissible cut* C on a *planar*, rooted tree T is an operation that acts as follows:

- Selects a number of edges of T with the property that every oriented path in T from the root to one of the leaves contains at most one of the selected edges
- Removes the selected edges

The result of an admissible cut is a disjoint union of a *planar* tree $\rho_C(T)$ that contains the root vertex and a *planar* forest $\pi_C(T)$; that is, a disjoint union $\pi_C(T) = \sqcup_i^{pl} T_i$ of *planar* trees, where each T_i has a unique source vertex (no incoming edges), which we select as the root vertex of T_i; and where the order of the T_i in $\pi_C(T)$ is fixed by the planar embedding (which makes the product $\sqcup^{pl}$ noncommutative). An *elementary* admissible cut (or simply an *elementary cut*) is a cut C consisting of a single edge.

2.2.3 The Loday–Ronco Hopf Algebra of Binary Rooted Trees

We can note right away that this algebraic structure is different from that we considered in chapter 1. Since older forms of Minimalism incorporate linear ordering, it is necessary to work with binary, rooted trees with an assigned choice of planar embedding (linear ordering of the leaves). Moreover, since they do not include workspaces, it is necessary to work only with trees, not with forests. This has immediate consequences on the type of algebraic structure that we need to work with, as the possible forms of product and coproduct operations are affected by the presence of this linear ordering and by the absence of forests. Thus, the main differences with respect to our setting in chapter 1 is that in New Minimalism, Merge acts on workspaces (a Hopf algebra of binary forests with no planar embeddings), while in Stabler's Computational Minimalism, Merge acts on planar binary rooted trees (and also admits a Hopf algebra interpretation, as we will show).

Consider the vector space $\mathcal{V}_k$ spanned by the planar binary rooted trees T with k internal vertices (or, equivalently, with $k+1$ leaves). See section 4.3 in chapter 4 for more details: in particular, the dimension of $\mathcal{V}_k$ is given in (4.3.1).

Definition 2.2.2. Let $\mathcal{V} = \oplus_{k\geq 0}\mathcal{V}_k$, with $\mathcal{V}_0 = \mathbb{Q}$. Let D_V denote the set of possible vertex labels. For a given label $d \in D_V$, the grafting operator $\wedge_d$ is defined as

$$\wedge_d : \mathcal{V} \otimes \mathcal{V} \to \mathcal{V}, \quad T_1 \otimes T_2 \mapsto T = T_1 \wedge_d T_2\,, \tag{2.2.1}$$

with $\wedge_d : \mathcal{V}_k \otimes \mathcal{V}_\ell \to \mathcal{V}_{k+\ell-1}$, which attaches the two roots (v_{r_1} of T_1 and v_{r_2} of T_2) to a single root vertex v, with label $d \in D_V$.

Note that this is *not* the same as with free symmetric Merge $\mathfrak{M}$, discussed in the previous chapter, because it acts on *planar* trees, producing new planar trees, so in particular it is *not* commutative, unlike free symmetric Merge $\mathfrak{M}$. It corresponds instead to the type of noncommutative Merge operation $\mathfrak{M}^{nc}$ that we considered in (1.12.2). However, in the case of the planar syntactic objects in $\mathcal{SO}^{nc}$ produced by $\mathfrak{M}^{nc}$, only the leaves carry labels (in the set $\mathcal{SO}_0$ of lexical items and syntactic features), while in the case of the operation (2.2.1), internal vertices of the tree also carry labels, so the operation itself depends on the label $d \in D_V$ that is assigned to the new root vertex. Thus, we use here the different notation $\wedge_d$ of (2.2.1), which keeps track of this labeling.

One also introduces the associative concatenation operations on planar binary rooted trees, given in Definition 2.2.3.

Definition 2.2.3. Given S and T, the tree $S\backslash T$ (S under T) is obtained by grafting the root of T to the rightmost leaf of S, while T/S (S over T) is the tree obtained by grafting the root of T to the leftmost leaf of S. The grafting operation (2.2.1) is related to these concatenations by

$$T_1 \wedge_d T_2 = T_1/S\backslash T_2, \tag{2.2.2}$$

where S is the planar binary tree with a single vertex decorated by $d \in D_V$.

Remark 2.2.4. Each planar binary rooted tree is described as a grafting,

$$T = T_\ell \wedge_d T_r ,$$

of the trees stemming to the left and right of the root vertex.

We remove the explicit mention of the vertex decorations when not needed.

Definition 2.2.5. The Loday–Ronco Hopf algebra $\mathcal{H}_{LR}$ of planar binary rooted trees is obtained from the vector space $\mathcal{V}$ by defining a multiplication and a comultiplication inductively by degrees. For trees $T = T_\ell \wedge T_r$ and $T' = T'_\ell \wedge T'_r$ the product defined in [124] can be built inductively using the property

$$T \star T' = T_\ell \wedge (T_r \star T') + (T \star T'_\ell) \wedge T'_r,$$

with the tree consisting of a single root vertex $\bullet$ as the unit. The coproduct similarly can be built from lower-degree terms by the property

$$\Delta(T) = \sum_{j,k} (T_{\ell,j} \star T_{r,k}) \otimes (T'_{\ell,n-j} \wedge T'_{r,m-k}) + T \otimes \bullet ,$$

where $T = T_\ell \wedge T_r$ and $\Delta(T_\ell) = \sum_j T_{\ell,j} \otimes T'_{\ell,n-j}$ and $\Delta(T_r) = \sum_k T_{r,k} \otimes T'_{r,m-k}$, for $T_\ell \in \mathcal{V}_n$ and $T_r \in \mathcal{V}_m$.

In [124], this Hopf algebra $\mathcal{H}_{LR}$ of planar binary rooted trees is described in terms of the Hopf algebra structure on the group algebra $\mathbb{Q}[S_\infty] = \oplus_n \mathbb{Q}[S_n]$ of the symmetric group. Namely, the inclusion of the Hopf algebra of noncommutative symmetric functions in the Malvenuto-Reutenauer Hopf algebra of permutations factors through the Loday–Ronco Hopf algebra of planar binary rooted trees; on this point, see also [1]. In [18], a version of the Loday–Ronco Hopf algebra of planar binary rooted trees was used for the renormalization of massless quantum electrodynamics and explicit isomorphisms between the Loday–Ronco Hopf algebra of planar binary rooted trees and the noncommutative Connes–Kreimer Hopf algebra of renormalization was constructed in [1], [94], and [66]. We discuss this relation to the Connes–Kreimer Hopf algebra in section 2.2.5, after a more explicit illustration of the Loday–Ronco product and coproduct in section 2.2.4.

2.2.4 Graphical Form of Coproduct and Product

It is shown in [1] that the coproduct and product of the Loday–Ronco Hopf algebra of planar binary rooted trees can be conveniently visualized in the following way:

A given tree T can be subdivided into two parts by cutting along the path from one of the leaves to the root (see Figure 2.1 and section 1.5 of [1]), and

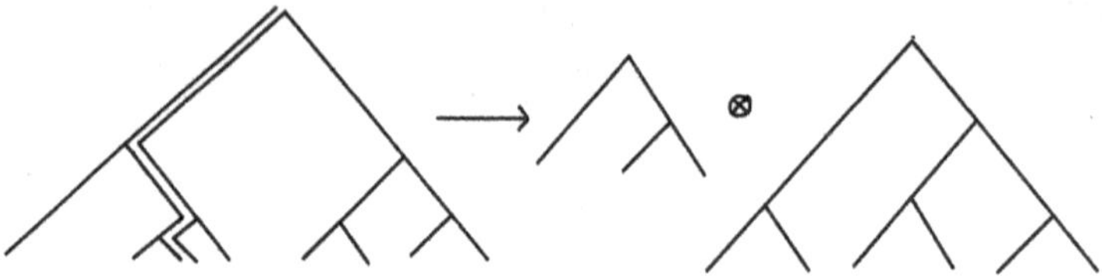

Figure 2.1
A term in the coproduct in the Loday–Ronco Hopf algebra.

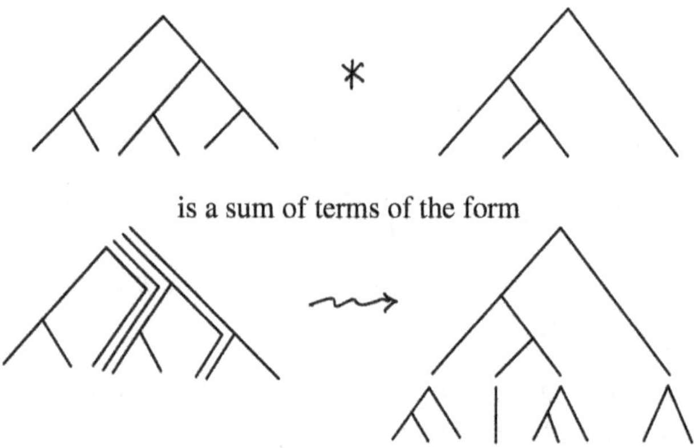

Figure 2.2
A product in the Loday–Ronco Hopf algebra, in a case with $n_1 = 5, n_2 = 3$.

the coproduct then takes the form of a sum over all such decompositions:

$$\Delta(T) = \sum T' \otimes T'' \tag{2.2.3}$$

where T', T'', respectively, are the parts of T to the left and right of the path from a leaf to the root, and the sum is over all choices of leaves.

To form the product $T_1 \star T_2$, suppose that T_1 has $n_1 + 1$ leaves and T_2 has $n_2 + 1$ leaves. Consider again subdivisions of the tree T_1 obtained by cutting along paths from the leaves to the root, including trees consisting of just the path itself, with the cuts chosen to obtain $n_2 + 1$ subtrees of T_1; see Figure 2.2.

We write $\mathcal{V}^{(k)}$ for the $\mathbb{Q}$-vector space spanned by the planar binary rooted trees with k leaves, and we write the usual operadic composition operation of grafting roots to leaves as before in the form

$$\gamma : \mathcal{V}^{(k_0)} \times \cdots \times \mathcal{V}^{(k_n)} \times \mathcal{V}^{(n+1)} \to \mathcal{V}^{(k_0+\cdots+k_n)}, \tag{2.2.4}$$

where $\gamma(T_0, \dots, T_n; T)$ is the tree obtained by grafting the trees $T_i \in \mathcal{V}^{(k_i)}$ in collection $(T_0, \dots, T_n)$ to the tree $T \in \mathcal{V}^{(n+1)}$, by attaching the root of T_i to the i-th leaf of T, as illustrated in Figure 2.2.

The product in $\mathcal{H}_{LR}$ is then written in the form

$$T \star T' = \sum_{(T_0,\dots,T_n)} \gamma(T_0, \dots, T_n; T'), \tag{2.2.5}$$

with $n+1$ being the number of leaves of T', and the sum over subdivisions into subtrees extracted according to the rule described previously.

One can observe here how the requirement of working only with *trees* and not with *forests* (i.e., no workspaces) makes it more complicated to obtain a viable form of compatible product and coproduct, compared to the case of the Hopf algebra of workspaces discussed in the previous chapter. For instance, if we compare the product structure in the two cases, we see that the product in the Hopf algebra of workspaces is simply the disjoint union that takes two workspaces and combines them into a single one. It has an immediately clear interpretation. If one wants to work only with trees, though, the product of two trees should produce a new tree, which requires that one of the trees is somehow distributed over the other (in particular, resulting in a noncommutative product, as it now does matter which of the two trees is distributed over the other). The operation illustrated in Figure 2.2 displays this procedure, which takes one of the two trees apart and distributes it over the other by grafting the resulting pieces to the leaves.

2.2.5 Hopf Algebra Comparisons

There is another notion, closely related to that of admissible cut that we used in chapter 1 for abstract binary rooted trees and in Definition 2.2.1 for *planar* binary rooted trees, which we will refer to as *admissible pruning* (see [1]). In the case of admissible cuts, it only happens for an elementary cut (a cut of a single edge) that both parts of the cut are trees. More generally, for nonelementary cuts, one side $\pi_C(T)$ is a forest, and only the other one $\rho_C(T)$ is a tree. Admissible prunings always produce two resulting trees.

Definition 2.2.6. An admissible pruning set S of a planar binary rooted tree T is a subset of the internal vertices of T, with the property that if $v \in S$, then the left-successor (child) vertex of v is also in S. Cutting each edge connecting a vertex in S to one not in S, one obtains two planar forests F'_S and F''_S, where one of the components of F''_S contains the root of T. A tree T'_S is then obtained by gluing together the component trees of F'_S and F''_S with the operations of Definition 2.2.3, as shown in Figure 2.3.

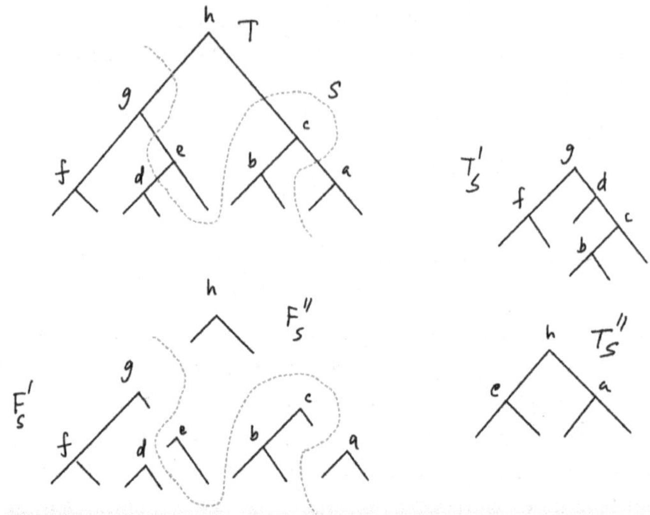

Figure 2.3
An admissible pruning set S of a planar binary rooted tree T and the resulting trees T'_S and T''_S.

It is shown in section 8 of [1] that there is a linear basis, which the authors denote by the notation M^*_T, of the Loday–Ronco Hopf algebra, labelled by the planar binary rooted trees T, with the property that, expressed in this basis, the Loday–Ronco product is the free associative, noncommutative algebra generated by the elements M^*_T, with the product

$$M^*_T \star M^*_{T'} = M^*_{T\backslash T'},$$

with $T\backslash T'$ the operation recalled in Definition 2.2.3. The Loday–Ronco coproduct, expressed in this basis, takes the form

$$\Delta_{LR}(M^*_T) = \sum_S M^*_{T'_S} \otimes M^*_{T''_S},$$

where the sum is over admissible pruning sets S, as in Definition 2.2.6, with the resulting trees T'_S and T''_S. Admissible pruning sets do not always correspond to admissible cuts (the example in Figure 2.3 shows a pruning set that is not an admissible cut), but if C is an admissible cut of T, then the set of vertices of $\pi_C(T)$ gives an admissible pruning set S. In particular, for the case of an elementary admissible cut, $T'_S = \pi_C(T) = T_v$ and $T''_S = T/^c\pi_C(T) = T/^cT_v$, so we can match the corresponding term of the coproduct to the coproduct term $T_v \otimes T/^cT_v$, which extracts an accessible term T_v for use by Internal Merge.

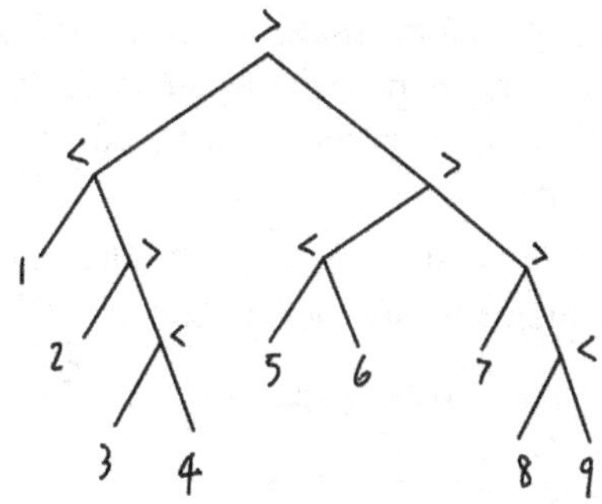

Figure 2.4
Planar binary rooted trees in Stabler's Computational Minimalism.

In [1], it is also proved that there is an explicit isomorphism between the Connes–Kreimer noncommutative Hopf algebra of planar, binary, rooted forests and the Loday–Ronco Hopf algebra of planar, binary, rooted trees. However, this cannot be used to match iterated applications of Internal Merge (as we will describe more explicitly in Proposition 2.2.25). Thus, in the formulation of Old Minimalism, the relation between the Merge operation and the Hopf algebra of binary rooted trees is less direct and less transparent than in New Minimalism based on free symmetric Merge.

2.2.6 Labeled Trees

The previous subsections only dealt with general mathematical formalism about planar binary rooted trees. We now consider more specifically the case of Old Minimalism as embodied in [192].

In this formulation, one considers planar binary rooted trees such as the example in Figure 2.4, where the ordering of the leaves determines and is determined by the planar embedding of the tree, and the labels at the inner vertices serve the purpose of pointing toward the head.

More precisely, all internal vertices and the root vertex are labeled by symbols in the set $\{>, <\}$. The purpose of the labels $>$ and $<$ is to identify where the *head* of the tree is: in the example of Figure 2.4, the head is the leaf with the vertex number 8. In addition to these labels, one considers a finite set of syntactic features $X \in \{N, V, A, P, C, T, D, \ldots\}$ and *selector* features denoted by the symbol σX for a head selecting a phrase XP. More generally, we can have labels that are strings (ordered finite sets) $\alpha = X_0 X_1 \cdots X_r$ of syntactic features. We also consider labels given by the letters ω and $\bar{\omega}$, which stand for *licensor* and *licensee*. (Note: in [192], the notation $= X$ is used to refer to the feature selector instead of our σX, and the notation $\pm X$ is used for licensor/licensee pairs, but to avoid confusion, we prefer to avoid using mathematical notation with a meaning different from its generally accepted one. Hence, we will instead use the letters σ and $\omega, \bar{\omega}$.)

We write the result of External Merge applied to the constituents α and β as in [192], as the labeled planar binary rooted tree

$$\overset{<}{\widehat{\alpha \quad \beta}} \qquad \text{rather than} \qquad \overset{\alpha}{\widehat{\alpha \quad \beta}}$$

This means that we consider as generators of the Loday–Ronco Hopf algebras the binary trees with labels that include the syntactic features, as well as symbols $<$ and $>$, as in [192], used to point toward the head when Merge is applied.

By (2.2.2), it is clear that the head of the tree $T_1 \wedge_> T_2$ is the head of T_2 and the head of the tree $T_1 \wedge_< T_2$ is the head of T_1.

Recall also that a *maximal projection* in T is a subtree of T (an accessible term) that is not a proper subtree of any larger subtree with the same head (see section 1.14). As pointed out in [192], in the example illustrated in Figure 2.4, the leaves $\{2, 3, 4\}$ determine a subtree with head the leaf 3, but any larger subtree in T would have a different head; hence, this subtree is a maximal projection. So is the subtree determined by the leaves $\{5, 6\}$, for instance.

Definition 2.2.7. We define $\mathcal{H}_{ling}$ to be the Loday–Ronco Hopf algebra $\mathcal{H}_{LR}$ with the choice of labels described here. We will write $\mathcal{V}_{ling}$ for the underlying $\mathbb{Q}$-vector space spanned by the binary trees with the labeling described here.

As we already pointed out, the choice of working with planar, binary, rooted trees corresponds to an assigned linear ordering of its leaves, which in turn is the linear ordering of the resulting sentence when spoken or read. Thus, by working with planar trees, we are assuming that linear ordering is determined for the result of Merge. This is the same assumption made in [192]. This is the crucial point in the comparison with New Minimalism, and we will discuss it more explicitly in Section 2.3.1.

2.2.7 Old Formulation of External Merge and Internal Merge

We now discuss the External Merge and Internal Merge operations in the old formulations of Minimalism. We first describe the operations at the level of the underlying combinatorial tree, and then we give the more precise constraints on when the operations can be applied, based on the labels of the trees involved.

Thus, we should view External Merge and Internal Merge as *partially defined operations*

$$\mathcal{E} : \mathcal{V}_{ling} \otimes \mathcal{V}_{ling} \to \mathcal{V}_{ling}$$

$$\mathcal{I} : \mathcal{V}_{ling} \to \mathcal{V}_{ling} ,$$

where we assume that the operations defined on generators are extended by linearity or bilinearity, so they are defined on the linear subspaces

$$\mathrm{Dom}(\mathcal{E}) \subset \mathcal{V}_{ling} \otimes \mathcal{V}_{ling}$$

$$\mathrm{Dom}(\mathcal{I}) \subset \mathcal{V}_{ling} ,$$

which we will describe more precisely shortly.

At the level of the underlying combinatorial trees, External Merge and Internal Merge are simply defined as the following operations.

The combinatorial structure of the External Merge is given by

$$\mathcal{E}(T_1 \otimes T_2) = \begin{cases} \bullet \wedge T_2 & T_1 = \bullet \\ \\ T_2 \wedge T_1 & \text{otherwise,} \end{cases} \tag{2.2.6}$$

where $\bullet$ denotes the tree consisting of a single vertex; and the $\wedge$ operation is the grafting operation $\wedge_d$ defined as in (2.2.1), with the specific label $d \in \{>, <\}$, according to a rule that will be discussed in more detail in section 2.2.8.

The combinatorial structure of Internal Merge is given by

$$\mathcal{I}(T) = \pi_C(T) \wedge \rho_C(T), \tag{2.2.7}$$

where C is an *elementary* admissible cut of T, where $\rho_C(T)$ is the remaining pruned tree that contains the root of T and $\pi_C(T)$ is the part that is severed by the cut (which in the case of an elementary cut is a tree rather than a forest). Again, $\wedge$ is as in (2.2.1), with a label that will be made more precise shortly, and the choice of the elementary cut will also depend on conditions on tree labels; see section 2.2.9.

We now look more precisely at the structure of External Merge and Internal Merge and their domains of definition. It should be noted how working with planar structures and with conditions of matching/related labels makes the underlying mathematical operations more difficult to describe, as they are defined only on specific domains and with additional conditions.

2.2.8 Old Form of External Merge

The more precise definition of the External Merge in the older forms of Minimalism is given in Definition 2.2.8 and Proposition 2.2.9 (where we are adapting the description of [192] to our terminology and notation).

Definition 2.2.8. As in [192], we use the notation $T[\alpha]$ to refer to a tree where the head is labeled by an ordered set of syntactic features starting with α. The label α consists of a string of syntactic features of the form

$$\alpha = X_0 X_1 \cdots X_r \quad \text{or} \quad \alpha = \sigma X_0 X_1 \cdots X_r,$$

with σ as the selector symbol.

We introduce the notation $\hat{\alpha}$ to refer to the string obtained from α after removing the first feature (other than the selection symbol). Namely, for $\alpha = X_0 X_1 \cdots X_r$ or $\alpha = \sigma X_0 X_1 \cdots X_r$, we have $\hat{\alpha} = X_1 \cdots X_r$.

Proposition 2.2.9. *The External Merge* $T = \mathcal{E}(T_1[\sigma\alpha], T_2[\alpha])$ *of two trees* $T_1[\sigma\alpha]$ *and* $T_2[\alpha]$ *is given by*

$$\mathcal{E}(T_1[\sigma\alpha], T_2[\alpha]) = \begin{cases} T_1[\widehat{\sigma\alpha}] \wedge_< T_2[\hat{\alpha}] & |T_1| = 1 \\ T_2[\hat{\alpha}] \wedge_> T_1[\widehat{\sigma\alpha}] & |T_1| > 1\,, \end{cases} \tag{2.2.8}$$

defined on the domain given by the vector space on the set of generators

$$\mathrm{Dom}(\mathcal{E}) = \mathrm{span}_{\mathbb{Q}}\{(T_1[\beta], T_2[\alpha]) \,|\, \beta = \sigma\alpha\}. \tag{2.2.9}$$

Proof. This clearly agrees with (2.2.6) at the level of the underlying combinatorial trees. The main difference with (2.2.6) is that here, the operation can be performed only if the heads are decorated with the syntactic features $\sigma\alpha$ and α, respectively. Thus, the domain of definition $\mathrm{Dom}(\mathcal{E}) \subset \mathcal{V}_{ling} \otimes \mathcal{V}_{ling}$ of the External Merge operation is the vector space with generators given by the pairs $(T_1[\beta], T_2[\alpha])$ satisfying $\beta = \sigma\alpha$, as in (2.2.9). □

2.2.9 Old Version of Internal Merge

In the older formulations of Minimalism, one considers a second Merge operation: *Internal Merge*, considered as distinct from External Merge. We again adapt to our notation and terminology the formulation given in [192]. We use the same notation as before for $T[\alpha]$, with $\alpha = X_0 X_1 \ldots X_r$ or $\alpha = \sigma X_0 X_1 \ldots X_r$ being a string of syntactic features possibly starting with a selector, and we write $\hat{\alpha} = X_1 \dots X_r$. Internal Merge is again modeled on the basic grafting operation of planar rooted binary trees $\wedge_d$, but this time, its input is a single tree $T[\alpha]$.

Definition 2.2.10. Given a planar binary rooted tree T containing a subtree T_1, and given another binary rooted tree T_2, we denote as in [192] by $T\{T_1 \to T_2\}$ the planar binary rooted tree obtained by removing the subtree T_1 from T and replacing it with T_2. In particular, we write $T\{T_1 \to \emptyset\}$ for the tree obtained by removing the subtree T_1 from T.

Remark 2.2.11. Note that, in order to ensure that the result of this operation is still a tree, we need to assume that if an internal vertex of T belongs to the subtree T_1, then all oriented paths in T from this vertex to a leaf also belong to T_1, where paths in the tree are oriented from root to leaves. We only consider subtrees that have this property. When it is necessary to stress this fact, we will refer to such subtrees as *complete subtrees*. Note that this matches the notion of *accessible term* used in New Minimalism, that we discussed in chapter 1.

Lemma 2.2.12. *The complete subtrees of a given tree T are exactly the subtrees that can be obtained by applying a single elementary admissible cut C to the tree T. The tree $T\{T_1 \to \emptyset\}$ is then the same as the tree $\rho_C(T)$.*

Proof. The property described in Remark 2.2.11 characterizing complete subtrees is that there are no further cuts along any path from the root of the subtree to any of the leaves, which is precisely the case of a subtree obtained by performing a single admissible cut on T. □

The following statement specifies the domain for Internal Merge, and it follows directly from [192].

Proposition 2.2.13. *Consider a tree $T[\alpha]$, where $\alpha = X_0 \cdots X_r$ or $\alpha = \sigma X_0 \cdots X_r$ or $\alpha = \omega X_0 \cdots X_r$ or $\alpha = \bar{\omega} X_0 \cdots X_r$. The domain* $\mathrm{Dom}(\mathcal{I}) \subset \mathcal{V}_{ling}$ *of the Internal Merge is given by the subspace*

$$\mathrm{Dom}(\mathcal{I}) = \mathrm{span}_{\mathbb{Q}}\{T[\alpha] \mid \exists T_1[\beta] \subset T[\alpha], \textit{with}\, \beta = \bar{\omega} X_0 \hat{\beta},\ \alpha = \omega X_0 \hat{\alpha}\}, \tag{2.2.10}$$

where $T_1 \subset T$ is a subtree as before, and

$$\mathcal{I}(T[\alpha]) = T_1^M[\hat{\beta}] \wedge_> T\{T_1[\beta]^M \to \emptyset\} = \pi_C(T) \wedge_> \rho_C(T), \tag{2.2.11}$$

where C is the elementary admissible cut specified by the subtree T_1^M (the maximal projection of the head of T_1) and the condition (given by the matching *labels ωX_0 and $\bar{\omega} X_0$) that $T[\alpha] \in \mathrm{Dom}(\mathcal{I})$. The head of $\pi_C(T) \wedge_> \rho_C(T)$ gets the label $\hat{\alpha}$.*

Proof. The domain (2.2.10) simply describes the label-matching conditions for performing Internal Merge in the formulation of [192]. The only thing that needs an explanation is the identification

$$T_1^M[\hat{\beta}] \wedge_> T\{T_1[\beta]^M \to \emptyset\} = \pi_C(T) \wedge_> \rho_C(T)$$

of (2.2.11). To see this, observe that the maximal projection T_1^M of the head of T_1 determines an associated elementary admissible cut that separates T_1^M from $T\{T_1[\beta]^M \to \emptyset\}$. Thus, with respect to this admissible cut $T_1^M = \pi_C(T)$, we obtain $T\{T_1[\beta]^M \to \emptyset\} = \rho_C(T)$, as observed in Lemma 2.2.12. □

In the tree $\rho_C(T)$ we maintain the "trace" (in the linguistic sense) of the deeper copy of T_1 through a labeling (a leaf node in $T/^c\pi_C(T)$ or a nonbranching node in $T/^\rho\pi_C(T)$ with the $\emptyset$ label), indicating the empty subtree left behind by the operation $T_1[\beta]^M \to \emptyset$; see section 2 of [192].

These versions of External Merge and Internal Merge, which we recalled in (2.2.8) and (2.2.11), have issues at the level of linguistic description, such as problems with potentially unlabelable exocentric constructions. For example (as observed by Riny Huijbregts), Internal Merge as in (2.2.11) can yield $\{XP, YP\}$ results, and a valid labeling cannot be performed, unless a further application of Internal Merge displaces XP or YP to a so-called *criterial position*; that is, where structural agreement, specifier-Head agreement, can take place.

In [192], two further variants of External Merge and Internal Merge are introduced to deal with so-called "persistent features." This results in what is referred to in [192] as "conflated minimalist grammars (CMGs)". These additional forms of External Merge and Internal Merge, however, do not solve the problem mentioned here and further complicate the structure, so we will focus on analyzing the algebraic structure determined just by the forms (2.2.8) and (2.2.11) for External and Internal Merge, respectively.

2.2.10 Old Version of Internal Merge and the Coalgebra Structure

As we have discussed here, Internal Merge is only partially defined on $\mathcal{V}_{ling}$ because it requires the existence of matching conditions on the label, which determine the domain $\mathrm{Dom}(\mathcal{I}) \subset \mathcal{V}_{ling}$. We now discuss how the domain $\mathrm{Dom}(\mathcal{I})$ and the Internal Merge operation $\mathcal{I}$ behave with respect to the coproduct of the Hopf algebra $\mathcal{H}_{ling}$.

Remark 2.2.14. Consider the coproduct $\Delta(T)$ of a tree $T \in \mathrm{Dom}(\mathcal{I})$. This is given by the sum of decompositions $\Delta(T) = \sum T' \otimes T''$ as illustrated in (2.2.3). In each of these decompositions, one side will contain the head of the tree T (or possibly both sides, if the leaf used to cut the tree also happens to be the head). If both the head of T and the head of $\pi_C(T)$ are contained in the same part of the partition, then that part still belongs to $\mathrm{Dom}(\mathcal{I})$. Thus, these pieces of the coproduct are in $\mathrm{Dom}(\mathcal{I}) \otimes \mathcal{H}_{ling} + \mathcal{H}_{ling} \otimes \mathrm{Dom}(\mathcal{I})$. The remaining terms, with the heads of T and of $\pi_C(T)$ on different sides of the decomposition only belong in general to $\mathcal{H}_{ling} \otimes \mathcal{H}_{ling}$ and fall outside the domain of Internal Merge.

This suggests a modification of the coproduct on $\mathcal{H}_{ling}$ with respect to the usual Loday–Ronco coproduct (2.2.3).

Definition 2.2.15. For a generator T that is not in $\mathrm{Dom}(\mathcal{I})$, we just set $\pi_C(T) = T$; that is, no cut is performed. For $T \in \mathrm{Dom}(\mathcal{I})$, let $h(T)$ and $h(\pi_C(T))$ be the leaves of T that are the head of T and the head of $\pi_C(T)$, respectively. If T is in $\mathrm{Dom}(\mathcal{I})$, then C is, as before, the elementary admissible cut determined by the label conditions. Let $\mathcal{P}_{\mathcal{I}}(T)$ denote the set of bipartitions:

$$\mathcal{P}_{\mathcal{I}}(T) = \left\{ T = (T', T'') \middle| \begin{array}{c} h(T) \in T' \text{ and } h(\pi_C(T)) \in T' \\ \text{or} \\ h(T) \in T'' \text{ and } h(\pi_C(T)) \in T'' \end{array} \right\}. \tag{2.2.12}$$

Remark 2.2.16. For trees outside the domain $\mathrm{Dom}(\mathcal{I})$, the set $\mathcal{P}_{\mathcal{I}}(T)$ just counts all bipartitions as we have set $\pi_C(T) = T$.

Proposition 2.2.17. *Consider the modified coproduct*

$$\Delta_{\mathcal{I}}(T) := \sum_{(T', T'') \in \mathcal{P}_{\mathcal{I}}(T)} T' \otimes T'', \tag{2.2.13}$$

which is the same as (2.2.3) *outside of* $\mathrm{Dom}(\mathcal{I})$. *With this modified coproduct,* $\mathrm{Dom}(\mathcal{I})$ *is a* coideal *of the coalgebra* $\mathcal{H}_{ling}$*:*

$$\Delta_{\mathcal{I}}(\mathrm{Dom}(\mathcal{I})) \subset \mathrm{Dom}(\mathcal{I}) \otimes \mathcal{H}_{ling} + \mathcal{H}_{ling} \otimes \mathrm{Dom}(\mathcal{I}). \tag{2.2.14}$$

Proof. The definition of the coproduct (2.2.13) using bipartitions $\mathcal{P}_{\mathcal{I}}(T)$ as in (2.2.12) is precisely designed to avoid the problem mentioned in Remark 2.2.14 of terms of the Loday–Ronco coproduct that are in $\mathcal{H}_{ling} \otimes \mathcal{H}_{ling}$ but outside of the domain of Internal Merge. □

2.2.11 Old Version of Internal Merge and Algebra Structure

We now discuss the behavior of Internal Merge with respect to the algebra structure of $\mathcal{H}_{ling}$ (i.e., the product operation). In this case, arguing in a similar way as for the previous coproduct above, there is a modification $\star_{\mathcal{I}}$ of the original product of $(\mathcal{H}_{LR}, \star)$ that remains the same outside of $\mathrm{Dom}(\mathcal{I})$ and takes into account the additional data in $\mathrm{Dom}(\mathcal{I})$, and that has the effect of making $\mathrm{Dom}(\mathcal{I})$ into a *right ideal* in the algebra $(\mathcal{H}_{ling}, \star_{\mathcal{I}})$. However, $\mathrm{Dom}(\mathcal{I})$ is not a left ideal.

We define the modified product $\star_{\mathcal{I}}$ on $\mathcal{H}_{ling}$ as in Definition 2.2.18.

Definition 2.2.18. Let $h(T)$ denote the head of T. Given trees T, T' where T' has $n{+}1$ leaves, consider the set $\mathcal{P}_{\mathcal{I}}(T, T')$ given by decompositions $(T_0, \ldots, T_n)$ of T as before, with

$$\mathcal{P}_{\mathcal{I}}(T, T') := \{(T_0, \ldots, T_n) \,|\, h(T) \text{ and } h(\pi_C(T)) \in T_{h(T')}\} . \tag{2.2.15}$$

Namely, these are the decompositions $(T_0, \ldots, T_n)$, with the property that the head of T lies in the component $T_{h(T')}$ that is grafted to the head of T'. Same for the head of $\pi_C(T)$, in the case where $T \in \mathrm{Dom}(\mathcal{I})$. In the case of $T \notin \mathrm{Dom}(\mathcal{I})$, the condition reduces to just $h(T) \in T_{h(T')}$.

Remark 2.2.19. Note that it is always possible to have such decompositions, as some of the components T_i can always be taken to be copies of just the path from a leaf to the root, so the relevant piece of the decomposition can be placed at the $h(T')$-position.

Proposition 2.2.20. *Consider the modified product $\star_{\mathcal{I}}$ on $\mathcal{H}_{ling}$ of the form*

$$T \star_{\mathcal{I}} T' = \sum_{(T_0,\ldots,T_n)\in\mathcal{P}_{\mathcal{I}}(T,T')} \gamma(T_0, \ldots, T_n; T') . \tag{2.2.16}$$

Then the domain of Internal Merge $\mathrm{Dom}(\mathcal{I})$ *is a right ideal (but not a left ideal) in the algebra $(\mathcal{H}_{ling}, \star_{\mathcal{I}})$; that is, it satisfies*

$$\mathrm{Dom}(\mathcal{I}) \star_{\mathcal{I}} \mathcal{H}_{ling} \subset \mathrm{Dom}(\mathcal{I}) . \tag{2.2.17}$$

For $T \in \mathrm{Dom}(\mathcal{I})$, we then have

$$\mathcal{I}(T \star_{\mathcal{I}} T') = \sum_{(T_0,\ldots,T_n)\in\mathcal{P}_{\mathcal{I}}(T,T')} \pi_C(T_{h(T')}) \wedge_{>} \gamma(T_0, \ldots, \rho_C(T_{h(T')}), \ldots, T_n; T') . \tag{2.2.18}$$

Proof. We consider the algebra $(\mathcal{H}_{ling}, \star_{\mathcal{I}})$ with the modified product (2.2.16). The head of each $\gamma(T_0, \ldots, T_n; T')$ is then the same as the head of T, which we write in shorthand as $h(T \star_{\mathcal{I}} T') = h(T)$. Moreover, by construction, the component $T_{h(T')}$ is in $\mathrm{Dom}(\mathcal{I})$ when $T \in \mathrm{Dom}(\mathcal{I})$, so $T \star_{\mathcal{I}} T'$ is itself in $\mathrm{Dom}(\mathcal{I})$. This shows that $\mathrm{Dom}(\mathcal{I})$ is a right ideal in the algebra $(\mathcal{H}_{ling}, \star_{\mathcal{I}})$. It is, however, not a left ideal due to the asymmetric form of the product in the Loday–Ronco Hopf algebra, which is inherited by the modified product (2.2.16). The relation (2.2.18) then follows from the form (2.2.11) of Internal Merge and the form (2.2.16) of the modified product. □

We can then combine the behavior with respect to the product with the previous observation about the coproduct, to obtain the structure of a *right module coalgebra.*

Proposition 2.2.21. *Internal Merge $\mathcal{I}$ as in [192] defines a right $(\mathcal{H}_{ling}, \star_{\mathcal{I}})$-module given by the coset*

$$\mathcal{M}_{\mathcal{I}} := \mathrm{Dom}(\mathcal{I}) \backslash \mathcal{H}_{ling} \,,$$

where $\mathcal{M}_{\mathcal{I}}$ is also a coalgebra with the coproduct induced by $(\mathcal{H}_{ling}, \Delta_{\mathcal{I}})$.

2.2.12 Iterated Internal Merge

As mentioned in the introduction to this chapter, since we need to consider domains that make Internal Merge and External Merge operations partially defined, this has the effect that, when one *composes* a chain of Merge operations, as is typical in the course of a derivation, the problem with domains also compounds: one needs to check multiple domain conditions. It is indeed a well-known phenomenon that computationally implementing such an approach can run into a problem with feature checking that rapidly grows with the length of the derivation. This phenomenon can be seen, for instance, in the construction of parsers for MGs in [104], where it is embedded into the logical formulas and interface conditions that account for the feature checking problem.

In our setting, as described previously, we see this problem of compounding effects of domains and feature checking in the following way: iterations of Internal Merge give rise to a nested family of right-ideal coideals organized into a corresponding projective system of right-module coalgebras. The growth of this projective system accounts for the rapidly growing computational complexity of feature checking, as we argue in section 2.2.13.

Definition 2.2.22. For a forest $F = T_1 \cdots T_\ell$, we define a grafting operation $\wedge^\ell$ that consists of the repeated application of the grafting $\wedge$ to the trees in F:

$$\bigwedge^{\ell} F = T_1 \wedge T_2 \wedge \cdots \wedge T_\ell \,. \tag{2.2.19}$$

Definition 2.2.23. For a tree $T[\alpha]$, where α is a finite list of features each of which can be either X_i or ωX_i of $\bar{\omega} X_i$ as above, we consider the following combinatorial conditions:

1. There are N subtrees $T_1, \ldots, T_N$ in T, each of which is a complete subtree, in the sense specified previously.
2. Let $T_1^M, \ldots, T_N^M$ be the maximal projections of the subtrees in the sense recalled previously. These are also complete subtrees of T.
3. The subtrees T_i^M are disjoint.

We say that $T[\alpha]$ is N-fold complete if it satisfies these three conditions.

Proposition 2.2.24. *The domain of the Nth iteration of Stabler's Internal Merge*

$$\mathcal{D}_N := \mathrm{Dom}(\mathcal{I}^N) \tag{2.2.20}$$

is given by the span of

$$\left\{ T[\alpha] \,\middle|\, \begin{array}{c} \exists \textit{ subtrees} \\ T_1[\beta^{(1)}], \ldots, T_N[\beta^{(N)}] \\ \textit{such that} \end{array} \left\{ \begin{array}{l} (i)\ (1), (2), (3) \textit{ are satisfied} \\ (ii)\ \beta_0^{(1)} = \bar{\omega} X_0, \ldots, \beta_0^{(N)} = \bar{\omega} X_{N-1} \\ (iii)\ \alpha = \omega X_0 \omega X_1 \cdots \omega X_{N-1} \cdots \end{array} \right. \right\}. \tag{2.2.21}$$

Proof. When we consider repeated applications of N Internal Merge operations, starting from a given tree $T[\alpha]$, with α a finite list of features as described previously, we need to assume that $T[\alpha]$ is N-fold complete (it satisfies the three combinatorial conditions of Definition 2.2.23), as these arise from the condition characterizing $\mathrm{Dom}(\mathcal{I})$ on each subsequent iteration. In addition to these combinatorial conditions, we have a matching labels condition that specifies the domain of the composite operation. Let $T[\alpha]$ be the given tree as a labeled tree, and let $T_1[\beta^{(1)}], \ldots, T_N[\beta^{(N)}]$ be the labeled subtrees, with the conditions listed here. Then the remaining conditions listed in (2.2.21) provide the label-matching conditions necessary for each successive iteration. □

Proposition 2.2.25. *On the domain* (2.2.21), *the iterated Internal Merge acts as*

$$\mathcal{I}^{\#C}(T[X]) = \bigwedge^{1+\#C} \left(\pi_C(T)[\hat{\mathbb{Y}}]\ \rho_C(T)[\hat{X}^N] \right), \tag{2.2.22}$$

where we use the notation $\pi_C(T)[\hat{\mathbb{Y}}]$ *for the forest* $\pi_C(T) = T_N^M \cdots T_1^M$, *where the label* $[\hat{\mathbb{Y}}]$ *means*

$$\pi_C(T)[\hat{\mathbb{Y}}] = T_N^M[\hat{\beta}^{(N)}] \cdots T_1^M[\hat{\beta}^{(1)}]$$

and the label $[\hat{\alpha}^N]$ *of the tree* $\rho_C(T)$ *stands for what remains of the original label* X *after the initial terms* $\omega X_0 \omega X_1 \cdots \omega X_{N-1}$ *are removed.*

Proof. Arguing as in Proposition 2.2.13, we see that conditions (1), (2), and (3) of Definition 2.2.23 ensure that the choice of the subtrees

$$T_1^M, \ldots, T_N^M$$

corresponds to an admissible cut C of the tree T, with the number of cut branches $\#C = N$. Using (2.2.11), we then see that the planar binary rooted tree obtained by the iterated Internal Merge operation takes the form (2.2.22). □

Note that the extraction of a forest $\pi_C(T)$ for the Internal Merge iteration, instead of a tree for a single elementary cut, is not directly describable in terms of the Hopf algebra coproduct here, as the coproduct of the Loday–Ronco Hopf algebra only has trees in both the left and the right channel, unlike the Hopf algebra of workspaces introduced in chapter 1.

Proposition 2.2.24 directly implies the generalization of Proposition 2.2.21 obtained in Proposition 2.2.26.

Proposition 2.2.26. *The domains* (2.2.20) *of the iterated Internal Merges determine a nested family of right-ideal coideals, given by the domains of the iterations of Internal Merge,* $\mathcal{D}_{N+1} \subset \mathcal{D}_N$ *with* $\mathcal{D}_N = \mathrm{Dom}(\mathcal{I}^N)$ *as before. Each* $\mathcal{D}_{N+1} \backslash \mathcal{D}_N$ *determines a coideal in the coalgebra* $\mathcal{D}_{N+1} \backslash \mathcal{H}_{ling}$*, and this gives a projective system of right-module coalgebras:*

$$\mathcal{M}_{\mathcal{I}^N} := \mathrm{Dom}(\mathcal{I}^N) \backslash \mathcal{H}_{ling} \,. \tag{2.2.23}$$

2.2.13 Feature Checking Complexity

The results of Propositions 2.2.24 and 2.2.26 suggest an interpretation of the modules $\mathcal{M}_{\mathcal{I}^N}$ of (2.2.23) as measuring, in a sense, the growing complexity of feature checking in Stabler's Computational Minimalism, with the length of the derivation. We can give a heuristic argument in the following way in terms of a simple count of vector space dimensions.

As we have mentioned, in the planar binary rooted trees of Stabler's Minimalism, the internal vertices are labeled by strings (ordered finite sets) $\alpha = X_0 X_1 \ldots X_r$ of syntactic features. We denote as Σ^* the set of all finite (but arbitrary-length) strings in the alphabet given by the set of syntactic features, and as Σ_ℓ the subset of strings of fixed length ℓ. Let $\mathfrak{s}$ denote the cardinality of the set of syntactic features so $\#\Sigma_\ell = \mathfrak{s}^\ell$. (For simplicity, since we are only interested in a rough estimate, we ignore here the possibility of having constraints on subsequent elements $X_i X_{i+1}$ in such sequences that could make this a smaller set.) We also denote as $\Sigma_\ell(a_0 \dots a_r)$, with $r < \ell$, the subset of sequences in Σ_ℓ where the first $r+1$ elements are fixed and equal to $a_0, \dots, a_r$.

The following observation is a well-known counting formula for planar binary rooted trees.

Remark 2.2.27. When we consider all the possible planar binary rooted trees with k internal vertices labeled by a set Σ_ℓ, we obtain a total counting of the form

$$d_{k,\ell} = \dim \mathcal{V}_{k,\Sigma_\ell} = (\#\Sigma_\ell)^k \frac{(2k)!}{k!(k+1)!} = \mathfrak{s}^{k\ell} \frac{(2k)!}{k!(k+1)!}, \tag{2.2.24}$$

where $\mathcal{V}_{k,\Sigma_\ell} = \mathcal{V}_{ling,k,\ell}$ denotes the vector space of planar binary rooted trees with internal vertices labeled by the set Σ_ℓ. Then, let $\mathcal{V}_{ling,k,\ell}(\alpha)$ be the span of those trees with a given label $\alpha \in \Sigma_\ell$ assigned at the root vertex and the other vertex labels assigned arbitrarily. We have

$$d_{k,\ell}(\alpha) = \dim \mathcal{V}_{ling,k,\ell}(\alpha) = (\#\Sigma_\ell)^{k-1} \frac{(2k)!}{k!(k+1)!} = \mathfrak{s}^{(k-1)\ell} \frac{(2k)!}{k!(k+1)!}. \tag{2.2.25}$$

Definition 2.2.28. Consider the set $\mathcal{R}_{\mathcal{I},\ell}$ defined as

$$\mathcal{R}_{\mathcal{I},\ell} := \{(\alpha,\beta) \in \Sigma_\ell \times \Sigma_\ell \,|\, \alpha = \omega X_0 \hat{\alpha}\,,\ \beta = \bar{\omega} X_0 \hat{\beta}\}\,. \tag{2.2.26}$$

Similarly, we define the set $\mathcal{R}_{\mathcal{I}^N,\ell}$ as

$$\mathcal{R}_{\mathcal{I}^N,\ell} = \left\{ (\alpha,\beta_1,\dots,\beta_N) \in \Sigma_\ell \times \Sigma_\ell^N \;\middle|\; \begin{array}{l} \alpha = \alpha = \omega X_0 \omega X_1 \cdots \omega X_{N-1} \cdots \\ \beta_1 = \bar{\omega} X_0, \dots \\ \cdots \\ \beta_N = \bar{\omega} X_{N-1} \end{array} \right\}. \tag{2.2.27}$$

We can assume here that k and ℓ are both large. We will also be interested in the case where N is large, and we will assume that $k, \ell > N$.

Proposition 2.2.29. *The dimension of $\mathcal{D}_{1,k,\ell}(\alpha) = \mathcal{D}_1 \cap \mathcal{V}_{ling,k,\ell}(\alpha)$ for $\mathcal{D}_1 =$ Dom$(\mathcal{I})$ is given by*

$$d_{\mathcal{I},k,\ell}(\alpha) = \dim \mathcal{D}_{1,k,\ell}(\alpha) = (\mathfrak{s}^{(k-1)\ell} - \mathfrak{s}^{(k-1)(\ell-2)}(\mathfrak{s}^2-1)^{k-1}) \frac{(2k)!}{k!(k+1)!}. \tag{2.2.28}$$

Proof. Consider a single application of Internal Merge $\mathcal{I}$. Note that, according to the description (2.2.26) of the set $\mathcal{R}_{\mathcal{I},\ell}$, we need at least $\ell \geq 3$. Out of all the $d_{k,\ell}$ planar binary rooted trees with k internal vertices labeled by Σ_ℓ, we can identify those that are the generators of the subspace $\mathcal{D}_{1,k,\ell}(\alpha) = \mathcal{D}_1 \cap$

$\mathcal{V}_{ling,k,\ell}(\alpha)$ for $\mathcal{D}_1 = \mathrm{Dom}(\mathcal{I})$ by considering all the possible ways to have one non-root, non-leaf vertex so the pair of the root and this vertex have a pair of labels $(\alpha,\beta) \in \mathcal{R}_{\mathcal{I},\ell}$, with arbitrary assignments of labels in all the remaining $k-2$ vertices. The condition $(\alpha,\beta) \in \mathcal{R}_{\mathcal{I},\ell}$ means that if we fix the value of α, then the value of β has the first two entries (which we just write as a_0a_1) fixed and only the remaining terms in the sequence $\beta \in \Sigma_\ell$ are free to be assigned. On the remaining points, we impose no constraints. As befor, $\mathfrak{s}^{(k-1)\ell} = \#(\Sigma_\ell)^{k-1}$, which counts all the arbitrary assignments of labels in Σ_ℓ to the $k-1$ non-root vertices. Also, we have

$$\mathfrak{s}^{\ell} - \mathfrak{s}^{\ell-2} = \mathfrak{s}^{\ell-2}(\mathfrak{s}^2 - 1) = \#(\Sigma_\ell \smallsetminus \Sigma_\ell(a_0a_1)). \tag{2.2.29}$$

Thus, when we count all the assignments of labels in Σ_ℓ such that not all of them are in the complement $\Sigma_\ell \smallsetminus \Sigma_\ell(a_0a_1)$, which is the same as counting all the possible ways of having at least one of the vertices labeled by $\Sigma_\ell(a_0a_1)$, we obtain (2.2.28). □

When we consider iterations of Internal Merge, we obtain a similar statement. We define

$$S_N(a,b) := \binom{k-1}{N} b^N (a-b)^{k-1-N}. \tag{2.2.30}$$

and

$$S_{N,k}(a,b) := S_N(a,b) + S_{N+1}(a,b) + \cdots + S_{k-1}(a,b) \le a^{k-1}, \tag{2.2.31}$$

where the inequality is a consequence of the binomial theorem $\sum_{i=0}^{k-1} S_i = a^{k-1}$. We then obtain the result of Proposition 2.2.30.

Proposition 2.2.30. *The dimension of* $\mathcal{D}_{N,k,\ell}(\alpha) = \mathcal{D}_N \cap \mathcal{V}_{ling,k,\ell}(\alpha)$, *with* $D_N = \mathrm{Dom}(\mathcal{I}^N)$ *is given by*

$$d_{\mathcal{I},k,\ell,N}(\alpha) = \dim \mathcal{D}_{N,k,\ell}(\alpha) = S_{N,k}(\mathfrak{s}^{\ell}, \mathfrak{s}^{\ell-2N}(\mathfrak{s}^{2N} - 1)) \frac{(2k)!}{k!(k+1)!}, \tag{2.2.32}$$

with $d_{\mathcal{I},k,\ell} = d_{\mathcal{I},k,\ell,1}$.

Proof. Consider N iterations of Merge, with $k\ell \ge N$. By the description of the set (2.2.27), for a given label $\alpha = a_0a_1 \ldots a_{2N-1}\hat{\alpha} \in \Sigma_\ell$ at the root vertex, we are counting the label assignments for which there are N non-root vertices whose labels are determined by the label α of the root, according to (2.2.27). Meanwhile, at all the remaining vertices, the labels can be arbitrarily assigned in Σ_ℓ. As before, we have $\mathfrak{s}^{(k-1)\ell} = \#(\Sigma_\ell)^{k-1}$ for arbitrary assignments of labels at all the non-root vertices and

$$\mathfrak{s}^{\ell} - \mathfrak{s}^{\ell-2N} = \mathfrak{s}^{\ell-2N}(\mathfrak{s}^{2N} - 1) = \#(\Sigma_\ell \smallsetminus \Sigma_\ell(a_0 \ldots a_{2N-1})). \tag{2.2.33}$$

We can identify among all the d_k trees those that are generators of the subspace $\mathcal{D}_{N,k,\ell}(\alpha) = \mathcal{D}_N \cap \mathcal{V}_{ling,k,\ell}(\alpha)$ as the subset of trees that have a choice of N non-root internal vertices which, together with the root, have an $(N+1)$–tuple of labels that belongs to the set $\mathcal{R}_{\mathcal{I}^N,\ell}$. This means that we are counting all the possible ways in which among the $k-1$ labels assigned to root vertices at least N are not in the complement of $\Sigma_\ell(a_0 \dots a_{2N-1})$. To do that, note that (2.2.30) counts the number of label assignments to a set of $k-1$ points where N of them have labels in a set $B \subset A$ with $b = \#B$, and the remaining $k-1-N$ points have labels in the complement $A \setminus B$. The result is the situation in which exactly N points out of the $k-1$ are in B. The counting of the assignments for which at least N of the points have labels in B is then given by (2.2.31). In our case, we are only imposing that at least N of the non-root vertices have labels in the subset $\Sigma_\ell(a_0 \dots a_{2N-1})$, so we find that the number of trees satisfying this property is given by (2.2.32). □

The explicit counting of (2.2.28) and (2.2.32) provides a measure of the growth in complexity, when N, k, and ℓ grow large, of the feature-checking requirements in Stabler's Minimalism.

2.2.14 Coideals, Recursion, and Symmetries

To interpret the results of Propositions 2.2.21 and 2.2.26, consider again the general fact that the Hopf-algebraic formalism occurs naturally in settings involving the construction of recursive and hierarchical structures due to the role of the coproduct and product in separating basic building blocks and combining them. In both Old Minimalism and New Minimalism, the Merge operator is the main mechanism for the construction of the recursive and hierarchical structures of syntax. So indeed, the fact that even in this older formulation of Stabler's Computational Minimalism, we find a Hopf algebra interpretation is not surprising.

It is more interesting to interpret the specific meaning of objects like the right-module coalgebras that we obtain in Propositions 2.2.21 and 2.2.26. It is helpful to again consider the comparison with the role that similar algebraic structures play in the context of theoretical physics, where Hopf algebras describe similar hierarchical structures (such as Feynman graphs) in fundamental physics. In the applications of Hopf algebras to physics, especially to renormalization in perturbative quantum field theories, Hopf ideals are closely related to the recursive implementation of symmetries (Ward identities) and the recursive construction of solutions to equations of motion (Dyson-Schwinger (DS) equations); see [65] and [203].

We have discussed in chapter 1 how the algebraic formulation of New Minimalism closely resembles the mathematical structure of DS equations in theoretical physics. However, the analogous mathematical structure describing Merge in the old forms of Minimalism differs significantly from that basic fundamental formulation we have in the case of New Minimalism.

The main difference, as previously shown, lies in the intrinsic asymmetry of the old form of the Merge operation, which only gives rise to a *weaker structure* than Hopf ideals (namely, the right-ideal coideals discussed in section 2.2.11). Thus, instead of a quotient Hopf algebra as in the case of implementation of symmetries in quantum field theory, one only obtains a quotient right-module coalgebra.

Such objects, quotient right-module coalgebras, sometimes referred to as "generalized quotients" of Hopf algebras, do indeed provide a suitable notion of quotients in the case of noncommutative Hopf algebras; see [149] and [194]. They are also studied in the context of the theory of Hopf–Galois extensions, designed to provide good geometric analogs in the theory of noncommutative geometry, of the notions of principal bundles and torsors in ordinary geometry, see, for instance, [184]. However, the type of structure involved, in order to accommodate the requirement of working with planar trees as well as the feature-checking requirements, becomes significantly more complicated than in the case of free symmetric Merge of New Minimalism.

We can think of the right-module coalgebras $\mathcal{M}_{\mathcal{I}^N}$ as a family of geometric spaces (in a noncommutative sense) that implement the recursive structures of syntax generated by the Merge operation, in a sense similar to how quotients by Hopf ideals in physics recursively implement the gauge symmetries or the recursively constructed solutions to the equations of motion. This, however, is a significantly weaker (and at the same time significantly more complicated) algebraic structure than the one that we see occurring in New Minimalism, as discussed in the previous chapter.

2.2.15 Partial Operated Algebra

We now discuss in more detail the mathematical structure of the old formulation of External Merge and show how it creates an independent structure of a very different nature from Internal Merge, as discussed in section 2.2.11. These very different mathematical formulations of Internal Merge and External Merge are another significant drawback in the algebraic properties of Old Minimalism in comparison with New Minimalism, since one expects that both forms of Merge arise as part of the same fundamental structure. We first need to introduce the notion of a partial operated algebra. We illustrate this context first in the case of Internal Merge.

We can consider the data $(\mathcal{H}_{ling}, \star_{\mathcal{I}}, \mathrm{Dom}(\mathcal{I}), \mathcal{I})$ discussed previously as a generalization of the notion of *operated algebra* (see [90] and [205]) where one considers data $(\mathcal{A}, \star, F)$ of an algebra with a linear operator $F : \mathcal{A} \to \mathcal{A}$. Here, because of the presence of domains, we need to extend this notion to the partially defined case.

Definition 2.2.31. A *partial operated algebra* is a pair $(\mathcal{A}, F)$ where $\mathcal{A}$ is an associative algebra and $F : \mathrm{Dom}(F) \to \mathcal{A}$ is a linear operator defined on a smaller domain $\mathrm{Dom}(F) \subset \mathcal{A}$, which is a right ideal of $\mathcal{A}$.

Remark 2.2.32. The pair $(\mathcal{H}_{ling}, \star_{\mathcal{I}}, \mathrm{Dom}(\mathcal{I}), \mathcal{I})$ is a partial operated algebra.

The setting of operated algebras was introduced by Rota [177] as a way of formalizing various instances of linear operators $F : \mathcal{A} \to \mathcal{A}$ on an algebra that satisfy polynomial constraints. The simplest example of such polynomial constraints is the identity $F(a \star b) = F(a) \star F(b)$ that makes F an actual algebra homomorphism. Another example is the Leibniz rule constraint $F(a \star b) = a \star F(b) + F(a) \star b$, which makes F a derivation. Other interesting polynomial constraints are Rota–Baxter relations, $F(a) \star F(b) = F(a \star F(b)) + F(F(a) \star b) + \lambda F(a \star b)$, which depending on the value of the parameter λ can represent various types of operations, such as integration by parts or extraction of the polar part of a Laurent series. These Rota–Baxter relations play an important role in the Hopf algebra formulation of renormalization in physics, and they will play a crucial role in chapter 3, in our model of the syntax–semantics interface.

In the case of Internal Merge considered here, not only is the linear operator only partially defined as $\mathcal{I} : \mathrm{Dom}(\mathcal{I}) \to \mathcal{H}_{ling}$, but the identity (2.2.18) that expresses the Internal Merge of a product is not directly of a simple polynomial form, as in Rota's operated algebra program. However, it appears to be an interesting question about whether the type of relation (2.2.18) can be accommodated in terms of the approach to Rota's program via *term-rewriting systems* (in the sense of [4], further developed in [75] and [90]).

2.2.16 Old Version of External Merge and Hopf Algebra Structure

We now focus on the binary operation $\mathcal{E}$ of External Merge, defined on the subdomain $\mathrm{Dom}(\mathcal{E}) \subset \mathcal{H}_{ling} \otimes \mathcal{H}_{ling}$ as described in (2.2.9). In Old Minimalism, the structure of External Merge is simpler than Internal Merge as discussed before, and it is closely related to the usual Hochschild cocycle that defines the grafting operations in the Hopf-algebraic construction of solutions of DS equations in physics.

The "operated algebra" viewpoint discussed briefly earlier is especially useful in analyzing External Merge, along the lines of extending the notion of operated algebra from unitary to binary operations discussed in [205]. We recall from [205] the notion of operated algebra based on binary operations. This structure is called "$\vee_\Omega$-algebra" in [205]. In our notation, we use $\wedge$ instead of $\vee$ for the grafting of binary trees used in Merge, following the convention of writing syntactic parsing trees with the root at the top and the leaves at the bottom. So we are going to refer to this structure as "$\wedge_\Omega$-algebra".

Definition 2.2.33. A triple $(\mathcal{A}, \star, \{\wedge_\alpha\}_{\alpha\in\Omega})$ is a $\wedge_\Omega$-algebra if $(\mathcal{A}, \star)$ is an algebra and $\{\wedge_\alpha\}_{\alpha\in\Omega}$ is a family of binary operations satisfying the identity

$$a \star b = a_1 \wedge_\alpha (a_2 \star b) + (a \star b_1) \wedge_\alpha b_2 \tag{2.2.34}$$

for $a = a_1 \wedge_\alpha a_2$ and $b = b_1 \wedge_\alpha b_2$. A $\wedge_\Omega$-bialgebra (or Hopf algebra) $(\mathcal{H}, \Delta, \star, \wedge_\Omega)$ is a bialgebra (or Hopf algebra) that is also a $\wedge_\Omega$-algebra. It is a *cocycle* $\wedge_\Omega$-Hopf algebra if the binary operations $\wedge_\Omega$ also satisfy the cocycle identity

$$\Delta(a \wedge_\alpha b) = (a \wedge_\alpha b) \otimes 1 + (\star \otimes \wedge_\alpha) \circ \tau(\Delta(a) \otimes \Delta(b)), \tag{2.2.35}$$

where $\tau : \mathcal{H} \otimes \mathcal{H} \otimes \mathcal{H} \otimes \mathcal{H} \to \mathcal{H} \otimes \mathcal{H} \otimes \mathcal{H} \otimes \mathcal{H}$ is the permutation that exchanges the two middle factors.

Again, because we have to use operations that are only partially defined on a smaller domain, we need to extend this notion to a partially defined version, as in the case of Definition 2.2.31.

Definition 2.2.34. A partially defined cocycle $\wedge_\Omega$-bialgebra (or Hopf algebra) is a bialgebra (or Hopf algebra) $(\mathcal{H}, \Delta, \star)$ together with a family $\{\wedge_\alpha\}_{\alpha\in\Omega}$ of binary operations defined on linear subspaces $\mathrm{Dom}(\wedge_\alpha) \subset \mathcal{H} \otimes \mathcal{H}$:

$$\wedge_\alpha : \mathrm{Dom}(\wedge_\alpha) \hookrightarrow \mathcal{H} \otimes \mathcal{H} \to \mathcal{H},$$

which satisfy (2.2.34) and (2.2.35) whenever all the terms are in $\mathrm{Dom}(\wedge_\alpha)$.

Proposition 2.2.35. *The Hopf algebra $\mathcal{H}_{ling}$ with the External Merge operator $\mathcal{E}$ is a partially defined cocycle $\wedge_\Omega$-Hopf algebra.*

Proof. It is shown in [205] that the Loday–Ronco Hopf algebra $\mathcal{H}_{LR}$ of planar binary rooted trees with the binary operations $\wedge_\alpha$ that graft two trees T_1, T_2 as the left and right subtrees at a new binary root with label α is a cocyle $\wedge_\Omega$-Hopf algebra. The range of $\wedge_\alpha$ is a coideal in $\mathcal{H}_{LR}$. It is also shown in [205] that the $\wedge_\Omega$-Hopf algebra $\mathcal{H}_{LR}$ is the free cocycle $\wedge_\Omega$-Hopf algebra; that is, the initial object in the category of cocycle $\wedge_\Omega$-Hopf algebras.

The External Merge operation is modeled on the grafting operations $\wedge_\alpha$ of the Loday–Ronco Hopf algebra, with the labels $\alpha \in \{<, >\}$. The main differences are that External Merge inverts the order when the first tree is nontrivial,

$$\mathcal{E}(T_1, T_2) = T_2 \wedge_> T_1 \quad \text{for } T_1 \neq \bullet$$

$$\mathcal{E}(\bullet, T_2) = \bullet \wedge_< T_2 ,$$

and the operation is only defined on the domain

$$\mathrm{Dom}(\mathcal{E}) \subset \mathcal{H}_{ling} \otimes \mathcal{H}_{ling}$$

of pairs of trees $(T_1[\beta], T_2[\alpha])$, with matching labels $\beta = \sigma\alpha$.

Both the $\wedge_\alpha$-algebra identity (2.2.34) and the cocycle identity (2.2.35) are still satisfied, whenever all the terms involved are in the domain of the Merge operator. This yields the description of External Merge as a structure of *partially defined cocycle $\wedge_\Omega$-Hopf algebra.* □

2.3 Summary Comparison of Old Minimalism and New Minimalism

New Minimalism, as presented in [28], [29], [33], and in our mathematical formulation in the previous chapter, avoids all the complications arising in the older formulation, with both issues arising from the presence of planar structure, and from the need for labeling and projection. By delegating these aspects to a later Externalization procedure, the core computational mechanism of Merge becomes very transparent and simple from the mathematical perspective and no longer leads to very different structures underlying Internal Merge and External Merge. The formulation is also no longer plagued by the problem of partially defined operations and corresponding domains.

We detailed in the previous chapter the whole mathematical structure of Merge in the New Minimalism, including a proposed model for Externalization. Here, we present a direct comparison with Old Minimalism, as discussed in the previous sections.

2.3.1 The Core Computational Structure

The first point is identifying the core computational structures behind the generative process in the old and the new versions of Minimalism and comparing them.

2.3.1.1 Core process of New Minimalism In New Minimalism, one starts with a core computational structure that is the recursive construction of binary rooted trees, where trees are now just abstract trees, not endowed with planar structure.

As described in section 1.10 of chapter 1, the set $\mathfrak{T}$ of finite binary rooted trees without a planar structure is obtained as the *free, nonassociative, commutative magma* whose elements are the balanced bracketed expressions in a single variable x, with the binary operation (binary set formation)

$$(\alpha,\beta) \mapsto \mathfrak{M}(\alpha,\beta) = \{\alpha,\beta\}, \tag{2.3.1}$$

where α and β are two such expressions.

This description of the generative process of binary rooted trees is immediate from the identification of these trees with balanced bracketed expressions in a single variable x, such as

$\{\{x\{xx\}\}x\} \longleftrightarrow$ 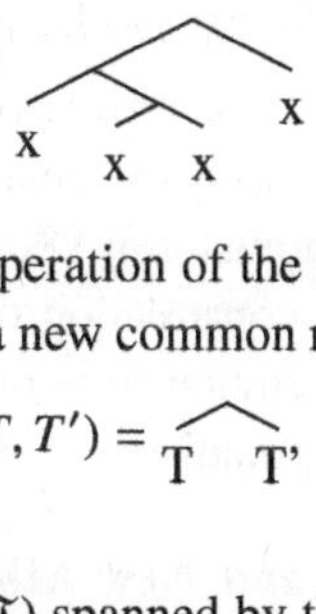

.

Under this identification, the binary operation of the magma takes two rooted binary trees and attaches the roots to a new common root:

$(T,T') \mapsto \mathfrak{M}(T,T') =$ 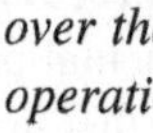

. (2.3.2)

If one takes the $\mathbb{Q}$-vector space $\mathcal{V}(\mathfrak{T})$ spanned by the set $\mathfrak{T}$, the magma operation $\mathfrak{M}$ on $\mathfrak{T}$ induces on $\mathcal{V}(\mathfrak{T})$ the structure of an algebra. More precisely, $\mathcal{V}(\mathfrak{T})$ is the *free, commutative, nonassociative algebra generated by a single variable x*, or equivalently, the *free algebra over the quadratic operad freely generated by the single commutative binary operation $\mathfrak{M}$* (see [96]). We will return to this description in terms of operads in the next chapter, where we discuss its relation to semantics.

2.3.1.2 Core process of Old Minimalism By comparison, the key generative process in Old Minimalism is governed by the grafting operators of

Definition 2.2.3 for *planar* binary rooted trees:

$$T = T_\ell \wedge^d T_r = \overset{d}{\widehat{T_\ell \quad T_r}} ,$$

where the planar trees now also carry labels at the internal vertices, not only labels at the leaves (as in the case of syntactic objects in New Minimalism). Thus, in particular, here one has such a grafting operation for every possible label $d \in D_V$ assigned to the new root vertex.

As we have seen, these grafting operations are involved in the inductive construction of the product and coproduct of the Loday–Ronco Hopf algebra of planar binary rooted trees.

This operations $\wedge^d$, with $d \in D_V$, inductively construct planar binary rooted trees. However, here one needs to impose further constraints, because the labeling of internal vertices of the planar trees also keeps track of the syntactic head of the tree and the subtrees (see the example in Figure 2.4).

These additional requirements lead to the operation $\wedge^d$ being defined on a subdomain of pairs of T_ℓ, T_r, with the property that $T_\ell \wedge^d T_r$ has a well-defined head. For comparison, we discussed the role of the head in the context of the New Minimalism in section 1.13 of chapter 1.

2.3.2 Relation to Combinatorial Dyson–Schwinger Equations

We have observed that the generative process of syntax in Minimalism shares strong formal similarities with the generative process of recursive constructions of solutions of combinatorial DS equations in physics, the mechanism underlying the inductive construction of solutions of quantum equations of motion. This formal resemblance manifests itself in different ways in Old Minimalism and New Minimalism.

2.3.2.1 DS equations and New Minimalism As we discussed in section 1.10 of chapter 1, the core computational mechanism of New Minimalism, based on free symmetric Merge $\mathfrak{M}$, is a very natural and simple mathematical object, and the generative process for the binary rooted trees can be seen as the simplest and most fundamental possible case of recursive solution of a fixed-point problem, of the kind that in physics is the DS equation.

Indeed, one can view the recursive construction of binary rooted trees through repeated application of the Merge operation (2.3.1) as the recursive construction of a solution to the fixed-point equation

$$X = \mathfrak{M}(X, X), \tag{2.3.3}$$

where $X = \sum_\ell X_\ell$ is a formal infinite sum of variables X_ℓ in $\mathcal{V}(\mathfrak{T})_\ell$, where ℓ is the grading by the number of leaves. The problem (2.3.3) can be solved recursively by degrees, with

$$X_n = \mathfrak{M}(X,X)_n = \sum_{j=1}^{n-1} \mathfrak{M}(X_j, X_{n-j}),$$

with solution $X_1 = x$, $X_2 = \{xx\}$, $X_3 = \{x\{xx\}\} + \{\{xx\}x\} = 2\{x\{xx\}\}$, $X_4 = 2\{x\{x\{xx\}\}\} + \{\{xx\}\{xx\}\}$, and so on. The coefficients with which the trees occur in these solutions count the different possible choices of planar embeddings.

Equation (2.3.3) is the simplest case (with the simplest quadratic polynomial) of the more general form of DS equations $X = \mathcal{B}(P(X))$ for rooted trees, where $P(X)$ is a polynomial in the formal variable X, and $\mathcal{B}$ is the operation (as in (1.1.1)) that takes a collection of rooted trees $T_1, T_2, \ldots, T_n$, seen as the forest $F = T_1 \sqcup T_2 \sqcup \ldots \sqcup T_n$, which appears as a monomial in $P(X)$, and constructs a new tree by attaching all the roots of the T_i's to a single new root (an n-ary Merge):

$$\mathcal{B} : T_1 \sqcup T_2 \sqcup \ldots \sqcup T_n \mapsto$$ 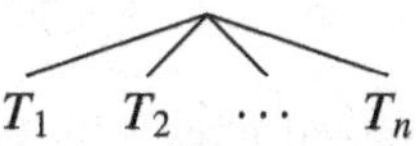 ; (2.3.4)

see Definition 1.3.2 and Lemma 1.3.3 in chapter 1. For a detailed discussion of the role of this grafting operator in physics, see, for instance, [65] and [203]. The case where $P(X)$ has a single quadratic term gives the binary Merge operation in its core generative process.

2.3.2.2 DS equations and Old Minimalism In the case of Old Minimalism, the recursive structure of combinatorial DS equations is not as directly visible as in New Minimalism. As discussed in section 2.2.14, the relation goes through a more subtle point. It is known (see [5], [65] and [66]) that solutions of combinatorial DS equations correspond to Hopf ideals in a Hopf algebra. In the case of the Old Minimalism formulation, Internal Merge does not quite define a Hopf ideal, but it does define a closely related structure: a right-ideal coideal. While a Hopf ideal determines an associated quotient Hopf algebra, noncommutative Hopf algebras in general have a weaker notion of quotient, which is given by the structure of quotient right-module coalgebras, which is indeed the structure that we have seen associated with Internal Merge in Old Minimalism. Thus, we can think of the right-ideal coideals defined by Internal Merge and its iterations as what remains visible in this setting of a recursive combinatorial DS equation (which corresponds to Hopf ideals).

2.3.3 Combinatorial Objects

We can now compare the main hierarchical structures involved in New Minimalism and Old Minimalism. New Minimalism has *syntactic objects* (abstract binary rooted trees with leaves decorated by lexical items and syntactic features) and *workspaces*, which are forests whose connected components are syntactic objects. Old Minimalism has planar binary rooted trees (no forests) with feature labels at both leaves and internal vertices and a well-defined syntactic head.

2.3.3.1 Syntactic objects and workspaces in New Minimalism As we discussed in the previous chapter, the core computational structure of section 2.3.1 introduces Merge in the most basic form of binary set formation (2.3.1). This can then be extended to the generative process that gives rise to *syntactic objects*. One starts with an assigned set $\mathcal{SO}_0$ of *lexical items* and *syntactic features* such as $N, V, A, P, C, T, D, \ldots$ The set $\mathcal{SO}$ of syntactic objects is then identified with the set (1.1.4):

$$\mathcal{SO} \simeq \mathfrak{T}_{\mathcal{SO}_0}$$

of finite binary rooted trees with no assigned planar embedding, and with leaves labeled by elements of $\mathcal{SO}_0$. Just as the set of binary rooted trees with no labeling of leaves has a magma structure with the binary set formation operation $\mathfrak{M}$ of (2.3.1), the set of syntactic objects also has a magma structure; namely, it is the *free, nonassociative, commutative magma over the set* $\mathcal{SO}_0$, as in (1.1.2):

$$\mathcal{SO} = \text{Magma}_{na,c}(\mathcal{SO}_0, \mathfrak{M}),$$

with the binary Merge operation $\mathfrak{M}$ defined as in (2.3.2) for pairs of trees with labeled leaves. In both the core magma $(\mathfrak{T}, \mathfrak{M})$ and the magma (1.1.2) of syntactic objects, one can introduce a multiplicative unit 1 that satisfies $\mathfrak{M}(T, 1) = T = \mathfrak{M}(1, T)$ for all trees by formally adding a trivial (empty) tree. (Note that this is different from the Loday–Ronco Hopf algebra in Old Minimalism, where one uses the "trivial tree" •, which consists of a single vertex.)

As discussed in section 1.1.3 of chapter 1, the description as elements of $\mathcal{SO} = \text{Magma}_{na,c}(\mathcal{SO}_0, \mathfrak{M})$ is usually expressed in the linguistics setting (see [28] and [29]) in terms of *sets* (in fact, *multisets*) rather than *trees*, and one expresses the elements of the magma in terms of brackets corresponding to the Merge operations, rather than in tree form, with corresponding notations:

$\{a, \{\{b, c\}, d\}\} \leftrightarrow$ 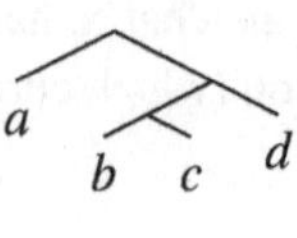

,

where the tree on the right should not be considered as being planar. As sets

$$\mathfrak{T}_{SO_0} \simeq \mathrm{Magma}_{na,c}(SO_0, \mathfrak{M})$$

are in bijection. Thus, both descriptions are equivalent. The description as magma is important in linguistics, as it encodes the generative process. The formulation in terms of bracketed expressions, as in the left-hand-side of the previous example, avoids confusion as to whether the trees have a planar embedding: working with (multi)sets (curly brackets notation) rather than with lists (round brackets notation) clearly means that no planarity is assumed. As argued in section 1.1.3 the tree notation is the standard mathematical terminology and certain mathematical operations that we are using have a simpler and more immediately visualizable description in terms of trees rather than in terms of the corresponding multisets of leaves.

In New Minimalism, Merge acts on workspaces, consisting of material (lexical items and syntactic objects) available for computation. The Merge operation updates the workspace for the next step of structure formation. This notion of workspace is formalized in [44]. Mathematically, as discussed in the previous chapter, workspaces are just finite, disjoint unions of binary rooted trees (i.e., *forests*) with leaves labeled by SO_0. Equivalently, workspaces are multisets of syntactic objects. Thus, the set of workspaces can be identified with the set $\mathfrak{F}_{SO_0}$ of *binary rooted forests with no assigned planar structure* with leaf labels in SO_0.

This is a key difference with respect to Old Minimalism, which does not use the notion of workspaces. Mathematically, the fact of using forests (the workspaces) instead of just trees (individual syntactic objects) has the advantage that it admits a very simple and natural product structure (namely, the disjoint union product $\sqcup$, that combines two given workspaces into a single one). This gives a very simple commutative product on the resulting Hopf algebra, to be compared with the more involved and noncommutative product of the Loday–Ronco Hopf algebra in the case of Old Minimalism. The coproduct on workspaces also simplifies with respect to the coproduct of the Loday–Ronco Hopf algebra, as we now discuss in more detail.

Given a workspace $F \in \mathfrak{F}_{SO_0}$, the material in F that is accessible for computation consists of the lexical items and all the trees that were obtained through previous applications of Merge. This inductive definition can be rephrased more directly by defining the *set of accessible terms* of F. As we discussed in the previous chapter, for a single binary rooted tree T with no assigned planar structure, the set $Acc(T)$ of accessible terms of T is the set of all subtrees $T_v \subset T$ given by all the descendants of a given non-root vertex of T;

see Definition 1.2.2 in chapter 1. Also, for a workspace given by a forest $F = \sqcup_{a \in I} T_a \in \mathfrak{F}_{SO_0}$, with T_a as the component trees and I as a finite indexing set, the set $Acc'(F)$ of accessible terms of the workspace consists of the syntactic objects, (i.e. the connected components T_a of F), together with all the accessible terms $\alpha \in Acc(T_a)$ of each component T_a (see Definition 1.2.2).

2.3.3.2 Labeled planar trees The fundamental combinatorial objects that describe the hierarchical structures in Old Minimalism differ from the setting of syntactic objects and workshops of New Minimalism in several important ways:

1. The use of *planar* trees
2. The use of labeling of the internal vertices (which reflects the features and head projection)
3. The fact that only trees are used and not forests (workspaces)

The first and third are responsible for the Hopf algebra being noncommutative and for the asymmetry of the resulting algebraic structure (the right-ideal coideals) associated with Internal Merge. The second is responsible for the partially defined action of Merge (the domains that correspond to the feature-checking conditions). These domains in turn determine the algebraic structure of right-ideal coideals, which one does not see in New Minimalism. The third is what makes the algebraic structure significantly more complicated, in terms of the form of the product and coproduct of the Hopf algebra, since, for example, the product is forced to be an asymmetric merging of two trees into a single one rather than just a simple combination of workspaces via disjoint union as in New Minimalism. Moreover, the accessible terms used by Internal Merge are not intrinsically part of the Loday–Ronco coproduct (see our previous discussion on the difference between admissible cuts and pruning sets), so that the operations of Internal Merge and External Merge require additional structure on top of the underlying Hopf algebra and are not fully determined by it as in the New Minimalism.

2.3.4 Action of Merge

Finally, we can directly compare the algebraic structures of Old Minimalism and New Minimalism at the level of the action of Merge, comparing the algebraic properties of the action of free symmetric Merge on workspaces of New Minimalism with the very different properties of the partial actions of Internal Merge and External Merge in Old Minimalism.

2.3.4.1 Action of Merge in New Minimalism With the same notation and terminology used in the previous chapter, we have seen that in New Minimalism, the Hopf algebra structure on workspaces can be seen as underlying the action of Merge. Indeed, the fundamental property of the coproduct in a Hopf algebra provides the list of all the possible ways of decomposing an object (the term of the algebra that the coproduct is applied to) into a pair consisting of a subobject and the associated quotient object. In other words, in the case of a tree T, one can write a coproduct in the form

$$\Delta(T) = \sum_{\underline{v}} F_{\underline{v}} \otimes T/F_{\underline{v}}, \tag{2.3.5}$$

where we include the special cases $T \otimes 1$ and $1 \otimes T$, and where, for $\underline{v} = \{v_i\}_{i=1}^k$, we write $F_{\underline{v}}$ for the forest consisting of a union of disjoint subtrees $T_{v_i} \subset T$. This can be equivalently described as a forest obtained from an *admissible cut* of the tree T. It represents a way of compiling a list of all the accessible terms of T with the corresponding cancellation of the deeper copy.

As we discussed in chapter 1, Internal Merge and External Merge use the terms of the coproduct where the subforest $F_{\underline{v}}$ in the left-channel has at most two components; see sections 1.3.3, 1.4.2, 1.4.3, and 1.9. In these terms, the action of Merge can be described as taking, for each of the two arguments of the binary operation $\mathfrak{M}_{S,S'}$ for a pair $S, S' \in \mathfrak{T}_{SO_0}$ of syntactic objects, the coproduct $\Delta(F)$ over the entire workspace, which extracts all the accessible terms, searching among them for matching copies of S and S', merging them if found, and keeping the other terms of the coproduct that perform the cancellation of the deeper copies (see section 1.8). As we saw in section 1.16, the remaining terms of the coproduct play a role in the FormSet operation.

The $\mathbb{Q}$-vector space $\mathcal{V}(\mathfrak{F}_{SO_0})$ spanned by the workspaces, with the product given by the disjoint union $\sqcup$ and the coproduct (2.3.5) extended from trees to forests by $\Delta(F) = \sqcup_a \Delta(T_a)$ for $F = \sqcup_a T_a$, has the structure of an associative, commutative, coassociative, and noncocommutative bialgebra $(\mathcal{V}(\mathfrak{F}_{SO_0}), \sqcup, \Delta)$. The vector space $\mathcal{V}(\mathfrak{F}_{SO_0}) = \oplus_\ell \mathcal{V}(\mathfrak{F}_{SO_0})_\ell$ is graded by the number of leaves, compatibly with the operations. Thus, an antipode compatible with the product and coproduct, making $\mathcal{V}(\mathfrak{F}_{SO_0})$ a Hopf algebra, can be constructed inductively by degrees (see the general discussion of the properties of Hopf algebras in section 4.2 in chapter 4). A more nuanced discussion of the different forms of the coproduct, Δ^c, Δ^ρ, Δ^d, and the effect on the coassociativity and the antipode properties were discussed in section 1.2.1 of chapter 1.

The action of Merge on workspaces recalled here is then formulated explicitly in terms of the product and coproduct structure of $(\mathcal{V}(\mathfrak{F}_{SO_0}), \sqcup, \Delta)$, as

$$\mathfrak{M}_{S,S'} = \sqcup \circ (\mathcal{B} \otimes \mathrm{id}) \circ \delta_{S,S'} \circ \Delta\,, \tag{2.3.6}$$

where the coproduct Δ produces the list of accessible terms and corresponding cancellations, and the operator $\delta_{S,S'}$ identifies the matching copies. These are then fed into Merge by the operation $\mathcal{B} \otimes \mathrm{id}$, which at the same time produces the new merged tree through the grafting operation $\mathcal{B}$ of (2.3.4) and performs the cancellation of the deeper copies by keeping in the new workspace the corresponding quotient terms produced by the coproduct. The resulting new workspace is then produced by taking all these components and the new component created by Merge together through the product operation $\sqcup$.

For the comparison with Old Minimalism, it is important to stress here that the entire action of Merge (including both Internal Merge and External Merge) arises in New Minimalism from the *same* mechanism associated with the coproduct and product operations of the Hopf algebra; namely, the operation (2.3.6). This is unlike what happens with Old Minimalism, where Internal Merge and External Merge are independently defined and are not directly constructed in terms of the coproduct and product of the Loday–Ronco Hopf algebra. Indeed the relation to the Hopf algebra structure is described in a more indirect way: through the case of an elementary cut and the relation between the Loday–Ronco and the Connes–Kreimer coproducts, and via the behavior of the domains of Merge with respect to the Hopf algebra operations. In New Minimalism, by contrast, Merge is globally defined so the domain is the full Hopf algebra (i.e., arbitrary workspaces).

The form (2.3.6) of the action of Merge includes other possible forms of Merge, in addition to External and Internal Merge: the forms of Sideward discussed in chapter 1. This again differs from Old Minimalism, where only External Merge and Internal Merge are defined. In New Minimalism, the additional forms of Sideward Merge can be eliminated, as subdominant terms, through a mechanism of Minimal Search that extracts the terms corresponding to External and Internal Merge as the least effort contributions, and also through Minimal Yield optimality constraints. While the presence of Sideward Merge in (2.3.6) may appear as a complication in New Minimalism, the presence of these suboptimal terms as part of the mathematical structure is useful, as discussed in section 1.9 (see also the forthcoming work [142]).

We have seen in the previous chapter that the usual form of Minimal Search can be implemented in this Hopf algebra setting by selecting the leading-order part of the coproduct with respect to a grading that weights the accessible terms

$T_v \subset T$ by the distance of the vertex v from the root vertex of T. Thus, searching among terms deeper into the tree becomes less efficient, with respect to this cost function, than searching near the top of the tree. At the same time the grading weights the cancellation of a deeper term as less costly than the cancellation of a term near the root. Taking the leading-order term with respect to this grading has exactly the same effect as implementing Minimal Search as usually described in the linguistics literature; see [28], [29], and [33]. More precisely, one has a weighted form of the Merge operation where this cost function is taken into account:

$$\mathfrak{M}^{\epsilon}_{S,S'} = \sqcup \circ (\mathcal{B}^{\epsilon} \otimes \mathrm{id}) \circ \delta_{S,S'} \circ \Delta\,;$$

and taking the leading term (namely, the only nonzero term in the limit $\epsilon \to 0$) of arbitrary compositions of operators $\mathfrak{M}^{\epsilon}_{S,S'}$, one finds only Internal Merge and External Merge, while all the remaining forms of Merge are of a lower order and disappear in the $\epsilon \to 0$ limit. In chapter 3, we will see a way of reinterpreting Minimal Yield using the formalism of Rota–Baxter algebras and Birkhoff factorization.

2.3.4.2 Action of Merge in Old Minimalism The action of Merge in Old Minimalism again differs in several significant ways from the newer version:

1. Internal Merge and External Merge are independently defined operators instead of being cases of the same operator as in New Minimalism.
2. No other forms of Merge occur (unlike the Sideward Merge).
3. Merge is partially defined due to feature-checking.
4. Internal Merge and External Merge lead to different algebraic structures.
5. Internal Merge can be described in terms of admissible cuts: the relation of these to the coproduct of the Loday–Ronco Hopf algebra is more involved than the direct relation of free symmetric Merge to the coproduct of the Hopf algebra of workspaces.

The first difference is a consequence of the fact that the combinatorial objects considered are only trees, not forests. This implies that External Merge requires the additional algebraic structure of $\wedge_{\Omega}$-algebra discussed in §2.2.16. Even when the two operations of Internal Merge and External Merge are compared in the way that most closely highlights the similarities, as explained in sections 2.2.15 and 2.2.16, these two operations cannot be reduced to special cases of a single algebraic structure. This is very different from what happens in New Minimalism, where the expression (2.3.6) covers all the possible cases of Merge in a single operation.

The second difference is that the lack of a unified Merge operation also implies that Old Minimalism does not have Sideward Merge, while in New Minimalism it is part of the algebraic structure, but is suppressed as subdominant and violating optimality, through the mechanism of Minimal Search and the Minimal Yield constraints.

The third difference, that Merge is partially defined in Old Minimalism, creates the compounding problem of domain checking in derivation chains. It is also responsible at the level of algebraic structures for the presence of an additional structure beyond Hopf algebras, in the form of associated right-ideal coideals. Finally, the description of Internal Merge in terms of admissible cuts in Old Minimalism is very similar to the description in terms of extraction of accessible terms using the Hopf algebra coproduct in New Minimalism. However, in Old Minimalism, the admissible cuts that occur in Internal Merge and its iterations do not come directly from the terms of the coproduct of the Loday–Ronco Hopf algebra. There is a known relation between the admissible cuts and the coproduct of the Loday–Ronco Hopf algebra, which is not immediately evident in the form described in sections 2.2.3 and 2.2.4. This is shown in Theorem 8.4 of [1], which demonstrates that there is a choice of basis for the Loday–Ronco Hopf algebra with respect to which the coproduct is expressible in terms of a collection of a version of the notion of admissible cuts, adapted to the decomposition of the tree into its left/right subtrees, $T = T_\ell \wedge^d T_r$ (see section 2.2.5). These, however, are not quite the same collection of admissible cuts that one needs in the iterations of Internal Merge. Moreover, a difference with respect to the Hopf algebra of workspaces is that, even when the Loday–Ronco coproduct is expressed in terms of admissible cuts, the forest part $\pi_C(T)$ of the admissible cut needs to be assembled into a tree, as only trees occur in the Loday–Ronco Hopf algebra and not forests, while one does need to use the forest $\pi_C(T)$ to describe the iterated Internal Merge.

2.3.4.3 Labeling in Old Minimalism and New Minimalism We discussed in sections 1.13 and 1.15 the notions of head function, planarization, and labeling algorithm in the setting of New Minimalism based on free symmetric Merge. In comparison, descriptions of Merge at the level of planar trees, as in Old Minimalism, involve a partially defined operations with restrictions on domains based on some label assignments. We discussed this explicitly earlier in this chapter, in the case of Stabler's formulation, but analogous conditions exist in other formulations of Old Minimalism. In view of the considerations given here, this can be read as evidence of the fact that one is trying to describe at the level of planar trees a Merge operation that is really taking place at the underlying level of nonplanar abstract trees, correcting for the incompatibility by restricting the domain of applicability. In particular, this means that the significant complications in the underlying algebraic structure of External Merge and Internal Merge that we have observed in the case of Stabler's formulation similarly apply to any other formulation that locates Merge *after* the planar embedding of trees.

This issue does not arise within the formulation of New Minimalism, since the planarization that happens in Externalization is simply a noncanonical (meaning dependent on syntactic parameters) choice of a section σ of the projection $\mathcal{SO}^{nc} \to \mathcal{SO}$. The only requirement for σ is to be compatible with syntactic parameters, not to be a morphism with respect to the magma operations $\mathfrak{M}$ and $\mathfrak{M}^{nc}$, since in this model, Merge only acts before Externalization as $\mathfrak{M}$ (not after Externalization as $\mathfrak{M}^{nc}$: the operation $\mathfrak{M}^{nc}$ is in fact never involved). In the previous chapter, we described Externalization in the form of a correspondence. This represents a two-step procedure that first chooses a noncanonical section σ and then quotients the image by eliminating those planar trees obtained in this way that are not compatible with further (language-specific) syntactic constraints. Merge is not applied anywhere in this Externalization process, which takes place *after* the results of Merge have been computed at the level of trees without planar structure.

The same issues we discussed in section 1.13.2 regarding planarization occur in Stabler's formalism [192] that we have been discussing in the previous sections of this chapter: it is easy to see the same issues arise in terms of the labels $>$ and $<$ assigned to the internal vertices of planar trees. The difference is that in [192], one starts with a tree that already has a planar assignment and uses heads, maximal projection, and c-command (see Definition 1.13.2), to obtain a new planar projection. In that case, as already discussed, the labels $>$ and $<$ are assigned to the root and the non-leaf vertices of a planar tree by pointing, at each vertex v, in the direction of the branch where the head of the tree T_v resides. If this assignment of labels is well defined, then a new planar structure, the Linear Correspondence Axiom (LCA) discussed in section 1.13.2, can be obtained simply by flipping subtrees about their root vertex every time the vertex is labeled $<$ until all the resulting vertices become labeled by $>$ (harmonic head-final LCA), or viceversa (harmonic head-initial LCA).

The implicit assumption that makes this possible is that every subtree T_v of a given tree T has a head (i.e., a marked leaf). This assumption regarding heads is necessary both for the labeling by $>$ and $<$ in Stabler's formalism and for the use of maximal projections for the definition of ordering in the LCA.

In the case of the labels $>$ and $<$, the problem of what label is assigned to the new root vertex when External Merge or Internal Merge is performed demands restrictions on the domain of applicability of these Merge operations, depending on the conditions on the labels at the heads of the trees used in the Merge operation (which makes them partially defined operations). Further restrictions of a similar nature arise from other labels attached to vertices in Old Minimalism, involving the licensor/licensee and selector features discussed in section 2.2.6. This also requires further steps, such as allowing Merge results to be unlabeled during derivation and labeled at the end so that successive cyclic raising can remove the elements responsible for un-labelability. (This refers to the same linguistic problem with the formalism that we discussed at the end of section 2.2.9, which was mentioned to us by Riny Huijbregts.)

In the case of LCA, the partial corrections to this problem (see [98], pp. 230–231) that we mentioned in section 1.13.2 of chapter 1 are similar in nature to the problem of un-labelable structures in Stabler's formalism mentioned here.

2.4 Conclusions

We have seen from this comparative analysis of the algebraic structures underlying one of the older versions of Minimalism (Stabler's Computational Minimalism) and Chomsky's newer version (New Minimalism) that the new version has a simpler mathematical structure with a unifying description of Internal Merge and External Merge, and with a core generative process that reflects the most fundamental magma of binary set formation, generating the binary rooted trees without any assigned planar structure.

The more complicated mathematical structure of Stabler's Minimalism is caused by several factors. The intrinsic asymmetry of Internal Merge is due to working with planar binary rooted trees. Abandoning the idea that planar embeddings should be part of the core computational structure of Minimalism is justified linguistically by the relevance of structures (abstract binary rooted trees) rather than strings (linearly ordered sets of leaves, or equivalently planar embeddings of trees) in syntactic parsing; see [62]. Thus, working with abstract trees without planar embeddings and with forests (workspaces) is one of the simplifying factors of New Minimalism. This advantage in the description of the fundamental generative process is accompanied by the difficult problem of developing an appropriate formalization of Externalization, discussed in section 1.12 of chapter 1, which will be addressed with a more detailed treatment in the forthcoming work [141].

The other main issue that complicates the mathematical structure of the older versions of Minimalism is the very different nature of the Internal Merge and External Merge operations: in the case of Stabler's Minimalism analyzed here, these two forms of Merge relate to two very different algebraic objects (operated algebras and right-ideal coideals), hence they cannot be reconciled as coming from the same operation, while in New Minimalism both Internal Merge and External Merge are cases of the same operation, and both arise as the leading terms with respect to the appropriate formulation of Minimal Search.

Finally, another main issue that makes the mathematical structure of older versions of Minimalism significantly more complicated is the presence of conditions on labels that need to be matched for Internal Merge and External Merge to be applicable, related to the problem of projections discussed in [26]. Mathematically, this makes all operations only partially defined on particular domains where conditions on labels are met. As we have discussed, this creates problems with having to work with partially defined versions of various algebraic structures and significantly complicates iterations of the Merge operations, where the conditions on domains compound. Since the conditions on labels are absent from the fundamental structure of New Minimalism, this problem of dealing with partially defined operations also disappears, leading to another simplification at the level of the mathematical structures involved. In New form of Minimalism, the labeling algorithm, which we discussed in section 1.15, does not run into this problem, because labeling takes place *after* the process of structure formation via free symmetric Merge, not as part of the Merge operation. In New Minimalism, all nonviable structures (in terms of heads, labels, theta roles, or syntactic parameters in Externalization) are *filtered out* and eliminated after the free structure formation produced by Merge.

Within New Minimalism, the planar structure of trees is introduced as the later step of Externalization, not at the level of the Merge action. The choice of planar structure in Externalization is done through a noncanonical (that is, dependent on syntactic parameters) section of the projection from planar to abstract trees. We showed in section 1.13 that the proposed construction of a unique canonical choice of planar embeddings, based on heads of trees, maximal projections, and c-command, can be only partially defined on the domain $\mathrm{Dom}(h) \subset \mathcal{SO}$. The difference with Old Minimalism is that this domain restriction does not interfere with the generative process of structure formation.

In all of these aspects, we see that the main change of perspective in New Minimalism is the *decoupling* of structure formation from filters that happens either at in Externalization or at the syntax–semantics interface.

2.5 Comparison with Other Models

In this chapter we have mostly discussed the comparison, at the level of algebraic properties, between New Minimalism, based on free symmetric Merge, and Old Minimalism, focusing on the formulation of Stabler's Computational Minimalism.

We finish this chapter by discussing briefly the comparison with other models of syntax, which may appear at first to have a very similar structure as the free symmetric Merge, but which in fact turn out to be significantly different in various respects.

2.5.1 Tree-Adjoining Grammars

Before we proceed to discuss further aspects of our model, we want to make a comment, regarding the formalism presented in chapter 1, that disambiguates between our setting based on Merge and other settings in generative linguistics. In particular, we add here a very brief clarification of the difference between the algebraic structure of Merge described in chapter 1 and that of *tree-adjoining grammars* (TAGs). In the setting of TAGs, one considers a generative process that depends on an initial choice of a given finite set of *elementary trees* with vertex labels. In TAGs, trees generally are not necessarily assumed to be binary. There are two composition rules: one composition operation (substitution rule) consists of grafting the root of a tree to the leaf of another tree; and a second composition operation (a so-called adjoining rule) inserts at an internal vertex of a tree with a label x another tree in which both the root and one of the leaves carry the same label x. The adjoining rule can be obtained as a suitable composition of grafting of roots to leaves, so the basic generative operations of TAGs are the *operad compositions* of rooted trees. Namely, if $O(n)$ denotes the set of trees in a given TAG with n leaves, then there are composition maps

$$\circ_i : O(n) \times O(m) \to O(n + m - 1) \tag{2.5.1}$$

that plug the root of a tree in $O(n)$ to the ith leaf of a tree in $O(m)$, resulting in a tree in $O(n + m - 1)$. Such operations, subject to associativity and unitarity conditions, define the algebraic structure of an *operad* (which we will return to in section 3.8.1 in chapter 3). Label matching conditions would require the notion of colored operads, but we will not discuss this here (see section 3.8.1 for a similar label-matching condition that arises in our setting).

To compare the TAG formalism with the algebraic formulation of Merge from chapter 1, one should note that there is an important relation between the two, as well as important differences: the latter show that these two formalisms do *not* constitute the same algebraic structure. This is why, in our view, algebraic structure is essential to the line of work presented here, rather than the formal language theoretic notion of weak generative capacity (such as mild context sensitivity). This is not to deny that notions of generative capacity might be useful to illuminate one or another aspect of human language; it is simply that the algebraic approach presented here does not draw on this more familiar formal language theory tradition, and can detect structures that are not visible within that framework.

The relationship between TAGs and Merge arises from the fact that, in the Merge formalism of chapter 1, the syntactic objects $T \in \mathcal{SO} = \mathfrak{T}_{\mathcal{SO}_0}$ are generated as elements of the free, nonassociative, commutative magma (3.1.1) on the Merge operation $\mathfrak{M}$. This *does* have an equivalent operad formulation, in terms of the quadratic operad freely generated by the single commutative binary operation $\mathfrak{M}$, see [96]. Thus, there is an equivalent way of formulating the generative process for the syntactic objects in terms of operad compositions (2.5.1), which makes this generative process appear similar to TAGs. Moreover, as we will discuss in section 3.8.1 of chapter 3, the formulation of Theta Theory in our setting does involve operads and colored operads. So there is some part of the algebraic structure that is common to both models. However, the main difference between the two lies in the fact that the Merge formalism does not *just* consist of the generation of syntactic objects through the magma operation, but also of the action of Merge on workspaces, given by the forests $F \in \mathfrak{T}_{\mathcal{SO}_0}$.

The action of Merge on workspaces is not only determined by the operad underlying the $\mathcal{SO}$ magma, but also requires the additional datum of the Hopf algebra structure on workspaces. This makes it possible to incorporate not only External Merge, which is involved in the magma $\mathcal{SO}$, but also Internal Merge, which requires an additional coproduct operation. The Hopf algebra coproduct has no direct analog in TAGs.

It is important to observe here that the operad underlying $\mathcal{SO}$ also determines a Hopf algebra, but the associative, commutative Hopf algebra on the vector space $\mathcal{V}(\mathfrak{F}_{\mathcal{SO}_0})$ spanned by forests of binary rooted trees, used in chapter 1 to formulate the action of Merge on workspaces, is *not* the same as the *nonassociative*, commutative Hopf algebra structure induced by the operad on the vector space $\mathcal{V}(\mathfrak{T}_{\mathcal{SO}_0})$ spanned by binary rooted trees; see [95] and [96]. This is a key algebraic difference between TAGs and Merge. The introduction of workspaces and the action of Merge on workspaces thus amounts to a key innovation in the modern Minimalist account.

Another main difference lies in the fact that the other main operation of TAGs, insertion at inner vertices of the tree, does not fit entirely with the operad formalism (if the insertion happens at an internal vertex and is not realized by repeated insertions at leaves) or with the Hopf algebra. We discussed in section 1.7 of chapter 1 how certain insertion operations at internal vertices of trees (as used, for instance, in Late Merge models) can be expressed in terms of a Lie algebra structure related to the Hopf algebra by duality (hence making these extensions of Merge expressible in terms of only the original Internal Merge and External Merge). Thus, an equivalent reformulation of TAGs in terms of Lie algebras appears possible (though we will not discuss it here). However, the insertion operation of TAGs does not fit the *same* Lie algebra structure that we have seen in section 1.7, so it is not an equivalent model.

2.5.2 Tensor Models

Smolensky proposed in [188] and [189] an *integrated connectionist symbolic architecture* for complex combinations of serial and parallel processing, aimed at modeling cognition based on a formalism of tensor products of vector spaces. In chapter 3, we will return in section 3.2.4 to discuss criticism of this approach and to explain why our model of syntax–semantics interface is *not* a tensor model in the sense of Smolensky. Tensor models of the kind described in [188] and [189] have been used in computational linguistics for parsing, both in the setting of context-free grammars and of TAGs, and in the case of Stabler's Computational Minimalism and MGs in [81] and [83], with an attempt to unify symbolic and connectionist approaches in language processing. The main idea is that, given a planar binary rooted trees with labeling by features as in Stabler's Minimalism, one encodes the set of features into a set of corresponding vectors and encodes the planar binary rooted tree structure using so-called "position roles" r_1, r_2, r_3 for the mother vertex and left/right daughter vertices and also encodes them as independent vectors V_{r_i}. Thus, one can encode a tree

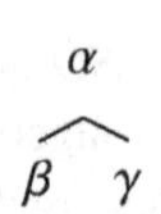

with a vector $V_\alpha \otimes V_{r_1} + V_\beta \otimes V_{r_2} + V_\gamma \otimes V_{r_3}$.

This type of tensor product relations $V_\alpha \otimes V_r$ is referred to as a "filler–role binding relation", where fillers stand for the features α. Here, we have illustrated position roles, which keep track of the tree structure, but more generally, roles are considered in this approach to represent the filler positions in the larger structures and can be represented by theta-roles, the argument of a predicate. We will discuss in section 3.2.4 that, when used in this more general sense, the tensor product pairing $V_\alpha \otimes V_r$ of fillers and roles would lead to assignments of tensor products of vectors when External Merge operations are performed that involve theta-role matching (see also our discussion of theta-roles in section 3.8.1), and this has problems with semantic interpretation, as pointed out in [147]. In [83], this approach is realized in terms of vector representations in Fock spaces, and it is shown that External Merge and Internal Merge (formulated in the older terminology of "merge" and "move") act as transformations on the Fock space.

As observed in [157], a fundamental construction in theoretical physics, the Fermionic Fock space, can be related to rooted trees via a procedure that labels in a one-to-one correspondence states in the Fermionic Fock space with rooted trees (which are not necessarily binary). This correspondence is based on two intermediate steps: the Matula number of a rooted tree [150], which gives a unique labeling of rooted trees by integers (using prime number decomposition), and an arithmetic construction of both Bosonic and Fermionic Fock spaces, which is also based on prime numbers. Combining these two parameterizations, one can match rooted trees to states in the Fermionic Fock space. This correspondence with Fock spaces makes it possible to express operations such as the coproduct of the Connes–Kreimer Hopf algebra of rooted trees and solutions to combinatorial DS equations in terms of states in the Fermionic Fock space (see [157]). Thus, it is possible that the formalism we are describing here, using the Hopf algebra of rooted trees for the action of Merge on workspaces, may admit a compatible encoding via Fock spaces of the kind used for Old Minimalism in [81] and [83]. Such representations should not assign a tensor product of vectors $V_{T_1} \otimes V_{T_2}$ to a Merge $\mathfrak{M}(T_1, T_2)$, due to the difficulties outlined in section 3.8.1; rather, they should rely on an appropriate encoding of binary rooted trees (with the magma operation $\mathfrak{M}$) in states of a Fock space (Bosonic or Fermionic). We will not discuss this topic further in this book.

2.5.3 Physics Methods in Minimalism

The algebraic model of Merge and Minimalism that we present in this book is directly inspired by the mathematical formalism of Hopf algebras used in theoretical physics to describe the generative process of Feynman diagrams in quantum field theory and the renormalization of the perturbative Feynman integrals.

There have been other approaches that considered adopting methods from theoretical physics to model aspects of generative linguistics, especially Merge and Minimalism. For example, in the approach of [164], Phase Theory and the "phase impenetrability conditions", formulated in terms of trees in X-bar theory, is considered as a dynamical process governed by certain creation and annihilation operators akin to those used in the context of quantum physics. This viewpoint is consistent with the Fock space approach outlined in section 2.5.2, where such operators are a natural part of the structure. While the approach of [164] is very different from the one that we follow here, they do make the observation that *copies* in generative syntax should be governed by a Hopf algebra structure, consistently with our formalism.

In [74], Merge is regarded as a "process of coarse graining" akin to a renormalization process in physics. The point of view illustrated there is similar to what physicists refer to as the "MERA renormalization in tensor networks", with the binary tree structures of syntactic objects regarded as a tensor network. The impenetrability of phases is then interpreted as the irreversibility of the renormalization group flow. A similar approach in terms of tensor networks renormalization was developed in [167].

Again, the approach described in [74] and [167] is different from the one that we follow in this book. In particular, while we model the syntax–semantics interface on a form of renormalization used in quantum field theory, we do not model this in a coarse graining process: in our model, no syntactic information is lost or averaged out, although phase impenetrability does hold in the form Δ_Φ of the Hopf algebra coproduct described in section 1.14. Thus, in particular, we expect that it should be possible to give a formulation similar to MERA renormalization for the phase algorithm and the coproduct Δ_Φ in section 1.14 of chapter 1, though we will not discuss it in this book.

3 The Syntax–Semantics Interface: An Algebraic Model

3.1 Modeling the Syntax–Semantics Interface

This chapter is based on the results of [140].[7]

As we have argued in the previous chapters, the modeling of generative syntax, based on the core computational structure of Merge, within the setting of the Minimalist Model, satisfies the following fundamental properties:

1. A concise conceptual framework
2. A precise mathematical formulation
3. A good explanatory power

For the most recent formulation of Minimalism, the first and third properties are articulated in [29], [30], and [33], as well as in *Elements* [40]. While requiring the existence of precise mathematical models has been traditionally associated with sciences like physics, since syntax is essentially a computational process, this suggests that a mathematical formulation might also be a desirable requirement in linguistics, and more specifically in the modeling of I-language.

In comparison with syntax, modeling semantics is presently in a less satisfactory state from the point of view of the same three properties listed here. Some main approaches to semantics include forms of compositional semantics (such as [168] and [169]), truth-conditional semantics, semiring semantics (such as [82]), and vector-space models (the latter especially in computational linguistics). General views of logic-oriented approaches to semantics, which we will not discuss here, can be seen, for instance, in [183], [195], and [204].

[7] Regarding the planning, writing, and authorship of this chapter, see the "Editor's Note" at the beginning of this book. Due to medical circumstances, Noam Chomsky was unable to review and approve the final version of this manuscript before it went to print.

We will discuss the version of [93]. In our view, each of these viewpoints has limitations of different nature.

Our purpose here is not to carry out a comprehensive comparative analysis and criticism of contemporary models of semantics. Rather, we want to approach the problem of modeling the syntax–semantics interface on the basis of a list of abstract properties and articulate a possible mathematical setting that such properties suggest. We can then compare existing models with the specific structure that we identify. We will show that one can remain, to some extent, agnostic about specific models of semantics, beyond some basic requirements, and still retain a fundamental functioning model of the interaction with syntax. This reflects a view of the syntax–semantics interface that is primarily syntax-driven.

As in the case of our mathematical formulation of Merge in terms of Hopf algebras, our guiding principle will be an analogy with conceptually similar structures that arise in theoretical physics. In particular, in the context of fundamental physical interactions described by quantum field theory, a fundamental problem is the assignment of "meaningful" physical values to the computation of the expectation values of the theory. This can be compared with the assignment of meaning—semantics—to syntactic objects. More precisely, assignment of meaning in the quantum field theory setting consists of the extraction of a finite (hence meaningful) part from Feynman integrals that are generally divergent (i.e., they produce meaningless infinities). This process is known in theoretical physics as *renormalization*. In the algebraic form that we consider here, it is known as "Connes–Kreimer renormalization" or as "Hopf algebra renormalization". While the renormalization problem and procedures leading to satisfactory solutions for it have been known to physicists since the development of quantum electrodynamics in the 1950s and 1960s (see [14]), a complete understanding of the underlying mathematical structure is much more recent (see [45] and [46]).

Even more recently, Yuri I. Manin showed that the same mathematical formalism can be applied in the theory of computation, in order to extract, in a similar way, computable "subfunctions" from undecidable problems, with undecidability being the analog in the theory of computation of the unphysical infinities; see [129], [130], and [50] and [112].

Assuming the conceptual standpoint that internal language (or I-language) is, in essence, a computational process, the extension of the mathematical framework of renormalization to the theory of computation suggests the existence of a similar possible manifestation in linguistics as well. In the case of linguistics, one does not have to deal with divergences of a physical or computational nature; rather, one has to carry out a consistent assignment of meaning to syntactic objects produced by the Merge mechanism, and reject inconsistencies and impossibilities. In the rest of this chapter, we plan to turn this heuristic comparison into a precise formulation.

There are several reasons why developing such a mathematical model of the syntax–semantics interface is desirable. Aside from general principles based on the three good properties of theoretical modeling stated at the beginning of this section, there are other possible applications of interest. For instance, a significant ongoing debate and controversy have ensued from the recent development of statistically based large language models (LLMs), with various claims of incompatibility with the generative linguistics framework itself. Since such theories, in our view, ultimately describe computational processes (albeit most likely also in our view of a different nature from those governing language in human brains), a viable computational and mathematical setting is helpful, so a specific comparative analysis can be carried out and such claims can be addressed.

3.1.1 Some Conceptual Requirements for a Syntax–Semantics Interface

We beging our analysis by setting out a simple list of what we regard as desirable properties of a model of the syntax-semantic interface:

1. Autonomy of syntax
2. Syntax supporting semantic interpretation
3. Semantic interpretation, to a large extent, independent of Externalization
4. Compositionality

The first requirement, the autonomy of syntax, expresses that the computational generative process of syntax described by Merge is independent of semantics. The second requirement, that syntax supports semantics, can be seen as positing that the syntax-semantic interface proceeds *from* syntax *to* semantics (a syntax-first view), while syntax itself is not semantic in nature. The third claim separates the interaction of the core computational mechanism of syntax and the Conceptual–Intensional (CI) system, which gives rise to the syntax-semantic interface, from the interaction with an Articulatory–Perceptual or Sensory–Motor (SM) system, which includes the process of Externalization. While it is reasonable to assume a certain level of interaction between these two interface channels, with "independence of Externalization" we emphasize that semantic interpretation depends primarily on *structural relations* and proximity in the syntactic structure rather than on linear proximity of words in a sentence. The compositional property is meant here simply as a requirement of consistency across syntactic substructures.

We also add another general principle that we will try to incorporate in our model, which may be at odds with some of the traditional approaches to semantics (such as the truth value-based approaches). We propose the following fundamental distinction between the roles of syntax and semantics in language (which we realize not everyone will agree with):

- Syntax is a computational process.
- Semantics is *not* a computational process and is in essence grounded in a notion of topological proximity.

The first statement is clear in the context of generative linguistics, and in particular in the setting of Minimalism, where the computational process is run by the fundamental Merge operation. The second assertion requires some contextual clarification. Saying that semantics is endowed only with a notion of proximity of a topological nature does *not* mean that it is not possible, or desirable, to consider models of semantics where additional structure is present, but rather that these other properties (metric, linear, and semiring structures, for instance, or forms of logic) mostly play a role to instantiate or quantify proximity relations, and the topological notion of proximity is the minimal basic requirement that is needed for a viable interface model. The compositionality of semantics does not require positing an additional computational structure on semantics itself: the computational structure of syntax suffices to induce it. In this view, semantics is not really a part of language itself, but rather an autonomous structure of "conceptual spaces" that mostly deals with proximity classifications.

3.1.2 Syntax

On the syntax side of the syntax-semantic interface, we assume the formulation of free symmetric Merge presented in chapter 1. This accounts for the properties (1) and (3) in our list from section 3.1.1: it provides a computational model of syntax that is independent of semantics, and where the interface with semantics takes place at the level of free symmetric Merge without requiring prior Externalization. Free symmetric Merge generates syntactic objects, described by binary rooted trees without any assigned planar embedding. Thus, our choice of modeling the syntax-semantic interface starting from the level of free symmetric Merge, as the syntactic part of the interface, has the effect of ensuring that the interface of syntax and semantics (also sometimes called the "Conceptual-Intensional system") is parallel and separate from the channel connecting the output of Merge to Externalization (the so-called "Sensory–Motor system"), although interactions between these two channels can be incorporated in the model (and will be discussed in section 3.4). It is useful to keep in mind that there are important differences in how one should model the syntax–semantics interface and its relation to Externalization, according to whether one focuses on the production of language or on the parsing of language at the receiving end. In the first case, the assumption that the products of Merge are checked for semantic parsing prior to Externalization is reasonable within this model, while in the second case one perceives language that has already been through Externalization and needs to parse it for semantic interpretation, so that some interaction between these two interfaces is required. We will discuss this in section 3.4.

Summarizing briefly the setting given in chapter 1, syntax is represented by the following:

- A (finite) set $\mathcal{SO}_0$ of *lexical items and syntactic features*.
- The set of *syntactic objects* $\mathcal{SO}$, identified with the set $\mathfrak{T}_{\mathcal{SO}_0}$ of finite binary rooted trees (with no planar structure), with leaves labeled by $\mathcal{SO}_0$, generated as the free, nonassociative, commutative magma over the set $\mathcal{SO}_0$,

$$\mathcal{SO} = \text{Magma}_{na,c}(\mathcal{SO}_0, \mathfrak{M}) = \mathfrak{T}_{\mathcal{SO}_0} . \tag{3.1.1}$$

- The set of *accessible terms* of a syntactic object $T \in \mathfrak{T}_{\mathcal{SO}_0}$, given by the set of all the subtrees $T_v \subset T$ with root a non-root vertex $v \in V(T)$.
- The commutative Hopf algebra of workspaces obtained by considering the vector space $\mathcal{V}(\mathfrak{F}_{\mathcal{SO}_0})$ spanned by the set $\mathfrak{F}_{\mathcal{SO}_0}$ of finite binary rooted forests with leaves decorated by elements of the set $\mathcal{SO}_0$, with the product given by the disjoint union $\sqcup$ and the coproduct determined by

$$\Delta(T) = T \otimes 1 + 1 \otimes T + \sum_{\underline{v}} F_{\underline{v}} \otimes T/F_{\underline{v}} , \tag{3.1.2}$$

  with $F_{\underline{v}} = T_{v_q} \sqcup \cdots \sqcup T_{v_n}$ being a collection of accessible terms.
- The action of Merge on workspaces

$$\mathfrak{M} = \sqcup \circ (\mathcal{B} \otimes \text{id}) \circ \Delta ,$$

  where $\mathcal{B}$ is the grafting of components of a forest to a common root vertex, or, for a fixed pair of syntactic objects S, S',

$$\mathfrak{M}_{S,S'} = \sqcup \circ (\mathcal{B} \otimes \text{id}) \circ \delta_{S,S'} \circ \Delta , \tag{3.1.3}$$

  where $\delta_{S,S'}$ selects matching pairs in the workspace (see chapter 1 for a more detailed description).

We will use the notation $\mathcal{H} = (\mathcal{V}(\mathfrak{F}_{SO_0}), \sqcup, \Delta, S)$ for this Hopf algebra, as in chapter 1. As discussed in chapter 1, the antipode S is defined inductively using the coproduct, so it suffices to specify the bialgebra part of the structure, $\mathcal{H} = (\mathcal{V}(\mathfrak{F}_{SO_0}), \sqcup, \Delta)$. When needed, we will consider the different forms of the coproduct discussed in chapter 1, and the corresponding variations of this basic algebraic structure. We comment on this briefly in section 3.1.2.1.

3.1.2.1 Remark on the Hopf Algebra coproduct We pointed out in Chapter 1 that there are different possible ways of interpreting the quotient T/T_v (or, more generally, $T/F_{\underline{v}}$) in the coproduct (1.2.8), which we wrote as $T/^cT_v$, performing a contraction of T_v to its root vertex (which becomes a new leaf labelled as $\not{T_v}$), or $T/^dT_v$, which deletes T_v (taking the unique maximal binary tree determined by the complement), as one sees in Externalization. The intermediate construction $T/^\rho T_v$ maintains an unlabeled nonbranching vertex marking the cancellation of the deeper copy T_v. This provides so-called *traces*, the empty categories left behind by movement implemented by Internal Merge. As is familiar from the long historical discussion of what is called "reconstruction", such traces are needed for semantic parsing.

In chapter 1, we argued that the two different forms $T/^cT_v$ and $T/^dT_v$ of quotient account for the fact that the deeper copy of the accessible term T_v is interpreted at the syntax–semantics interface but not expressed in Externalization.

Thus, when it comes to interfacing syntax with semantics, it is generally better to retain the root vertex v of T_v in the quotient $T/^cT_v$ with the $\not{T_v}$ label. Similarly, in quotients $T/F_{\underline{v}}$, each tree component T_{v_i} of the forest $F_{\underline{v}}$ is contracted to its root vertex v_i with the label $\not{T_{v_i}}$. (This label $\not{T_{v_i}}$ should be understood as retaining the features of T_{v_i} that suffice for semantic parsing.)

However, there will be cases, in our discussion of the syntax–semantics interface, where it will still be useful to consider also the quotient of the form $T/^dT_v$ with the deletion of T_v. This will be the case when we want to separately compare parsing on a substructure T_v and on a complementary structure where T_v is absent. We will encounter such cases, for instance, in sections 3.2.5 and 3.9.

To formulate more precisely property (2) of our list from section 3.1.1, we will rely on the notion of an *abstract head function* that we introduced in section 1.13.3 in chapter 1, considering the role of the notion of *head* in syntax and semantics. So we will generally work with a subset of the set $\mathcal{SO}$ of syntactic objects, which is the domain Dom(h) of a head function.

3.1.3 Consistency and Substructures: Preliminary Discussion

A fundamental aspect of how we want to approach the question of modeling the interface between syntax and semantics is the idea that the assignment of semantic values to syntactic objects requires a recursive checking of consistency across substructures. This is the key guiding principle that dictates the algebraic properties of our model.

As we discussed in chapter 1, where we presented the mathematical formulation of Merge as described within the Strong Minimalist Thesis (SMT), a syntactic object $T \in \mathcal{SO} \simeq \mathfrak{T}_{\mathcal{SO}_0}$ comes endowed with its family of accessible terms $T_v \in \mathrm{Acc}(T)$. These are *substructures*; namely, they are themselves syntactic objects $T_v \in \mathcal{SO}$ contained in T by way of its recursive construction through the magma operation $\mathfrak{M}$. As discussed in chapter 1, the accessible terms T_v are exactly the substructures that are available for computation via the action of Merge.

The key idea here is that, given the hierarchical, recursive nature of the construction of syntactic objects (i.e., the generation of the magma $\mathcal{SO}$ from the set $\mathcal{SO}_0$ via repeated application of the magma operation $\mathfrak{M}$), one should expect that an assignment of semantic values will have to be compatible with this recursive computational structure. This means that one should be able to assign values first to the lexical items in $\mathcal{SO}_0$, and then (whenever possible) to the smaller objects built from these in the magma, and then to larger ones, with the constraint that when one assigns a value to a larger object T, this value assignment should be compatible with the values already assigned, in this recursive procedure, to the smaller substructures T_v that it contains.

As we will discuss briefly in section 3.1.4, this idea about recursive assignment of values with compatibility across substructures is very familiar in physics, where one also deals with recursive generative processes and a very similar problem of value assignments.

The way that this consistency problem is addressed mathematically is by replacing a simple assignment of values (for example, a function $\phi : \mathcal{SO} \to \mathcal{S}$ with values in some—for the moment unspecified—target space $\mathcal{S}$) with a more elaborate function that incorporates the consistency checking for substructures into its own recursive definition. This more elaborate function, known as the *Bogolyubov preparation*, is extensively used to solve this same problem in the context of quantum physics.

To make this Bogolyubov preparation work and modify a given map ϕ : $\mathcal{SO} \to \mathcal{S}$ recursively, in such a way that it incorporates the needed consistency checking, one needs two fundamental ingredients, one of which we already had from our setting in chapter 1:

- A way of extracting substructures T_v of a given structure T and separately considering the substructure T_v by itself, as well as the part T/T_v that remains of T when the substructure is removed
- A way of separating out, in the target space $\mathcal{S}$, agreement and disagreement (or to filter by levels of agreement and disagreement)

The first requirement is provided to us by the coproduct of the Hopf algebra of workspaces that we introduced in chapter 1. The second will be the fundamental requirement that we impose on whatever model of semantics we use as the target space $\mathcal{S}$. It is important to notice here how the first requirement lies entirely on the side of syntax and is very closely tied to the fundamental computational structure of syntax based on Merge, while the second requirement lies entirely on the side of semantics and has to do with whatever notion of proximity and similarity one can introduce for semantic values.

Once these two ingredients are available, one can recursively identify places where the agreement of values between a substructure T_v and its complement T/T_v fails. These reveal exactly where and why the overall consistency of semantic interpretation runs into possible problems. This recursive separation between inconsistencies and consistencies is a mathematical procedure known as *Birkhoff factorization*. We will discuss this in more specific detail in the rest of this chapter.

3.1.4 Algebraic Renormalization: A Short Summary

The physical procedure of *renormalization* can be formulated in algebraic terms (see [45], [46], [56], and [57]) using Hopf algebras and Rota–Baxter algebras, as in the setting of Connes–Kreimer renormalization. In this formulation, the procedure describes a very general form of Birkhoff factorization, which separates out an initial (unrenormalized) mapping into two parts of a convolution product, with one term describing the desirable (*meaningful*) part and one term describing the *meaningless* part, which needs to be removed (divergences, in the case of Feynman integrals in quantum field theory).

The mathematical setting that describes renormalization in physics, which we summarize here, may seem far-fetched as a model for linguistics, but the point here is that mathematical structures exist as flexible templates for the description of certain types of universal fundamental processes in nature, which are likely to manifest themselves in similar mathematical form in a variety of different contexts.

The Hopf algebra datum $\mathcal{H} = (\mathcal{V}, \cdot, \Delta, S)$, a vector space with compatible multiplication, comultiplication (with unit and counit), and antipode, takes care of describing the underlying combinatorial data and their generative process. In the case of quantum field theory, these are Feynman graphs with their subgraphs. The Feynman graphs of a given quantum field theory can be described as a generative process in two ways: one in terms of graph grammars (see [143]), which is similar to the older approach to generative linguistics based on formal languages; another in terms of Hopf algebras (see [45], [46], [56], and [57]). The comparison between these two generative descriptions of Feynman graphs shows direct similarities with what happens in the case of syntax, with the difference between the old formal languages approach and the new Merge approach in generative linguistics, where syntactic objects and the workspaces with the action of Merge can also be described in terms of Hopf algebras, as in chapter 1.

The Hopf algebra structure is central to the renormalization process, and the coproduct operation is the key part of the structure that is responsible for implementing the renormalization procedure, as we will recall next. The other algebraic datum, the Rota-Baxter algebra $(\mathcal{R}, R)$, represents what in physics is called a "regularization scheme". This is the choice of a model space where factorization into meaningful and meaningless parts takes place. There is an important conceptual difference between these two algebraic objects $\mathcal{H}$ and $\mathcal{R}$, in the sense that $\mathcal{H}$ is essentially intrinsic to the process while $\mathcal{R}$ is an accessory choice; and in principle, many different regularization schemes can be adopted to achieve the same desired renormalization. In terms of our linguistic model, one should think of this choice of a regularization scheme $\mathcal{R}$ as the choice of *some* model of semantics. As in the case of regularization in physics, we view the specifics of such a model as accessories to the interface that we are describing, while we view the role of the syntactic structures encoded in $\mathcal{H}$ as the essential part. This again reflects the view of a primarily syntax-driven interface between syntax and semantics.

In our context, this reflects the fact that there are several approaches to the construction of possible models of semantics, which are, in our view, not entirely satisfactory and not entirely compatible. However, we argue that this is not as serious an obstacle as it might first appear, in the sense that this also is very much the situation with regularization schemes in the physics of renormalization, where one has dimensional, cutoff, zeta function regularizations, among others, and yet one can still extract a viable procedure for the assignment of meaningful physical values, consistently across the choices of regularization. We will argue that indeed, a viable model of the interface between syntax and semantics rests on specific formulations of semantics only through some very simple abstract properties that can be satisfied within several different models.

We have briefly recalled the detailed definition of the Hopf algebra structure introduced in chapter 1; for more details, we refer readers to our discussion there and the short review of Hopf algebras in chapter 4. Here the Hopf algebra datum $\mathcal{H} = (\mathcal{V}, \cdot, \Delta, S)$ is assumed to be a commutative, associative, coassociative, graded, and connected Hopf algebra, but in general *not* cocommutative. This reflects the setting of New Minimalism (Merge and SMT) that we follow in this chapter (unlike the Old Minimalism discussed for comparison in chapter 2). We will fix this datum to be $\mathcal{H} = (\mathcal{V}(\mathcal{F}_{SO_0}), \sqcup, \Delta)$, with $\mathcal{F}_{SO_0}$ being the set of finite binary rooted forests (with no planar structure), and with the coproduct Δ as in (1.2.8), with the grading by the number of leaves, and with the unique, inductively defined antipode S. As mentioned in section 3.1.2.1, we will use the coproducts Δ^c or Δ^d, and modify this basic algebraic structure accordingly, as discussed in chapter 1.

For the Rota-Baxter part of the structure, we can distinguish two cases: the algebra case and the semiring case. The algebra case is the one that was originally introduced in the physics setting.

Definition 3.1.1. A Rota-Baxter algebra $(\mathcal{R}, R)$ of weight -1 is a commutative associative algebra $\mathcal{R}$ endowed with a linear operator $R : \mathcal{R} \to \mathcal{R}$ satisfying the identity

$$R(a)R(b) = R(aR(b)) + R(R(a)b) - R(ab),$$

for all $a, b \in \mathcal{R}$.

The prototype example (relevant to physics) is the algebra of Laurent series with the operator R of projection onto their polar (divergent) part. We will see an explicit linguistic example in section 3.5.2.

The case of a semirings (where addition is no longer invertible), more closely related to settings like the theory of computation, was introduced in [144], see also [135] and [145]. It will apply to various forms of semiring parsing in linguistics, which we will discuss in this chapter.

Definition 3.1.2. A Rota-Baxter semiring of weight $+1$ is a semiring $\mathcal{R}$ endowed with a Rota-Baxter operator R of weight $+1$. This is an additive (with respect to the semiring addition) map $R : \mathcal{R} \to \mathcal{R}$ satisfying

$$R(a) \odot R(b) = R(a \odot R(b)) \boxplus R(R(a) \odot b) \boxplus R(a \odot b),$$

with $(\boxdot, \odot)$ designating the semiring addition and multiplication operations. A Rota-Baxter semiring of weight -1 similarly satisfies the identity

$$R(a) \odot R(b) \boxdot R(a \odot b) = R(a \odot R(b)) \boxdot R(R(a) \odot b)\,.$$

Note that since semiring addition is not invertible, in this case we cannot move the term $R(a \odot b)$ to the other side of the identity. The purpose of the Rota-Baxter operator R is to project onto the "part of interest" (e.g., divergencies in physics). We will discuss in section 3.1.5 how to adapt Rota-Baxter data of the form $(\mathcal{R}, R)$ to semantic models, where the operator R will typically play the role of setting a threshold of agreement/disagreement.

Definition 3.1.3. A character of a commutative Hopf algebra $\mathcal{H}$ with values in a commutative algebra $\mathcal{R}$ is a map

$$\phi : \mathcal{H} \to \mathcal{R}$$

which is assumed to be a *morphism of algebras*, and hence it satisfies $\phi(xy) = \phi(x)\phi(y)$, as well as being a linear map of the underlying vector spaces. In the case where $\mathcal{R}$ is a semiring, we will consider two cases of semiring-valued characters:

1. *Semiring maps:*
$$\phi : \mathcal{H}^{semi} \to \mathcal{R}$$
defined on a subdomain $\mathcal{H}^{semi}$ of $\mathcal{H}$ that is a commutative semiring, with ϕ a morphism of commutative semirings.
2. *Maps on cones:* assuming that $\mathcal{H}$ is defined over the field $\mathbb{R}$, we consider maps
$$\phi : \mathcal{H}^{cone} \to \mathcal{R}$$
where the subdomain $\mathcal{H}^{cone}$ of $\mathcal{H}$ is a cone that is closed under convex linear combinations and under multiplication in $\mathcal{H}$, with ϕ compatible with convex combinations and products, $\phi(xy) = \phi(x) \odot \phi(y)$, with $\odot$ the semiring product.

In physics, such a datum $\phi : \mathcal{H} \to \mathcal{R}$ describes the Feynman rules for computing Feynman integrals in an assigned regularization scheme (given by the Rota-Baxter datum). In our setting, the map $\phi : \mathcal{H} \to \mathcal{R}$ is a map from syntactic objects to a semantic space, which includes the possibility of meaninglessness, when consistent semantics cannot be assigned. By "consistent," we mean that the assignment of semantic values to larger hierarchical structures has to be compatible with assignments to substructures: this is exactly the same consistency requirement that is used in the physics of renormalization and determines the required algebraic structure. The multiplicativity condition here just means that, in a workspace containing many different syntactic objects, the image of each of them in the semantic model $\mathcal{R}$ is independent of the others. Of course, when different syntactic objects are assembled by the action of Merge, their images in $\mathcal{R}$ need to be compared for consistency: this is indeed the crucial part of the interpretive process, which corresponds to the *compositionality* requirement, number (4) on our list of desired properties for the syntax–semantics interface.

Remark 3.1.4. It is important to stress the fact that a character $\phi : \mathcal{H} \to \mathcal{R}$ is only a map of algebras: it does not know anything about the fact that $\mathcal{H}$ also has a coproduct Δ and $\mathcal{R}$ also has a Rota-Baxter operator R. In particular, the target $\mathcal{R}$ does not carry a coproduct operation and ϕ is *not* a morphism of Hopf algebras.

The observation made in Remark 3.1.4 will play an important role in our linguistic model. It is in fact closely related to the statement at the beginning of this chapter: the computational structure of syntax—which, as we explained in chapter 1 depends on the coproduct structure of $\mathcal{H}$—does not require an analogous computational counterpart in semantics. We will discuss this point in more detail in the following sections, where we will show that in our model, the compositional properties of semantics are entirely governed by the computational structure of syntax, along with the topological nature of semantics (as a classifier of proximity relations). This is a very strong statement on the relative roles of syntax and semantics, presenting what can be viewed as a strong "syntax-first" model. While several of the examples presented in this chapter will be simplified mathematical models aimed at illustrating the fundamental algebraic properties, we will discuss at some length how the principle stated here can be understood in the case of Paul Pietroski's model of semantics, which we compare with our framework in section 3.6. We will also discuss in section 3.9 how our model can be applied if one adopts the Heim–Kratzer (HK) model of semantics.

In the physics setting, as well as in our linguistics model, the interaction between the two additional data, Δ and R, is used to implement *consistency* across substructures (our desired property of compositionality). This happens by recursively constructing a factorization (over the grading of the Hopf algebra), in the following way.

Definition 3.1.5. A *Birkhoff factorization* of a character $\phi : \mathcal{H} \to \mathcal{R}$ is a decomposition

$$\phi = (\phi_- \circ S) \star \phi_+ , \tag{3.1.4}$$

with S being the antipode and $\star$ being the convolution product determined by the coproduct Δ:

$$(\phi_1 \star \phi_2)(x) = (\phi_1 \otimes \phi_2)\,\Delta(x)\,.$$

One interprets the term ϕ_+ as the meaningful renormalized part, and the term ϕ_- as the meaningless part that needs to be removed. The semiring case is similar.

Definition 3.1.6. A Birkhoff factorization of a semiring character $\phi : \mathcal{H}^{semi} \to \mathcal{R}$ is a factorization of the form

$$\phi_+ = \phi_- \star \phi.$$

A Birkhoff factorization as in Definition 3.1.5 is constructed inductively using R and Δ in the following way.

Proposition 3.1.7. (See [45] and [56].) *Consider a Rota–Baxter algebra $(\mathcal{R}, R)$ of weight -1 and a commutative graded connected Hopf algebra $\mathcal{H}$. if $\phi : \mathcal{H} \to \mathcal{R}$ is a character, then there is (uniquely up to normalization) a Birkhoff factorization of the form in* (3.1.4), *obtained inductively (on the Hopf algebra degree) as*

$$\begin{aligned} \phi_-(x) &= -R(\phi(x) + \textstyle\sum \phi_-(x')\phi(x'')) \\ \phi_+(x) &= (1-R)(\phi(x) + \textstyle\sum \phi_-(x')\phi(x''))\,, \end{aligned} \tag{3.1.5}$$

where $\Delta(x) = 1 \otimes x + x \otimes 1 + \sum x' \otimes x''$, with the x', x'' having a lower degree. The $\phi_\pm : \mathcal{H} \to \mathcal{R}_\pm$ are algebra homomorphisms to the range of R and $(1-R)$. These are subalgebras (not just vector subspaces) because of the Rota–Baxter identity satisfied by R.

Remark 3.1.8. One usually refers to the expression

$$\tilde{\phi}(x) := \phi(x) + \sum \phi_-(x')\phi(x'') \tag{3.1.6}$$

as the *Bogolyubov preparation* of ϕ and writes $\phi_- = -R(\tilde{\phi})$ and $\phi_+ = (1-R)(\tilde{\phi})$.

The case of semirings is similar.

Proposition 3.1.9. (See [144].) *If $(\mathcal{R}, R)$ is a Rota–Baxter semiring of weight $+1$ and $\mathcal{H}$ is a commutative graded connected Hopf algebra with a semiring character $\phi : \mathcal{H}^{semi} \to \mathcal{R}$, where $\mathcal{H}^{semi}$ has an induced grading, one has the* *following factorization:*

$$\begin{aligned}\phi_-(x) = &\quad R(\tilde{\phi}(x)) &= R(\phi(x) \boxdot \phi_-(x') \odot \phi(x''))\\ \phi_+(x) = &\quad (\phi_- \star \phi)(x) &= \phi(x) \boxdot \phi_-(x) \boxdot \phi_-(x') \odot \phi(x'')\\ = &\quad \phi_- \boxdot \tilde{\phi}\,, &\end{aligned} \tag{3.1.7}$$

where $\phi_\pm$ are also multiplicative with respect to the semiring product, $\phi_\pm(xy) = \phi_\pm(x) \odot \phi_\pm(y)$.

Remark 3.1.10. In the case where $(\mathcal{R}, R)$ is a Rota–Baxter semiring of weight -1, one still obtains a Birkhoff factorization of the form (3.1.7). In this case, both $\phi_\pm$ still satisfy the multiplicative property if R has the additional property that

$$R(x \odot y) \boxdot R(x) \odot R(y) = R(x) \odot R(y)\,; \tag{3.1.8}$$

see [144]. This happens, for instance, if R satisfies $R(x + y) \leq R(x) + R(y)$ in the case with $\mathcal{R} = (\mathbb{R} \cup \{-\infty\}, \max, +)$.

3.1.5 Semantic Spaces

If we follow the idea described here of a syntax–semantics interface modeled after the formalism of algebraic renormalization in physics, we encode the syntactic side of the interface in terms of the Hopf algebra model of Merge and Minimalism as described in chapter 1. We then need a general description of what types of mathematical objects should be feasible on the semantic side, such that a Birkhoff factorization mechanism as described previously can be used to implement the assignment of semantic values to syntactic objects. As we will be illustrating in a series of different examples in the following sections, Birkhoff factorizations of the form (3.1.5) or (3.1.7) will serve the purpose, in our model, of checking and implementing the consistency of semantic assignments throughout all substructures of given syntactic hierarchical structures. Consistency checking will be realized through the use of a combination of values on substructures, provided by the Bogolyubov preparation in (3.1.6), as well as the use of a Rota–Baxter operator as a way of checking the possible failures of consistency across substructures. Since the target of our map from the Hopf algebra of syntax has to be a model of a "semantic space", again we proceed first by trying to identify certain key formal properties that we would like to have for such "semantic spaces" (thought of in similar terms to the regularization schemes in the physics of renormalization).

We first discuss in section 3.1.5.1 some analogies of the type of model that we have in mind, originating in neuroscience. The first is a neuroscience model that is somewhat controversial (and will play no direct role in this chapter, except as an analogy), while the second is a well-established result on neural codes and homotopy types.

3.1.5.1 Neuroscience data and syntax–semantics interface models Neuroscience data that study the human brain's handling of syntax and semantics in response to auditory or other signals (see [71], [9], and [72]), e.g., in experiments measuring event-related brain potential waveform components and in functional magnetic resonance imaging studies, display rapid recognition of syntactic violations and activation in the middle and posterior superior temporal gyrus for both semantic and syntactic violations. In contrast, the anterior superior temporal gyrus and the frontal operculus are activated by syntactic violations. Syntax and semantics have been claimed to be disentangled in such experiments by using artificial grammars: syntactic errors in simple cases that do not involve significant hierarchical syntactic structures appear to be related to activation of the frontal operculum, while the type of syntactic structure building that is modeled by the Merge operation appears related to activation in the most ventral anterior portion of the BA 44 part of Broca's area.

Additional semantic information shows involvement of other areas of the brain, in particular the BA 45 area. This suggests a possible "syntax-first" model of language processing in the brain, with an initial structure building process taking place at the syntactic level and an interface with semantics through the connectome involving the frontal operculus, BA 44, and BA 45. It should be noted, though, that this proposal regarding brain regions implicated in syntax and semantics has been strongly contested, such as according to the results in [63], which dispute the disentanglement and partitioning into areas of the syntax-first proposal in [71]. We only mention this proposal here as an analogy that can help illustrate some of our modeling of the syntax–semantics interface according to the list of properties outlined in §3.1.1. While we understand that this view is considered controversial by some, it does furnish a suggestive analogy for some of the basic geometric requirements that we will be assuming about the semantic side of the interface that we wish to model.

Another insight from neuroscience that we would like to carry into our modeling is the idea of information encoded via "open coverings" (i.e., overlapping open neighborhoods in a topological space and their pattern of intersections) and the topological notion of "homotopy types". This is well known in the setting of visual stimuli, when hyppocampal place cells, which fire in response to a restricted area of the spatial stimulus, are analyzed to address the question of how neuron-spiking activity encodes and relays information about the stimulus space. In such settings, one can show that patterns of neuron firing and their receptive fields determine an open covering that (under a convexity hypothesis) can reconstruct the stimulus space up to homotopy; see [49] and the mathematical survey in [132]. While this picture is specific to visual stimuli, an important idea that can be extracted from it is the role of open coverings[8] (in particular, open coverings associated with binary codes) in encoding proximity relations, as well as the role of convexity in such coverings. We will incorporate these ideas into a general basic picture of a notion of "semantic spaces" that can be compatible with how semantic information may effectively be stored in human brains. It was already observed in [133] that this structure should be part of modeling of semantic spaces.

3.1.5.2 Formal properties of semantic spaces

As the basic structure for an adequate parametrizing space for semantics, we focus on two compositional aspects: measuring degrees of proximity and a notion of agreement and disagreement. Our setting here is similar to the notion of semantic spaces of Manin–Marcolli [133]. We argue that, at least, semantics should be able to compare different semantic values (points in a semantic space) in terms of their level of agreement/disagreement, and to form new semantic points by some form of combination/interpolation of previously achieved ones. The type of interpolation considered may vary with specific models, but in general, we can think of it in the following related forms:

- Geodesic paths
- Convex combinations
- Overlapping open neighborhoods

A typical example that would combine these forms of interpolation is provided by a geodesically convex Riemannian manifold (a smooth space with a notion of distance and where convex interpolation can be achieved along the shortest paths, see section 3.2.2). Another aspect to take into consideration is the idea that, for instance, one can usually associate with a lexical item a collection of different "semes," and hence of points in a semantic space. In other words, the target of a map from lexical items and syntactic objects should allow for such "lists of semes."

[8] Open coverings (or just "coverings") of a topological space X are collections $\mathcal{U} = \{\mathcal{U}_\alpha\}$ of open sets $\mathcal{U}_\alpha \subset X$, such that their union covers the entire space $X = \cup_\alpha \mathcal{U}_\alpha$.

A very simple mathematical structure where notions of agreement or disagreement, proximity, and lists are simultaneously present, and combination operations are possible, is a vector space structure, and for this reason, it happens that frequently used elementary computational models of semantics tend to be based on vectors and vector space operations. More sophisticated geometric models of semantics based on spaces with properties of convexity, local coordinates representing semic axes, and realizations of notions of similarity were presented in [76] and [77], for example. Such geometric models also incorporate the possibility of open coverings, intersections of open sets, and homotopy as a way of realizing a "meeting of minds" model of Gärdenfors ([76] and [77]), where different observers may produce somewhat different sets of semantic associations with the same linguistic items (see the corresponding discussion in [133]).

An additional structure can be incorporated, if one wants to include a notion akin to that of "independent events." This can be achieved by working with spaces that also have a product operation, such as algebras, rings, or semirings, or that can be mapped to a space with this kind of structure, where such independence hypotheses can be tested. Thus, for example, elementary operations like assignments of truth values or of probabilities/likelihood estimates fall within this category and are usually performed by mapping to some ring or semiring structure. More generally, rings, algebras, and semirings can be seen as repositories for comparisons with specific test hypotheses, probing agreement/disagreement, or likelihood of representation along a chosen semic axis. We will discuss a few such examples in the following sections.

3.1.5.3 Concept spaces outside of language In this viewpoint, the type of fundamental structure that we associate with semantic spaces is not strictly dependent on their role in language. Indeed, the idea of extracting classifications from certain kinds of sensory data and associating with them some representation where proximity and difference can be evaluated is common to other cognitive processes. Conceptual spaces associated with vision are intensely studied in the context of neuroscience, computer vision, and artificial intelligence, and in that case, certainly the most relevant structures involved are topological in nature (see, e.g., the theory of perceptual manifolds in [41]). This suggests that it is possible to consider a model in language where the conceptual spaces that syntax interfaces with would have an essentially similar nature as other conceptual spaces, and not necessarily endowed with any additional structure specific to their role in language, with all the required structure that is of a specifically linguistic nature provided by syntax.

Formulated in such terms, this leads to a view of semantics that is essentially external to language and becomes a part of linguistics through the presence of a map *from* syntax. A more nuanced position, as we will illustrate in specific examples in this chapter, endows the semantic conceptual spaces with just enough additional structure extending the topological notion of proximity to make the mapping from syntax sufficiently robust to induce a compositional structure on semantics, modeled on the Merge operation in syntax.

For ease of computation, we will be using examples where such additional structure, aimed at quantifying proximity relations, consists of metrics with convexity properties and/or evaluations in semirings. This viewpoint will bring us close to Pietroski's model of compositional semantics [169], where a compositional structure in semantics is modeled on the Merge operation of syntax. One significant difference in our setting, though, is that we do not need to posit a separate compositional/computational operation on semantics itself (why should a Merge-type operation develop twice, once for syntax and once for semantics?). In our model, the compositionality of semantics is directly induced by the computational structure of syntax through the Birkhoff factorization mechanism described previously. This will constitute the key to our interface model.

Of course, one should allow for enough structure on the semantics side to incorporate the possibility of conjunctions of predicates, as well as a way of distinguishing the possibilities of mapping to conjunctions, predicate saturation, and existential closure. We will discuss more of this in section 3.6, where we analyze Pietroski's semantics, and in section 3.9, where we analyze Heim–Kratzer semantics. The main point that we want to stress here is that one does not need two parallel generative computational processes, one on the side of syntax and one on the side of semantics (as would be the case if we assumed that our maps $\phi : \mathcal{H} \to \mathcal{R}$ were Hopf algebra homomorphisms; see Remark 3.1.4). What one has instead is a map between two different kinds of mathematical structures, only one of which (syntax) is constructed by a recursive generative process.

3.2 Syntax–Semantics Interface as Renormalization: Toy Models

In this section, we outline some of the mathematical properties of the type of model we are proposing. To do that, we focus on what we refer to as "toy models", following the standard physics use of the term, where we present an overly simplified form of the type of structure we are describing, which will make it possible to outline more easily all the essential ingredients of the construction and the essential resulting algebraic properties. We will then proceed to gradually modify these toy models to incorporate more realistic features.

3.2.1 A Simple Toy Model: Head-Driven Syntax–Semantics Interface

We discuss, as a first illustrative example, a very simple-minded toy model of the type of syntax–semantics interface we are proposing. The example that we present in this section is intentionally oversimplified to more easily illustrate its main formal aspects.

Consider the semiring $(\mathbb{R} \cup \{-\infty\}, \max, +)$, where the addition is the maximum (with $-\infty$ as the unit of addition), and where the product is the usual sum of real numbers (with the rule that $-\infty + x = -\infty$), with 0 as the unit of the semiring multiplication $+$. This is also known as the "tropical semiring".

Lemma 3.2.1. *The ReLU (rectified linear unit) operator*

$$R : x \mapsto x^+ = \max\{x, 0\}$$

is a Rota–Baxter operator of weight $+1$ *on* $\mathcal{R} = (\mathbb{R} \cup \{-\infty\}, \max, +)$.

Proof. To see this, we need to check that the Rota–Baxter relation

$$x^+ + y^+ = \max\{(x^+ + y)^+, (x + y^+)^+, (x + y)^+\}$$

is verified for all $x, y \in \mathbb{R} \cup \{-\infty\}$. The following table shows that this is indeed the case:

	$x \leq 0, y \leq 0$	$x \geq 0, y \leq 0$	$x \leq 0, y \geq 0$	$x \geq 0, y \geq 0$
$x^+ + y^+$	0	x	y	$x + y$
$(x^+ + y)^+$	0	$\begin{cases} x + y & x + y \geq 0 \\ 0 & x + y \leq 0 \end{cases}$	y	$x + y$
$(x + y^+)^+$	0	x	$\begin{cases} x + y & x + y \geq 0 \\ 0 & x + y \leq 0 \end{cases}$	$x + y$
$(x + y)^+$	0	$\begin{cases} x + y & x + y \geq 0 \\ 0 & x + y \leq 0 \end{cases}$	$\begin{cases} x + y & x + y \geq 0 \\ 0 & x + y \leq 0 \end{cases}$	$x + y$
max	0	x	y	$x + y$

□

Remark 3.2.2. The identity operator $R = \mathrm{id}$ on the same semiring $(\mathbb{R} \cup \{-\infty\}, \max, +)$ is a Rota–Baxter operator of weight -1.

Definition 3.2.3. Consider a semantic space $\mathcal{S}$ with a map $s : \mathcal{SO}_0 \to \mathcal{S}$, which assigns a meaning (a point in $\mathcal{S}$) to the lexical items and syntactic features in $\mathcal{SO}_0$. Given a tree $T \in \mathfrak{T}_{\mathcal{SO}_0}$ and a leaf $\ell \in L(T)$, we write $\lambda(\ell) \in \mathcal{SO}_0$ for the label (lexical item and/or syntactic feature) assigned to that leaf. Given a head function h, defined on a domain $\mathrm{Dom}(h) \subset \mathfrak{T}_{\mathcal{SO}_0}$, we obtain a map

$$s \circ h : \mathrm{Dom}(h) \subset \mathfrak{T}_{\mathcal{SO}_0} \to \mathcal{S}, \quad T \mapsto s(\lambda(h(T))),$$

where $h(T) \in L(T)$ is the head. For simplicity of notation, we will also just write $s(h(T))$ for $s(\lambda(h(T)))$, leaving the label assignment λ implicit.

We now assume that the semantic space $\mathcal{S}$ has *probes*, given by functions $\Upsilon : \mathcal{S} \to \mathbb{R}$, that check the degree of agreement or disagreement with some particular semantic hypothesis. We assume that for $s \in \mathcal{S}$, a value $\Upsilon(s) < 0$ means that there is disagreement between the semantic object s and the semantic hypothesis Υ, while a value $\Upsilon(s) > 0$ signifies agreement, with the magnitude $|\Upsilon(s)|$ signifying the amount of agreement or disagreement. A value $\Upsilon(s) = 0$ signifies indifference.

Example 3.2.4. In the case of the familiar vector space model of semantics, such a probe can be obtained by taking the inner product with a specified hypothesis-vector:

$$\Upsilon(s) = \langle s, v_\Upsilon \rangle,$$

where the semantic hypothesis being tested is semantic proximity to a chosen vector v_Υ.

Lemma 3.2.5. *Consider a semantic space $\mathcal{S}$, a probe $\Upsilon : \mathcal{S} \to \mathbb{R}$, a map $s : \mathcal{SO}_0 \to \mathcal{S}$ assigning semantic values to lexical items and syntactic features, and a head function h defined on its domain* $\mathrm{Dom}(h) \subset \mathfrak{T}_{\mathcal{SO}_0}$. *Let $\mathcal{V}(\mathfrak{F}_{\mathcal{SO}_0})^{semi} \subset \mathcal{V}(\mathfrak{F}_{\mathcal{SO}_0})$ denote the semiring of linear combinations $\sum_i c_i F_i$, with $c_i \geq 0$. Then the data (Υ, s, h) determine a semiring homomorphism:*

$$\phi_{\Upsilon,s,h} : \mathcal{V}(\mathfrak{F}_{\mathcal{SO}_0})^{semi} \to \mathbb{R} \cup \{-\infty\}.$$

Proof. The data (Υ, s, h) determine a map:

$$\Upsilon_{s,h} : \mathfrak{T}_{\mathcal{SO}_0} \to \mathbb{R} \cup \{-\infty\}$$

$$\Upsilon_{s,h} : T \mapsto \begin{cases} \Upsilon(s(\lambda(h(T))) & T \in \mathrm{Dom}(h) \\ -\infty & T \notin \mathrm{Dom}(h). \end{cases} \tag{3.2.1}$$

The value $-\infty$ in the case of $T \notin \mathrm{Dom}(h)$ represents the case where the comparison with the hypothesis in the probe cannot be performed due to the lack of a well-defined head in T. This map can be extended from trees to forests by setting

$$\phi_{\Upsilon,s,h} : \mathfrak{F}_{\mathcal{SO}_0} \to \mathbb{R} \cup \{-\infty\}, \quad \phi_{\Upsilon,s,h}(F) = \sum_a \Upsilon_{s,h}(T_a), \quad \text{for } F = \sqcup_a T_a.$$

We can further extend this map to the subdomain $\mathcal{V}(\mathfrak{F}_{SO_0})^{semi} \subset \mathcal{V}(\mathfrak{F}_{SO_0})$ by setting

$$\phi_{\Upsilon,s,h}(\sum_i c_i F_i) = \boxdot_i \phi_{\Upsilon,s,h}(F_i) \odot \log(c_i) = \max_i \{\phi_{\Upsilon,s,h}(F_i) + \log(c_i)\} .$$

□

The extension to linear combinations is only needed for formal consistency. In the case of sums where all the coefficients are 1, the corresponding $\log(c_i)$ terms vanish.

Remark 3.2.6. One reason why this simple-minded toy model is too oversimplified is that the assignment $\phi_{\Upsilon,s,h}$ only follows the semantic value of the head of the tree, and hence it only uses the semantic values already attached to the leaves of the tree. However, in general, we want to obtain new points in semantic space, as the lexical items attached to the leaves are related and combined inside more elaborate syntactic objects. We will show in section 3.2.2 how to correct this problem and obtain more refined toy models.

To see how our interface model works in this simplified example, we first perform the Birkhoff factorization with respect to the Rota–Baxter operator $R = \mathrm{id}$ of weight -1, and then with respect to the ReLU Rota–Baxter operator $R = (\cdot)^+$ of weight $+1$.

Lemma 3.2.7. *For a semiring homomorphism $\phi : \mathcal{V}(\mathfrak{F}_{SO_0})^{semi} \to \mathcal{R} = (\mathbb{R} \cup \{\infty\}, \max, +)$, where the values $\phi(T)$ signify agreement or disagreement between a semantic values assigned to trees T and a semantic probe, the Birkhoff factorization with $R = \mathrm{id}$ has the effect of checking, for a given syntactic object $T \in \mathfrak{T}_{SO_0}$ and all chains of subforests $F_{\underline{v}_N} \subset F_{\underline{v}_{N-1}} \subset \cdots \subset F_{\underline{v}_1} \subset T$, when the* combined *agreement with the semantic probe of the parts*

$$\phi(F_{\underline{v}_N}) + \phi(F_{\underline{v}_{N-1}}/F_{\underline{v}_N}) + \cdots + \phi(T/F_{\underline{v}_1})$$

is greatest, and this is at least as good as the overall agreement $\phi(T)$.

Proof. The Birkhoff factorization with respect to the Rota–Baxter operator $R = \mathrm{id}$ of weight -1 simply gives $\phi_- = \tilde{\phi}$, so we have

$$\phi_-(T) = \tilde{\phi}(T) = \max\{\phi(T), \sum_{i=1}^{N} \phi(F_{\underline{v}_i}) + \phi(F_{\underline{v}_{i-1}}/F_{\underline{v}_i})\} ,$$

where $F_{\underline{v}_N} \subset F_{\underline{v}_{N-1}} \subset \cdots F_{\underline{v}_0} = T$ is a nested sequence of subforests (collections of accessible terms), and the maximum is taken over all such sequences of arbitrary length $N \geq 1$. □

Corollary 3.2.8. *For the case of $\phi = \phi_{\Upsilon,s,h}$, the Birkhoff factorization as in Lemma 3.2.7 has the effect of checking, for a given syntactic object $T \in \mathfrak{T}_{SO_0}$ and all chains of subtrees (subforests) $T_{v_N} \subset T_{v_{N-1}} \subset \cdots \subset T_{v_1} \subset T$, when the* combined *agreement with the semantic probe is maximal. If $\phi_{\Upsilon,s,h}(T) > 0$, this maximum is bounded below by the sum of values on the chain of subtrees with $h(T_{v_i}) = h(T)$, which is $N \cdot \phi_{\Upsilon,s,h}(T)$, with N being the length of the path from the root of T to the leaf $h(T)$. If $\phi_{\Upsilon,s,h}(T) < 0$, on the other hand, the maximum is bounded below by the values $\phi_{\Upsilon,s,h}(T) + M_v \cdot \phi_{\Upsilon,s,h}(T_v)$, where T_v is an accessible term with $\phi_{\Upsilon,s,h}(T_v) > 0$ and M_v is the length of the path from v to the leaf $h(T_v)$.*

Proof. Observe that we have $\phi_{\Upsilon,s,h}(T/T_v) = \phi_{\Upsilon,s,h}(T)$ since if $h(T) \notin T_v$, then quotienting the subtree T_v will not affect the head, and if $h(T) \in T_v$, then $h(T) = h(T_v)$, by the properties of head functions; and we label the leaf of T/T_v with a *trace* carrying the semantic value that was assigned to the leaf $h(T_v)$. The case of $T/F_{\underline{v}}$ is treated similarly. Note that here, we take quotients $T/^cT_v$ and $T/^cF_{\underline{v}}$, with contraction of each component of the subforest, as discussed in section 3.1.2.1.

For simplicity, we write out in full only the case where each $F_{\underline{v}_k}$ consists of a single subtree T_{v_k}, as the more general case of forests is analogous. In this case, we are computing

$$\begin{aligned}\phi_{\Upsilon,s,h,-}(T) = \max\{ &\phi_{\Upsilon,s,h}(T), \phi_{\Upsilon,s,h}(T) + \phi_{\Upsilon,s,h}(T_1), \cdots, \\ &\phi_{\Upsilon,s,h}(T) + \phi_{\Upsilon,s,h}(T_1) + \cdots + \phi_{\Upsilon,s,h}(T_N) \quad \},\end{aligned}$$

where N is the longest chain of nested accessible terms in T. The maximum is achieved at sequences $T_k \subset \cdots \subset T_1 \subset T$, where all $\phi_{\Upsilon,s,h}(T_i) > 0$ and as large as possible; that is, at the chains of nested accessible terms that achieve the *combined* maximal agreement with the probe.

For example, for a chain of length $N = 1$ (i.e., a single accessible term $T_v \subset T$), we are comparing $\phi_{\Upsilon,s,h}(T)$ and $\phi_{\Upsilon,s,h}(T) + \phi_{\Upsilon,s,h}(T_v)$, and hence we are checking whether $\phi_{\Upsilon,s,h}(T_v) > 0$ or $\phi_{\Upsilon,s,h}(T_v) < 0$; that is, whether individual accessible terms of T_v have heads $h(T_v)$ that semantically agree or disagree with the probe Υ. Clearly, among all subtrees T_v, one can always find some for which $\phi_{\Upsilon,s,h}(T) + \phi_{\Upsilon,s,h}(T_v) > \phi_{\Upsilon,s,h}(T)$: the subtrees for which $h(T_v) = h(T)$. The case of longer chains is analogous.

It is then clear that a lower bound in the case $\phi_{\Upsilon,s,h}(T) > 0$ is obtained by following the path from the root of T to the head $h(T)$, while in the case $\phi_{\Upsilon,s,h}(T) < 0$, one maximizes over collections of accessible terms with positive values $\phi_{\Upsilon,s,h}(T_{v_i}) > 0$, and one such collection is obtained by following the head of any T_v that has $\phi_{\Upsilon,s,h}(T_v) > 0$. □

We compare this to taking the Birkhoff factorization with respect to the ReLU Rota–Baxter operator $R(x) = x^+ = \max\{x, 0\}$ of weight $+1$. This shows that using different Rota–Baxter structures on the target semiring corresponds to performing different tests of semantic compositionality.

Lemma 3.2.9. *For the semiring homomorphism $\phi_{\Upsilon,s,h} : \mathcal{V}(\mathfrak{F}_{SO_0})^{semi} \to \mathcal{R} = (\mathbb{R} \cup \{\infty\}, \max, +)$, consider the Birkhoff factorization with respect to the ReLU Rota–Baxter operator $R(x) = x^+ = \max\{x, 0\}$ of weight $+1$. In this case, the value of $\phi_{\Upsilon,s,h,-}(T)$ is computed as a maximum value $\phi_{\Upsilon,s,h}(F_{\underline{v}_N}) + \phi_{\Upsilon,s,h}(F_{\underline{v}_{N-1}}) + \cdots + \phi_{\Upsilon,s,h}(F_{\underline{v}_1}) + \phi_{\Upsilon,s,h}(T)$ over all nested sequences, with the property that all $\phi_{\Upsilon,s,h}(F_{\underline{v}_i}) > 0$ and, in the case where $\phi_{\Upsilon,s,h}(T) < 0$, satisfying $\sum_i \phi_{\Upsilon,s,h}(F_{\underline{v}_i}) > |\phi_{\Upsilon,s,h}(T)|$. The maximum computing $\phi_{\Upsilon,s,h,-}(T)$ is bounded below by $N\phi_{\Upsilon,s,h}(T)$, with N being the length of the path from the root of T to the leaf $h(T)$, in the case with $\phi_{\Upsilon,s,h}(T) > 0$, and by $\phi_{\Upsilon,s,h}(T) + M_v \cdot \phi_{\Upsilon,s,h}(T_v)$, where T_v is any accessible term with $\phi_{\Upsilon,s,h}(T_v) > |\phi_{\Upsilon,s,h}(T)|$ and M_v is the length of the path from v to the leaf $h(T_v)$, when $\phi_{\Upsilon,s,h}(T) < 0$.*

Proof. We obtain in this case

$$\begin{aligned}\phi_{\Upsilon,s,h,-}(T) = \quad & \max\{\phi_{\Upsilon,s,h}(T), (\cdots(\phi_{\Upsilon,s,h}(F_{\underline{v}_N})^+ + \cdots \\ & + \phi_{\Upsilon,s,h}(F_{\underline{v}_{i-1}}/F_{\underline{v}_i}))^+ + \cdots + \phi_{\Upsilon,s,h}(T/F_{\underline{v}_0}))^+\}^+ ,\end{aligned}$$

with the maximum taken over all nested sequences of subforests of arbitrary length $N \geq 1$, as before. By the same argument as in Lemma 3.2.7 about heads of subtrees T_v and quotient trees T/T_v, in the case of chains of subtrees $T_{v_N} \subset T_{v_{N-1}} \subset \cdots \subset T_{v_1} \subset T$, this gives

$$(\cdots((\phi_{\Upsilon,s,h}(T_{v_N})^+ + \phi_{\Upsilon,s,h}(T_{v_{N-1}}))^+ \cdots + \phi_{\Upsilon,s,h}(T_{v_1}))^+ + \phi_{\Upsilon,s,h}(T))^+ .$$

The case of forests is similar (with sums over the component trees). Then ReLU is applied to the maximum taken over all these sums.

For example, for a chain of length $N = 1$, one compares $\phi_{\Upsilon,s,h}(T)$ with $\phi_{\Upsilon,s,h}(T) + \phi_{\Upsilon,s,h}(T_1)$. Thus, $\max\{\phi_{\Upsilon,s,h}(T), (\phi_{\Upsilon,s,h}(T) + \phi_{\Upsilon,s,h}(T_1)^+)^+\}^+$ has value $\phi_{\Upsilon,s,h}(T)$ if $\phi_{\Upsilon,s,h}(T) > 0$ and $\phi_{\Upsilon,s,h}(T_1) < 0$; value $\phi_{\Upsilon,s,h}(T) + \phi_{\Upsilon,s,h}(T_1)$ if $\phi_{\Upsilon,s,h}(T) > 0$ and $\phi_{\Upsilon,s,h}(T_1) > 0$, or if $\phi_{\Upsilon,s,h}(T) < 0$ and $\phi_{\Upsilon,s,h}(T_1) > 0$ with $\phi_{\Upsilon,s,h}(T) + \phi_{\Upsilon,s,h}(T_1) > 0$; and value 0 if $\phi_{\Upsilon,s,h}(T) < 0$ and $\phi_{\Upsilon,s,h}(T_1) < 0$, or if $\phi_{\Upsilon,s,h}(T) < 0$ and $\phi_{\Upsilon,s,h}(T_1) > 0$ with $\phi_{\Upsilon,s,h}(T) + \phi_{\Upsilon,s,h}(T_1) < 0$.

Thus, we see that when $\phi_{\Upsilon,s,h}(T) > 0$, the value $\phi_{\Upsilon,s,h,-}(T)$ is bounded below by $N\phi_{\Upsilon,s,h}(T)$, where N is the length of the path from the root of T to the leaf $h(T)$, as in Corollary 3.2.8. However, when $\phi_{\Upsilon,s,h}(T) < 0$, the Birkhoff factorization with respect to ReLU gives a more refined test than the Birkhoff factorization with respect to $R = \text{id}$ of Lemma 3.2.7 and Corollary 3.2.8. Indeed, in this case, we not only search over nested sequences with $\phi_{\Upsilon,s,h}(T_{v_N}) + \phi_{\Upsilon,s,h}(T_{v_{N-1}}) \cdots + \phi_{\Upsilon,s,h}(T_{v_1}) > 0$, but also further require that individual terms are positive and $\phi_{\Upsilon,s,h}(T_{v_N}) + \phi_{\Upsilon,s,h}(T_{v_{N-1}}) \cdots + \phi_{\Upsilon,s,h}(T_{v_1}) > |\phi_{\Upsilon,s,h}(T)|$, because of applying ReLU to the resulting sum. In particular, one obtains such a lower bound by following the head of any accessible term T_v with $\phi_{\Upsilon,s,h}(T_v) > |\phi_{\Upsilon,s,h}(T)|$, as stated. □

A case where $\phi_{\Upsilon,s,h}(T) < 0$, with the maximum realized by a sequence of positive terms with $\phi_{\Upsilon,s,h}(T_{v_N}) + \phi_{\Upsilon,s,h}(T_{v_{N-1}}) \cdots + \phi_{\Upsilon,s,h}(T_{v_1}) > |\phi_{\Upsilon,s,h}(T)|$, signals a situation where the semantic value assigned to the head $h(T)$ *disagrees* with the semantic probe used, but there are accessible terms in T that individually agree with the semantic probe and whose combined agreement is greater than the magnitude of the disagreement for $h(T)$.

Remark 3.2.10. The construction illustrated in Lemma 3.2.7, Corollary 3.2.8, and Lemma 3.2.9 can be seen as a way of extracting substructures where agreement or disagreement with a given semantic value is concentrated.

As mentioned at the beginning of this section and in Remark 3.2.6, the example of semiring homomorphism $\phi_{\Upsilon,s,h}(T)$ used in Lemma 3.2.7, Corollary 3.2.8, and Lemma 3.2.9 is unsatisfactory because it only uses the semantic values assigned to the leaves of the syntactic objects T through the map $s : \mathcal{SO}_0 \to \mathcal{S}$ and does not create new semantic values assigned to the syntactic objects T themselves that go beyond the value already assigned to the leaf that is the head $h(T)$.

We can easily see the problem outlined here and in Remark 3.2.6 by considering some simple examples where this setting works and where it runs into problems. Consider first a sentence like

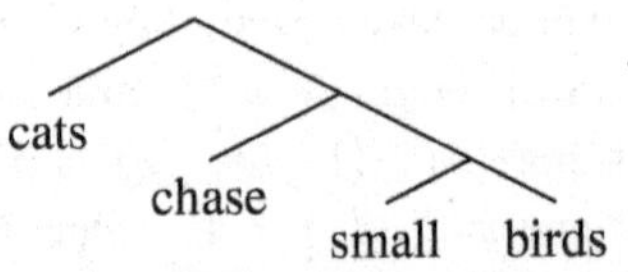

Let's assume that the lexical items λ are mapped to values $s(\lambda)$ in a vector-space model of semantics. In this vector space, consider a probe v_Υ, which is the vector associated with "predation". The vectors v assigned to the lexical items "cats", "chase", and "birds" will all positively correlate (as predator, prey, and hunting action) to the chosen semantic probe. Extracting coproduct terms and building the recursive factorization give

$$\phi_{\Upsilon,s,h,-}(T) = (\max\{\phi_{\Upsilon,s,h}(T), \phi_{\Upsilon,s,h}(F_{\underline{v}})^{+}\phi_{\Upsilon,s,h}(T/F_{\underline{v}}),$$

$$\ldots, \phi_{\Upsilon,s,h}(F_{\underline{v}_N})^+ \phi(F_{\underline{v}_{N-1}}/F_{\underline{v}_N}) \cdots \phi(T/F_{\underline{v}_1})\})^+$$

$\phi_{\Upsilon,s,h,-}($ 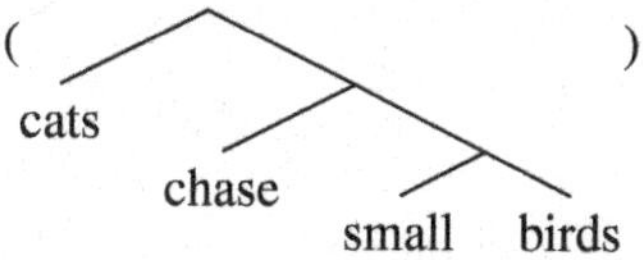

$) =$

$(\max\{\phi_{\Upsilon,s,h}($ 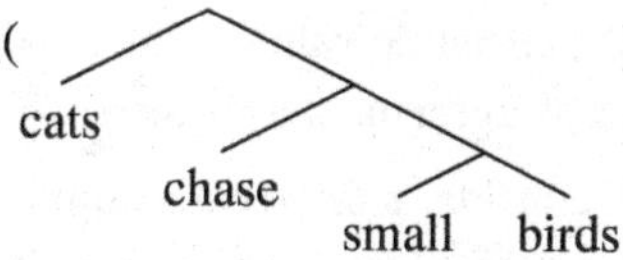

$),$

$\phi_{\Upsilon,s,h}(\text{cats})^+ \phi_{\Upsilon,s,h}($ 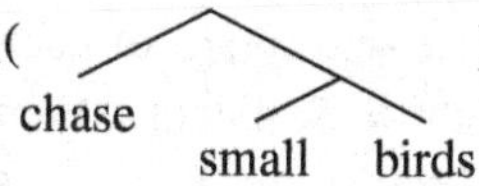

$),$

$\phi_{\Upsilon,s,h}(\text{small})^+ \phi_{\Upsilon,s,h}($ 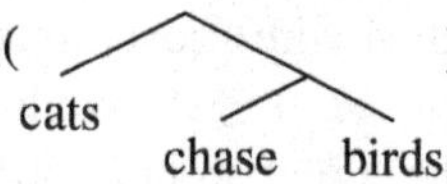

$),$

$\phi_{\Upsilon,s,h}(\text{birds})^+ \phi_{\Upsilon,s,h}($ 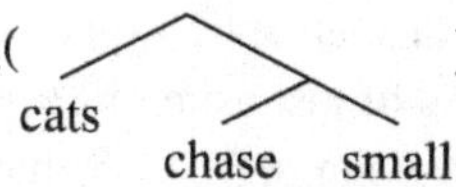

$),$

$\phi_{\Upsilon,s,h}($ 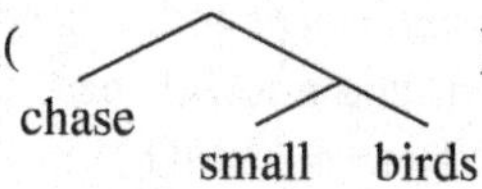

$)^+ \phi_{\Upsilon,s,h}(\text{cats}),$

$\phi_{\Upsilon,s,h}($ 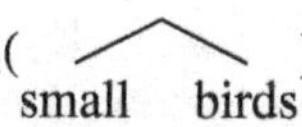

$)^+ \phi_{\Upsilon,s,h}($ 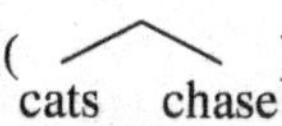

$), \ldots\})^+,$

where ... stands for the further terms of the coproduct involving a forest rather than just a tree in the left-channel. Here, *head* is either the verb "chase" or the noun "bird" on the subtrees, and if the Rota–Baxter operator R is ReLU (or some threshold), one gets agreement with the probe on the substructures. Then, consider the same sentence, but replace "chase" with "nurture" and use the same probe. Now some substructures are filtered out by ReLU because of disagreement with the probe:

$\phi_{\Upsilon,s,h}($ 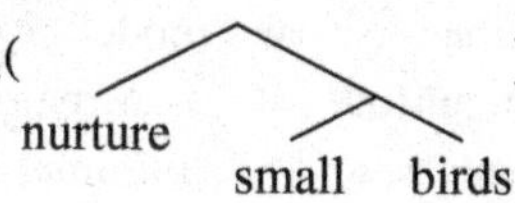

$)^+ = 0.$

Also, the final application of $R = ReLU$ to the maximum in $\phi_{\Upsilon,s,h,-} = R(\tilde{\phi}_{\Upsilon,s,h})$ cuts off all structures with *head* given by the verb "nurture" because of the disagreement with the probe:

$$\phi_{\Upsilon,s,h}([\text{cats}\ \text{nurture}])^+ = 0, \quad \phi_{\Upsilon,s,h}([\text{cats}\ [\text{nurture}\ \text{birds}]])^+ = 0\,,$$

and similarly for other terms of this kind. Thus, in this simple example, comparing probes with the semantic value of the heads of the structures suffices to detect agreement and disagreement as desired.

On the other hand, consider a different example, given by the same sentence "France is a hexagonal republic," which we already discussed in the introduction, and which we will return to later in this chapter. In a vector-space model of semantics, the lexical items "France", "hexagon(al)", and "republic" have corresponding vectors, $v_{Fr}, v_{hex}, v_{rep} \in \mathcal{S}$. Then choose a probe v_Υ, such as the vector associated with "government". This would give $\langle v_\Upsilon, v_{Fr}\rangle > 0$, $\langle v_\Upsilon, v_{rep}\rangle > 0$, but $\langle v_\Upsilon, v_{hex}\rangle \leq 0$. We have the same tree structure, but now just looking at the head leaf would not help to detect these agreements and disagreements with the semantic probe. So one can see by comparing these examples that, instead of just using the semantic value $s(h(T))$ associated with the head, we want a point $s(T) \in \mathcal{S}$ that displaces the position of the head $s(h(T))$ in the direction of the *complement* of the head: just slightly, if there is large agreement between them, and a lot, if there is disagreement. Thus, $\phi_{\Upsilon,s,h}(T)$ can differ significantly from $\phi_{\Upsilon,s,h}(h(T))$ in the case where there is disagreement inside T.

So we can conclude that this toy model construction does not represent correctly how an assignment of semantic values to sentences should work. We discussed it here mostly as a way to show, in the simplest possible form, how Birkhoff factorizations work. We now move on to more realistic models, as suggested by the comparison of the previous examples. These will also be simplified toy models, but we will gradually introduce more realistic features.

3.2.2 Head-Driven Interfaces and Convexity

We now assume that our semantic space model $\mathcal{S}$ is a geodesically convex region inside a Riemannian manifold (M, g). A region $\mathcal{S} \subset M$ is geodesically convex if, for any given points $s, s' \in \mathcal{S}$, minimal-length geodesic arcs $\gamma : [0, 1] \to M$, with $\gamma(0) = s$ and $\gamma(1) = s'$, are contained in the region: $\gamma(t) \in M$ for all $t \in [0, 1]$.

This includes in particular when $\mathcal{S}$ is a vector space or a simplex. In these cases, we write $\{\lambda s + (1 - \lambda)s' \,|\, \lambda \in [0, 1]\}$ for the segment connecting s, s' in $\mathcal{S}$ (the convex combinations of s and s'). With a slight abuse of notation, in the more general case of geodesically convex regions inside a Riemannian manifold, we will still write $\lambda s + (1 - \lambda)s'$ to indicate the point $\gamma(\lambda)$ along a given minimal geodesic arc $(\gamma(t))_{0\leq t\leq 1}$ in $\mathcal{S}$.

We assume, as before, that there is a map $s : \mathcal{SO}_0 \to \mathcal{S}$ that assigns semantic values to the lexical items and syntactic features.

3.2.2.1 Comparison functions We assume that the semantic space $\mathcal{S}$ is endowed with one of the following additional data:

1. On the product $\mathcal{S} \times \mathcal{S}$, there is a function

$$\mathbb{P} : \mathcal{S} \times \mathcal{S} \to [0, 1] \tag{3.2.2}$$

that evaluates the probability that two points s, s' are semantically associated (interpreted as the frequency with which they are semantically associated within a specified context). We assume that $\mathbb{P}$ is symmetric, $\mathbb{P}(s, s') = \mathbb{P}(s', s)$; that is, it factors through the symmetric product

$$\mathbb{P} : \mathrm{Sym}^2(\mathcal{S}) \to [0, 1] \,.$$

One can also assume that $\mathbb{P}$ is a probability measure on $\mathcal{S} \times \mathcal{S}$, although this is not strictly necessary in what follows. If the underlying space $\mathcal{S}$ is convex, we always assume that $\mathbb{P}$ is a biconcave function.

2. On the product $\mathcal{S} \times \mathcal{S}$, there is a function

$$\mathfrak{C} : \mathcal{S} \times \mathcal{S} \to \mathbb{R} \tag{3.2.3}$$

that evaluates the level of semantic agreement or disagreement between two points s, s', where

   - The function $|\mathfrak{C}(s, s')|$ measures the magnitude of agreement or disagreement
   - The function $\mathrm{sign}(\mathfrak{C}(s, s')) = \mathfrak{C}(s, s')/|\mathfrak{C}(s, s')| \in \{\pm 1\}$ measures whether there is agreement or disagreement

   Again, we assume that the function $\mathfrak{C}$ is symmetric. In the case of a semantic vector space $\mathcal{S}$, one can assume that $\mathfrak{C}$ is obtained from a symmetric bilinear form by

$$\mathfrak{C}(s, s') = \frac{\langle s, s' \rangle}{\|s\| \, \|s'\|} \,, \tag{3.2.4}$$

which gives the usual cosine similarity, but in general, it is not necessary for $\mathfrak{C}(s, s')$ to be as in (3.2.4).

This type of comparison functions $\mathbb{P} : \mathcal{S} \times \mathcal{S} \to [0, 1]$ as in (3.2.2), or $\mathfrak{C} : \mathcal{S} \times \mathcal{S} \to \mathbb{R}$ as in (3.2.3), should really be thought of, more generally, as a collection $\mathbb{P} = \{\mathbb{P}_\sigma\}$ or $\mathfrak{C} = \{\mathfrak{C}_\sigma\}$, where the index σ runs over certain syntactic functions (in the sense of functional relations between constituents in a clause). For example, suppose that one looks at the two sentences "Dog bites man" and "Man bites dog." In the first case, the phrase determines a point on the geodesic arc in $\mathcal{S}$ between the points s(bite) and s(man), at a distance $\mathbb{P}_\sigma(s(\text{bite}), s(\text{man}))$ from the vertex s(bite), where σ is the verb-object relation, while in the second case σ is the subject-verb relation. The value $\mathbb{P}_\sigma(s(\text{bite}), s(\text{man})) \in [0,1]$ evaluates the degree of likelihood of this association in the specified relation; see the discussion of this example in section 3.2.2.3.

3.2.2.2 Threshold Rota-Baxter operators As in the cases discussed in section 3.2.2.1, we can consider a semiring $\mathcal{P}$ endowed with a Rota–Baxter structure. Here, we take $\mathcal{P} = ([0,1], \max, \cdot, 0, 1)$, the Viterbi semiring, for probabilistic parsing.

Lemma 3.2.11. *Consider the Viterbi semiring* $\mathcal{P} = ([0,1], \max, \cdot, 0, 1)$. *Then the threshold operators*

$$c_\lambda : \mathcal{P} \to \mathcal{P} \quad \text{with} \quad \lambda \in [0,1],$$

given by

$$c_\lambda(x) = \begin{cases} x & x < \lambda \\ 1 & x \geq \lambda, \end{cases} \tag{3.2.5}$$

are Rota–Baxter operators of weight -1 *that satisfy the property* (3.1.8).

Proof. We can compare the values in the Rota–Baxter identity as follows:

	$x < \lambda, y < \lambda$	$x \geq \lambda, y < \lambda$	$x < \lambda, y \geq \lambda$	$x \geq \lambda, y \geq \lambda$
$c_\lambda(xy)$	xy	xy	xy	$\begin{cases} xy & xy < \lambda \\ 1 & xy \geq \lambda \end{cases}$
$c_\lambda(x)c_\lambda(y)$	xy	y	x	1
$c_\lambda(c_\lambda(x)y)$	xy	y	xy	1
$c_\lambda(xc_\lambda(y))$	xy	xy	x	1

Indeed, we have $x, y, \lambda \in [0,1]$; hence, if either $x < \lambda$ or $y < \lambda$, then $xy < \lambda$. The maximum of the first two rows is $\max\{c_\lambda(xy), c_\lambda(x)c_\lambda(y)\} = c_\lambda(x)c_\lambda(y)$, which shows that the identity (3.1.8) holds. Moreover, the maximum between the last two rows of this table is also equal to $c_\lambda(x)c_\lambda(y)$ so the Rota–Baxter identity of weight -1 holds. □

3.2.2.3 Viterbi-valued semiring character We then consider the construction of characters. For our target semiring, given by the Viterbi semiring $\mathcal{P} = ([0,1], \max, \cdot, 0, 1)$, we can consider characters $\phi : \mathcal{H}^{cone} \to \mathcal{P}$ with the domain given by a convex cone inside $\mathcal{H}$, which ensures that if generators $F \in \mathfrak{F}_{SO_0}$ are mapped to $\mathcal{P}$, linear combinations that are in the cone will also map to $\mathcal{P}$.

Lemma 3.2.12. *Consider a semantic space $\mathcal{S}$ that is geodesically convex, endowed with a function $s : \mathcal{SO}_0 \to \mathcal{S}$ and a function $\mathbb{P} : \mathrm{Sym}^2(\mathcal{S}) \to [0,1]$ as in section 3.2.2.1. Also, consider a head function h defined on a domain $\mathrm{Dom}(h) \subset \mathfrak{T}_{SO_0}$. The function $s : \mathcal{SO}_0 \to \mathcal{S}$ extends to a map $s : \mathrm{Dom}(h) \to \mathcal{S}$, and these data determine a character given by a map*

$$\phi_{s,\mathbb{P},h} : \mathcal{H}^{cone} \to \mathcal{P},$$

where $\mathcal{H}^{cone}$ is the cone of convex linear combinations $\sum_i a_i F_i$ with $0 \leq a_i$ and $\sum_i a_i = 1$, and with forests $F_i \in \mathfrak{F}_{SO_0}$. The character is defined on the generators by $\phi_{s,\mathbb{P},h}(T) = 0$ for $T \notin \mathrm{Dom}(h)$, while for $T \in \mathrm{Dom}(h)$, the value $\phi_{s,\mathbb{P},h}(T)$ is inductively determined by the description of T as iterations of the Merge operation $\mathfrak{M}$ in the magma (3.1.1). It is extended to $\mathcal{H}^{cone}$ by $\phi_{s,\mathbb{P},h}(F) = \prod_k \phi_{s,\mathbb{P},h}(T_k)$, for $F = \sqcup_k T_k$, as well as $\phi_{s,\mathbb{P},h}(\sum_i a_i F_i) = \max_i a_i \phi_{s,\mathbb{P},h}(F_i)$.

Proof. To an unordered pair $\mathfrak{M}(\alpha,\beta) = \{\alpha,\beta\}$ of $\alpha,\beta \in \mathcal{SO}_0$, we assign a value in $\mathcal{P}$ in the following way. If the tree $T = \mathfrak{M}(\alpha,\beta) \in \mathcal{SO} = \mathfrak{T}_{SO_0}$ is not in $\mathrm{Dom}(h)$, we assign the value $\phi_{s,\mathbb{P},h}(T) = 0$. If $T \in \mathrm{Dom}(h)$, consider the value

$$p_{\alpha,\beta} := \mathbb{P}(s(\alpha), s(\beta))$$

and define $s(T) \in \mathcal{S}$ as

$$s(T) = p s(\alpha) + (1-p) s(\beta), \tag{3.2.6}$$

where $p \in [0,1]$ is

$$p = \begin{cases} p_{\alpha,\beta} & \alpha = h(T) \\ 1 - p_{\alpha,\beta} & \beta = h(T). \end{cases} \tag{3.2.7}$$

We then set

$$\phi_{s,\mathbb{P},h}(\mathfrak{M}(\alpha,\beta)) = p_{\alpha,\beta}. \tag{3.2.8}$$

We then proceed inductively. If $T = \mathfrak{M}(T_1, T_2)$ is not in $\mathrm{Dom}(h)$, we set $\phi_{s,\mathbb{P},h}(T) = 0$. If it is in $\mathrm{Dom}(h)$, then by the properties of head functions, T_1 and T_2 are also in $\mathrm{Dom}(h)$. So we can assign to T the point $s(T) \in \mathcal{S}$, given by

$$s(T) = p\, s(T_1) + (1-p)\, s(T_2),$$

where

$$p = \begin{cases} p_{s(T_1),s(T_2)} & h(T) = h(T_1) \\ 1 - p_{s(T_1),s(T_2)} & h(T) = h(T_2), \end{cases} \quad (3.2.9)$$

with

$$p_{s(T_1),s(T_2)} = \mathbb{P}(s(T_1), s(T_2)) .$$

We then set

$$\phi_{s,\mathbb{P},h}(T) = p_{s(T_1),s(T_2)} .$$

It is clear that this determines a map

$$\phi_{s,\mathbb{P},h} : \mathcal{H}^{cone} \to \mathcal{P},$$

with $\phi_{s,\mathbb{P},h}(\sum_i a_i F_i) = \max_i\{a_i \phi_{s,\mathbb{P},h}(F_i)\}$ and $\phi_{s,\mathbb{P},h}(F) = \prod_k \phi_{s,\mathbb{P},h}(T_k)$, for $F = \sqcup_k T_k \in \mathfrak{F}_{\mathcal{SO}_0}$. □

Remark 3.2.13. The semiring-valued character $\phi_{s,\mathbb{P},h}$ that we constructed in Lemma 3.2.12 improves on the construction of the character $\phi_{\Upsilon,s,h}$ given in Lemma 3.2.5, in the sense that the values $\phi_{s,\mathbb{P},h}(T)$ assigned to syntactic object do not depend just on the semantic values of the lexical items, but also on other points of semantic space $\mathcal{S}$, obtained as convex combinations of values assigned to lexical items. However, it should still be regarded as a toy model case, as the way in which these combinations are obtained and the corresponding value of $\phi_{s,\mathbb{P},h}(T)$ is computed is still overly simplistic. We show in section 3.2.3 another similar simplified toy model example, with a choice of a semiring-valued character that combines the properties of $\phi_{s,\mathbb{C},h}$ of Lemma 3.2.12 and $\phi_{\Upsilon,s,h}$ of Lemma 3.2.5.

Note that, in principle, we have two simple choices of how to extend inductively (3.2.7) from the cherry tree case $T = \mathfrak{M}(\alpha, \beta)$ to the more general case $T = \mathfrak{M}(T_1, T_2)$. One is to define $p_{s(T_1),s(T_2)}$ as in (3.2.9), with $p_{s(T_1),s(T_2)} = \mathbb{P}(s(T_1), s(T_2))$, inductively using the previously constructed points $s(T_1)$ and $s(T_2)$. Another possibility, more similar to our previous example $\phi_{\Upsilon,s,h}$ from Lemma 3.2.5, is to define it using the heads, $\mathbb{P}(h(T_1), h(T_2))$. To see why the option of (3.2.9) is clearly preferable, consider the following example. Take the three sentences "Man bites dog", "Man bites apple", and "Dog bites man". Denoting by M, B, D, and A the respective points in $\mathcal{S}$ associated with these lexical items, the points associated to the respective sentences are shown in the diagram in Figure 3.1.

In the sentence "Dog bites man", the verb phrase (VP) determines a point on the geodesic arc in $\mathcal{S}$ between the points B and M at a distance $\mathbb{P}_\sigma(B, M)$ from the vertex B, where in this case, σ is the verb-object relation and the

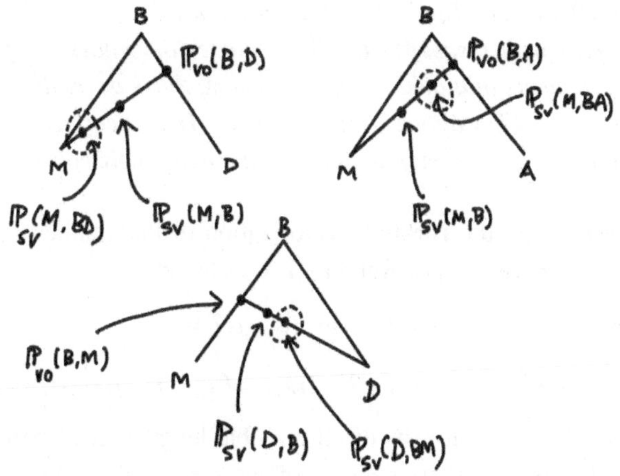

Figure 3.1
Sketch of different semantic points constructed by geodesic arcs for the three sentences "Man bites dog", "Man bites apple", and "Dog bites man", and with the two different choices of $p_{s(T_1),s(T_2)} = \mathbb{P}(s(T_1), s(T_2))$ (circled) or $\mathbb{P}(h(T_1), h(T_2))$.

value $\mathbb{P}_\sigma(B, M) \in [0, 1]$ expresses the degree of likelihood of this association in the relation σ. One then considers, on the geodesic arc in $\mathcal{S}$ between this point associated to the VP and the point D, a new point. In the case of the choice $p_{s(T_1),s(T_2)} = \mathbb{P}(s(T_1), s(T_2))$ as in (3.2.9), this point is located at a distance of either $\mathbb{P}_{\sigma'}(D, BM)$, where we write BM for the point $s(\mathfrak{M}(B, M))$ associated with the VP by the procedure just described, and σ' is the subject-verb relation between D and $h(\mathfrak{M}(B, M))$. In the case where we use $\mathbb{P}(h(T_1), h(T_2))$, this point is located at a distance $\mathbb{P}_{\sigma'}(D, B)$, where σ' is the subject-verb relation. The cases of the second and third sentences are analogous, as sketched in Figure 3.1. One can see in a simple example like this why the choice $p_{s(T_1),s(T_2)} = \mathbb{P}(s(T_1), s(T_2))$ is preferable to $\mathbb{P}(h(T_1), h(T_2))$ by comparing the location of points in the first two cases in Figure 3.1. If one uses $\mathbb{P}(h(T_1), h(T_2))$, the length of the geodesic arc between M and the points BD or BA is determined in both cases by the same value $\mathbb{P}_{\sigma'}(M, B)$, while in the case of $\mathbb{P}(s(T_1), s(T_2))$ one has different lengths $\mathbb{P}_{\sigma'}(M, BD) << \mathbb{P}_{\sigma'}(M, BA)$.

3.2.2.4 Birkhoff factorization with threshold operators The Birkhoff factorization of the character $\phi_{s,\mathbb{P},h}$ with respect to the threshold Rota–Baxter operators provides a way of searching for substructures with large semantic agreement between constituent parts. This is formulated more precisely in the following statement:

Proposition 3.2.14. *The Birkhoff factorization of the character $\phi_{s,\mathbb{P},h}$ defined in Lemma 3.2.12 with respect to the Rota–Baxter operators c_λ of weight -1 identifies, as elements that achieve the maximum, those accessible terms $T_v \subset T$ with values $\phi_{s,\mathbb{P},h}(T_v)$ above a threshold λ. This identifies substructures within T that carry large semantic agreement among their constituent parts.*

Proof. If we perform the Birkhoff factorization of the character $\phi_{s,\mathbb{P},h}$ using the Rota–Baxter operator c_λ of weight -1, we obtain

$$\phi_{s,\mathbb{P},h,-}(T) = c_\lambda(\tilde{\phi}_{s,\mathbb{P},h}(T)) =$$

$$c_\lambda(\max\{\phi_{s,\mathbb{P},h}(T), c_\lambda(\cdots c_\lambda(\phi_{s,\mathbb{P},h}(F_{\underline{v}_N}))\phi_{s,\mathbb{P},h}(F_{\underline{v}_{N-1}}/F_{\underline{v}_N}))\cdots\phi_{s,\mathbb{P},h}(T/F_{\underline{v}_0})\})$$

over nested chains of subforests of all possible lengths N, as before. Again, we can consider for simplicity the case of subtrees, as the value for forests is the semiring product of the values on the tree components. When we look at chains of length $N = 1$ with subtrees, we are comparing $\phi_{s,\mathbb{P},h}(T)$ to the value $c_\lambda(\phi_{s,\mathbb{P},h}(T_v)) \cdot \phi_{s,\mathbb{P},h}(T/T_v)$. Arguing as before, we have

$$c_\lambda(\max\{\phi_{s,\mathbb{P},h}(T), c_\lambda(\phi_{s,\mathbb{P},h}(T_v)) \cdot \phi_{s,\mathbb{P},h}(T/T_v)\}) =$$

$$c_\lambda(\max\{p_{s(T_1),s(T_2)}, c_\lambda(p_{s(T_{v,1})s(T_{v,2})}) \cdot p_{s(T_1),s(T_2)}\}) = c_\lambda(p_{s(T_1),s(T_2)}),$$

where this time, the maximal value is realized by all the terms $T_v \subset T$ that have $p_{s(T_{v,1})s(T_{v,2})} \geq \lambda$ and $p_{s(T_1),s(T_2)} \geq \lambda$. Note that longer sequences will have products with intermediate terms $\phi_{s,\mathbb{P},h}(F_{\underline{v}_{i-1}}/F_{\underline{v}_i}) < 1$, hence will not achieve the same maximum. Thus, the maximizers are accessible terms that carry large semantic agreement among their constituent parts. □

For example, suppose that we consider again the two sentences "Dog bites man" and "Man bites dog". As shown previously, the resulting semantic points associated with these two sentences are, as they should be, in different locations in $\mathcal{S}$. Moreover, the fact that one will have $\mathbb{P}_{\sigma'}(M, BD) << \mathbb{P}_{\sigma'}(D, BM)$ when σ' is the subject-verb relation, implies that the threshold operators c_λ discussed in Lemma 3.2.11 will filter out the second sentence before the first.

3.2.2.5 Convex neighborhoods The construction of the character $\phi_{s,\mathbb{P},h}$ of Lemma 3.2.12 is also a toy model. It is better than the initial oversimplified toy model of Lemma 3.2.5 (see Remark 3.2.6), because it does not use only the points in the semantic space $\mathcal{S}$ associated with the head leaf, but it still uses only geodesic arcs in the semantic space $\mathcal{S}$. Passing from a zero-dimensional to a one-dimensional representation of syntactic relations is an improvement, and as we will discuss in section 3.3, it is already sufficient to obtain an embedded image of syntax inside semantics (in essence, because the syntactic objects are themselves one-dimensional tree structures). However, this representation can be improved by considering, along with geodesic arcs, higher dimensional convex structures like simplexes and geodesic neighborhoods of points.

While we will not expand this approach in this chapter, it is worth mentioning some ideas that relate to some of what we will be discussing in the following sections. Given a syntactic object $T \in \mathcal{SO}$, with $T \in \mathrm{Dom}(h)$, a geodesically convex semantic space $\mathcal{S}$, and a mapping $s : \mathrm{Dom}(h) \to \mathcal{S}$ constructed as in Lemma 3.2.12, we can consider the points $s(T_v) \in \mathcal{S}$ associated with all the accessible terms of T. (See section 3.3 for the embedding properties of this map.) Now consider geodesic balls $B_v(\epsilon)$ in $\mathcal{S}$, centered at the points $s(T_v)$ with radius $\epsilon > 0$. Here, by geodesic ball, we mean the image under the exponential map of a ball in the tangent space. We assume that the injectivity radius of $\mathcal{S}$ is larger than the maximal distance between the points $s(T_v)$ for all $v \in V(T)$. In terms of the semantic space, a geodesic neighborhood around a given point $s \in \mathcal{S}$ represents all the close semantic associations with the semantic point s recorded in $\mathcal{S}$. We can then vary the scale ϵ of the geodesic balls and form a simplicial complex (a Vietoris–Rips complex) associated with the intersections of these geodesic balls (see Figure 3.2). As the scale $\epsilon > 0$ varies, one obtains a filtered complex, according to the familiar construction of *persistent topology* (see [58]). The scale ϵ provides another form of filtering that generalizes what we previously described in terms of the threshold operator c_λ. In this case, the persistent structures that arise can be seen as detecting "collections of substructures that carry higher semantic relatedness" inside the given hierarchical structure T. A possible way to obtain an open covering in semantic space, consisting of a system of overlapping open neighborhoods of the semantic values of the heads of the substructures, would be to construct such open coverings based on the decomposition of the syntactic object into *phase*, according to Phase Theory, as we discussed in section 1.14 of chapter 1.

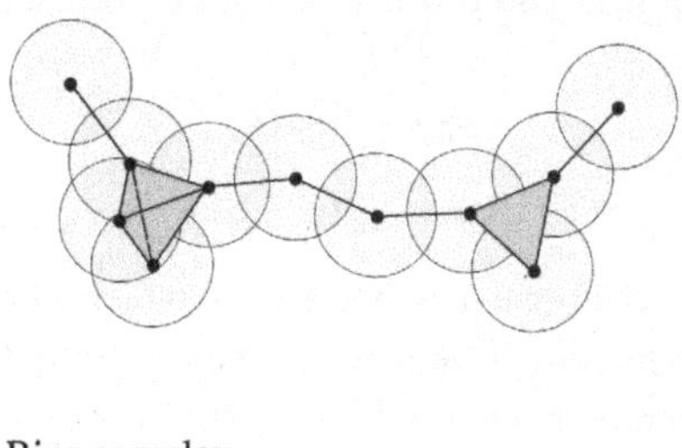

Figure 3.2
Example of a Vietoris–Rips complex.

3.2.3 Head-Driven Interfaces and Vector Models

Now consider the case where the semantic space $\mathcal{S}$ is modeled by a vector space, and assume that it is endowed with a function $\mathfrak{C} : \mathcal{S} \times \mathcal{S} \to \mathbb{R}$ that describes the level of semantic agreement, as in section 3.2.2.1. This may be based on cosine similarity or on other methods: the detailed form of $\mathfrak{C}$ is not important in what follows, beyond the basic properties described in section 3.2.2.1.

3.2.3.1 Max-plus-valued semiring character We discuss here an example where we consider again the max-plus semiring $\mathcal{R} = (\mathbb{R} \cup \{-\infty\}, \max, +)$ and a semantic comparison function of the form $\mathfrak{C} : \mathcal{S} \times \mathcal{S} \to \mathbb{R}$, as discussed in section 3.2.2.1.

Lemma 3.2.15. *Consider the semiring $\mathcal{R} = (\mathbb{R} \cup \{-\infty\}, \max, +)$. The data of a function $\mathfrak{C} : \mathcal{S} \times \mathcal{S} \to \mathbb{R}$ as before, a function $s : \mathcal{SO}_0 \to \mathcal{S}$, and a head function defined on a domain $\mathrm{Dom}(h) \subset \mathfrak{T}_{\mathcal{SO}_0}$ determine a semiring-valued character*

$$\phi_{s,\mathfrak{C},h} : \mathcal{H}^{semi} \to \mathcal{R},$$

where $\mathcal{H}^{semi}$ is the semiring of linear combinations $\sum_i a_i F_i$, with $a_i \geq 0$.

Proof. For any tree $T \notin \mathrm{Dom}(h)$, we set $\phi_{s,\mathfrak{C},h}(T) = -\infty$. We then consider only trees that are in $\mathrm{Dom}(h)$. As in Lemma 3.2.12, we start by considering the case of a tree of the form $T = \mathfrak{M}(\alpha,\beta) = \{\alpha,\beta\}$, with $\alpha,\beta \in \mathcal{SO}_0$. We assign to this tree a value in $\mathcal{R}$ obtained by computing $\mathfrak{C}(s(\alpha), s(\beta)) \in \mathbb{R}$ and considering the line $L_{\alpha,\beta}$, in the vector space $\mathcal{S}$, through the points $s(\alpha)$ and $s(\beta)$:

$$L_{\alpha,\beta} = \{t\alpha + (1-t)\beta = \beta + t(\alpha - \beta) \,|\, t \in \mathbb{R}\},$$

if $\beta = h(T)$ (exchanging α and β if $\alpha = h(T)$; i.e., replacing t with $1-t$). We then define

$$t_{\alpha,\beta} = \mathfrak{C}(\alpha,\beta)$$

$$s(T) := \beta + t_{\alpha,\beta}(\alpha - \beta) \in L_{\alpha,\beta}. \tag{3.2.10}$$

This has the effect of creating a new point $s(T)$, which moves the value $s(h(T))$ along the line $L_{\alpha,\beta}$ in the direction α (or in the opposite direction), depending on the agreement/disagreement sign of $\mathfrak{C}(\alpha,\beta)$. We then set

$$\phi_{s,\mathfrak{C},h}(\mathfrak{M}(\alpha,\beta)) = \begin{cases} \mathfrak{C}(\alpha,\beta) & \beta = h(T) \\ 1 - \mathfrak{C}(\alpha,\beta) & \alpha = h(T). \end{cases}$$

We can then proceed inductively, setting, for $T = \mathfrak{M}(T_1, T_2) \in \mathrm{Dom}(h)$,

$$t_T = \begin{cases} \mathfrak{C}(s(T_1), s(T_2)) & h(T) = h(T_2) \\ 1 - \mathfrak{C}(s(T_1), s(T_2)) & h(T) = h(T_1) \end{cases}$$

$$s(T) = t_T s(T_1) + (1 - t_T) s(T_2)$$

$$= \begin{cases} s(T_2) + t_T(s(T_1) - s(T_2)) & h(T) = h(T_2) \\ s(T_1) + t_T(s(T_2) - s(T_1)) & h(T) = h(T_1) \end{cases}$$

$$\phi_{s,\mathfrak{C},h}(T = \mathfrak{M}(T_1, T_2)) = t_T \,.$$

Setting $\phi_{s,\mathfrak{C},h}(F) = \sum_k \phi_{s,\mathfrak{C},h}(T_k)$ for $F = \sqcup_k T_k$ and

$$\phi_{s,\mathfrak{C},h}(\sum_i a_i F_i) = \max\{a_i \phi_{s,\mathfrak{C},h}(F_i)\}$$

then completely determines $\phi_{s,\mathfrak{C},h}$ on $\mathcal{H}^{semi}$. □

3.2.3.2 Hyperplane arrangements The observation in Lemma 3.2.16 follows from Lemma 3.2.15, rephrased in a more geometric way.

Lemma 3.2.16. *Let $S_{\mathfrak{C}}$ denote the multiplicative subsemigroup of $\mathbb{R}^*$ generated by the set of nonzero elements in $\mathfrak{C}(s(\mathcal{SO}_0) \times s(\mathcal{SO}_0))$. For $T \in \mathfrak{T}_{\mathcal{SO}_0}$ in* Dom(*h*)*, let $L(T)$ be the set of leaves of the tree. We write, for simplicity of notation, $s(L(T))$ for the set of vectors $s(\lambda(L(T))) \subset S$, with λ the labeling in $\mathcal{SO}_0$. Let $S_T \subset S_{\mathfrak{C}} \subset \mathbb{R}^*$ be the multiplicative semigroup generated by the set $\mathbb{R}^* \cap \mathfrak{C}(s(L(T)) \times s(L(T)))$. The vector $s(T)$ of* (3.2.10) *is in the linear span of the set $s(L(T))$, with coefficients in S_T.*

Proof. Consider a binary rooted tree $T \in \mathrm{Dom}(h) \subset \mathfrak{T}_{\mathcal{SO}_0}$, with $L(T)$ being its set of leaves. By the recursive procedure of Lemma 3.2.15, based on the construction of T by repeated application of free symmetric Merge $\mathfrak{M}$, as an element in the magma (3.1.1), the resulting point $s(T)$ in the vector space $\mathcal{S}$ is a linear combination of the vectors $s(\ell)$, with $\ell \in L(T)$, where we write $s(\ell)$ as a shorthand notation for $s(\lambda(\ell))$:

$$s(T) = \sum_{\ell \in L(T)} a_\ell \, s(\ell) \in \mathrm{span}(L(T)) \,,$$

with coefficients a_ℓ in the multiplicative subsemigroup $S_T \subset S_{\mathfrak{C}}$. □

Lemma 3.2.17. *If $\mathfrak{C}$ on the vector space $\mathcal{S}$ is given by a cosine similarity as in* (3.2.4)*, then the set of vectors $s(\mathcal{SO}_0) \subset \mathcal{S}$ determines an associated* *hyperplane arrangement* $\mathcal{HA}_{SO_0}$ *of hyperplanes:*

$$\mathcal{HA}_{SO_0} = \{H_\lambda = \{v \in \mathcal{S} \,|\, \langle v, s(\lambda) \rangle = 0\} \,|\, \lambda \in \mathcal{SO}_0,\ s(\lambda) \neq 0\}\,, \tag{3.2.11}$$

where the hyperplane H_λ *describes all semantic vectors that are neutral with respect to* $s(\lambda)$*; namely, vectors* $v \neq 0$ *with* $\mathfrak{C}(v, s(\lambda)) = 0$.

This is immediate, as the set of hyperplanes here is simply given by the normal hyperplanes to a specified set of vectors under the inner product that also defines the cosine similarity.

One can then see the construction of the character $\phi_{s,\mathfrak{C},h}$ of Lemma 3.2.15 as presented in Lemma 3.2.18.

Lemma 3.2.18. *The vectors* $s(T)$*, for* $T \in \mathrm{Dom}(h) \subset \mathfrak{T}_{SO_0}$*, give a refinement of the hyperplane arrangement* $\mathcal{HA}_{SO_0}$ *of Lemma 3.2.17, with the resulting arrangement*

$$\mathcal{HA}_{SO} = \{H_T = \{v \in \mathcal{S} \,|\, \langle v, s(T) \rangle = 0\} \,|\, T \in \mathfrak{T}_{SO_0},\ s(T) \neq 0\}\,, \tag{3.2.12}$$

where the values $t_{T_v} = \phi_{s,\mathfrak{C},h}(T_v)$*, for* $v \in V(T)$*, determine which chambers of the complement of the arrangement* $\mathcal{HA}_{SO_0}$ *the hyperplane* H_T *crosses.*

Proof. The inductive construction of $\phi_{s,\mathfrak{C},h}$ in Lemma 3.2.15 shows that, for $\alpha, \beta \in \mathcal{SO}_0$ the value $\phi_{s,\mathfrak{C},h}(\mathfrak{M}(\alpha,\beta)) = t_{\alpha,\beta}$ determines which chambers of the complement of $H_\alpha \cup H_\beta$ the hyperplane $H_{\mathfrak{M}(\alpha,\beta)}$ crosses, depending on the sign of $t_{\alpha,\beta}$ and of $1 - t_{\alpha,\beta}$. Inductively, the same applies to the role of $t_T = \phi_{s,\mathfrak{C},h}(T)$ in determining the position of H_T with respect to H_{T_1} and H_{T_2}, and hence the role of the values t_{T_v}, for the accessible terms $T_v \subset T$, in determining the position of H_T with respect to $\mathcal{HA}_{SO_0}$. □

3.2.3.3 ReLU Birkhoff factorization We then consider, in this model, the effect of taking the Birkhoff factorization with respect to the ReLU Rota-Baxter operator of weight +1. Note that this gives an instance of a situation that is quite familiar from the theory of neural networks, where a ReLU function is applied to certain linear combinations and an optimization is performed over the result.

Proposition 3.2.19. *The Birkhoff decomposition of the character* $\phi_{s,\mathfrak{C},h}$ *defined in Lemma 3.2.15, with respect to the ReLU Rota–Baxter operator of weight* +1 *selects, for a given tree* T*, chains* $T_{v_N} \subset T_{v_{N-1}} \subset \cdots \subset T_{v_1} \subset T$ *of accessible terms of* T*, where each* $\phi_{s,\mathfrak{C},h}(T_{v_i}) > 0$ *and of maximal values among all accessible terms of* $T_{v_{i-1}}$*; that is, every* T_{v_i} *optimizes the value of the character among the available accessible terms.*

Proof. As in Lemma 3.2.9, we consider

$$\phi_{s,\mathfrak{C},h,-}(T) = \max\{\phi_{s,\mathfrak{C},h}(T), (\cdots(\phi_{s,\mathfrak{C},h}(F_{\underline{v}_N})^+ + \cdots + \phi_{s,\mathfrak{C},h}(F_{\underline{v}_{i-1}}/F_{\underline{v}_i}))^+ + \cdots)^+ + \phi_{s,\mathfrak{C},h}(T/F_{\underline{v}_0})\}^+$$

over all nested sequences of subforests of arbitrary length $N \geq 1$. For chains of length $N = 1$, considering the case of subtrees $T_v \subset T$, we are comparing $\phi_{s,\mathfrak{C},h}(T)$ and $\phi_{s,\mathfrak{C},h}(T_v)^+ + \phi_{s,\mathfrak{C},h}(T/T_v)$. Again, we have $h((T/T_v)_1) = h(T_1)$ and $h((T/T_v)_2) = h(T_2)$, with $T/T_v = \mathfrak{M}((T/T_v)_1, (T/T_v)_2)$, so $\phi_{s,\mathbb{P},h}(T/T_v) = \phi_{s,\mathbb{P},h}(T)$. Thus, the maximum $\max\{\phi_{s,\mathfrak{C},h}(T), \phi_{s,\mathfrak{C},h}(T_v)^+ + \phi_{s,\mathfrak{C},h}(T/T_v)\}^+ = (\phi_{s,\mathfrak{C},h}(T_v)^+ + \phi_{s,\mathfrak{C},h}(T/T_v))^+$ is achieved at the largest positive value $\phi_{s,\mathfrak{C},h}(T_v)$ over all the accessible terms $T_v \subset T$. The next step then compares this maximal value with the values $(\phi_{s,\mathfrak{C},h}(T_w)^+ + \phi_{s,\mathfrak{C},h}(T_v))^+ + \phi_{s,\mathfrak{C},h}(T)$ over all accessible terms $T_w \subset T_v$ and the maximum is again realized at the largest positive $\phi_{s,\mathfrak{C},h}(T_w)$ among them. This shows that the overall maximum is achieved at the longest chain $T_{v_N} \subset T_{v_{N-1}} \subset \cdots \subset T_{v_1} \subset T$ of accessible terms where each T_{v_i} has $\phi_{s,\mathfrak{C},h}(T_{v_i}) > 0$ and of maximal values among all the accessible terms of $T_{v_{i-1}}$. □

3.2.4 Not a Tensor-Product Semantics

While the examples of characters, Rota–Baxter structures, and Birkhoff factorizations considered here are just a simplified model, they are already detailed enough to illustrate some important points. Consider the property, mentioned in Remark 3.1.4, that characters are *not* morphisms of coalgebras, but only morphisms of algebras (or semirings). This has important consequences, such as the fact that we are *not* dealing here with what is often referred to as "tensor product-based" connectionist models of computational semantics, such as [188] (se also our comparative discussion in section 2.5). The compositional structure of such Smolensky tensor-product models has in our view been rightly criticized (see, for instance, [147]) for not being compatible with human behavior. Indeed one can easily see the problem with such models: the idea of "tensor-product based" compositionality is that, given vectors $s(\alpha), s(\beta) \in \mathcal{S}$ for lexical items α, β, one would assign to a *planar* tree $T = \mathfrak{M}^{nc}(\alpha, \beta)$ a vector $s(\alpha) \otimes s(\beta) \in \mathcal{S} \otimes \mathcal{S}$, and correspondingly evaluate cosine similarity between T and another $T' = \mathfrak{M}_{nc}(\gamma, \delta)$ in the form $\mathfrak{C}(\alpha, \gamma) \cdot \mathfrak{C}(\beta, \delta)$.

There are several obvious problems with such a proposal. In a simple example with lexical items $\alpha = \gamma =$*light* and $\beta =$*blue* and $\delta =$*green*, the planar trees $T =$*light blue* and $T' =$*light green* should have *closer* semantic values $s(T)$ and $s(T')$ than the values $s(\beta)$ and $s(\delta)$ (since both colors share the property of being light), but a measure of similarity of the product form $\mathfrak{C}(\alpha, \gamma) \cdot \mathfrak{C}(\beta, \delta)$

would just be equal to $\mathfrak{C}(\beta,\delta)$. A further issue with these tensor-models, from our perspective, is that this type of model would require previous planarization of trees and cannot be defined at the level of the products of free symmetric Merge.

In contrast, in the type of model that we are discussing, these issues do *not* arise. While we have described in chapter 1 the Merge operation on workspaces in terms of a coproduct on a Hopf algebra of binary rooted forests, which maps to a tensor product $\Delta : \mathcal{H} \to \mathcal{H} \otimes \mathcal{H}$ (since comultiplication has two outputs), the characters used for mapping to semantic spaces have no requirement of compatibility with the coproduct structure. Indeed, in our setting, we would *not* assign to a tree $T = \mathfrak{M}(\alpha,\beta)$ a tensor product of vectors and a product of cosine similarities, but rather a *linear combination* $s(T) = t_T s(\alpha) + (1 - t_T)s(\beta)$, which is indeed seemingly more directly compatible with the empirically observed human behavior, as described in [147].

3.2.5 Boolean Semiring

As a final example of a simple toy model of syntax–semantics interface, in preparation for the discussion of section 3.3, we consider the simplest choice of semiring—namely, the Boolean semiring

$$\mathfrak{B} = (\{0,1\}, \vee, \wedge) = (\{0,1\}, \max, \cdot)\,. \tag{3.2.13}$$

Assignments of values in the Boolean semiring can be regarded as a form of truth-valued semantics, in which one assigns a 0/1 (F/T) value to (parts of) sentences or to syntactic objects.

A map $\phi : \mathfrak{T}_{SO_0} \to \mathfrak{B}$ is an assignment of truth values, extended to $\phi : \mathfrak{F}_{SO_0} \to \mathfrak{B}$ by $\phi(F) = \prod_i \phi(T_i)$ for $F = \sqcup_i T_i$. We use the identity as a Rota–Baxter operator.

The Bogolyubov preparation $\tilde{\phi}$ is then given by

$$\tilde{\phi}(T) = \max\{\phi(T), \phi(F_{\underline{v}})\phi(T/F_{\underline{v}}), \ldots, \phi(F_{\underline{v}_N})\phi(F_{\underline{v}_{N-1}}/F_{\underline{v}_N}) \cdots \phi(T/F_{\underline{v}_1})\}\,, \tag{3.2.14}$$

with the maximum taken over all chains of nested forests of accessible terms. Thus, $\tilde{\phi}$ detects, in cases where the truth value assigned to T may be False ($\phi(T) = 0$), the longest chains of decompositions into accessible terms and their complements, which separately evaluate as True, hence identifying where the truth-value changes from T to F when substructures are combined in the full structure.

While we will return in section 3.9 to a more specific discussion of truth-conditional semantics in the context of Heim–Kratzer semantics, we can use an example we already discussed to illustrate here some known difficulties with that model and possibly some way of reconsidering some of the issues involved. This simple example was mentioned in the context of criticism of truth-conditional semantics in Pietroski's work [168]. It is based on the observation that, while the truth conditions of "*France is a republic*" and "*France is hexagonal*" are satisfied, the sentence "*France is a hexagonal republic*" seems weird due to the semantic mismatch in the expression "*hexagonal republic*".

We view this example in the light of an assignment $\phi : \mathcal{H} \to \mathfrak{B}$ and the corresponding Birkhoff factorization with the identity Rota–Baxter operator, as written previously. We can assume that ϕ assigns value $\phi(T) = 1$ when T has a well-determined associated truth condition, and $\phi(T) = 0$ when it does not. Thus, the trees corresponding to "*France is a republic*" and "*France is hexagonal*" would have value 1 because a country can be a republic and can have a certain type of shape on a map, while the tree corresponding to "*hexagonal republic*" would have value 0 if we agree that a polygonal shape is not one of the attributes of a form of state governance. The tree T that corresponds to "*France is a hexagonal republic*" contains an accessible term T_v, which corresponds to "*hexagonal republic*," and accessible terms (in this case leaves) ℓ and ℓ', which correspond to the lexical items "*hexagonal*" and "*republic*". Each accessible term T_v has a corresponding quotient T/T_v. The Bogolyubov preparation $\tilde{\phi}$ of (3.2.14) then takes the form

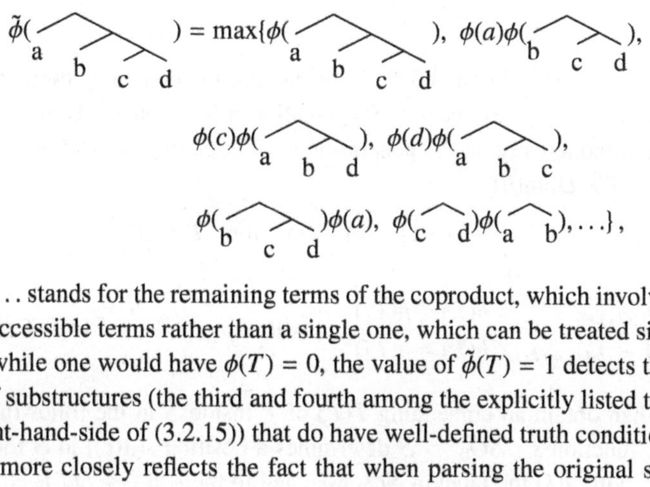

(3.2.15)

where ... stands for the remaining terms of the coproduct, which involve a forest of accessible terms rather than a single one, which can be treated similarly. Thus, while one would have $\phi(T) = 0$, the value of $\tilde{\phi}(T) = 1$ detects the presence of substructures (the third and fourth among the explicitly listed terms on the right-hand-side of (3.2.15)) that do have well-defined truth conditions.

This more closely reflects the fact that when parsing the original sentence for semantic assignments, one does indeed detect the presence of the two substructures that have unproblematic truth conditions, and the fact that these do not combine to assign a truth condition to the full tree T, causing a mismatch between the values of $\phi(T)$ and $\tilde{\phi}(T)$. This manifests itself in the weird impression resulting from the parsing of the full sentence.

3.3 The Image of Syntax Inside Semantics

The examples illustrated here also demonstrate one additional property of this model of syntax–semantics interface: syntactic objects are mapped, together with their compositional structure under Merge, inside semantic spaces, and so they are, at least in principle, reconstructible from this syntactic "shadow" projected on the model used for the representation of semantic proximity relations. This observation is in fact directly relevant to the current controversy about the relationship between large language models (LLMs) and generative linguistics, as we discuss more explicitly in section 3.10. For now, let us add some detail to this picture.

Consider again the setting of Lemma 3.2.12.

Proposition 3.3.1. *Let $\mathcal{S}$ be a semantic space that is a geodesically convex Riemannian manifold, endowed with a semantic proximity function $\mathbb{P}$: $\mathrm{Sym}^2(\mathcal{S}) \to [0,1]$, with the property that for $s \neq s'$, one has $\mathbb{P}(s,s') \in (0,1)$. Also assume that, as before, there is a map $s : \mathcal{SO}_0 \to \mathcal{S}$ that assigns semantic values to lexical items and syntactic features. Let h be a head function with domain $\mathrm{Dom}(h) \subset \mathfrak{T}_{\mathcal{SO}_0}$. These data determine embeddings of trees $T \in \mathrm{Dom}(h)$ inside the semantic space $\mathcal{S}$.*

Proof. Arguing as in Lemma 3.2.12, we can use the convexity property of $\mathcal{S}$ and the function $\mathbb{P}$ to extend $s : \mathcal{SO}_0 \to \mathcal{S}$ to a function $s : \mathrm{Dom}(h) \to \mathcal{S}$, constructed inductively on the generation via Merge of objects $T \in \mathfrak{T}_{\mathcal{SO}_0}$, by setting, for $T \in \mathrm{Dom}(h)$,

$$s(T) = p\, s(T_1) + (1-p)\, s(T_2) \quad \text{for} \quad T = \overset{\frown}{T_1 \quad T_2} \tag{3.3.1}$$

$$p = \begin{cases} p_{s(T_1),s(T_2)} & h(T) = h(T_1) \\ 1 - p_{s(T_1),s(T_2)} & h(T) = h(T_2) \end{cases} \quad \text{with} \quad p_{s(T_1),s(T_2)} = \mathbb{P}(s(T_1), s(T_2)) . \tag{3.3.2}$$

We can then obtain an embedding $\mathcal{I}(T)$ of T inside $\mathcal{S}$ in the following way. First, the function $s : \mathcal{SO}_0 \to \mathcal{S}$ determines a position $s(\lambda(\ell))$ in $\mathcal{S}$ for every leaf of T, with $\lambda(\ell)$ the label in $\mathcal{SO}_0$ assigned to the leaf $\ell \in L(T)$. Note that the same lexical item $\lambda \in \mathcal{SO}_0$ may be assigned to more than one leaf in $L(T)$, resulting in an assignment $\ell \mapsto s(\lambda(\ell))$ that is not always an embedding of $L(T)$ in $\mathcal{S}$. For each pair $\ell, \ell' \in L(T)$ that are adjacent in T, consider the syntactic object

$$T_{v_{\ell,\ell'}} = \overset{\frown}{\ell \quad \ell'}$$

where we denote by $v_{\ell,\ell'}$ the vertex above the leaves ℓ, ℓ'. The subtree $T_{v_{\ell,\ell'}}$ is in Dom(h), since T is. We can assume that $\ell = h(T_{v_{\ell,\ell'}})$: the case with ℓ' is analogous. Then (3.3.1) assigns to $T_{v_{\ell,\ell'}}$ a point in $\mathcal{S}$ on the geodesic arc between $s(\lambda(\ell))$ and $s(\lambda(\ell'))$, where these two points are distinct since $T_{v_{\ell,\ell'}} \in$ Dom(h). We then obtain embeddings of all the subtrees $T_{v_{\ell,\ell'}}$ in $\mathcal{S}$ by taking the image $\mathcal{I}(T_{v_{\ell,\ell'}})$ to consist of the geodesic arc $ts(\lambda(\ell)) + (1-t)s(\lambda(\ell'))$, with $t \in [0,1]$ with a marked point (the root $s(T_{v_{\ell,\ell'}})$ of $\mathcal{I}(T_{v_{\ell,\ell'}})$) placed at the position $t = p_{s(\ell),s(\ell')}$, as in (3.3.1) and (3.3.2).

We proceed similarly for the subsequent steps of the construction of T in the $\mathcal{SO}$ magma by obtaining the image $\mathcal{I}(T_v)$ of a subtree T_v as the union of the images $\mathcal{I}(T_{v,1})$ and $\mathcal{I}(T_{v,2})$, where $T_v = \mathfrak{M}(T_{v,1}, T_{v,2})$, and an additional geodesic arc between the points $s(T_{v,1})$ and $s(T_{v,2})$, with a marked point (the root vertex $s(T_v)$ of $\mathcal{I}(T_v)$) placed along this geodesic arc according to (3.3.1) and (3.3.2). The images $\mathcal{I}(T)$ of trees $T \in$ Dom(h) constructed in this way are in general immersions rather than simply embeddings because of the possible coincidence of the points assigned to some of the leaves, as well as because of possible intersections of the geodesic arcs at points that are not tree vertices. Both of these issues can be readily resolved to obtain embeddings. Indeed, the semantic space $\mathcal{S}$ will be generally high dimensional. So long as it has dimension greater than two, crossings of strands of a diagram can be eliminated by a very small perturbation. In the case of leaves carrying the same lexical item, one can argue that the different context (in the sense of the different subtree) in which the item appears will naturally slightly modify its semantic location in $\mathcal{S}$. This can be modeled by a small movement of the endpoints of the geodesic arc to the interior of the arc (which functions as modifier of the semantic proximity relations). This deforms the immersions to embeddings. □

The resolution of possible coincident assignments of semantic points at the leaves, as mentioned in the proof, can also be seen from the fact that repeated lexical items at the leaves are, in the linguistic sense, "repetitions" not "copies", which means that they do not play identical roles in the sentence represented by the structure T. This difference in roles makes it possible to resolve the ambiguity of the noninjective assignment of points $\ell \mapsto s(\lambda(\ell))$, by slightly moving the points $s(\lambda(\ell))$ in $\mathcal{S}$ in different directions marked by the different roles of the lexical items $\lambda(\ell)$. Thus, it is always possible to assume, without loss of generality, that the map $\ell \mapsto s(\lambda(\ell))$ is an embedding.

The assumption that the function $\mathbb{P}$ that measures semantic relatedness has values $0 < \mathbb{P}(s, s') < 1$ whenever $s \neq s'$ means that we want to model a situation where different points in the semantic space $\mathcal{S}$ are never completely semantically disjoint or entirely coincident. In such a semantic space model, even an apparently "nonsensical" pair would not score 0 under the function $\mathbb{P}$, so, for example, different locations in $\mathcal{S}$ would distinguish "colorless green" from "colorless red", as different (mental) images of absence of green rather than red. The fact that the expression is semantically awkward would correspond to a small (but nonzero) value of $\mathbb{P}(s, s')$ that affects the metric shape of the resulting image tree (which in the geometric setting we describe in section 3.4, will end up located very near a boundary stratum of the relevant moduli space).

On the other hand, if we allow the possibility that $\mathbb{P}(s, s') = 0$ or $\mathbb{P}(s, s') = 1$ for some pairs $s \neq s'$, the construction of Proposition 3.3.1 would no longer yield an embedding since for lexical items mapped to such pairs, the root of the associated Merge tree would map to one leaf or the other, rather than to an intermediate point. Such models will result in certain syntactic trees being mapped to degenerate image trees in $\mathcal{S}$, which are located not just near, but on the boundary strata of the moduli spaces we introduce in section 3.4. Such cases should also be taken into consideration. (We will see the relevance of this in the context of Pietroski's semantics in section 3.6.) Here, we focus on models where this situation can be avoided.

It is important to note that the image of the syntactic trees $T \in \text{Dom}(h) \subset \mathfrak{T}_{\mathcal{SO}_0}$ inside the semantic space $\mathcal{S}$ is like a static photographic image, rather than a dynamical computational process. Indeed, *all* computational manipulations of syntactic objects are performed by Merge on the syntax side of the interface, not inside the space $\mathcal{S}$, which does not have on its own a computational structure. The only properties of $\mathcal{S}$ that are used to obtain an embedded copy of the syntactic trees are proximity relations (here realized in the form of geodesic convexity).

In particular, given that this construction determines an embedding of syntactic trees in semantic spaces, one can consider the *inverse problem* of reconstructing syntactic objects and the action of Merge from their image under this embedding. In other words, given enough measurements of semantic proximities in text, can we reconstruct the underlying generative process of syntax? Since the computational mechanism of syntax is not directly acting on semantic spaces, and one is only able to see the embedding of the syntactic objects, it is reasonable to expect that this inverse problem (reconstructing the map $\tilde{\phi}$ of the syntax–semantics interface from the embedding $\mathcal{I}$) could be, and we suspect probably is, computationally hard. (See [134] for recent work that in a certain sense attempts to solve this problem, but not within the explicit framework described here.) We will discuss another instance of this problem, in the context of LLMs, in section section 3.10. In the next section, we further discuss the image of syntax inside semantics and its relation to the Externalization of free symmetric Merge.

3.4 Head Functions, Moduli Spaces, Associahedra, and Externalization

We now revisit the simple model of section 3.2.2.3, with the recursive construction of semantic values associated with trees in the domain of a head function. We view here the same construction in terms of points in a moduli space of metric trees introduced in [53], related to moduli spaces of real curves of genus zero with marked points. We will show that this viewpoint provides further insight into the geometry of an Externalization process that introduces language-dependent planarization of the syntactic trees, and the interaction between the core generative process of free symmetric Merge and the Conceptual–Intensional system (the syntax–semantics interface), and an Externalization mechanism that interfaces the same core computational process with the Articulatory-Perceptual or Sensory–Motor system. This will provide a more careful and elaborate explanation of the viewpoint sketched in the introduction regarding independence of the syntax–semantics interface and Externalization. The relation between these two mechanisms can also be approached in a geometric form.

3.4.1 Preliminary Discussion

In the formulation of Minimalism in terms of free symmetric Merge as the core computational mechanism, as presented in [40] and formalized mathematically in the previous chapters, the generative process of syntax produces hierarchical structures through syntactic objects and the action of Merge on workspaces in terms of the Hopf algebra $\mathcal{H}$ of binary rooted forests (with no assigned planar structure). A mechanism of *Externalization* takes place after this generative process. This mechanism describes the connection to the Sensory–Motor system, which due to its physical and physiological nature externalizes language in the form of a temporally ordered sequence of words, realized as sounds or signs or writing (or, inversely, for parsing). The necessity of temporal ordering in the Externalization of language requires a *planarization* of the binary rooted trees (syntactic objects), as the choice of a planar structure is equivalent to the choice of an ordering of the leaves. This choice of planarization is subject to language-dependent constraints, through the syntactic parameters of languages. In chapter 1 we proposed a mathematical formalism for Externalization based on a suitable notion of correspondences.

In the previous sections of this chapter, we analyzed possible models (some of them highly simplified) of how the products of the free Merge generative process of syntax can be mapped to semantic spaces, where the main property of semantic space that we have used is a notion of topological/metric proximity. This type of mapping of syntax to semantics is designed to directly apply to the hierarchical structures produced by free symmetric Merge without having to first pass through the choice of a planar structure as is done in the Externalization process. This mapping to semantic spaces represents the interaction between the core computational mechanism of Merge with the Conceptual–Intensional system.

These two mechanisms are illustrated in the form of the two top arrows as depicted in Figure 3.3. This part of the picture corresponds to property (3) on the list in section 3.1.1—namely, that semantic interpretation is, to a large extent, independent of Externalization. However, obviously the Externalization process and the mapping to semantic spaces need to be compatibly combined, as Figure 3.3 suggests. The goal of the rest of this section is to introduce the mathematical framework in which both processes simultaneously coexist.

It is important to stress here that what we refer to as a "combined process" in Figure 3.3 and later in this section is not incorporated into the modeling of the generative process of Merge, as formulated in the *Elements* text [40], as for that purpose, it suffices to regard as separate the two interface channels with the Conceptual–Intensional system (syntax–semantics interface) and with the Sensory–Motor system (Externalization). Where it becomes useful to consider a common parameterizing space where these two channels can be simultaneously accessed is at the *receiver's end*, or, in other words, when one models *parsing* rather than production. In particular, compatibility is necessary since one receives a product of Externalization and, in proceeding to its syntactic and semantic parsing, one needs to be able to relate it to the productions of the free symmetric Merge and their mapping to semantic spaces. We will see that indeed, there is a good geometric setting for such comparisons, and an "origami folding" projection map (discussed in section 3.4.4) that undoes the planarization of Externalization while maintaining compatible mapping to a semantic space.

We proceed in the following way. First, we introduce a framework designed for the comparison of different planar structures on the same abstract binary rooted tree. Since the planarization of Externalization is language dependent, we need a space where different planarization can be considered. Such a space is well studied in mathematics and is called the *associahedron*. We recall its properties in section 3.4.2. At the same time, we want to keep track of the fact

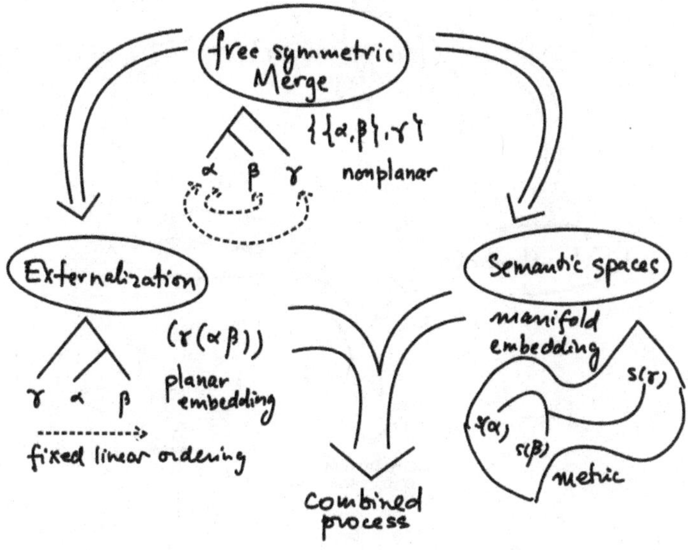

Figure 3.3
Free symmetric Merge, Externalization, and semantic spaces.

that the hierarchical structures produced by free symmetric Merge have also acquired a metric structure through its mapping to semantic spaces, where this metric structure keeps track of information about semantic relatedness across substructures. This assignment of metric data on nonplanar binary rooted trees is also described by a well known mathematical object, the Billera–Holmes–Vogtmann (BHV) moduli space, which we also discuss in section 3.4.2.

Thus, we present a formulation where, taken separately (as in the top arrows of Figure 3.3) the Externalization and the mapping to semantic spaces result in the assignment to a given syntactic object $T \in \mathcal{SO}$ with n leaves of a vertex in the K_n associahedron, and of a point in the BHV_n moduli space, respectively.

These two geometric objects, the associahedron and the BHV moduli space, naturally combine into another space, which accounts for what happens when we enrich the combinatorial associahedron with metric data. This is again a geometric object that is very well known in mathematics, where it is identified with a certain moduli space of curves, $\bar{M}^{or}_{0,n}(\mathbb{R})$. We review in section 3.4.2 the relation between these three fundamental spaces K_n, BHV_n, and $\bar{M}^{or}_{0,n}(\mathbb{R})$; see Figure 3.4.

In the sections 3.4.3 and 3.4.4, we explain more in detail how the mapping to semantic spaces and Externalization can be seen in this perspective. We

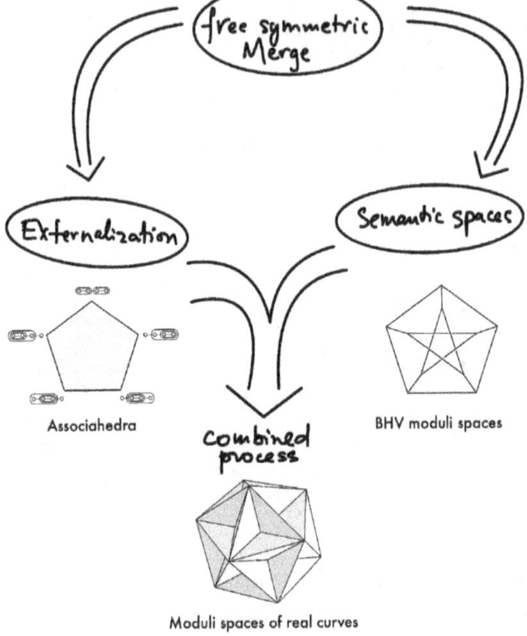

Figure 3.4
Free symmetric Merge, Externalization, the syntax–semantics interface, and the respective moduli spaces.

include a discussion of how Richard Kayne's Linear Correspondence Axiom (LCA) algorithm, Guglielmo Cinque's Abstract Functional Lexicon, and constraints implemented by syntactic parameters appear in this formulation.

3.4.2 Associahedra and Moduli Spaces

We recall here some general facts about moduli spaces of abstract and planar binary trees, and their relation to the moduli space of genus zero real curves with marked points. For a more detailed account, we refer the reader to [11], [13], and [53].

The Stasheff associahedron K_n is a convex polytope of dimension $n - 2$, where the vertices correspond to all the balanced parentheses insertions on an ordered string of n symbols (equivalently, all planar binary rooted trees on n leaves) and the edges are given by a single application of the associativity rule. For example, the one-dimensional associahedron K_3 is the graph with a single edge and two vertices:

$$((ab)c) \longleftrightarrow (a(bc)).$$

The two-dimensional associahedron K_4 is similarly a pentagon, while the three-dimensional associahedron K_5 is illustrated in Figure 3.5. Faces of the associahedron K_n are products of lower-dimensional associahedra. These strata K_{n_i} correspond to the degeneration of a binary tree where some of the internal vertices acquire higher valencies. The description in terms of planar binary rooted trees has an equivalent formulation in terms of triangulations of an $(n+1)$–gon by drawing diagonals.

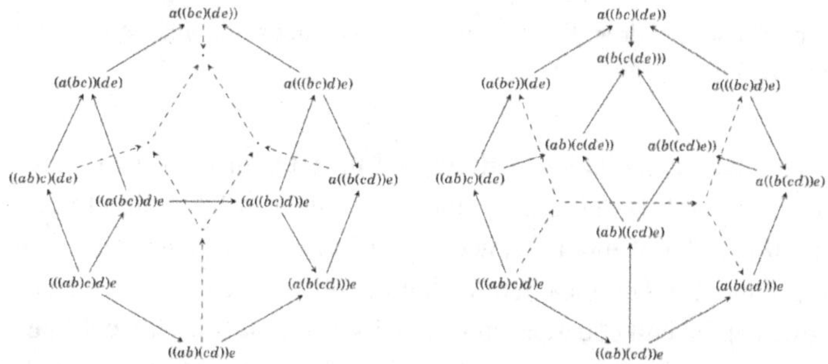

Figure 3.5
The Stasheff associahedron K_5, front and back view.

Boardman and Vogt [13] showed that the associahedron K_n can be decomposed into C_{n-1} cubes of dimension $n-2$, where C_{n-1} is the Catalan number

$$C_{n-1} = \frac{1}{n}\binom{2n-2}{n-1}.$$

The decomposition of the associahedron K_4 is illustrated in Figure 3.6.

Each vertex of the associahedron can be identified with a *planar* binary rooted tree. A way to interpret the polytope points here is as metric structures on planar binary rooted trees that assign weights in $\mathbb{R}_{\geq 0}$ to the internal edges of the tree, with degeneracies along the faces and vertices of the cubic decomposition; see Figure 3.6 for K_4 (and the corresponding discussion in [53]). Each cube in the decomposition parametrizes the normalized choices of weights for the internal edges for the planar tree structure associated with that cube, and the faces are glued according to the transitions from one tree structure to an adjacent one, as dictated by the associahedron structure.

It was further shown in [53], that the Devadoss–Morava tree spaces give a parametrization of planar binary rooted trees with weights on the internal

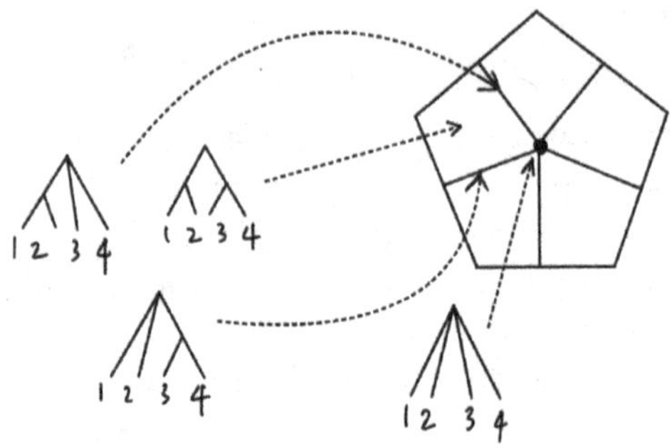

Figure 3.6
The Stasheff associahedron K_4 with its cubic decomposition and parametrization of planar metric binary rooted trees.

edges. This parameterization is described in terms of the (open cells of the) associahedron, and its cubic decomposition can then be related to compactifications $\overline{M}_{0,n+1}(\mathbb{R})$ of moduli spaces of real curves of genus zero with $n + 1$ marked points. The key idea here is that the ordered leaves of a planar binary rooted tree can be embedded as an ordered set of points in the real line, where the coordinates of the points are obtained from the weights assigned at the internal edges of the tree as a function e^{-W} of the sum W of the weights along the path from the root to one of the leaves (see the example in Figure 3.7). Note that, while the open cells of the associahedron correspond to binary trees, the boundary strata of these cells contain trees with higher valences (corresponding to the limits of binary trees when one or more of the edge lengths go to zero). Since the trees coming from syntax are binary (see section 1.11 of chapter 1 for why Merge operators with higher arity are excluded), the image from syntax will lie inside the open cells. The boundary structure is still important, though, because the boundaries of cells in the associahedron encode all the possible structural changes to the underlying hierarchical structures (syntactic objects). Moreover, all boundary strata will become relevant when FormSet is also included in the structure formation process, as discussed in section 1.16 of chapter 1.

As shown by Devadoss in [52], the orientation double cover $\overline{M}^{or}_{0,n+1}(\mathbb{R})$ of $\overline{M}_{0,n+1}(\mathbb{R})$ can be decomposed into a collection of $n!$ copies of the associahedron K_n, where the $n!/2$ associahedra of $\overline{M}_{0,n+1}(\mathbb{R})$ correspond to the permutations of the $n + 1$ points on the real line preserving the cyclic order of $\{0, 1, \infty\}$, with gluings corresponding to certain twist operations on the triangulated $(n + 1)$–gons (see Figure 3.10 for the example of $n = 3$). Note that for $n \leq 3$, the moduli space $\overline{M}_{0,n+1}(\mathbb{R})$ is orientable, so one does not see the role

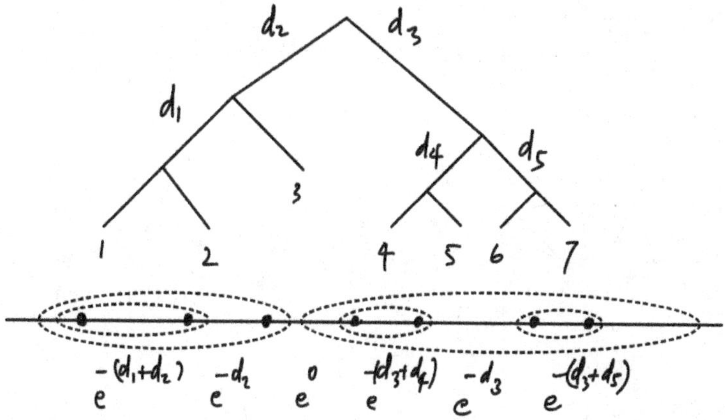

Figure 3.7
A planar binary rooted tree with weighted internal edges and the associated ordered configuration of points on the real line, as shown in [53].

of the orientation double cover; see [53] for a more detailed discussion of the more general case.

One can also consider the moduli space BHV_n of abstract binary rooted trees with n leaves (with no assigned planar structure), along with weighted internal edges, and their one-point compactification BHV^+, constructed by Billera, Holmes, and Vogtmann [11]. The moduli space BHV_n is obtained by considering all the $(2n-3)!!$ abstract binary rooted trees with n labeled leaves. All these trees have $n-2$ internal edges. For each tree, one considers an orthant $\mathbb{R}^{n-2}_{\geq 0}$, which represents all the possible choices of a weight (length) for the internal edges. These orthants are glued along the common faces (which correspond to shrinking one of the internal edges), and this gives the space BHV_n. The link $\mathcal{L}_n$ of the origin in BHV_n is an $(n-3)$–dimensional simplicial complex. In the case $n = 3$ it consists of three points. For $n = 4$ it is the Peterson graph of Figure 3.9. In general, there are $(2n-3)!!$ top $(n-3)$–dimensional simplexes of $\mathcal{L}_n$ (e.g., 15 edges in the case of $\mathcal{L}_4$) that correspond to the different trees, and two of them share a face when the corresponding trees give rise to the same quotient tree when contracting an internal edge.

It is shown in [53] that there is a projection map between these moduli spaces:

$$\Pi_n : \overline{M}^{or}_{0,n+1}(\mathbb{R}) \twoheadrightarrow \mathrm{BHV}^+_n\,, \tag{3.4.1}$$

with a finite projection that is generically 2^{n-1}-to-1, obtained by folding of the cubes of the cubical decomposition of the associahedra in the moduli space

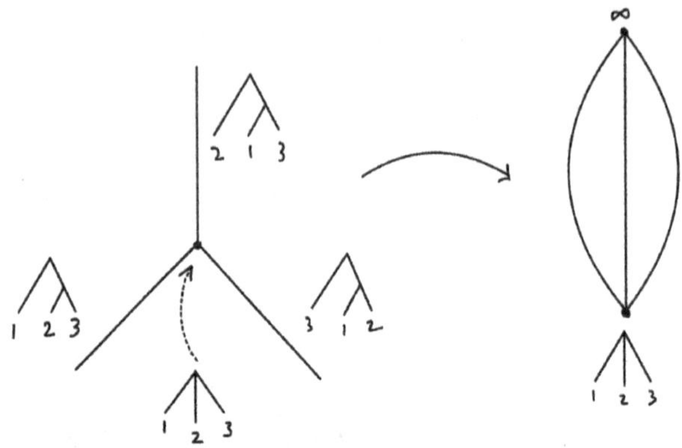

Figure 3.8
The moduli space BHV_3 of abstract binary rooted trees and its one-point compactification BHV_3^+.

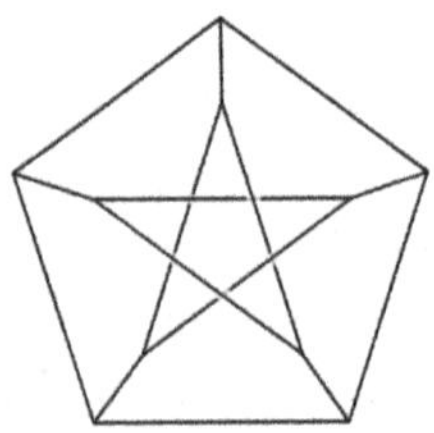

Figure 3.9
The Peterson graph is the link $\mathcal{L}_4$ of the origin in BHV_4.

$\overline{M}^{or}_{0,n+1}(\mathbb{R})$ (this is referred to as "origami folding" in [53]). The folding projection is done according to the formula

$$n! \cdot C_{n-1} = 2^{n-1} \cdot (2n-3)!!\,,$$

where the left-hand-side lists the C_{n-1} cubes of the $n!$ associahedra inside $\overline{M}^{or}_{0,n+1}(\mathbb{R})$, and the right-hand-side lists the $(2n-3)!!$ simplexes of dimension $(n-3)$ of $\mathcal{L}_n$, and 2^{n-1} is the multiplicity of the generic fibers of the projection map. Note that 2^{n-1} is the number of different planar structures for a given abstract binary rooted tree on n leaves, since such a tree has $n-1$ non-leaf vertices and the total number of planar embeddings can be obtained by choosing one of two possible planar embeddings (left/right) for each pair of edges below a given non-leaf vertex. The origami folding quotient takes each $(n-2)$–dimensional cube and folds it in half in each direction, obtaining 2^{n-2} foldings, with two copies of each cube in the orientation double cover, so that one obtains 2^{n-1} points in each general fiber. We will see this more explicitly in section 3.4.3, applied to our setting.

3.4.3 Head Functions, Convex Semantic Spaces, and Metric Trees

With these facts in hand, now consider again the setting discussed in our simple example of section 3.2.2.3.

Consider the set of all $(2n-3)!!$ abstract binary rooted trees with n labeled leaves. Suppose that the leaves are labeled by a given multiset $\{\lambda_i\}_{i=1}^n$ of lexical items and syntactic features in $\mathcal{SO}_0$. If this is a multiset instead of a set, we still interpret the multiple copies of a given item in $\mathcal{SO}_0$ as *repetitions*, not as *copies*, in the sense that they can play different roles in structure formation via applications of Merge—hence, we will still regard them as distinct labels. Thus, we have the following geometric description of our data:

- For any choice of the lexical items associated to the leaves, we obtain a corresponding copy of the moduli space BHV_n.
- The link of the origin $\mathcal{L}_n \subset \mathrm{BHV}_n$ can be seen as an assignment of weights to the internal edges that is normalized (e.g., by requiring that the total sum of weights is equal to 1).
- We write $\mathrm{BHV}_n(\Lambda)$ for $\Lambda = \{\lambda_i\}_{i=1}^n$ for the copy of BHV_n that corresponds to the given choice Λ of the lexical items assigned to the leaves.
- We similarly write $\mathcal{L}_n(\Lambda)$ for the associated copy of $\mathcal{L}_n$.

Proposition 3.4.1. *The choice of a head function h determines simplicial subcomplexes* $\mathcal{L}_n(\Lambda, h) \subset \mathcal{L}_n(\Lambda)$, $\mathrm{BHV}_n(\Lambda, h) \subset \mathrm{BHV}_n(\Lambda)$, $M_n(\Lambda, h) \subset \overline{M}^{or}_{0,n+1}(\mathbb{R})$, *compatible with the maps relating these moduli spaces. It also identifies a lift of* $\mathcal{L}_n(\Lambda, h)$ *and* $\mathrm{BHV}_n(\Lambda, h)$ *inside* $M_n(\Lambda, h)$, *which is determined by the planar structure* π_h *associated with the head function (i.e., the harmonic head-initial LCA planarization).*

Proof. The choice of the head function h selects, for each of these copies $\mathcal{L}_n(\Lambda) \subset \mathrm{BHV}_n(\Lambda)$, a simplicial subcomplex $\mathcal{L}_n(\Lambda, h) \subset \mathcal{L}_n(\Lambda)$ and the associated cone $\mathrm{BHV}_n(\Lambda, h) \subset \mathrm{BHV}_n(\Lambda)$, where the set of top $(n-3)$–dimensional simplexes of $\mathcal{L}_n(\Lambda, h)$ corresponds to the subset of the given $(2n-3)!!$ trees that belong to $\mathrm{Dom}(h)$.

Let $M_n(\Lambda, h) \subset \overline{M}^{or}_{0,n+1}(\mathbb{R})$ denote the locus in $\overline{M}^{or}_{0,n+1}(\mathbb{R})$ obtained as a image under the projection map of the image $\mathrm{BHV}_n(\Lambda, h)^+$ in the one-point compactification BHV_n^+ of the cone $\mathrm{BHV}_n(\Lambda, h)$:

$$M_n(\Lambda, h) := \Pi_n^{-1}(\mathrm{BHV}_n(\Lambda, h)^+) . \tag{3.4.2}$$

A point in $\mathrm{BHV}_n(\Lambda, h)$ is a pair $(T, \underline{\ell})$ of an abstract binary rooted tree on n leaves labeled by the points of Λ together with a set $\underline{\ell} = (\ell_k)_{k=1}^{n-2}$ of weights $\ell_i \in \mathbb{R}_{\geq 0}$ assigned to the internal edges of T. The 2^{n-1} points in the fiber $\Pi_n^{-1}(T, \underline{\ell}) \subset M_n(\Lambda, h)$ are given by the points $(T^\pi, \underline{\ell})$, where T^π ranges over all the possible planarizations π of T and the lengths of the internal edges stay the same.

We have seen that the choice of a head function h determines an associated planar structure π_h for all trees $T \in \mathrm{Dom}(h)$. Thus, the choice of a head function determines a lift of the subcomplex $\mathrm{BHV}_n(\Lambda, h)$ (and in particular of $\mathcal{L}_n(\Lambda, h) \subset \mathcal{L}_n(\Lambda)$) to a subcomplex of $M_n(\Lambda, h) \subset \overline{M}^{or}_{0,n+1}(\mathbb{R})$. □

Consider then, as in section 3.2.2.3, a semantic space $\mathcal{S}$ that is a geodesically convex subspace of a Riemannian manifold, together with a map $s : \mathcal{SO}_0 \to \mathcal{S}$. Assume that, for points in $\mathcal{S}$, we can evaluate the frequency of semantic relatedness in a specified context in terms of a biconcave function $\mathbb{P} : \mathrm{Sym}^2(\mathcal{S}) \to [0, 1]$.

Proposition 3.4.2. *Let $T \in \mathrm{Dom}(h) \subset \mathfrak{T}_{\mathcal{SO}_0}$ be a tree with n leaves. The data $(s : \mathcal{SO}_0 \to \mathcal{S}, \mathbb{P})$ determine a set $\underline{\ell} = (\ell_k)_{k=1}^{n-2} \in \mathbb{R}^{n-2}_{\geq 0}$ of weights assigned to the internal edges of T. Thus, the data $(s : \mathcal{SO}_0 \to \mathcal{S}, \mathbb{P})$ determine a point $(T, \underline{\ell}^{(h,s,\mathbb{P})}) \in \mathcal{L}_n(\Lambda, h)$ and a point in the corresponding fiber of the projection from $M_n(\Lambda, h)$.*

Proof. To see this, we proceed as in section 3.2.2.3. For each of the $n - 2$ vertices v of T that is neither the root nor one of the leaves, consider the two subtrees $T_{v,1}$ and T_{v_2} that have root vertices v_1, v_2 immediately below v, and compute $p_v := \mathbb{P}(s(T_{v,1}), s(T_{v,2}))$, where $h(T_{v,i})$ is the head leaf of the subtree $T_{v,i}$, used to determine the position of the points $s(T_{v,i})$. We label the $(n - 2)$ internal edges by the target vertex v (where the tree is oriented away from the root), and we take $\ell_v = p_v$.

Thus, we have that, for a tree $T \in \mathrm{Dom}(h) \subset \mathfrak{T}_{\mathcal{SO}_0}$ on n leaves labeled by Λ, the choice of a head function h, together with the choice of a map $s : \mathcal{SO}_0 \to \mathcal{S}$ and of the function $\mathbb{P} : \mathrm{Sym}^2(\mathcal{S}) \to [0, 1]$, determines, after an overall normalization of the weights, a point $(T, \underline{\ell}^{(h,s,\mathbb{P})}) \in \mathcal{L}_n(\Lambda, h)$, and a corresponding point $(T^{\pi_h}, \underline{\ell}^{(h,s,\mathbb{P})}) \in \mathcal{L}_n(\Lambda, h)$ in the fiber above $(T, \underline{\ell}^{(h,s,\mathbb{P})})$ in $M_n(\Lambda, h) \subset \overline{M}^{or}_{0,n+1}(\mathbb{R})$. □

Remark 3.4.3. Note that in section 3.2.2.3, we used the same coordinates:

$$\mathbb{P}(s(T_{v,1}), s(T_{v,2}))$$

to assign points $s(T_v) = p_v s(T_{v,1}) + (1 - p_v)s(T_{v,2})$ or $s(T_v) = p_v s(T_{v,2}) + (1 - p_v)s(T_{v,1})$ (according to whether the head $h(T_v)$ matches the head of either of the two subtrees). Thus, according to this construction, the weight of an internal edges of T obtained as in Proposition 3.4.2 reflects the positions in the semantic space $\mathcal{S}$ of the accessible term below that edge.

As a result, we can view the construction of the character $\phi_{s,\mathbb{P},h}$ of section 3.2.2.3 equivalently as the construction of a section.

Corollary 3.4.4. *The construction of the character $\phi_{s,\mathbb{P},h}$ of section 3.2.2.3 is equivalent to the construction of a partially defined section:*

$$\sigma_{s,\mathbb{P},h,n} : \mathrm{BHV}_n \to \overline{M}^{or}_{0,n+1}(\mathbb{R}), \tag{3.4.3}$$

which is defined over

$$\mathrm{Dom}(\sigma_{s,\mathbb{P},h,n}) = \mathrm{BHV}_n(\Lambda, h),$$

and a partially defined map

$$s_{\mathbb{P},h} : \mathfrak{T}_{\mathcal{SO}_0} \to \cup_n \mathcal{L}_n(\Lambda, h) \tag{3.4.4}$$

with $\mathrm{Dom}(s_{\mathbb{P},h}) = \mathrm{Dom}(h)$. *The construction of the character $\phi_{s,\mathbb{P},h}$ of section 3.2.2.3 is equivalent to the construction of the composite map $\sigma_{s,\mathbb{P},h} \circ s_{\mathbb{P},h}$.*

For the case $n = 3$, the projection maps are illustrated in Figure 3.10 (see also [53]).

We have described the construction here in terms of the simple model of assignment of semantic values to syntactic objects described in section 3.2.2.3. This can be adapted to other models, so we can incorporate, as part of the modeling of the syntax–semantics interface, the construction of a partially defined section:

$$\sigma_{\mathcal{S},n} : \mathrm{Dom}(\sigma_{\mathcal{S}}) \subset \mathrm{BHV}_n \to \overline{M}^{or}_{0,n+1}(\mathbb{R}), \tag{3.4.5}$$

which depends on the model of semantic space $\mathcal{S}$ used and on its properties. Similarly, the map (3.4.4) can be generalized as a map

$$s_{\mathcal{S},h,n} : \mathfrak{T}_{\mathcal{SO}_0} \to \mathcal{L}_n \cap \mathrm{Dom}(\sigma_{\mathcal{S},n}). \tag{3.4.6}$$

3.4.4 Origami Folding and Externalization

In chapter 1, we gave an account of Externalization as a section of the projection from planar to abstract binary rooted trees, where the section is language dependent and is chosen such that the resulting planar structure is compatible

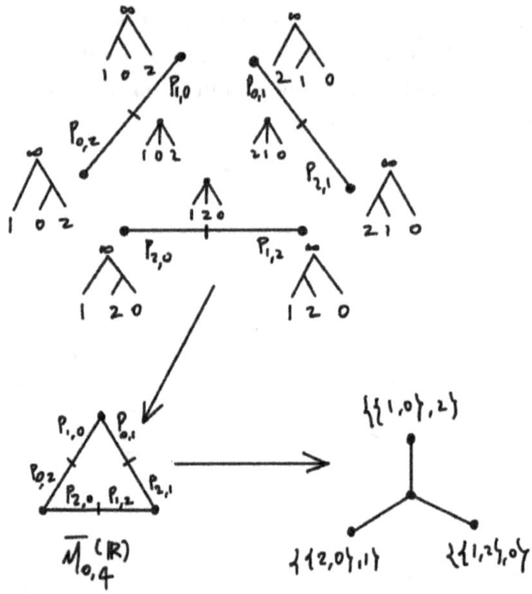

Figure 3.10
The projections from three associahedra K_3 to the moduli space $\overline{M}_{0,4}(\mathbb{R})$ and to the BHV_3^+ moduli space and the embedding map $\mathcal{I}$ of syntactic trees to semantic space $\mathcal{S}$ seen from the point of view of moduli spaces.

with certain syntactic parameters through the effect that these have on word order.

In terms of the geometry of moduli spaces described here, one can similarly view Externalization as the choice of a *section* of the origami folding projection (3.4.1) (see [53]):

$$\sigma_{L,n} : \mathrm{BHV}_n \to \overline{M}^{or}_{0,n+1}(\mathbb{R}), \tag{3.4.7}$$

depending on a specified language L through its syntactic parameters. The origami folding map (3.4.1) is then the projection that undoes the effect of Externalization and relates the moduli space of curves describing the combined effect of Externalization and mapping to semantic space to the mapping of the products of free symmetric Merge, before Externalization, to semantic space. The section (3.4.7) is defined at the level of the combinatorial trees, as a choice of a section $\sigma_{L,n} : \mathfrak{T}_{SO_0,n} \to \mathfrak{T}^{pl}_{SO_0,n}$ that assigns a planar structure, as discussed in chapter 1. It is extended to metric trees as the identity on the metric datum $\underline{\ell}$, since Externalization is decoupled from the metric structure, reflecting our initial assumption on the independence of semantic values from Externalization. This independence assumption only affects this independence of $\sigma_{L,n}$ on the metric structure. It does not mean that there would be *no* interaction with the semantics channel. One way to model such interaction is by comparing the two sections $\sigma_{L,n}$ and $\sigma_{S,n}$ on the subdomain $\mathrm{Dom}(\sigma_{S,n}) \subset \mathrm{BHV}_n$, where both are defined, and in particular on the target of the map $s_{S,h,n}$ of (3.4.6).

3.4.5 An Example

All these discussions of the relation between Externalization and the syntax–semantics interface in terms of moduli spaces are quite abstract. Let us illustrate what is happening with a very simple example. Consider a sentence such as

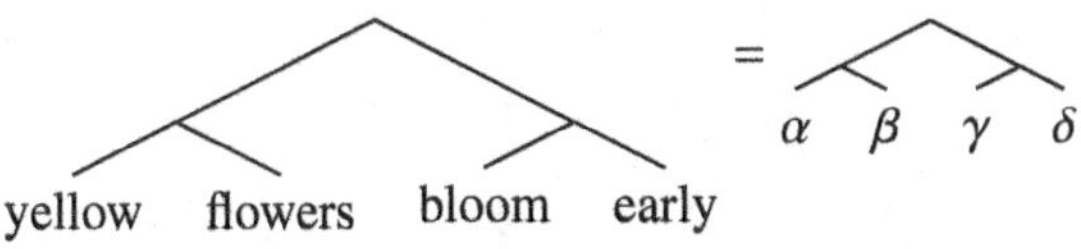

$= ((\alpha\,\beta)\,(\gamma\,\delta))$

In the form depicted, this is represented by a planar binary rooted tree on four leaves labeled by the lexical items in the set $\Lambda = \{\alpha, \beta, \gamma, \delta\}$. The tree does not contain exocentric constructions and has a well-defined syntactic head. Thus, we have the associated data (Λ, h) as in Proposition 3.4.1. The underlying syntactic object, as produced by a free symmetric Merge. is the *nonplanar* abstract binary rooted tree

$\{\{\alpha,\beta\},\{\gamma,\delta\}\} =$

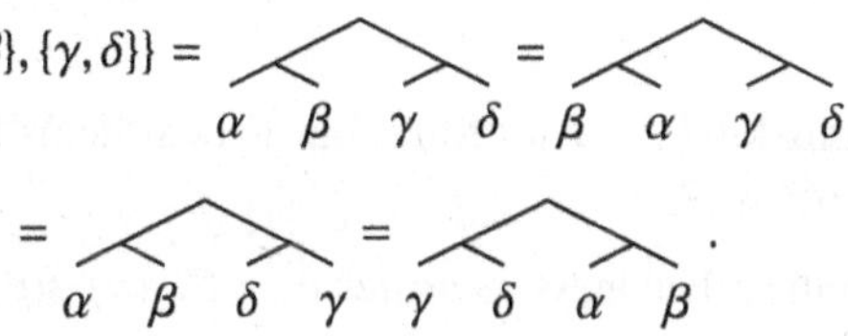

The planar tree $((\alpha\,\beta)\,(\gamma\,\delta))$ corresponds to a vertex of the associahedron K_4, as in Figure 3.11. The associahedron considered is one of the $4! = 24$ associahedra that correspond to the 4! permutations of the leaves' labels. This assignment of a vertex on one of the 24 associahedra corresponds to the left arrow (free symmetric Merge to Externalization) in the top part of Figure 3.4.

The abstract tree $\{\{\alpha,\beta\},\{\gamma,\delta\}\}$ produced by free symmetric Merge, on the other hand, is one of the $15 = (2n-3)!!$, for $n = 4$, possible abstract binary rooted trees on four labeled leaves. These 15 possible trees correspond to the 15 edges of the link $\mathcal{L}_4$ of the origin in the moduli space BHV_4. Thus, the syntactic object $\{\{\alpha,\beta\},\{\gamma,\delta\}\}$ selects one of these edges; see Figure 3.12.

Now suppose that we have chosen a semantic space $\mathcal{S}$ (for simplicity of discussion, consider using a vector space model, although it is not necessary

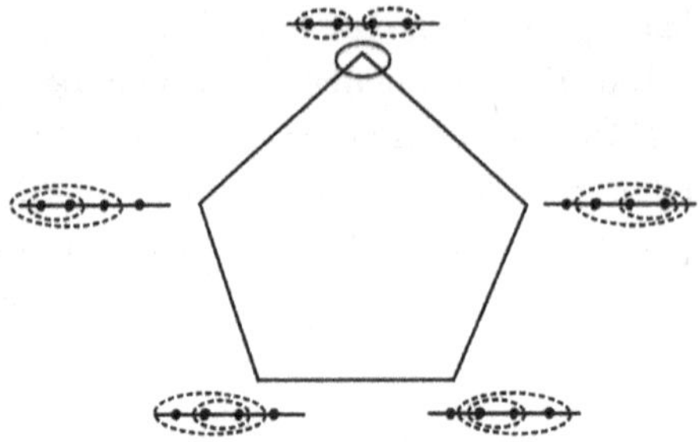

Figure 3.11
The selected vertex of the associahedron K_4 corresponding to the planar tree $((\alpha\ \beta)\ (\gamma\ \delta))$, as shown in [53].

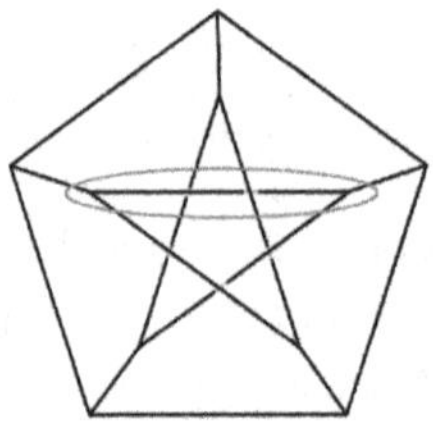

Figure 3.12
The selected edge in the link $\mathcal{L}_4$ of the origin in the moduli space BHV_4 corresponding to the abstract tree $\{\{\alpha,\beta\},\{\gamma,\delta\}\}$.

for $\mathcal{S}$ to be of this kind). Each of the four lexical items has a representation $s(\alpha), s(\beta), s(\gamma), s(\delta) \in \mathcal{S}$.

The two semantic relatedness measures $u_1 = \mathbb{P}(s(\alpha), s(\beta))$ (relating "yellow" and "flower") and $u_2 = \mathbb{P}(s(\gamma), s(\delta))$ (relating "blooming" and "early") in $\mathcal{S}$ provide two real coordinates associated with the accessible terms $\{\alpha,\beta\}$ and $\{\gamma,\delta\}$, respectively. These two coordinates fix a point $(u_1, u_2) \in [0,1]^2$ in a square (see Figure 3.13). The selected edge of Figure 3.12 corresponds to the diagonal of the square given by $u_1 + u_2 = 1$. Thus, the mapping of the result of the free symmetric Merge to semantic space determines a point in the moduli space BHV_4^+. This completes the right arrow (free symmetric Merge to semantic spaces) in the top part of Figure 3.4 for the example of this simple sentence.

We next see in this example the bottom part of Figure 3.4, which describes the compatibility between Externalization and the syntax–semantics interface. First, note that the associahedra K_4 are tiled with squares (quadrangles), as in

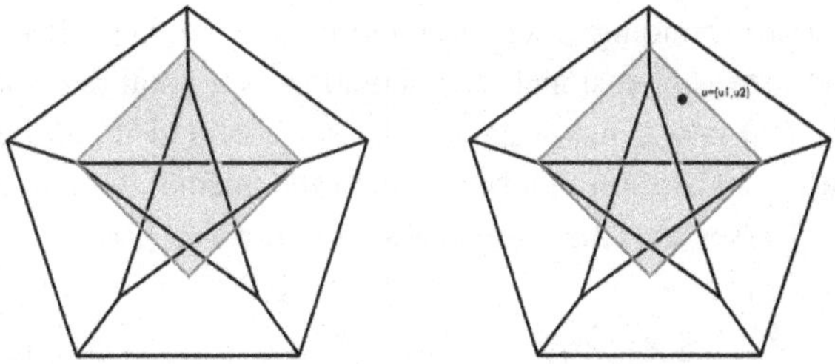

Figure 3.13
The square and the selected edge in the link $\mathcal{L}_4$ for corresponding to the abstract tree $T = \{\{\alpha,\beta\},\{\gamma,\delta\}\}$: the mapping of T to semantic space selects a point $u = (u_1, u_2)$ in this square, as in the right part of the figure.

Figure 3.14. The two vertices of the square adjacent to the marked vertex of the pentagon correspond to degenerate trees where one or the other of the internal edges is shrunk to zero length, while the other has a normalized length of 1. Thus, we see that we can map this square to the square of Figure 3.13 through the same coordinates (u_1, u_2) describing the lengths of the two internal edges (compare with Figure 3.10 for the case $n = 3$).

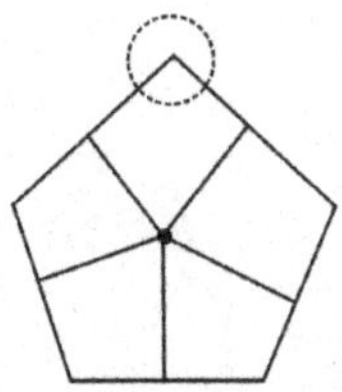

Figure 3.14
The associahedron K_4 tiled with squares (quadrangles), with the selected vertex associated with $((\alpha\ \beta)\,(\gamma\ \delta))$.

This lifting of the point associated with T in the square of Figure 3.13 to a corresponding point in the square of Figure 3.13 is the effect of the section $\sigma_{L,4}$ described in (3.4.7). To see this, we need to take into consideration the fact that the 24 associahedra combine into a single geometric space, obtained by gluing them along their boundaries. This is done in two steps: first, 12 associahedra are glued along their boundaries as in the left side of Figure 3.15, forming the space $\bar{M}_{0,5}(\mathbb{R})$. Then the orientation double cover is formed: in a self-intersecting three-dimensional visualization, this resulting space $\bar{M}^{or}_{0,5}(\mathbb{R})$ can be identified with the *great dodecahedron* in the right side of Figure 3.15

(see also the corresponding discussion and figures in [53]). It is not easy to see it from its three-dimensional representation as a great dodecahedron, but the space $\bar{M}^{or}_{0,5}(\mathbb{R})$ is a genus 4 hyperbolic surface, and it can be seen more directly from its description in terms of the fundamental domain given in [2], as in Figure 3.16 (see also the corresponding figure 8 in [2]).

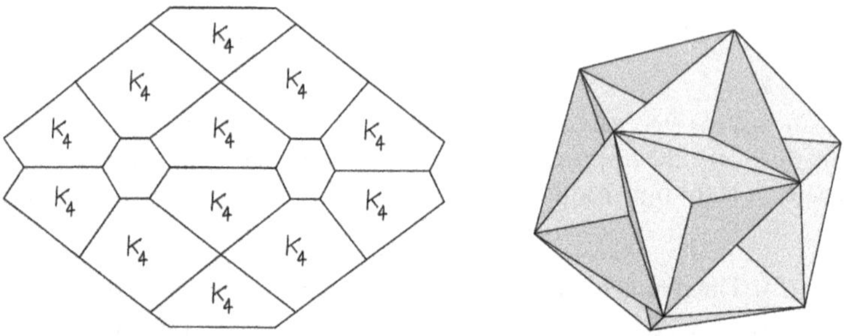

Figure 3.15
Here, 12 associahedra K_4 assemble into the space $\bar{M}_{0,5}(\mathbb{R})$ and its orientation double cover gives 24 associahedra assembled into the space $\bar{M}^{or}_{0,5}(\mathbb{R})$ identified with the great dodecahedron, as shown in [53].

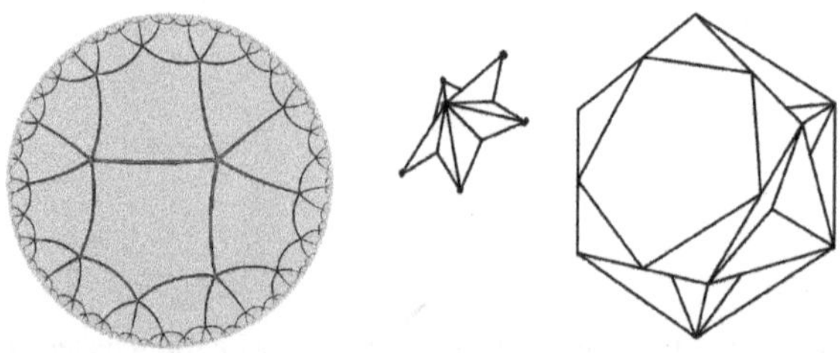

Figure 3.16
The great dodecahedron $\bar{M}^{or}_{0,5}(\mathbb{R})$ as a hyperbolic genus 4 surface, and the two different forms of the 24 associahedron tiles, as shown in [2].

The origami folding projection map $\Pi_4 : \overline{M}^{or}_{0,5}(\mathbb{R}) \twoheadrightarrow \mathrm{BHV}^+_4$ of (3.4.1) folds together and identifies eight squares in $\overline{M}^{or}_{0,5}(\mathbb{R})$ to each square in BHV^+_4. Thus, when we lift to $\overline{M}^{or}_{0,5}(\mathbb{R})$ the point assigned to the tree T in one of the squares of BHV^+_4 by the mapping of T to semantic space, the lifted point lies on one of the eight preimages of the given square of BHV^+_4. This choice that singles out one of these eight preimages is the choice of planar structure of the syntactic object determined by Externalization, and this indeed gives the section $\sigma_{L,4}$ described in (3.4.7), where here, L = English.

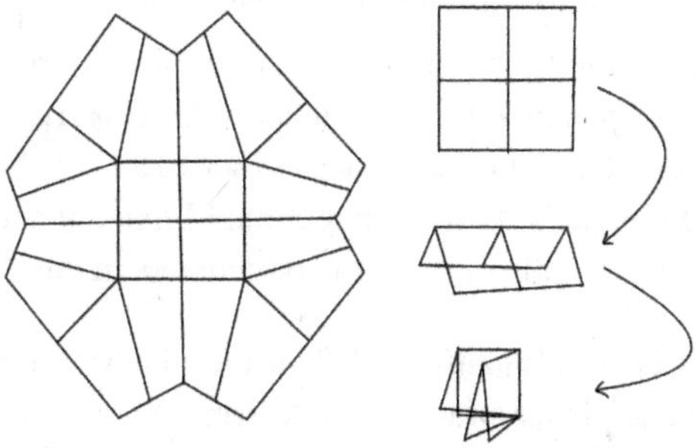

Figure 3.17
Four squares in adjacent associahedra K_4 are folded together (origami folding) in the projection to BHV_4^+ so that 8 squares in the double cover $\overline{M}^{or}_{0,5}(\mathbb{R})$ are identified in the projection $\Pi_4 : \overline{M}^{or}_{0,5}(\mathbb{R}) \twoheadrightarrow \mathrm{BHV}_4^+$, as shown in [53].

One can then see in this same simple example that if instead of taking the planarization $T^{\pi_L} = ((\alpha\ \beta)(\gamma\ \delta))$ of the syntactic object $T = \{\{\alpha,\beta\},\{\gamma,\delta\}\}$, one takes the planarization T^{π_h} determined by the head function, one would end up with a *different* planar tree:

$T^{\pi_h} =$ 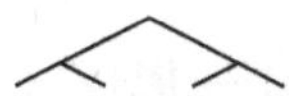 $= ((\gamma\delta)(\beta\alpha))\,.$

$\gamma \quad \delta \quad \beta \quad \alpha$

This means that one ends up on a different one of the 24 associahedra and a square inside that associahedron, which is still one of the eight squares that project to the same unchanged square in BHV_4^+. The same point in this square in BHV_4^+ determined by mapping T to semantic space is then lifted to a corresponding point in a different square inside $\overline{M}^{or}_{0,5}(\mathbb{R})$. This means that we are considering a different section of the projection Π_4. This is the section $\sigma_{S,4}$ described in (3.4.5). The difference between these two sections is measured by a transformation $\gamma_{L,4} \circ \sigma_{L,4} = \sigma_{S,4}$, which, applied to our syntactic object T, gives the permutation $\gamma_{L,4}(T) = (3421)$. As in (3.4.8), this transformation $\gamma_{L,4}$ is Kayne's LCA algorithm for this example.

We have considered here a very simple example with $n = 4$, where the geometry is straightforward to visualize. The spaces $\overline{M}^{or}_{0,n+1}(\mathbb{R})$ and BHV_n^+ grow significantly in combinatorial complexity as n becomes larger, but they are still very well understood and widely studied geometric spaces. Other more complicated geometries are likely to arise if the mapping of syntactic objects to semantic spaces is done in a more sophisticated way than the very simple type of mappings considered in this chapter as illustrative examples.

3.4.6 Geometric View of Planarization Questions

We conclude this section by briefly commenting on how certain frameworks, where the question of planarization of syntactic objects arises, also can be seen in terms of the geometry of moduli spaces described here. We discuss briefly Kayne's LCA and Cinque's Abstract Functional Lexicon, and we also outline how one can describe the role of syntactic parameters in this geometric setting.

3.4.6.1 Kayne's LCA algorithm When the planar structure assigned by the section $\sigma_{S,n}$ is the planar structure π_h determined by a head function, as in the case of (3.4.3), this means comparing the planar structure π_h, for h defined on $\mathrm{Dom}(h) \subset \mathfrak{T}_{SO_0}$, with the planar structure π_L defined by the section $\sigma_{L,n}$ through the constraints imposed by the syntactic parameters of the language L. This comparison can be seen as a version of Kayne's LCA, [109] and [110]. As we observed in section 1.13.2 of chapter 1, the LCA cannot be defined *globally* on $\mathfrak{T}_{SO_0}$, but it is only partially defined on the domain $\mathrm{Dom}(h)$ of a head function (the syntactic head). Thus, it does not play the same role as Externalization, which is a choice of a globally defined (noncanonical) section $\sigma_{L,n}$. However, on the domain $\mathrm{Range}(s_{S,h,n}) \subset \mathcal{L}_n \cap \mathrm{Dom}(h)$, where both $\sigma_{L,n}$ and $\sigma_{S,n}$ are defined, there is a covering transformation $\gamma_{L,n}$ of the projection map:

$$\overline{M}^{or}_{0,n+1}(\mathbb{R}) \to \mathrm{BHV}^+_n ,$$

which satisfies

$$\gamma_{L,n} \circ \sigma_{L,n} = \sigma_{S,n} \tag{3.4.8}$$

at all points in $\mathrm{Range}(s_{S,h,n})$. This covering transformation $\gamma_{L,n}$ plays the role of the partially defined LCA algorithm.

3.4.6.2 Cinque's Abstract Functional Lexicon There are other constructions that one can fit into this geometric picture with partially defined sections of the projection map

$$\overline{M}^{or}_{0,n+1}(\mathbb{R}) \to \mathrm{BHV}^+_n$$

and covering transformations permuting the 2^{n-1} points of the fibers of the projection map. For example, one can view this as the Abstract Functional Lexicon described by Cinque [42].

In [42], Cinque considers the problem of comparing word order relations imposed on individual languages by particular syntactic parameters, with a certain base ordering relation of proximity to the verbal properties of different morphemes (in a structural sense, rather than in terms of linear ordering), such as mood, tense, modality aspect, and voice. In [42], a general hierarchy of functional morphemes and adverbial classes is identified (see (6) and (7) of [42]). As observed in [42], with verbal morphemes as heads and corresponding classes of adverbs as phrases in so-called specifier position, this hierarchy determines a planar embedding (in the way that it appears in (6) and (7) of [42]). Syntactic parameters, on the other hand, also determine a planar embedding. While this could be a priori arbitrary, the variability across languages is far less than the space of combinatorial possibilities would allow. Also, different word order constraints appear not to be independent, but to exhibit a significant degree of relatedness. This can be seen both at the theoretical level (see [84]) and at the level of database analysis of syntactic parameters (see [158], [162], and [172]).

In terms of the geometry of moduli spaces, one can describe the difference between the ordering (planar structure) described by Cinque in [42] and the deviation from it in the word order of specific languages in terms of *covering transformations* $\gamma_{L,n}$ of the projection map $\overline{M}^{or}_{0,n+1}(\mathbb{R}) \to \mathrm{BHV}^+_n$ that act as permutations of the planar structures, and are language specific. The degree to which word order constraints deviate from the base structural hierarchy described in Cinque can then be measured in terms of how far the $\gamma_{L,n}$ are from the identity in the group of covering transformations of the projection map.

3.4.6.3 Geometric view of syntactic parameters Syntactic parameters fix constraints on the planar structure of Externalization. For an extensive recent account of syntactic parameters, see [176]. In chapter 1, we interpreted the role of syntactic parameters as constraints on the choice of a language-dependent section $\sigma_{L,n}$ for the Externalization of free symmetric Merge.

This discussion shows that in our setting, we can also interpret the role of syntactic parameters in a geometric way, as the choice, for a given language L, of a collection $L \mapsto \{\gamma_{L,n}\}_n$ of covering transformations of the origami folding projections $\overline{M}^{or}_{0,n+1}(\mathbb{R}) \to \mathrm{BHV}^+_n$, as in (3.4.8). Comparison of syntactic parameters across languages can be formulated in various computational forms. This includes the difficult problem of understanding the relation among parameters, as well as the much lower-dimensional space occupied by actual languages inside the high-dimensional space of possible values of the hundreds of parameters currently studied (see, e.g., [88], [128], [137], [113], [172], and [187]). In particular, one can focus on the effect of syntactic parameters on word order constraints. In this case, using the framework considered here, one can view this comparison across languages as the comparison between sections

$\sigma_{L,n}$ for different languages L, or equivalently as the properties of the collection of elements $\gamma_{L,n}\gamma_{L',n}^{-1}$, for $L \neq L'$ in the group of covering transformations of $\overline{M}^{or}_{0,n+1}(\mathbb{R}) \to \mathrm{BHV}_n^+$.

3.5 Birkhoff Factorization and (Semi)ring Parsing

This section is more technical, at the mathematical level, and is not required in the successive parts of the book. It can be skipped by readers interested in continuing directly to the discussion of other models of semantics: in particular, in section 3.6, we will analyze how our model relates to Pietroski's theory of *minimalist meaning*, [168], and in section 3.9, we will discuss the relation to Heim–Kratzer semantics.

In this section, we discuss the problem of parsing in semirings in the setting of Merge derivations. We focus on *semiring parsing*, as in [82]. Compared to the semantics models that we will be discussing in the following sections, semiring semantics incorporates the idea of truth values and generalizes it to values in arbitrary rings or semirings (which, for simplicity, we write collectively as "(semi)rings"), not necessarily Boolean. In a different direction, which we will discuss in section 3.6, Pietroski's approach provides an alternative that bypasses the idea of truth values entirely and is based on a compositional structure modeled on Minimalism's Merge operation. We will return to discuss another model of semantics based on truth values assignments in the Heim–Kratzer setting in section 3.9, which uses a different formulation, not based on the semiring parsing that we consider in this section but in terms of spaces of functions.

We extend here the setting introduced in the previous sections, with the interface between syntax and semantics modeled by the Birkhoff factorization of characters of a Hopf algebra, with values in a Rota-Baxter algebra or semiring, by developing a more refined description of the characters and the factorization procedure, which incorporate more detailed properties of semantic parsing (namely, chains of Merge derivations) and compositionality.

In this section, we analyze (semi)ring parsing, introducing a version that is adapted to Minimalism formulated in terms of free symmetric Merge.

Since this is the most mathematics-heavy section in this chapter, we provide a preliminary outline of the content and a more heuristic explanation of what is covered in the various subsections before starting to discuss the more precise details.

3.5.1 Preliminary Discussion

The relation between grammars and semirings was first observed by Chomsky–Schützenberger in [39]. Semiring parsing (see, e.g., [82]), when formulated in the setting of context-free grammars, considers deduction rules of the form

$$\frac{A_1 \dots A_k}{B} C_1 \dots C_\ell ,$$

where the terms A_i (main conditions) are rules R of the grammar or input non-terminals, the C_i are non-probabilistic Boolean side conditions, and the fraction notation means that if the numerator terms hold, then the denominator term also does. To the main condition terms, one assigns values in a semiring, combined with the semiring operations, to obtain a value for the deduced output. The target semiring varies according to the parsing algorithm considered. The main choices include:

- The Boolean semiring
- The tropical semiring
- The probability semiring (i.e., the familiar case of Viterbi parsing)
- The noncommutative derivation forest semiring, which collects all the possible derivations, with concatenation as multiplication and union as semiring addition.

Often, parsing with values in other semirings factor through the derivations semiring. This setting is specifically constructed in the formal language context, and specifically for context-free grammars, although some generalizations exist in mildly context-sensitive classes like those produced by tree-adjoining grammars (TAGs).

A natural question arises about what type of algebraic structure replaces this form of semiring parsing in the setting of Minimalism, and more specifically the form of Minimalism based on free symmetric Merge.

The main goal of this section is to provide an answer to this question, in a form that is again based on the Birkhoff factorization procedure that we present throughout this chapter as a natural formalism for different forms of assignments of semantic values in the context of a free symmetric Merge model of syntax.

Developing this form of "semiring parsing" (where semirings in fact will be replaced by more general algebraic objects) requires several steps, which we now briefly summarize.

In section 3.5.2, we introduce a ring of Merge derivations, formed by considering chains of Merge operations, with Merge acting on workspaces. These are assembled into a ring structure, where the linear structure is obtained by taking the vector space spanned by the derivations (i.e., including formal linear combinations) and the multiplication operation is the union of the workspaces with the corresponding Merge actions. These are the same operations on the algebra part of the Hopf algebra of workspaces that we introduced in chapter 1 and used earlier in this chapter. We only consider the product structure on this ring, not the coproduct, as we have for the Hopf algebra of workspaces. However, this does not lead to a loss of structure in this case, because the coproduct is built into the Merge operation on workspaces, so it is still encoded into the data of this ring of Merge derivations.

In order to illustrate more clearly the properties of this ring of Merge derivations, in section 3.5.2 we discuss again the notion of Minimal Search and Minimal Yield in the Merge model of syntax. In chapter 1, we gave an account of how Minimal Search is implemented as an extraction of leading-order term in the action of Merge on workspaces, and how constraints of Minimal Yield and Resource Restriction can exclude presumptively unwanted forms of Sideward Merge, while retaining only Internal Merge and External Merge. The idea of extraction of the leading-order term used in the Minimal Search argument is closely related to Birkhoff factorization, as originally observed in the context of renormalization in physics.

Here, we show that after extending the ring of Merge derivations to a ring of Laurent series with coefficients in this ring, Minimal Search and Minimal Yield and the "no complexity loss" constraints analyzed in chapter 1 can be reformulated as a Birkhoff factorization in this ring. This factorization is applied to a character from the Hopf algebra of workspaces that assigns to a workspace its Merge derivation and a power that counts the effect of that derivation on the size of the workspace. The Birkhoff factorization separates out, on one side, the unwanted forms of Merge, while retaining on the other side the fundamental ones (namely, External Merge and Internal Merge). This case of Birkhoff factorization happens to be the one that is closest to the original form used in physics.

This result on Minimal Yield as Birkhoff factorization in section 3.5.2 is not required for the following parts of this section; it is included to provide a more direct understanding of the ring of Merge derivations and to connect it to our original formulation in chapter 1. This section can be skipped (except for Definition 3.5.1) by readers interested in directly accessing the discussion of how to extend the semiring parsing framework.

The next step is to formulate a parsing where the ring of Merge derivations is the source rather than the target of a character. This requires endowing the ring of Merge derivations with an additional algebraic structure that is more closely related to the Hopf algebra structure of workspaces.

The main construction for the generalization of semiring parsing starts in section 3.5.3. The main viewpoint here is that, in order to formulate semiring parsing for Merge derivations based on the action of Merge on workspaces, one needs to replace the setting of Hopf algebras and semirings, which we used in the previous sections of this chapter to describe simple models of syntax–semantics interface, with a slightly more flexible form, where the algebraic structures of Hopf algebra and semiring are replaced by their "categorified" forms, which we refer to, respectively, as Hopf algebroids and semiringoids. The reason for this extension is very simple. Merge derivations given by actions of Merge on workspaces only compose when the target workspace of one derivation agrees with the source workspace of the next one. This differs from the situation considered in the previous sections where we only considered the Hopf algebra of workspaces, where the product is the disjoint union (combination of workspaces), which is always defined without source/target matching conditions.

Thus, just as in passing from the multiplication in a group to the multiplication in a groupoid, which precisely accounts for the fact that arrows compose only when the target of the first is the source of the second, one can obtain similar generalizations of the structures of Hopf algebra and Rota–Baxter algebra (or semiring) that we used in the formulation of mapping from syntax to semantics as Birkhoff factorization in the previous sections.

An extension of the notion of Hopf algebra that accommodates for the need for source/target matching conditions in the product was developed in the context of algebraic topology with the notion of commutative Hopf algebroid and bialgebroid. We take that as the starting point in section 3.5.3 by constructing a bialgebroid associated with the ring of Merge derivations introduced in section 3.5.2. This bialgebroid replaces the Hopf algebra of workspaces on the syntax side by encoding the Merge derivations in syntax.

In section 3.5.4, we then consider the other side of the Birkhoff factorization (namely, the semantics side), where we wish to replace the algebraic datum of a Rota-Baxter algebra or Rota-Baxter semiring with an analogous categorified version. We use a notion of algebroid that is compatible with the notion of Hopf algebroids and bialgebroids introduced in section 3.5.3, and we show that the notion of algebroid that we consider is dual to directed graphs, with the cases of bialgebroids being dual to directed graphs that are reflexive and transitive (small categories) and with Hopf algebroids being dual to groupoids.

We also extend the generalization of algebras to algeboids to an analogous generalization of semirings to a similar categorified structure of a semiringoid. (Note that some other different notions of algebroids and semiringoids exist in the mathematical literature, but they should not be confused with the version adopted here.)

In section 3.5.4.2, we describe how the notion of a Rota–Baxter operator of weight -1 on an algebra can be generalized to the case of an algebroid, and similarly in section 3.5.4.3, we show the analogous generalization of Rota–Baxter semirings of weight ± 1 to semiringoids.

With this, we have both sides of the mapping ready for the case of Merge derivations with their composition structure. We prove in section 3.5.4.4 the existence of Birkhoff factorizations of characters from Hopf algebroids with values in Rota–Baxter algebroids and from bialgebroids to Rota–Baxter semiringoids. The characters and the factorization can here be described dually in terms of maps of directed graphs.

We conclude in section 3.5.5 by showing that, with the algebraic setting constructed in the previous subsections, one obtains a form of semiringoid parsing that simultaneously generalizes the various semiring parsings of [82] and the Birkhoff factorizations described in section 3.2.

3.5.2 Minimal Yield as Birkhoff Factorization

In chapter 1, we presented a way to implement Minimal Search and eliminate unwanted forms of Merge (Sideward Merge) and retain only the Internal and External forms of Merge. (We put aside here the question of whether these excluded forms of Merge are indeed undesirable, and simply assume that this is so. We refer the readers to the forthcoming [142] for a further discussion of this question.) In section 1.6.1 of chapter 1, we also discussed the Minimal Yield condition (see Definition 1.6.1) and its effect on certain counting of size and complexity. In that formulation, Minimal Search is implemented by extracting the leading-order term with respect to a specific grading function imposed on the terms of the coproduct of the Hopf algebra, and we defined Minimal Yield as a condition on different counting functions keeping track of size and complexity. We show that there is another natural way of thinking about these minimality conditions, by formulating them as a Birkhoff factorization, very similar in form to the one used in quantum field theory with respect to a character with values in a ring of Laurent series.

3.5.2.1 Effect of Merge on workspaces Since here it is not important to keep track of traces in the effect of Internal Merge, we can either consider the quotients $T/^c F_{\underline{v}}$ in the coproduct or the quotient $T/^d F_{\underline{v}}$ (see section 1.2 of chapter 1). We have the same counting of the effect of Merge on the various measures of workspace size (number of components, number of accessible terms, number of vertices) as described in section 1.6.2 of chapter 1.

The different cases of Merge are given by External Merge (EM), Internal Merge (IM), and forms of Sideward Merge(SM):

- **EM:** $F = T \sqcup T' \sqcup \hat{F} \mapsto F' = \mathfrak{M}(T, T') \sqcup \hat{F}$
- **IM:** $F = T \sqcup \hat{F} \mapsto F' = \mathfrak{M}(T_v, T/T_v) \sqcup \hat{F}$
- **SM(i):** $F = T \sqcup T' \sqcup \hat{F} \mapsto F' = \mathfrak{M}(T_v, T'_w) \sqcup T/T_v \sqcup T'/T'_w \sqcup \hat{F}$
- **SM(ii):** $F = T \sqcup T' \sqcup \hat{F} \mapsto F' = \mathfrak{M}(T, T'_w) \sqcup T'/T'_w \sqcup \hat{F}$
- **SM(iii):** $F = T \sqcup \hat{F} \mapsto F' = \mathfrak{M}(T_v, T_w) \sqcup T/(T_v \sqcup T_w) \sqcup \hat{F}$

where $\hat{F}$ denotes the part of the workspace that is not affected, and

$$\mathfrak{M}(T, T') = \widehat{T \quad T'} \,.$$

The effect of these Merge operations on various forms of size counting is discussed in chapter 1, in section 1.6.1. Recall that we denote by $b_0(F)$ the number of connected components of a workspace $F \in \mathfrak{F}_{SO_0}$ (number of syntactic objects), by $\alpha(F) = \#\mathrm{Acc}(F)$ the number of accessible terms in F (the total number of non-root vertices), by $\sigma(F) = \#\mathrm{Acc}'(F) = \#V(F)$ the total number of vertices. For a chain of Merge derivations,

$$\Phi = \mathfrak{M}_{S_N, S'_N} \circ \cdots \circ \mathfrak{M}_{S_1, S'_1} \,,$$

we set

$$\delta b_0 := b_0(F) - b_0(\Phi(F)), \quad \delta\alpha = \alpha(\Phi(F)) - \alpha(F), \text{ and } \delta\sigma = \sigma(\Phi(F)) - \sigma(F),$$

so the Minimal Yield conditions of Definition 1.6.1 are equivalent to

$$\delta b_0 \geq 0, \quad \delta\alpha \geq 0, \text{ and } \delta\sigma = 1 \,,$$

describing the "no divergence", "no information loss", and "minimality of yield" conditions of Definition 1.6.1, respectively. We also consider the weaker condition of "positive yield," $\delta\sigma \geq 0$. Similarly, we consider the "no complexity loss" principle discussed in section 1.6.1. We compute the associated function $\Phi_0 : \pi_0(F) \to \pi_0(\Phi(F))$ and take, as in Definition 1.6.2,

$$\delta \deg_a := (\deg(\Phi_0(a)) - \deg(a)) \quad \text{for } a \in \pi_0(F) \,.$$

The "no complexity loss" principle corresponds to $\delta \deg_a \geq 0$ for all $a \in \pi_0(F)$. We will just write δ when we mean either δb_0 or $\delta\alpha$ or $\delta\sigma$ or $\delta \deg_a$. The measure δ will be used in the construction of section 3.5.2.2. Note that values of $\delta \geq 0$ eliminate some of the undesirable forms of Sideward Merge, and allow Internal Merge and External Merge.

We show here that the elimination of the forms of Merge, described in terms of Minimal Yield in section 1.6.1, can also be formulated as a Birkhoff factorization where one eliminates divergences as in the physical setting.

3.5.2.2 Laurent series ring of Merge derivations Here, we introduce a ring that organizes derivations in the Minimalist generative grammar defined by free symmetric Merge, weighted by their effect on the workspace.

Definition 3.5.1. The algebra of free Merge derivations $\mathcal{DM}$ is the commutative associative $\mathbb{Q}$-algebra with the underlying $\mathbb{Q}$-vector space spanned by elements of the form φ_A, where $A \subset \mathcal{SO} \times \mathcal{SO}$ is a set of pairs (S, S') of syntactic objects, and

$$\varphi_A = (F \overset{\mathfrak{M}_A}{\to} F') \tag{3.5.1}$$

consists of all possible chains of Merge operations:

$$F \overset{\mathfrak{M}_{S_1,S_1'}}{\to} F_1 \to \cdots F_{N-1} \overset{\mathfrak{M}_{S_N,S_N'}}{\to} F' , \tag{3.5.2}$$

with $(S_i, S_i') \in A$. Since the source and target workspaces are assigned, there are finitely many possible chains. The algebra multiplication is given by the operation

$$\varphi_A \sqcup \varphi_B = (F \sqcup \tilde{F} \overset{\mathfrak{M}_{A \sqcup B}}{\to} F' \sqcup \tilde{F}') \tag{3.5.3}$$

for $\varphi_A = (F \overset{\mathfrak{M}_A}{\to} F')$ and $\varphi_B = (\tilde{F} \overset{\mathfrak{M}_B}{\to} \tilde{F}')$, with the unit given by the empty forest mapped to itself. Let $\mathcal{DM}[t^{-1}][[t]]$ denote the associative commutative $\mathbb{Q}$-algebra of Laurent power series with coefficients in $\mathcal{DM}$.

The meaning of the product (3.5.3) is to perform in parallel different Merge operations that affect different parts of a workspace. Such operations, if conducted sequentially, would commute with each other, and hence would be independent of the order of execution (unlike operations that affect the same components of the workspace), so composition can be regarded as simultaneous and parallel rather than sequential, and it can be grouped together as a single operation.

The fact recalled in Proposition 3.5.2 is well known (see [45], [46], [56], and [57]).

Proposition 3.5.2. *Given a commutative associative algebra $\mathcal{A}$ and the algebra of Laurent series $\mathcal{A}[t^{-1}][[t]]$, the linear operator $R : \mathcal{A}[t^{-1}][[t]] \to$*

$\mathcal{A}[t^{-1}][[t]]$ that projects onto the polar part,

$$R(\sum_{i=-N}^{\infty} a_i t^i) = \sum_{i=-N}^{-1} a_i t^i\,, \tag{3.5.4}$$

makes $(\mathcal{A}[t^{-1}][[t]], R)$ a Rota–Baxter algebra of weight -1.

Proposition 3.5.3. *Consider the map $\phi : \mathcal{H} \to \mathcal{DM}$,*

$$\phi(F) = (L(F) \stackrel{\mathfrak{M}_{A(L(F),F)}}{\longrightarrow} F)\,, \tag{3.5.5}$$

which assigns to a forest F the set $A(L(F), F)$ of all Merge derivations from the multiset of individual lexical items and syntactic features that form the set of leaves $L(F)$, to the forest F (the generative process for F). This defines a character (a morphism of commutative algebras) from the Hopf algebra $\mathcal{H}$ of nonplanar binary rooted forests to the algebra of free Merge derivations $\mathcal{DM}$. Then the assignment

$$\phi_t(F) = (L(F) \stackrel{\mathfrak{M}_{A(L(F),F)}}{\longrightarrow} F)\; t^{\delta(\mathfrak{M}_{A(L(F),F)})}\,, \tag{3.5.6}$$

where δ is δb_0, $\delta\alpha$, or $\delta\sigma$, defines a morphism of commutative algebras $\phi_t : \mathcal{H} \to \mathcal{DM}[t^{-1}][[t]]$.

Proof. It suffices to check that $\phi(F \sqcup F') = \phi(F) \sqcup \phi(F')$; namely,

$$(L(F) \sqcup L(F') \stackrel{\mathfrak{M}_{A(L(F)\sqcup L(F'),F\sqcup F')}}{\longrightarrow} F \sqcup F') = (L(F) \sqcup L(F') \stackrel{\mathfrak{M}_{A(L(F),F)\sqcup A(L(F'),F')}}{\longrightarrow} F \sqcup F')\,.$$

This is the case since, if the end result of a chain of Merge operations contains a disjoint union $F \sqcup F'$ of two trees, then all the individual Merge operations $\mathfrak{M}_{T_v,T_w}$ in the chain will use syntactic objects T_v T_w, where both sets of leaves $L(T_v)$ and $L(T_w)$ are subsets of $L(F)$ or where both are subsets of $L(F')$, as otherwise, $\mathfrak{M}_{T_v,T_w}$ would create a connected component T with $L(T) \cap L(F) \neq \emptyset$ and $T \cap L(F') \neq \emptyset$ so the end result would not contain $F \sqcup F'$. Moreover, for δ equal to δb_0, $\delta\alpha$, or $\delta\sigma$, as discussed in section 3.5.2.1, we define $\delta(\mathfrak{M}_A)$, where $\mathfrak{M}_A$ is a finite set of Merge derivations $\{\phi = \mathfrak{M}_{S_N,S'_N} \circ \cdots \circ \mathfrak{M}_{S_1,S'_1}\}$ with fixed source F and target F', by taking

$$\delta\alpha(F \stackrel{\mathfrak{M}_A}{\longrightarrow} F') = \alpha(F') - \alpha(F)\,, \quad \delta b_0(F \stackrel{\mathfrak{M}_A}{\longrightarrow} F') = b_0(F) - b_0(F')\,,$$

$$\delta\sigma(F \stackrel{\mathfrak{M}_A}{\longrightarrow} F') = \sigma(F') - \sigma(F)\,,$$

where $\alpha(F \sqcup \tilde{F}) = \alpha(F) + \alpha(\tilde{F})$, and similarly for b_0 and σ; hence, (3.5.6) is an algebra homomorphism. □

As we will see in Lemma 3.5.5, the character $\phi_t : \mathcal{H} \to \mathcal{DM}[t^{-1}][[t]]$ of Proposition 3.5.3 is not good enough to detect the difference between Internal/External Merge and Sideward/Countercyclic Merge. However, one can consider similar characters more suitable for this purpose. A simple modification of ϕ_t that works can be obtained in the following way, where the statement follows as in Proposition 3.5.3.

Corollary 3.5.4. *For $T \in \mathfrak{T}_{SO_0}$ let $\mathcal{F}_T \subset \mathfrak{F}_{SO_0} \times \mathfrak{F}_{SO_0}$ denote the set of pairs (F, F') of forests F with $L(F) = L(F') = L(T)$ that are intermediate derivations for T; namely, those pairs for which there is a chain of free symmetric Merge derivations*

$$L(T) \stackrel{\mathfrak{M}_{S_1,S'_1}}{\longrightarrow} \cdots \stackrel{\mathfrak{M}_{S_i,S'_i}}{\longrightarrow} F \stackrel{\mathfrak{M}_{S_{i+1},S'_{i+1}}}{\longrightarrow} \cdots \stackrel{\mathfrak{M}_{S_j,S'_j}}{\longrightarrow} F' \stackrel{\mathfrak{M}_{S_{j+1},S'_{j+1}}}{\longrightarrow} \cdots \stackrel{\mathfrak{M}_{S_m,S'_m}}{\longrightarrow} T,$$

for some $m \geq 1$, including the case with $F = L(T)$ and $F' = T$. Consider the assignment

$$\psi_t(T) = \sum_{(F,F') \in \mathcal{F}_T} (F \stackrel{\mathfrak{M}_{A(F,F')}}{\longrightarrow} F')\, t^{\delta(\mathfrak{M}_{A(F,F')})}, \tag{3.5.7}$$

where $\mathfrak{M}_{A(F,F')}$ is the set of all possible Merge derivations from F to F' and δ is δb_0, $\delta\alpha$, or $\delta\sigma$. This determines a morphism of commutative algebras $\psi_t : \mathcal{H} \to \mathcal{DM}[t^{-1}][[t]]$.

The reason why the choice of the character ψ_t of (3.5.7) is preferable to the choice of ϕ_t of (3.5.6) is explained by the simple property in Lemma 3.5.5.

Lemma 3.5.5. *The character $\phi_t : \mathcal{H} \to \mathcal{DM}[t^{-1}][[t]]$, where δ is δb_0, $\delta\alpha$, or $\delta\sigma$, takes values in the subring*

$$\mathcal{DM}[[t]] = (1 - R)\,\mathcal{DM}[t^{-1}][[t]]$$

of formal power series.

Proof. Consider the case of a tree $T \in \mathfrak{T}_{SO_0}$. The value

$$\phi_t(T) = (L(T) \stackrel{\mathfrak{M}_{A(L(T),T)}}{\longrightarrow} T)\, t^{\delta(T)}$$

represents the complete set of all possible chains of free symmetric Merge derivations that construct the syntactic object T starting from a multiset $L = L(T)$ of lexical items and syntactic features. If $\#L = \ell \geq 2$, then $\#V(T) = 2\ell - 1 = \sigma(T)$ and $\alpha(T) = 2\ell - 2$, with $b_0(T) = 1$, so we have $\delta b_0 = \ell - 1 \geq 0$, $\delta\alpha = \ell - 2 \geq 0$, and $\delta\sigma = \ell - 1 \geq 0$. Thus, notice that $\phi_t(T)$ is always in the nonpolar part $\mathcal{DM}[[t]]$ for any tree T, regardless of which types of Merge operation have been used along the chain of derivations. This means that $(1-R)\phi_t(T) = \phi_t(T)$ for all T. The case of forests is then immediate since $\phi_t(F) = \prod_a \phi_t(T_a)$, for $F = \sqcup_a T_a$ and $\delta(F) = \sum_a \delta(T_a) \geq 0$. □

Thus, the character ϕ_t does not suffice to separate Internal Merge and External Merge from Sideward Merge operations on the basis of the counting given by δ. On the other hand, the character ψ_t, which also considers all the intermediate derivations from $L(T)$ to T, each weighted according to the corresponding value of δ, will have a nontrivial polar part, when certain Sideward Merge operations are present somewhere in the chain of derivations.

However, even when using the character ψ_t, which detects the presence of the so-called undesirable forms of Merge in a derivation, simply applying the projection onto the regular part

$$\psi_t(F) \mapsto (1-R)\psi_t(F)$$

does not suffice to eliminate those Sideward/Countercyclic Merge operations and only retain Internal/External Merge. This is a consequence of the fact that the projection R onto the polar part is not an algebra homomorphism but a Rota–Baxter operator. The failure of the Rota–Baxter operator R of (3.5.4) to be an algebra homomorphism:

$$R((\sum_{i=-N}^{\infty} a_i t^i)(\sum_{j=-M}^{\infty} b_j t^j)) = R(\sum_{n=-(N+M)}^{\infty} \sum_{i+j=n} a_i b_j\, t^n) = \sum_{n=-(N+M)}^{-1} \sum_{i+j=n} a_i b_j\, t^n$$

$$\neq R(\sum_{i=-N}^{\infty} a_i t^i) R(\sum_{j=-M}^{\infty} b_j t^j) = \sum_{n=-(N+M)}^{-1} \sum_{i+j=n, i<0, j<0} a_i b_j\, t^n$$

reflects the fact that terms in a product of series can end up in the polar (or nonpolar) part of the product without being in the polar (or nonpolar) part of the individual factors because of the sum t^{i+j} of the exponents. This means that simply applying $(1-R)$ to $\phi(F)$ will not suffice to get rid of Merge derivations that violate Minimal Yield constraints. However, Birkhoff factorization achieves that result.

Proposition 3.5.6. *The inductively constructed Birkhoff factorization* (3.1.5) *of the character ψ_t of* (3.5.7) *implements a form of Minimal Search, in the sense that it inductively eliminates from the derivations those forms of Sideward Merge that violate the Minimal Yield properties, while retaining the compositions of Internal and External Merge.*

Proof. This is a direct consequence of Proposition 3.1.7. Taking $\psi_{t,+}(T) = (1-R)\tilde{\psi}_t(T)$, with $\tilde{\psi}_t$ the Bogolyubov preparation of $\psi_t(T)$ gives an algebra homomorphism

$$\psi_{t,+} : \mathcal{H} \to \mathcal{DM}[[t]] .$$

In the inductive construction of

$$\tilde{\psi}_t(T) = \psi_t(T) + \sum \psi_{t,-}(F_{\underline{v}})\psi_t(T/F_{\underline{v}}),$$

one analyzes in parallel the Merge derivations of accessible terms of T, ensuring that the so-called undesirable forms of Merge are progressively removed and the only ones retained at each step are derivations with $\delta \geq 0$ (i.e., containing Internal Merge and External Merge, or forms of Sideward Merge that do not violate Minimal Yield constraints; see Remark 1.6.9). More precisely, if there is a term in $\psi_t(T)$ of the form $(F \to F')t^{\delta}$, where the derivation is a form of Sideward Merge that violates the Minimal Yield conditions, then the forest F' will occur as a collection of accessible terms $F' = F_{\underline{v}}$ in T. Thus, in $\tilde{\psi}_t(T)$, the term $\psi_{t,-}(F_{\underline{v}})\psi_t(T/F_{\underline{v}})$ will contain a term $R(\psi_t(F'))\psi_t(T/F_{\underline{v}})$, which will contain a summand equal to $-(F \to F')t^{\delta}$, which has the effect of removing the unwanted derivation. Meanwhile, any term $(F \to F')t^{\delta}$ in $\psi_t(T)$ that only contains derivations that do not violate the Minimal Yield constraints is not cancelled by anything coming from the terms $\psi_{t,-}(F_{\underline{v}})\psi_t(T/F_{\underline{v}})$ because such terms are eliminated when applying R in the inductive construction of $\psi_{t,-}(F_{\underline{v}})$. □

In a similar way, we can consider the "no complexity loss" constraints of Definition 1.6.2, discussed in sections 1.6.1 and 1.6.3 of chapter 1, and the function $\delta \deg_a$ for $a \in \pi_0(F)$ discussed in section 3.5.2.1. In order to formulate these contraints as a Birkhoff factorization, consider a set of variables t_λ for $\lambda \in \mathcal{SO}_0$. This is a finite (albeit possibly large) set. We consider the algebra of Laurent series $\mathbb{Q}[[t_\lambda]][t_\lambda^{-1}]$ in all these variables, as well as the algebra $\mathcal{DM}[[t_\lambda]][t_\lambda^{-1}]$. Assume that we consider the subspace of $\mathcal{V}(\mathfrak{F}_{\mathcal{SO}_0})$ spanned by the forests $F = \sqcup_a T_a$, where all $T_a \in \mathrm{Dom}(h)$, for a head function h. We call this subspace $\mathcal{V}_h$. The product and coproduct $(\sqcup, \Delta^c)$ of the Hopf algebra of workspaces induce a product and coproduct on the subspace $\mathcal{V}_h$, respectively. We can then label the internal vertices of the syntactic objects in $\mathcal{V}_h$ using the head function, labeling $v \in V(T)$ by $h_T(v)$ (meaning the lexical item $\lambda = \lambda(h_T(v))$ in $\mathcal{SO}_0$ labeling the leaf $h_T(v)$). Thus, we can correspondingly assign to an internal vertex of T a variable t_λ. We can identify the components T_a of F with their root vertices and the latter with the correponding labeling by the head $h(T_a)$; hence, we can assign to each $a \in \pi_0(F)$ a variable $t_a := t_{\lambda(h(T_a))}$. We then set

$$\delta \deg_a(F \stackrel{\Phi}{\to} F') = \deg(\Phi_0(a)) - \deg(a)$$

$$\phi_t(F) = \sum_{\Phi: F \to F'} (F \xrightarrow{\Phi} F') \prod_{a \in \pi_0(F)} t_{h(T_a)}^{\delta \deg_a(\Phi)},$$

and we can apply the same procedure to this character that takes values in $\mathcal{DM}[[t_\lambda]][t_\lambda^{-1}]$.

3.5.3 Birkhoff Factorization in Algebroids

The construction of the ring (or algebra) $\mathcal{DM}$ of Merge derivations in the previous sections can be seen as an adaptation to the case of free symmetric Merge (in the form presented in chapter 1) of the idea of the *derivation forest semirings* of [82], where the original case treated in [82] was based on derivations in context-free grammars. We now show how to extend this notion from the setting of context-free semiring parsing to the Minimalist account. To see the analogy more directly, instead of the algebra used in section 3.5.2.2, one can construct a slightly different algebraic object encoding the same set of free symmetric Merge derivations. This will include the data of the Hopf algebra $\mathcal{H}$, while incorporating not just the workspaces, but also the explicit Merge derivations acting on them.

Next, we recall the notion of commutative bialgebroid and Hopf algebroid, originally introduced in the context to algebraic topology (see Appendix A1 of [173]). As elsewhere in this book, we will always assume that all algebras and vector spaces are over the field $\mathbb{Q}$ of rational numbers, unless otherwise stated.

To clarify the following definition of Hopf algebroids, it is helpful to recall the relation between commutative Hopf algebras and *affine group schemes*. The basic idea here is the fact that the set of characters of a commutative Hopf algebras; namely, the *morphisms of commutative algebras*

$$\phi \in \mathrm{Hom}_{\mathrm{CAlg}}(\mathcal{H}, \mathcal{R}),$$

where $\mathcal{R}$ is a chosen commutative algebra, is not just a set, but is in fact a *group*, because the coproduct operation of $\mathcal{H}$ determines a product operation

$$\phi_1 \star \phi_2 \in \mathrm{Hom}_{\mathrm{CAlg}}(\mathcal{H}, \mathcal{R}) \quad \text{with } (\phi_1 \star \phi_2)(x) := (\phi_1 \otimes \phi_2)\Delta(x),$$

and the antipode S determines the inverse: $\phi^{-1} := \phi \circ S$. Thus, a commutative Hopf algebra determines an assignment

$$G : \mathcal{R} \mapsto G(\mathcal{R}) = \mathrm{Hom}_{\mathrm{CAlg}}(\mathcal{H}, \mathcal{R})$$

of a group $G(\mathcal{R})$ to any commutative algebra $\mathcal{R}$, which is a functor between the respective categories. Such G is called an *affine group scheme*. Assigning a commutative Hopf algebra $\mathcal{H}$ is equivalent to assigning the associated affine group scheme $G = G_{\mathcal{H}}$, so one says that a commutative Hopf algebra is dual to an affine group scheme. The notion of a Hopf algebroid developed in [173], which we recall here, generalizes this relation between commutative Hopf algebras and affine group schemes to the case where, instead of groups, one works with groupoids, where the product operation is a composition of arrows that is possible only when the target of the first arrow agrees with the source of the next. The reason why we need this generalization is that chains of Merge derivations between workspaces are indeed composable only when the target workspace of one Merge application is the same as the source workspace of the next Merge application.

Definition 3.5.7. A commutative Hopf algebroid is dual to an *affine groupoid scheme*; namely, it consists of a pair of commutative algebras $\mathcal{A}^{(0)}$ and $\mathcal{H}^{(1)}$ with the property that, for any other commutative algebra $\mathcal{R}$, the sets $\mathcal{G}^{(0)}(\mathcal{R}) = \mathrm{Hom}(\mathcal{A}^{(0)}, \mathcal{R})$ and $\mathcal{G}^{(1)}(\mathcal{R}) = \mathrm{Hom}(\mathcal{H}^{(1)}, \mathcal{R})$ are the objects and morphisms of a groupoid $\mathcal{G}$. Equivalently, the pair of algebras $(\mathcal{A}^{(0)}, \mathcal{H}^{(1)})$ is endowed with homomorphisms $\eta_s, \eta_t : \mathcal{A}^{(0)} \to \mathcal{H}^{(1)}$, which give $\mathcal{H}^{(1)}$ the structure of a $\mathcal{A}^{(0)}$-bimodule (dual to source and target maps of the groupoid), a coproduct (dual to composition of arrows in the groupoid) given by a morphism of $\mathcal{A}^{(0)}$-bimodules:

$$\Delta : \mathcal{H}^{(1)} \to \mathcal{H}^{(1)} \otimes_{\mathcal{A}^{(0)}} \mathcal{H}^{(1)},$$

a counit $\epsilon : \mathcal{H}^{(1)} \to \mathcal{A}^{(0)}$, which is also a morphism of $\mathcal{A}^{(0)}$-bimodules (dual to the inclusion of identity morphisms), and a conjugation $S : \mathcal{H}^{(1)} \to \mathcal{H}^{(1)}$ (dual to the inverse of morphisms in the groupoid). These maps satisfy $\epsilon\eta_s = \epsilon\eta_t = 1$ (identity morphisms have the same source and target), $(1 \otimes \epsilon)\Delta = (\epsilon \otimes 1)\Delta = 1$ (composition with the identity morphism), $(1 \otimes \Delta)\Delta = (\Delta \otimes 1)\Delta$ (associativity of the composition of morphisms), and $S^2 = 1$ and $S\eta_s = \eta_t$ (inversion is an involution and exchanges the source and target of morphisms), and the property that composition of a morphism with its inverse gives the identity morphism; namely, that

$$\eta_t\epsilon = \mu(S \otimes 1)\Delta \quad \text{and} \quad \eta_s\epsilon = \mu(1 \otimes S)\Delta,$$

with $\mu : \mathcal{H}^{(1)} \otimes_{\mathcal{A}^{(0)}} \mathcal{H}^{(1)} \to \mathcal{H}^{(1)}$ extending the algebra multiplication $\mu : \mathcal{H}^{(1)} \otimes_{\mathbb{Q}} \mathcal{H}^{(1)} \to \mathcal{H}^{(1)}$. Also, one has $\Delta\eta_s = 1 \otimes \eta_s$, $\Delta\eta_t = \eta_t \otimes 1$ (the source of the composition of arrows is the source of the first and the target of the composition is the target of the second). A morphism of Hopf algebroids,

$$f : (\mathcal{A}_1^{(0)}, \mathcal{H}_1^{(1)}) \to (\mathcal{A}_2^{(0)}, \mathcal{H}_2^{(1)}),$$

is a pair of algebra homomorphisms:

$$f^{(0)} : \mathcal{A}_1^{(0)} \to \mathcal{A}_2^{(0)} \quad \text{and} \quad f^{(1)} : \mathcal{H}_1^{(1)} \to \mathcal{H}_2^{(1)},$$

satisfying $f^{(0)} \circ \epsilon_1 = \epsilon_2 \circ f^{(1)}$, $f^{(1)} \circ \eta_{s,1} = \eta_{s,2} \circ f^{(0)}$, $f^{(1)} \circ \eta_{t,1} = \eta_{t,2} \circ f^{(0)}$, $f^{(1)} \circ S_1 = S_2 \circ f^{(1)}$, $\Delta_2 \circ f^{(1)} = (f^{(1)} \otimes f^{(1)}) \circ \Delta_1$.

A commutative bialgebroid is a structure as described here, where one does not assume the invertibiliy of morphisms; namely, $C^{(0)}(\mathcal{R}) = \mathrm{Hom}(\mathcal{A}^{(0)}, \mathcal{R})$ and $C^{(1)}(\mathcal{R}) = \mathrm{Hom}(\mathcal{H}^{(1)}, \mathcal{R})$ are the objects and morphisms of a small category C (a semigroupoid) instead of a groupoid, so one has the same structure as before, but without the conjugation map S. Thus, a commutative bialgebroid is dual to an *affine semigroupoid scheme*.

Examples of Hopf algebroids arise, for instance, when the field of definition of a Hopf algebra $\mathcal{H}$ is replaced by the ring of functions $\mathcal{A}$ of some underlying space. In our setting, the natural modification of the Hopf algebra $\mathcal{H}$ of workspaces is a version where arrows corresponding to the action of Merge are also incorporated as part of the same algebraic structure. Since these will in general not necessarily be invertible arrows, the resulting structure will be a bialgebroid rather than a Hopf algebroid.

Remark 3.5.8. We assign a grading to a bialgebroid $(\mathcal{A}^{(0)}, \mathcal{H}^{(1)})$ by defining, for an arrow γ in the semigroupoid the degree as the maximal length of a factorization of γ, $\deg(\gamma) = \max\{n \geq 1 \mid \exists \gamma = \gamma_1 \circ \cdots \circ \gamma_n\}$. In the dual algebra, we assign $\deg(\delta_\gamma) = \deg(\gamma)$, with δ_γ being the Kronecker delta, and $\deg(\prod_i \delta_{\gamma_i}) = \sum_i \deg(\delta_{\gamma_i})$. The coproduct $\Delta(\delta_\gamma) = \delta_\gamma \otimes 1 + 1 \otimes \delta_\gamma + \sum_{\gamma=\gamma_1\circ\gamma_2} \delta_{\gamma_1} \otimes \delta_{\gamma_2}$ has the terms δ_{γ_1}, δ_{γ_2} of lower degrees. So we set $\mathcal{H}^{(1)} = \oplus_{n\geq 0} \mathcal{H}^{(1)}_n$ with $\mathcal{H}^{(1)}_0 = \mathbb{Q}$ and $\mathcal{H}^{(1)}_n$ spanned by the elements of degree n, compatibly with product and coproduct operations.

Lemma 3.5.9. *The data* $\mathcal{A}^{(0)} = (\mathcal{V}(\mathfrak{F}_{SO_0}), \sqcup)$ *and* $\mathcal{H}^{(1)} = (\mathcal{DM}, \sqcup)$*, define a bialgebroid.*

Proof. The algebra $\mathcal{H}^{(1)} = (\mathcal{DM}, \sqcup)$ dual to the arrows $C^{(1)}$ is the same algebra of Merge derivations introduced in Definition 3.5.1. We can identify elements $X = \sum_i a_i \varphi_{A_i}$ in $\mathcal{DM}$ with finitely supported functions $X = \sum_i a_i \delta_{\varphi_{A_i}}$ on the set of derivations of the form (3.5.1), (3.5.2), with $\delta_{\varphi_{A_i}}$ being the Kronecker delta. The left and right $\mathcal{A}^{(0)}$-module structures that correspond to the source and target maps are determined by

$$\eta_s(F)\varphi_A = \begin{cases} \varphi_A & s(\varphi_A) = F \\ 0 & \text{otherwise} \end{cases} \qquad \eta_t(F)\varphi_A = \begin{cases} \varphi_A & t(\varphi_A) = F \\ 0 & \text{otherwise.} \end{cases}$$

The coproduct $\Delta : \mathcal{H}^{(1)} \to \mathcal{H}^{(1)} \otimes_{\mathcal{A}^{(0)}} \mathcal{H}^{(1)}$ is given by

$$\Delta(\delta_{\phi_A}) = \delta_{\phi_A} \otimes 1 + 1 \otimes \delta_{\phi_A} + \sum_{\phi_A = \phi_{A_1} \circ \phi_{A_2}} \delta_{\phi_{A_1}} \otimes \delta_{\phi_{A_2}} ,$$

where for $\phi_{A_2} = (F \overset{\mathfrak{M}_{A_2}}{\to} F')$ and $\phi_{A_1} = (F' \overset{\mathfrak{M}_{A_1}}{\to} F'')$, the composition is given by

$$\phi_{A_1} \circ \phi_{A_2} = (F \overset{\mathfrak{M}_{A_1 \circ A_2}}{\to} F''),$$

where $\mathfrak{M}_{A_1 \circ A_2} = \mathfrak{M}_{A_1} \circ \mathfrak{M}_{A_2}$ denotes the set of all compositions of a chain of Merge derivations in the set A_2 followed by one in the set A_1. □

Remark 3.5.10. Note that the bialgebroid of Lemma 3.5.9 only uses the multiplication $(\mathcal{V}(\mathfrak{F}_{SO_0}), \sqcup)$ of the Hopf algebra $\mathcal{H}$ of workspaces, and the comultiplication of $\mathcal{H}$ does not appear in the expression for the coproduct on $\mathcal{H}^{(1)}$. The coproduct of $\mathcal{H}$, however, is also encoded in the bialgebroid, as it is built into the arrows of $\mathcal{H}^{(1)}$, since the Merge operations $\mathfrak{M}_{S,S'}$ that occur in the arrows are of the form (3.1.3), so that terms of the coproduct of $\mathcal{H}$ will contribute to arrows.

3.5.4 Bialgeroids and Rota-Baxter Algebroids

In order to simultaneously extend our setting with Rota–Baxter algebras (and semirings) and Birkhoff factorization of maps from Hopf algebras, as well as the setting of semiring parsing in semantics, we now introduce a version of Birkhoff factorization for algebroids.

3.5.4.1 Algebroids and directed graph schemes In our setting, we will take a different viewpoint on the notion of *algebroid* than what is more commonly used in mathematics. The common definition of an algebroid (over a field K) is just a K-linear category where the operation of morphism composition is the multiplication part of the algebroid and the linear structure on the spaces of morphisms provides the addition part. However, in view of our use of the notions of Hopf algebroid and bialgebroid in Definition 3.5.7, it is natural to think of a commutative algebroid in the way given in Definition 3.5.11.

Definition 3.5.11. An algebroid is a pair of commutative algebras $(\mathcal{A}, \mathcal{E})$ with two morphisms, $\eta_s, \eta_t : \mathcal{A} \to \mathcal{E}$, which give $\mathcal{E}$ the structure of bimodule over $\mathcal{A}$ and a morphism of $\mathcal{A}$-bimodules $\epsilon : \mathcal{E} \to \mathcal{A}$ with $\epsilon\eta_s = \epsilon\eta_t = 1_{\mathcal{A}}$. A morphism $f : (\mathcal{A}_1, \mathcal{E}_1) \to (\mathcal{A}_2, \mathcal{E}_2)$ is a pair of morphisms of commutative algebras $f_V : \mathcal{A}_1 \to \mathcal{A}_2$ and $f_E : \mathcal{E}_1 \to \mathcal{E}_2$, with $\eta_{s,2} \circ f_V = f_E \circ \eta_{s,1}$, $\eta_{t,2} \circ f_V = f_E \circ \eta_{t,1}$ and $f_V \circ \epsilon_1 = \epsilon_2 \circ f_E$.

A way of thinking of this notion of algebroid is as being dual to directed graphs. In other words, our algebroids are dual to *affine directed graph schemes* in the same way that Hopf algebras are dual to affine group schemes, Hopf algebroids are dual to affine groupoid schemes, and bialgebroids are dual to affine semigroupoid schems, as discussed in section 3.5.3. This can be seen immediately as stated in Lemma 3.5.12.

Lemma 3.5.12. *Let $(\mathcal{A}, \mathcal{E})$ be a commutative algebroid in the sense of Definition 3.5.11. Then for every other commutative algebra $\mathcal{R}$, the sets $V(\mathcal{R}) = \mathrm{Hom}(\mathcal{A}, \mathcal{R})$ and $E(\mathcal{R}) = \mathrm{Hom}(\mathcal{E}, \mathcal{R})$ are the sets of vertices and edges of a directed graph $G(\mathcal{R})$ with source and target maps $s, t : E(\mathcal{R}) \to V(\mathcal{R})$, determined by the morphisms $\eta_s, \eta_t : \mathcal{A} \to \mathcal{E}$, and where each vertex $v \in V(\mathcal{R})$ has a looping edge $e_v \in E(\mathcal{R})$, with $s(e_v) = t(e_v) = v$. A morphism of algebroids induces a morphism of directed graphs.*

Proof. A directed graph G is a functor from the category $\mathbf{2}$ with two objects V, E and two non-identity morphisms $s, t : E \to V$, with values in the category Sets. The assignment $G(\mathcal{R}) : V \mapsto \mathrm{Hom}(\mathcal{A}, \mathcal{R})$ and $G(\mathcal{R}) : E \mapsto \mathrm{Hom}(\mathcal{E}, \mathcal{R})$ and $G(\mathcal{R}) : s \mapsto \eta_s^*$ $G(\mathcal{R}) : t \mapsto \eta_t^*$, with $\eta_i^*(\phi) = \phi \circ \eta_i$, for $\phi \in \mathrm{Hom}(\mathcal{E}, \mathcal{R})$, determines such a functor. The inclusion of the looping edges e_v in $\mathrm{Hom}(\mathcal{E}, \mathcal{R})$ is given by $e_v = v \circ \epsilon$, with $v \in \mathrm{Hom}(\mathcal{A}, \mathcal{R})$. A morphism of directed graph $\alpha : G_2 \to G_1$ is a natural transformation of the functors from $\mathbf{2}$ to Sets; namely, a pair of maps:

$$\alpha_V : \mathrm{Hom}(\mathcal{A}_2, \mathcal{R}) \to \mathrm{Hom}(\mathcal{A}_1, \mathcal{R}) \quad \text{and} \quad \alpha_E : \mathrm{Hom}(\mathcal{E}_2, \mathcal{R}) \to \mathrm{Hom}(\mathcal{E}_1, \mathcal{R}),$$

such that $s \circ \alpha_E = \alpha_V \circ s$ and $t \circ \alpha_E = \alpha_V \circ t$. A morphism of algebroids,

$$f : (\mathcal{A}_1, \mathcal{E}_1) \to (\mathcal{A}_2, \mathcal{E}_2),$$

determines such a natural transformation with $\alpha_V = f_V^*$ and $\alpha_E = f_E^*$. The additional property $f_V \circ \epsilon_1 = \epsilon_2 \circ f_E$ ensures that a looping edge e_v in $\mathrm{Hom}(\mathcal{E}_2, \mathcal{R})$ is mapped to $\alpha_E(e_v) = e_{\alpha_V(v)}$ in $\mathrm{Hom}(\mathcal{E}_1, \mathcal{R})$. □

Corollary 3.5.13. *A bialgebroid $(\mathcal{A}, \mathcal{E})$ is a commutative algebroid with the property that the graphs $G(\mathcal{R})$ are categories (i.e., they are directed graphs satisfying reflexivity and transitivity).*

Proof. A directed graph G is a category (with objects given by the vertices and morphisms given by the directed edges) iff it is the directed graph of a preorder (namely, if it satisfies reflexivity and transitivity). In other words, it is a directed graph where every vertex has a looping edge attached to it, and if there is a pair of edges e, e' with $s(e) = v$, $t(e) = s(e')$ and $t(e') = v'$, then there is an edge $\tilde{e}$ with $s(\tilde{e}) = v$ and $t(\tilde{e}) = v'$. The coproduct of the bialgebroid ensures that the graphs $G(\mathcal{R})$ are transitive, while reflexivity is already a property of directed graphs determined by algebroids. □

3.5.4.2 Rota–Baxter algebroids The notion generalizing the Rota–Baxter algebra structure in this setting is given in Definition 3.5.14.

Definition 3.5.14. A commutative Rota–Baxter algebroid of weight -1 is a commutative algebroid $(\mathcal{A}, \mathcal{E})$, as in Definition 3.5.11, together with a pair of maps $R = (R_V, R_E)$, where $R_V \in \text{End}(\mathcal{A})$ is an algebra homomorphism and $R_E : \mathcal{E} \to \mathcal{E}$ is a linear map that satisfies

$$R_E(\eta_s(a) \cdot \xi) = \eta_s(R_V(a)) \cdot R_E(\xi) \quad R_E(\eta_t(a) \cdot \xi) = \eta_t(R_V(a)) \cdot R_E(\xi), \quad (3.5.8)$$

for all $a \in \mathcal{A}$ and $\xi \in \mathcal{E}$, with $\cdot$ the algebra product in $\mathcal{E}$, and $\epsilon \circ R_E = R_E \circ \epsilon$, and that satisfies the Rota–Baxter relation of weight -1:

$$R_E(\xi) \cdot R_E(\zeta) = R_E(R_E(\xi) \cdot \zeta) + R_E(\xi \cdot R_E(\zeta)) - R_E(\xi \cdot \zeta). \quad (3.5.9)$$

We moreover require a normalization condition—namely, that $R_E(1_\mathcal{E}) = 0$ or $R_E(1_\mathcal{E}) = 1_\mathcal{E}$, where $1_\mathcal{E}$ is the unit of the algebra $\mathcal{E}$.

Lemma 3.5.15. *The Rota–Baxter structure of Definition 3.5.14 has the following properties:*

1. *The condition* (3.5.8) *replaces the conditions $\eta_s R_V = R_E \eta_s$ and $\eta_t R_V = R_E \eta_t$ and is implied by these conditions in the case where R_E is an algebra homomorphism.*
2. *The normalization condition that $R_E(1) \in \{0, 1\}$, together with the conditions* (3.5.8) *and* (3.5.9)*, imply that R_E also satisfies*

$$\begin{aligned} R_E(R_V(\eta_s(a)) \cdot \xi) &= R_V(\eta_s(a)) \cdot R_E(\xi) \\ R_E(R_V(\eta_t(a)) \cdot \xi) &= R_V(\eta_t(a)) \cdot R_E(\xi) \end{aligned} \quad (3.5.10)$$

   for all $a \in \mathcal{A}$ and $\xi \in \mathcal{E}$; that is, R_E is a bimodule homomorphism when $\mathcal{E}$ is viewed as a bimodule over the subalgebra $R_V(\mathcal{A})$.
3. *If $R_V \in \text{Aut}(\mathcal{A})$ is an algebra automorphism, then* (3.5.8) *and* (3.5.9) *with $R_E(1_\mathcal{E}) \in \{0, 1\}$ imply that R_E is a bimodule homomorphism of $\mathcal{E}$ as a $\mathcal{A}$-bimodule.*

Proof. 1. If R_E is an algebra homomorphism, then the conditions $\eta_s R_V = R_E \eta_s$ and $\eta_t R_V = R_E \eta_t$ imply that

$$R_E(\eta_s(a) \cdot \xi) = R_E(\eta_s(a)) \cdot R_E(\xi) = \eta_s(R_V(a)) \cdot R_E(\xi),$$

and it is similar for η_t.

2. If R_E satisfies (3.5.9), then the subspaces $R_E(\mathcal{E})$ and $(1 - R_E)(\mathcal{E})$ of $\mathcal{E}$ are (possibly nonunital) subalgebras. If $R_E(1) \in \{0, 1\}$, then either $R_E(\mathcal{E}) \subset \mathcal{E}$ is unital and $(1 - R_E)(\mathcal{E})$ is not, or vice versa. If, moreover, R_E also satisfies (3.5.8), then $(\mathcal{A}, R_E(\mathcal{E}))$ and $(\mathcal{A}, (1 - R_E)(\mathcal{E}))$ are subalgebroids of $(\mathcal{A}, \mathcal{E})$ with the induced maps η_s, η_t, ϵ. Indeed, the Rota–Baxter identity (3.5.8) ensures that the product $R_E(\xi) \cdot R_E(\zeta)$ is in the range $R_E(\mathcal{E})$ for all $\xi, \zeta \in \mathcal{E}$, and hence $R_E(\mathcal{E}) \subset \mathcal{E}$ is a (possibly nonunital) subalgebra. The case of $(1 - R_E)\mathcal{E}$ is similar. If $R_E(1) = 0$, then $(1 - R_E)\mathcal{E}$ is unital and $R_E(\mathcal{E})$ is not, and vice versa if $R_E(1) = 1$. Note then that the conditions $R_E(1_\mathcal{E}) \in \{0, 1\}$ and (3.5.9) imply that the linear map R_E is a projector; namely, $R_E^2 = R_E$. In fact, by (3.5.9), we have

$$\begin{aligned} R_E(R_E(\xi)) &= R_E(R_E(\xi) \cdot 1)) = R_E(\xi) \cdot R_E(1) + R_E(\xi \cdot 1) - R_E(\xi \cdot R_E(1)) \\ &= R_E(\xi) \cdot (1 + R_E(1)) - R_E(\xi \cdot R_E(1)), \end{aligned}$$

where if $R_E(1) = 0$ or $R_E(1) = 1$, we get $R_E^2(\xi) = R_E(\xi)$. Applying condition (3.5.9) to a pair with $\xi = \eta_s(a)$ gives (using condition (3.5.8))

$$R_E(\eta_s(a)) \cdot R_E(\zeta) = R_E(R_E(\eta_s(a)) \cdot \zeta) + R_E(\eta_s(a) \cdot R_E(\zeta)) - R_E(\eta_s(a) \cdot \zeta),$$

which gives

$$\eta_s(R_V(a)) \cdot R_E(\zeta) = R_E(\eta_s(R_V(a)) \cdot \zeta) + \eta_s(R_V(a)) R_E^2(\zeta) - \eta_s(R_V(a)) R_E(\zeta).$$

Since we are also assuming that $R_E(1_\mathcal{E}) \in \{0, 1\}$, we have $R_E^2(\zeta) = R_E(\zeta)$, so we obtain

$$\eta_s(R_V(a)) \cdot R_E(\zeta) = R_E(\eta_s(R_V(a)) \cdot \zeta),$$

and similarly with η_t, so that (3.5.10) holds for all $a \in \mathcal{A}$ and $\zeta \in \mathcal{E}$.

3. If R_V is an automorphism of $\mathcal{A}$ rather than just an endomorphism, then this also implies

$$R_E(\eta_s(a) \cdot \xi) = \eta_s(a) \cdot R_E(\xi) \quad R_E(\eta_t(a) \cdot \xi) = \eta_t(a) \cdot R_E(\xi), \tag{3.5.11}$$

for all $a \in \mathcal{A}$ and $\zeta \in \mathcal{E}$. □

A simple source of examples of Rota–Baxter algebroids is obtained by considering functions on the edges of a directed graph, which take values in a

Rota–Baxter algebra. This means that, in these examples, the Rota–Baxter operator is acting only on the coefficients of functions. Lemma 3.5.16 is a direct consequence of the definition of Rota–Baxter algebroids.

Lemma 3.5.16. *Let G be a directed graph, and let $(\mathcal{R}, R)$ be a Rota–Baxter algebra of weight* -1. *Consider the pair of algebras $(\mathcal{A}, \mathcal{E})$ with $\mathcal{A} = \mathbb{Q}[V_G]$ (finitely supported $\mathbb{Q}$-valued functions on the set V_G of vertices of G), and $\mathcal{E} = \mathbb{Q}[V_G] \otimes_{\mathbb{Q}} \mathcal{R}$, with morphisms $\eta_s, \eta_t : \mathcal{A} \to \mathcal{E}$ given by composition with the source and target maps $s, t : E_G \to V_G$. The maps $R_V = \mathrm{id}$ on $\mathcal{A}$ and $R_E = 1 \otimes R$ give $(\mathcal{A}, \mathcal{E})$ the structure of a Rota–Baxter algebroid of weight* -1.

3.5.4.3 Rota–Baxter semiringoids There is a direct generalization of this notion of Rota–Baxter algebroids, as well as the class of examples discussed in Lemma 3.5.16, to the case where algebras are replaced by semirings. We will refer to those as *Rota–Baxter semiringoids*. The definition and properties are analogous to the algebroid case, in the same way in which we generalized from Rota–Baxter algebras to Rota–Baxter semirings in section 3.1.4. We will focus in particular on the analog of the examples in Lemma 3.5.16.

The category of commutative semirings, with the initial object given by the semiring $\mathbb{Z}_{\geq 0}$ of nonnegative integers, is dual to the category of semiring schemes; that is, affine schemes over Spec($\mathbb{Z}_{\geq 0}$). The full subcategory of idempotent commutative semirings, with the initial object given by $\mathfrak{B}$, the Boolean semiring of (3.2.13), is dual to the category of affine schemes over Spec($\mathfrak{B}$).

Definition 3.5.17. A semiringoid is the datum $(\mathcal{A}, \mathcal{E})$ of two commutative semirings with semiring homomorphisms $\eta_s, \eta_t : \mathcal{A} \to \mathcal{E}$ that give $\mathcal{E}$ the structure of a bi-semimodule over the semiring $\mathcal{A}$ and with a bi-semimodule homomorphism $\epsilon : \mathcal{E} \to \mathcal{A}$ with $\epsilon\eta_s = \epsilon\eta_t = 1_{\mathcal{A}}$. A morphism $(\mathcal{A}_1, \mathcal{E}_1) \to (\mathcal{A}_2, \mathcal{E}_2)$ of semiringoids is a pair of semiring homomorphisms $f_V : \mathcal{A}_1 \to \mathcal{A}_2$ and $f_E : \mathcal{E}_1 \to \mathcal{E}_2$, with $\eta_{s,2} \circ f_V = f_E \circ \eta_{s,1}$, $\eta_{t,2} \circ f_V = f_E \circ \eta_{t,1}$ and $f_V \circ \epsilon_1 = \epsilon_2 \circ f_E$. A Rota–Baxter semiringoid of weight $+1$ is a semiringoid $(\mathcal{A}, \mathcal{E})$ endowed with a semiring endomorphism $R_V : \mathcal{A} \to \mathcal{A}$ and an $R_E : \mathcal{E} \to \mathcal{E}$ a $\mathbb{Z}_{\geq 0}$-linear map (morphism of $\mathbb{Z}_{\geq 0}$-semimodules) that satisfies

$$R_E(\eta_s(a) \odot \xi) = \eta_s(R_V(a)) \odot R_E(\xi), \quad R_E(\eta_t(a) \odot \xi) = \eta_t(R_V(a)) \odot R_E(\xi) \quad (3.5.12)$$

for all $a \in \mathcal{A}$ and $\xi \in \mathcal{E}$, with $\odot$ denoting the semiring product in $\mathcal{E}$, and $\epsilon \circ R_E = R_E \circ \epsilon$, and that satisfies the Rota–Baxter relation of weight $+1$:

$$R_E(\xi) \odot R_E(\zeta) = R_E(R_E(\xi) \odot \zeta) \boxdot R_E(\xi \odot R_E(\zeta)) \boxdot R_E(\xi \odot \zeta), \quad (3.5.13)$$

with $\boxdot$ and $\odot$ denoting the semiring sum and product in $\mathcal{E}$. The case of a Rota–Baxter structure of weight -1 is similar, with (3.5.13) being replaced by

$$R_E(\xi) \odot R_E(\zeta) \boxdot R_E(\xi \odot \zeta) = R_E(R_E(\xi) \odot \zeta) \boxdot R_E(\xi \odot R_E(\zeta)). \qquad (3.5.14)$$

We moreover require the normalization condition $R_E(1_{\mathcal{E}}) = 0_{\mathcal{E}}$ or $R_E(1_{\mathcal{E}}) = 1_{\mathcal{E}}$, where $1_{\mathcal{E}}$ is the unit of the multiplicative monoid and $0_{\mathcal{E}}$ is the unit of the additive monoid of $\mathcal{E}$.

When considering semiringoids with commutative idempotent semirings, one can drop the $\mathbb{Z}_{\geq 0}$-linearity requirement for R_E and only require that R_E be a morphism of $\mathfrak{B}$-semimodules (Boolean semimodules).

Remark 3.5.18. Note that the notion of semiringoid used in Definition 3.5.17 differs from another commonly used notion, in which a semiringoid is a small category C where all the Hom sets $\mathrm{Hom}_C(X, Y)$, for $X, Y \in \mathrm{Obj}(C)$, are commutative monoids with bilinear composition of morphisms, and all the End sets $\mathrm{End}_C(X) = \mathrm{Hom}_C(X, X)$ are semirings.

We have then an analog for semiringoids of the class of Rota–Baxter algebroids of Lemma 3.5.16. Again, this follows directly from Definition 3.5.17.

Lemma 3.5.19. *Let G be a directed graph, and let $(\mathcal{R}, R)$ be a Rota–Baxter semiring of weight ± 1. Consider the pair of semirings $(\mathcal{A}, \mathcal{E})$ with $\mathcal{A} = \mathbb{Z}_{\geq 0}[V_G]$ (finitely supported $\mathbb{Z}_{\geq 0}$-valued functions on the set V_G of vertices of G) and $\mathcal{E} = \mathbb{Z}_{\geq 0}[V_G] \otimes_{\mathbb{Z}_{\geq 0}} \mathcal{R}$, with morphisms $\eta_s, \eta_t : \mathcal{A} \to \mathcal{E}$ given by precomposition with the source and target maps $s, t : E_G \to V_G$. The maps $R_V = \mathrm{id}$ on $\mathcal{A}$ and $R_E = 1 \otimes R$ give $(\mathcal{A}, \mathcal{E})$ the structure of a Rota–Baxter semiringoid of weight ± 1. In the case where $\mathcal{R}$ is a commutative idempotent semiring, we can replace this construction with $\mathcal{A} = \mathfrak{B}[V_G]$ (Boolean functions on V_G) and $\mathcal{E} = \mathfrak{B}[V_G] \otimes_{\mathfrak{B}} \mathcal{R}$, to obtain a Boolean Rota–Baxter semiringoid (a semiringoid over commutative idempotent semirings).*

3.5.4.4 Birkhoff factorization in algebroids and semiringoids We then consider morphisms of algebroids $\Phi : (\mathcal{A}^{(0)}, \mathcal{H}^{(1)}) \to (\mathcal{A}, \mathcal{E})$ from a Hopf algebroid to an an algebroid with a Rota–Baxter structure (R_V, R_E) of weight -1. The target algebroid $(\mathcal{A}, \mathcal{E})$ does *not* have a compositional structure in the sense that the directed graph (graph scheme) G dual to the algebroid does not generally have the transitive property: given two directed edges where the target of the first is the source of the second, it is not necessarily the case that there is also a edge from the source of the first to the target of the second.

The source $(\mathcal{A}^{(0)}, \mathcal{H}^{(1)})$ has a compositional structure that is encoded in the coproduct as bialgebroid, which is the convolution product of the groupoid algebra $\mathcal{H}^{(1)}$. As in the case of algebras, the convolution structure on $\mathcal{H}^{(1)}$ and the Rota–Baxter structure on $(\mathcal{A}, \mathcal{E})$ will perform the factorization of Φ : $(\mathcal{A}^{(0)}, \mathcal{H}^{(1)}) \to (\mathcal{A}, \mathcal{E})$, which accounts for the induced compositional structure of the image.

Lemma 3.5.20. *Let $(\mathcal{A}^{0)}, \mathcal{H}^{(1)})$ be a Hopf algebroid, and let $(\mathcal{A}, \mathcal{E})$ be an algebroid with a Rota–Baxter structure (R_V, R_E) of weight -1. Given a morphism $\Phi : (\mathcal{A}^{(0)}, \mathcal{H}^{(1)}) \to (\mathcal{A}, \mathcal{E})$ of algebroids, there is a pair $\Phi_\pm$ with $\Phi_{\pm,V} = \Phi_V$ and $\Phi_{+,E}(f) = (\Phi_{-,E} \star \Phi_E)(f) = (\Phi_{-,E} \otimes \Phi_E)(\Delta f)$ for all $f \in \mathcal{H}^{(1)}$, where we have*

$$\Phi_{-,E}(f) = -R_E(\tilde{\Phi}_E(f)) \quad \textit{with} \quad \tilde{\Phi}_E(f) = \Phi_E(f) + \sum \Phi_{-,E}(f')\Phi_E(f''),$$

for $\Delta(f) = f \otimes 1 + 1 \otimes f + \sum f' \otimes f''$, and with $\Phi_{+,E}(f) = (1 - R_E)(\tilde{\Phi}_E(f))$.

Proof. The argument for showing that the maps $\Phi_{\pm,E} : (\mathcal{A}^{0)}, \mathcal{H}^{(1)}) \to (\mathcal{A}, \mathcal{E}_\pm)$ with $\mathcal{E}_+ = (1 - R_E)(\mathcal{E})$ and $\mathcal{E}_- = R_E(\mathcal{E})$ are algebroid homomorphisms closely follows the same argument for Rota–Baxter algebras of weight -1, as in Theorem 1.39 of [46]. The factorization identity $\Phi_{+,E} = \Phi_{-,E} \star \Phi_E$ follows from $\Phi_{+,E} = (1 - R_E)\tilde{\Phi}_E$ and $\Phi_{-,E} = -R_E\tilde{\Phi}_E$ and the expression for $\tilde{\Phi}_E$ in terms of the coproduct Δ. □

We consider in particular the case where the Rota–Baxter algebroids are as in Lemma 3.5.16.

Lemma 3.5.21. *Let $(\mathcal{A}, \mathcal{E})$ be a Rota–Baxter algebroid as in Lemma 3.5.16, and $(\mathcal{A}^{(0)}, \mathcal{H}^{(1)})$ be a bialgebroid. The Birkhoff factorization of an algebroid homomorphism $\Phi : (\mathcal{A}^{(0)}, \mathcal{H}^{(1)}) \to (\mathcal{A}, \mathcal{E})$ consists of a map of directed graphs (graph schemes) $\alpha : G \to \mathcal{G}$, with G dual to $(\mathcal{A}, \mathcal{E})$ and $\mathcal{G}$ dual to $(\mathcal{A}^{(0)}, \mathcal{H}^{(1)})$, such that $\Phi_E(f) = f \circ \alpha$ for $f \in \mathcal{H}^{(1)}$, with the factorization $\Phi_{E,-}$ mapping $f = \delta_\gamma$, where γ is an arrow in $\mathcal{G}$, to the function $\Phi_{E,-}(\delta_\gamma)$, which acts on a combination $\sum_i a_i e_i$ with $e_i \in E_G$ as follows:*

$$\Phi_{E,-}(\delta_\gamma)(\sum_i a_i e_i) = -(\sum_{\alpha(e)=\gamma} R_E(a_e) + \sum_{\alpha(e_1)\circ\alpha(e_2)=\gamma} R_E(R_E(a_{e_1})a_{e_2}) + \cdots$$
$$+ \sum_{\alpha(e_1)\circ\cdots\circ\alpha(e_n)=\gamma} R_E(\cdots(R_E(a_{e_1})\cdots)a_{e_n})) . \tag{3.5.15}$$

Proof. The algebroid $(\mathcal{A}, \mathcal{E})$ is associated with a directed graph (graph scheme) G and the bialgebroid $(\mathcal{A}^{0)}, \mathcal{H}^{(1)})$ is associated with a semigroupoid $\mathcal{G}$ (equivalently, a graph that is reflexive and transitive). A morphism $\Phi : (\mathcal{A}^{(0)}, \mathcal{H}^{(1)}) \to$

$(\mathcal{A}, \mathcal{E})$ of algebroids is equivalent to the datum of a map of directed graphs $\alpha : G \to \mathcal{G}$. The map $\Phi_E : \mathcal{H}^{(1)} \to \mathcal{E}$ then is given by $\Phi_E(f) = f \circ \alpha$. It suffices to consider the case of $f = \delta_\gamma$ for some $\gamma \in \mathcal{G}^{(1)}$, as in general $f \in \mathbb{Q}[\mathcal{G}]$ will be a product of linear combinations of delta functions δ_γ. In the case where the Rota–Baxter operator of weight -1 is thc identity, the Bogolyubov preparation takes the form

$$\tilde{\Phi}_E(\delta_\gamma) = \delta_\gamma \circ \alpha + \sum_{\gamma=\gamma_1\circ\gamma_2} \delta_{\gamma_1} \circ \alpha \cdot \delta_{\gamma_2} \circ \alpha + \cdots + \sum_{\gamma=\gamma_1\circ\cdots\circ\gamma_n} \delta_{\gamma_1} \circ \alpha \cdots \delta_{\gamma_n} \circ \alpha ,$$

with $n = \deg(\gamma)$, which is then equal to

$$\tilde{\Phi}_E(\delta_\gamma) = \sum_{e\in E_G : \alpha(e)=\gamma} \delta_e + \cdots + \sum_{e_1,\dots,e_n\in E_G : \gamma=\alpha(e_1)\circ\cdots\circ\alpha(e_n)} \delta_{e_1} \cdots \delta_{e_n} , \quad (3.5.16)$$

so we have, for a collection of edges $e_i \in E_G$,

$$\tilde{\Phi}_E(\delta_\gamma)(\sum_i a_i e_i) = \sum_{\alpha(e)=\gamma} a_e + \sum_{\alpha(e_1)\circ\alpha(e_2)=\gamma} a_{e_1} a_{e_2} + \cdots + \sum_{\alpha(e_1)\circ\cdots\circ\alpha(e_n)=\gamma} a_{e_1} \cdots a_{e_n} .$$

In the case of a Rota–Baxter operator R_E of weight -1 that is not the identity, we similarly get (3.5.21). □

In the case of the bialgebroid $(\mathcal{A}^{0)} = \mathcal{V}(\mathfrak{T}_{SO_0}), \mathcal{H}^{(1)} = \mathcal{DM})$ of Merge derivations as in Lemma 3.5.9, where $\mathcal{G}$ is the associated semigroupoid, we can regard the choice of a map of directed graphs $\alpha : G \to \mathcal{G}$ from some graph G as a chosen *diagram of Merge derivations* modeled on G. The algebroid homomorphism $\Phi_E(f) = f \circ \alpha$ describes all the ways of obtaining a certain Merge derivation γ in $\mathcal{DM}$ as an arrow in G, $\Phi_E(\delta_\gamma) = \sum_{e : \alpha(e)=\gamma} \delta_e$. The Bogolyubov preparation with the identity Rota–Baxter operator lists all the possible ways of obtaining γ as a composition of Merge derivations through arrows in G, as in (3.5.16). Consider an element $\sum_i \lambda_i e_i$ as a weighted combination of edges in the diagram G. For example, if the coefficients $\Lambda = (\lambda_e)_{e\in E}$ are a probability distribution on the edges of G, then the value (using the identity as a Rota–Baxter operator) of the Bogolyubov preparation,

$$\tilde{\Phi}_E(\delta_\gamma)(\sum_e \lambda_e \, e) = \sum_{\alpha(e)=\gamma} \lambda_e + \cdots + \sum_{\alpha(e_1)\circ\cdots\circ\alpha(e_n)=\gamma} \lambda_{e_1} \cdots \lambda_{e_n} ,$$

is the total probability of realizing γ through the diagram E as a sum of the probabilities of all the possible ways of obtaining γ as a composition of arrows in the image of edges of E drawn the assigned probabilities λ_e.

The setting for algebroids generalizes to semiringoids as in the case of the generalization from Rota–Baxter algebras to Rota–Baxter semirings.

Corollary 3.5.22. *The Birkhoff factorization of Lemma 3.5.20 extends to the case of Rota–Baxter semiringoids of weight +1, in the form of a morphism of semiringoids* $\Phi : (\mathcal{A}^{(0)}, \mathcal{H}^{(1)})^{semi} \to (\mathcal{A}, \mathcal{E})$ *from a subdomain of a bialgebroid* $(\mathcal{A}^{(0)}, \mathcal{H}^{(1)})$ *that has a semiringoid structure and is closed under coproduct* Δ. *The terms of the factorization are as in the case of semirings (see Proposition 3.1.9), with* $\Phi_{E,-}(f) = R(\tilde{\Phi}_E(f)) = R(\Phi_E(f) \boxdot \phi_-(f') \odot \phi(f''))$, *with* $\Delta(f) = f \otimes 1 + 1 \otimes f + \sum f' \otimes f''$.

3.5.5 Parsing Semirings and Merge Derivations

After this preparatory work, we can now formulate the analog of parsing semirings in the setting of Merge derivations, replacing the usual formulation for context-free grammars, as in [82].

Here, we consider a map $\Phi : (\mathcal{A}^{(0)}, \mathcal{H}^{(1)})^{semi} \to (\mathcal{A}, \mathcal{E})$, where $(\mathcal{A}, \mathcal{E})$ is a Rota–Baxter semiringoid and $(\mathcal{A}^{(0)}, \mathcal{H}^{(1)})^{semi}$ is a subdomain of the bialgebroid $(\mathcal{A}^{(0)} = \mathcal{V}(\mathfrak{F}_{SO_0}), \mathcal{H}^{(1)} = \mathcal{DM})$ of Merge derivations that has a semiringoid structure, such that Φ is a morphism of semiringoids. We assume that the target $(\mathcal{A}, \mathcal{E})$ is of the form as in Lemma 3.5.19, where $(\mathcal{R}, R)$ is a Rota–Baxter semiring, such as the max-plus semiring $(\mathbb{R} \cup \{-\infty\}, \max, +)$ (also known as the "tropical semiring"), with R given by the ReLU operator, or the Viterbi semiring $([0, 1], \max, \cdot)$, with the threshold Rota–Baxter operators c_λ that we considered before. Then the map Φ may be viewed as assigning a diagram of Merge derivations through a map $\alpha : G \to \mathcal{G}$ as before, and checking all the possible ways of realizing some chain of Merge derivations γ through compositions coming from the chosen diagram, weighted by elements in the given semiring and filtered by the Rota–Baxter operator, which acts as a threshold.

Thus, as before, we start with a chosen diagram $\alpha : G \to \mathcal{G}$ of Merge derivations where we have assigned weights $\lambda_e \in \mathcal{R}$ with values in the parsing semiring $\mathcal{R}$, for each edge $e \in E_G$. For example, if $\mathcal{R} = ([0, 1], \max, \cdot)$, we can think of λ_e as a probability (or frequency counting) of the occurrence of e in the diagram of derivations. If $\mathcal{R} = (\mathbb{R} \cup \{-\infty\}, \max, +)$, we can think of λ_e as being real weights assigned to the edges e of the diagram G. Then the resulting factorization

$$\Phi_{E,-}(\delta_\gamma)(\sum_e \lambda_e e) = \sum_{\alpha(e)=\gamma} R_E(a_e) + \sum_{\alpha(e_1)\circ\alpha(e_2)=\gamma} R_E(R_E(a_{e_1}) a_{e_2}) + \cdots$$
$$+ \sum_{\alpha(e_1)\circ\cdots\circ\alpha(e_n)=\gamma} R_E(\cdots(R_E(a_{e_1})\cdots) a_{e_n}) \qquad (3.5.17)$$

measures all the possible ways of obtaining the Merge derivation γ via compositions in the chosen diagram with combined weights filtered by R.

- In the case of $\mathcal{R} = (\mathbb{R} \cup \{-\infty\}, \max, +)$, with R =ReLU, (3.5.5) lists all the possibilities with weights of the substructures involved that are above the ReLU threshold.
- In the case of $\mathcal{R} = ([0,1], \max, \cdot)$, with the threshold $R = c_\lambda$, (3.5.5) lists all the possible realizations of the derivation γ in the diagram that have probabilities above the threshold λ in the substructures involved.
- In the case of the Boolean semiring $\mathfrak{B} = (\{0,1\}, \max, \cdot)$, with $R = \mathrm{id}$, the factorization (3.5.5) evaluates the truth value (truth conditions) for the realization of a derivation γ through the diagram G, given that the arrows of G have assigned truth values (truth conditions), in such a way that the composition of arrows in the derivation corresponds to the AND operation on the respective truth values and the choices of different paths of derivations to obtain the same γ correspond to the OR operation on the respective truth values.

3.5.6 Summary of Semiringoid Parsing

Here, we summarize quickly the main outcome of the construction obtained in this section. The semiringoid parsing strategy can be outlined as the following steps:

1. An algebroid $(\mathcal{A}, \mathcal{E})$ used as the target of parsing specifies a directed graph G: this should be seen as a "chosen graphical template" for a possible scheme of derivations.
2. We can also regard this template as a "probabilistic" datum, where we can consider a distribution $\sum_i \lambda_i e_i$ on the arrows of the diagram G.
3. A character $\Phi : (\mathcal{A}^{(0)}, \mathcal{H}^{(1)}) \to (\mathcal{A}, \mathcal{E})$ from a bialgebroid (or Hopf algebroid) to the target algebroid induces a map of graphs $\alpha : G \to \mathcal{G}$ (where $\mathcal{G}$ is also a category): this can be seen as a way of realizing the chosen template G as actual derivations (i.e., composable arrows of the semigroupoid).
4. For a chosen arrow γ of the semigroupoid $\mathcal{G}$, identified with its dual function $f = \delta_\gamma$, the Bogolyubov preparation (for the trivial Rota–Baxter structure $R = \mathrm{id}$),

$$\tilde{\Phi}_E(\delta_\gamma) = \sum_{e \in E_G \,:\, \alpha(e) = \gamma} \delta_e + \cdots + \sum_{e_1, \ldots, e_n \in E_G \,:\, \gamma = \alpha(e_1) \circ \cdots \circ \alpha(e_n)} \delta_{e_1} \cdots \delta_{e_n},$$

lists all the possible ways of obtaining γ as compositions of images of arrows in G (i.e., realizing it in any possible way through the template diagram of derivations).

5. For a distribution $\sum_i \lambda_i e_i$ on the diagram G and a nontrivial Rota–Baxter structure R_E, the Birkhoff factorization
$$\Phi_{E,-}(\delta_\gamma)(\sum_i \lambda_i e_i) = -(\sum_{\alpha(e)=\gamma} R_E(\lambda_e) + \sum_{\alpha(e_1)\circ\alpha(e_2)=\gamma} R_E(R_E(\lambda_{e_1})\lambda_{e_2}) + \cdots$$
$$+ \sum_{\alpha(e_1)\circ\cdots\circ\alpha(e_n)=\gamma} R_E(\cdots(R_E(\lambda_{e_1})\cdots)\lambda_{e_n}))$$
selects only those realizations through the template diagram that meet a certain threshold (for example of probability).

In the specific case of Merge derivations, the process outlined here is realized in the following way:

- The targets of parsing are seimiringoids consisting of functions on a template graph G of derivations, with values in a parsing semiring, which we can take as the tropical semiring $(\mathbb{R} \cup \{-\infty\}, \max, +)$, with the Rota–Baxter structure R given by the ReLU threshold, or the Viterbi semiring $([0,1], \max, \cdot)$, with a Rota–Baxter structure given by a threshold $R = c_\lambda$.
- The character Φ from the bialgebroid of Merge derivations to the target semiringoid determines the assignment $\alpha : G \to \mathcal{G}$ of a specified diagram of Merge derivations, checking all the possible ways of realizing a given chain γ of Merge derivations in $\mathcal{G}$ through compositions coming from the chosen diagram G, weighted by elements in the given semiring and filtered by the Rota–Baxter threshold R.

This parsing strategy is designed to address questions of this kind: suppose that one selects a set of arrows $\mathfrak{M}_{S,S'}$ (individual Merge applications) or chains $\mathfrak{M}_{S_N,S'_N} \circ \cdots \circ \mathfrak{M}_{S_1,S'_1}$ with certain probabilities (or more complex diagrams of such derivations). Then, can one realize a certain specific transformation $F \to F'$ of workspaces via Merge, using only the given class of transformations, in such a way that all the probabilities for all the necessary compositions of these remain above an assigned threshold?

With this, we have shown that we can obtain a form of semiring parsing for Merge derivations that simultaneously generalizes the semiring parsings of [82], such as with values in the Boolean or the Viterbi semiring, and also the Birkhoff factorizations of our initial toy models of syntax–semantics interface discussed in section 3.2.

It should be possible to provide an equivalent reformulation, for the parsing ofr Merge derivations based on Hopf algebroids and Birkhoff factorizations in Rota–Baxter semiringoids, in the more familiar language of automata, though we will not develop this here.

3.6 Pietroski's Compositional Semantics

Among the various proposed models of semantics, Pietroski's compositional model (see, e.g., [168] and [169]) is closely linked to the structure of syntax as described by Merge. We discuss how this approach relates to our model of the syntax–semantics interface. Our main observation here is that, in our model, it is not necessary to assume an independent existence *within* semantics of what Pietroski refers to in [168] as the *Combine* binary operation, which mimics the functioning of Merge in syntax. The type of compositional structure postulated by Pietroski in [168] for semantics *follows* in our case from Merge itself acting on the syntax side of the interface, along with the map $\phi : \mathcal{H} \to \mathcal{R}$ with its Birkhoff factorization.

To see this, we recall briefly the setting of [168], focusing in particular on the discussion of the *Combine* operation, which is the aspect more directly connected to our setting. The general principles for the compositional structure of semantics articulated in [168] include the basic idea that "meanings are instructions to build concepts," which can be articulated in the following way, adapting the arguments of [168] to the terminology used in this chapter. Lexical items are seen as "instructions to fetch concepts." This corresponds to the assumption made in various examples in the previous sections of the existence of a map $s : \mathcal{SO}_0 \to \mathcal{S}$ from lexical items to a semantic space $\mathcal{S}$. One then considers i-expressions, generated by I-language, as building instructions for the construction of i-concepts, with principles that govern the combination of i-expressions.

This fits nicely with our proposal of a syntax-driven syntax–semantics interface, in which the i-expressions are provided, in our setting, by the syntactic objects $T \in \mathrm{Dom}(h) \subset \mathfrak{T}_{\mathcal{SO}_0}$. The corresponding i-concepts are provided in our setting by their images $s(T) \in \mathcal{S}$, under an extension of the map $s : \mathcal{SO}_0 \to \mathcal{S}$ from $\mathcal{SO}_0$ to $\mathrm{Dom}(h) \subset \mathcal{SO}$, as discussed in section 3.3, together with the corresponding $\phi(T) \in \mathcal{R}$, where $\mathcal{R}$ is an algebraic structure of the Rota–Baxter type associated to the topological/metric space $\mathcal{S}$.

On the side of syntax, the free commutative, nonassociative magma $\mathcal{SO} = \mathrm{Magma}_{nc,c}(\mathcal{SO}_0, \mathfrak{M})$ of (3.1.1) is the main *computational structure*, with Merge $\mathfrak{M}$ as the main *binary operation* of structure formation. The resulting hierarchical structures are the syntactic objects $T \in \mathcal{SO} = \mathfrak{T}_{\mathcal{SO}_0}$, identified with abstract (nonplanar) binary rooted trees with leaves labeled by lexical items in $\mathcal{SO}_0$. In Pietroski's formulation of [168], one considers a parallel form of binary structure formation operation, acting on the side of semantics.

The compositional rules for the building of i-concepts via i-expressions are described in [168] in terms of one basic noncommutative binary operation, *Combine*. This in turn consists of the composition of two operations, *Combine* = *Label* ∘ *Concatenate*, where, given two i-concepts α, β that can be combined in the I-language, one first forms a concatenation

$$Concatenate(\alpha,\beta) = \{\alpha,\beta\} = \overset{\frown}{\alpha \quad \beta} \tag{3.6.1}$$

of the two and then labels the resulting expression by one of the constituents α or β that plays the role of the head $h(\alpha,\beta)$ of the combined expression

$$Combine(\alpha,\beta) = Label \circ Concatenate(\alpha,\beta) = Label(\overset{\frown}{\alpha \quad \beta}) = \begin{matrix} h(\alpha,\beta) \\ \overset{\frown}{\alpha \quad \beta} \end{matrix} . \tag{3.6.2}$$

The binary operation *Combine* is not symmetric because of the head label. Note that this operation is closely modeled on the Merge operation, where, given two syntactic objects T_1 and T_2, with the property that

$$T = \mathfrak{M}(T_1, T_2) = \overset{\frown}{T_1 \quad T_2} \in \mathrm{Dom}(h) \subset \mathcal{SO} = \mathfrak{T}_{\mathcal{SO}_0},$$

where $\mathfrak{M}$ is free symmetric Merge and h is a head function, one can assign to the abstract tree T a planar structure T^{π_h} determined by the head function, resulting in a planar tree

$$T^{\pi_h} = \mathfrak{M}^{nc}(T_{h(T)}, T') \in \mathfrak{T}^{\text{planar}}_{\mathcal{SO}_0},$$

where $T' \in \{T_1, T_2\}$ is the one that does not contain $h(T)$.

In [168], the operation (3.6.2) is presented, in principle, as a compositional operation that takes place in the semantic space $\mathcal{S}$, hence requiring this space to be endowed with its own computational system (at least partially defined), analogous to the Merge operation in syntax. As a result, we would have two systems that each have a Merge–type operation, one for syntax and one for semantics. Besides an issue here with parsimony (we can get by, given the model presented here, with just one), this would be different from the case of other types of conceptual spaces, such as the perceptual manifolds associated with vision (see, e.g., [41]).

The most widely studied conceptual spaces and perceptual manifolds are in the context of vision. It should be noted that there have been significant attempts by mathematicians at formulating a compositional computational model for vision: among these in particular Pattern Theory, developed by Grenander,

Mumford, et al. (see, e.g., [85], [86], and [156]), which has found various applications, especially in computer vision. The original approach to Pattern Theory was based on importing ideas from the theory of formal languages, especially probabilistic context-free grammars. This was further articulated in Grenander's proposed "mathematical theory of semantics" [87]. We will not be discussing this viewpoint in this book, but it is important to stress that it is still *topological and geometric* properties of the relevant semantic spaces that play a fundamental role in that setting, and that there are serious limitations to the extent to which a generative model can be adapted to vision in comparison to language.

3.6.1 Combine Operation in Pietroski's Semantics

In terms of the syntax–semantics interface, using the terminology in this chapter, a setting such as that proposed by Pietroski in [168] would seem to correspond to postulating the existence of a morphism $\phi : \mathcal{H}^{nc} \to \mathcal{H}_S$, that maps a noncommutative version of the Hopf algebra structure of $\mathcal{H}$ (responsible for the action of Merge on syntax, formulated after planarization) to a (possibly partially defined) noncommutative Hopf algebra structure $\mathcal{H}_S$ on the side of the semantic space $\mathcal{S}$. At least this would be the case if one desired a *Combine* operation that fully mimicked the Merge operation, including the Internal Merge action. However, this would be a much stronger requirement than what is strictly needed to achieve the desired type of compositionality of i-concepts.

Indeed, one can view the construction of i-concepts postulated by [168] not as the result of a compositional structure on the semantic space $\mathcal{S}$ itself, but simply as the extension of the map $s : \mathcal{SO}_0 \to \mathcal{S}$ to a map $s : \mathrm{Dom}(h) \to \mathcal{S}$, built along the lines of what discussed in Lemma 3.2.12. In other words, in this formulation, the i-concept *Combine*(α,β), where $\alpha = s(T_1)$ and $\beta = s(T_2)$, is well defined if $T = \mathfrak{M}(T_1, T_2) \in \mathrm{Dom}(h)$; and in that case, is simply the image

$$\mathit{Combine}(\alpha,\beta) := s(\mathfrak{M}(T_1, T_2)) \in \mathcal{S}, \qquad (3.6.3)$$

where the construction of the point $s(T)$ depends on $s(T_1)$, $s(T_2)$, and on whether the head function satisfies $h(T) = h(T_1)$ or $h(T) = h(T_2)$. In other words, it depends on $\mathcal{S}$ only through the existence of a geodesically convex Riemannian structure and a semantic proximity function $\mathbb{P}$, without having to require any Merge-like computational mechanism in $\mathcal{S}$ itself. It suffices that syntax has such an operation, and $\mathcal{S}$ has a topological proximity relation (expressed in the case of the construction we presented in Lemma 3.2.12 in terms of a more specific metric property of convexity).

3.6.2 Concatenate Operation

Our form (3.6.3) of the Combine operation is not based on the concatenation of strings of leaves. Indeed, the *Concatenate* operations in (3.6.1) is defined as a nonplanar binary tree, not as an ordered set. This differs from the formulation often used in the setting of Pietroski's semantics, where *Concatenate* is understood as the usual operation of concatenation of ordered sets. We show here why it is necessary to use a different construction, not based on concatenation, as the latter is not directly compatible with a free symmetric Merge.

To ordered strings of elements in the set $\mathcal{SO}_0$ of lexical items and syntactic features, one can apply the usual concatenation operation of ordered sets. This can be formulated, as in Definition 3.6.1. We use the notation "Conc" for this operation to distinguish it from (3.6.1).

Definition 3.6.1. Let $\Sigma^*(\mathcal{SO}_0)$ be the set of ordered sequences of elements in $\mathcal{SO}_0$ of arbitrary (finite) length. The concatenation operation is the associative noncommutative binary operation

$$\mathrm{Conc} : \Sigma^*(\mathcal{SO}_0) \times \Sigma^*(\mathcal{SO}_0) \to \Sigma^*(\mathcal{SO}_0)$$

$$(\alpha, \beta) \mapsto \mathrm{Conc}(\alpha, \beta) = \alpha^\wedge\beta = \alpha\beta\,, \tag{3.6.4}$$

which combines the ordered sets α and β into a new ordered set where the string β follows the string α.

To a planar binary rooted tree $\tilde{T} \in \mathfrak{T}^{pl}_{\mathcal{SO}_0}$, with leaves labeled by $\mathcal{SO}_0$, one can associate the corresponding ordered set of leaves (ordered by the planar structure), $L(\tilde{T}) \in \Sigma^*(\mathcal{SO}_0)$. Let $\mathfrak{M}^{nc}$ denote the noncommutative, nonassociative binary operation of the free noncommutative, nonassociative magma

$$\mathfrak{T}^{pl}_{\mathcal{SO}_0} = \mathcal{SO}^{nc} = \mathrm{Magma}_{na,nc}(\mathcal{SO}_0, \mathfrak{M}^{nc})\,.$$

As before, we denote by Π the canonical projection (morphism of magmas)

$$\Pi : \mathfrak{T}^{pl}_{\mathcal{SO}_0} \to \mathfrak{T}_{\mathcal{SO}_0}\,,$$

which forgets the planar embedding. Given an abstract tree $T \in \mathfrak{T}_{\mathcal{SO}_0}$, we write $\tilde{T}$ for any choice of a planar tree $\tilde{T} \in \Pi^{-1}(T)$ in the fiber $\Pi^{-1}(T)$ of this projection. Then the concatenation operation of (3.6.4) is related to the noncommutative Merge of planar trees as shown in Lemma 3.6.2.

Lemma 3.6.2. *The map* $L : \mathfrak{T}^{pl}_{\mathcal{SO}_0} \to \Sigma^*(\mathcal{SO}_0)$ *with* $L : \tilde{T} \mapsto L(\tilde{T})$ *satisfies*

$$L(\mathfrak{M}^{nc}(\tilde{T}_1, \tilde{T}_2)) = \mathrm{Conc}(L(\tilde{T}_1), L(\tilde{T}_2))\,.$$

Proof. This is immediate, as the noncommutative Merge $\mathfrak{M}^{nc}(\tilde{T}_1, \tilde{T}_2)$ is the planar tree that has the planar tree $\tilde{T}_1$ as the left branch below the root vertex and $\tilde{T}_2$ as the right branch. Thus, the ordered set of leaves $L(\mathfrak{M}^{nc}(\tilde{T}_1, \tilde{T}_2))$ consists of the ordered set $L(\tilde{T}_1)$, followed by the ordered set $L(\tilde{T}_2)$ so it agrees with the result of the concatenation operation of (3.6.4). Note that $\mathfrak{M}^{nc}$ is nonassociative while Conc is associative, so the map L kills the associators of $\mathfrak{M}^{nc}$. □

When we consider abstract trees that are produced by the free symmetric Merge $\mathfrak{M}$—namely, the syntactic objects in

$$\mathcal{SO} = \mathrm{Magma}_{na,c}(\mathcal{SO}_0, \mathfrak{M}) = \mathfrak{T}_{\mathcal{SO}_0},$$

we know that there is no possible *morphism of magmas* $\mathcal{SO} \to \mathcal{SO}^{nc}$ (see Lemma 1.13.1 in chapter 1). Namely, given a map of sets

$$\sigma : \mathcal{SO} \to \mathcal{SO}^{nc}$$

that is a section of the projection $\Pi : \mathcal{SO}^{nc} \to \mathcal{SO}$ (i.e., such that $\Pi \circ \sigma =$ id, meaning that σ is an assignment of planar structure to abstract trees), we generally have

$$\mathfrak{M}^{nc}(\sigma(T_1), \sigma(T_2)) \neq \sigma(\mathfrak{M}(T_1, T_2)). \tag{3.6.5}$$

This means that there must necessarily be some T_1, T_2 for which one has the inequality (3.6.5).

Definition 3.6.3. A "linearization algorithm" is a map of sets $\sigma : \mathcal{SO} \to \mathcal{SO}^{nc}$ that is a section of the canonical projection $\Pi : \mathcal{SO}^{nc} \to \mathcal{SO}$ (i.e., one that satisfies $\Pi \circ \sigma = \mathrm{id}$).

The observation in (3.6.5) has the consequence that, in principle, there may be pairs T_1, T_2 for which

$$L(\sigma(\mathfrak{M}(T_1, T_2))) \neq \mathrm{Conc}(L(\sigma(T_1)), L(\sigma(T_2))) = L(\mathfrak{M}^{nc}(\sigma(T_1), \sigma(T_2))),$$

where the second equality is Lemma 3.6.2. Thus, there is a possible obstruction to a consistent definition of Conc on the image of a linearization algorithm $\sigma : \mathcal{SO} \to \mathcal{SO}^{nc}$. We need to check when an obstruction of this kind does or does not arise.

Definition 3.6.4. A linearization algorithm given by a section $\sigma : \mathcal{SO} \to \mathcal{SO}^{nc}$ of the projection $\Pi : \mathcal{SO}^{nc} \to \mathcal{SO}$ has a well-defined concatenation if, for all $T_1, T_2 \in \mathcal{SO}$,

$$L(\sigma(\mathfrak{M}(T_1, T_2))) = \mathrm{Conc}(L(\sigma(T_1)), L(\sigma(T_2))). \tag{3.6.6}$$

It has partial concatenation on $\mathcal{D}$ if there is a domain $\mathcal{D} \subset \mathcal{SO}$ such that, for all $T_1, T_2 \in \mathcal{D}$, the identity (3.6.6) holds.

Correspondingly, a linearization algorithm does not have well defined concatenation if there is a pair of $T_1, T_2 \in \mathcal{SO}$ for which, in $\Sigma^*(\mathcal{SO}_0)$,

$$L(\sigma(\mathfrak{M}(T_1, T_2))) \neq \mathrm{Conc}(L(\sigma(T_1)), L(\sigma(T_2)));$$

hence also necessarily, in $\mathfrak{T}^{pl}_{\mathcal{SO}_0}$,

$$\sigma(\mathfrak{M}(T_1, T_2)) \neq \mathfrak{M}^{nc}(\sigma(T_1)), L(\sigma(T_2)).$$

Lemma 3.6.5. *There is no linearization algorithm given by a section σ : $\mathcal{SO} \to \mathcal{SO}^{nc}$ of the canonical projection Π : $\mathcal{SO}^{nc} \to \mathcal{SO}$ that has a well-defined concatenation.*

Proof. We need to show that, given any section $\sigma : \mathcal{SO} \to \mathcal{SO}^{nc}$, there is a pair T_1, T_2 such that

$$L(\sigma(\mathfrak{M}(T_1, T_2))) \neq \mathrm{Conc}(L(\sigma(T_1)), L(\sigma(T_2))).$$

Since σ cannot be a morphism of magmas, we know that there is a pair T_1, T_2 such that

$$\sigma(\mathfrak{M}(T_1, T_2)) \neq \mathfrak{M}^{nc}(\sigma(T_1), \sigma(T_2)).$$

The two planar trees $\sigma(\mathfrak{M}(T_1, T_2))$ and $\mathfrak{M}^{nc}(\sigma(T_1), \sigma(T_2))$ are always in the same fiber over the abstract tree $T = \mathfrak{M}(T_1, T_2)$ of the projection map Π that collapses the 2^{n-1} different planar structures, with $n = \#L(T)$, where $L(T)$ is the unordered set of leaves of T. On the other hand, the two planar trees $\sigma(\mathfrak{M}(T_1, T_2))$ and $\mathfrak{M}^{nc}(\sigma(T_1), \sigma(T_2))$ will have the same ordered set of leaves iff they differ by an associator (a change of bracketing on the same ordered set). Observe that a given ordered set of leaves is thus realized by C_{n-1} possible planar structures (with C_n being the Catalan numbers), which correspond to the vertices of the associahedron K_n. These planar structures are exactly those that differ by associators. Consider the standard identity involving Catalan numbers:

$$n!\, C_{n-1} = 2^{n-1}\, (2n-3)!! \,, \tag{3.6.7}$$

where $(2n-3)!!$ counts the number of different possible abstract (i.e., nonplanar) binary rooted trees on n labeled leaves (those abstract trees T that have the same unordered set $L(T)$), and 2^{n-1} is the number of possible planar structures for each of these trees. This total number is equal to the number C_{n-1} of possible bracketing on a fixed ordered set (the total number of planar trees $\tilde{T}$ with the same ordered set $L(\tilde{T})$) times the total number $n!$ of possible orderings of an unordered set $L(T)$ of n-leaves. So the identity says that one can start with the unordered set S and count all possible orderings S^σ (these are counted by $n!$ permutations), and for each S^σ consider all possible planar trees $\tilde{T}$ with the ordered set $L(\tilde{T}) = S^\sigma$ (these are counted by C_{n-1} different bracketing); or equivalently, one can consider all possible abstract trees T with unordered sets $L(T) = S$ (these are counted by the double factorial $(2n-3)!!$) and then all possible planar structures $\tilde{T}$ for these trees (counted by the 2^{n-1} factor). The counting $n!C_{n-1}$ can be seen, as discussed in section 3.4.2, as $n!$ associahedra glued together to form the orientation double cover of $\bar{M}_{0,n+1}(\mathbb{R})$, with its projection to the moduli space BHV_n^+ of abstract metric trees, with 2^{n-1} being the size of the fibers. Here, we can think of it in the following way.

Given a planar binary rooted tree $\tilde{T}$ on n labeled leaves, the symmetric group S_n acts on the leaves producing other planar trees $\tau : \tilde{T} \mapsto \tau(\tilde{T})$ with the same tree structure but the labels of leaves permuted. If $\tilde{T}$ is one of the vertices of one of these $n!$ associahedra K_n counted on the left-hand-side of (3.6.7), then the orbit under S_n determines a corresponding vertex in each of the other associahedra. Moreover, for each abstract binary rooted tree T on n labeled leaves and a choice of planar structure $\tilde{T} \in \Pi^{-1}(T)$, we can consider a group $G_{\tilde{T}}$ of transformations generated by the involutions γ_v that flip the two subtrees $\tilde{T}_{v,L}$ and $\tilde{T}_{v,R}$ below v in the planar tree $\tilde{T}$, producing a new planar tree $\gamma_v(\tilde{T})$ with the same underlying abstract T. This has $\#G_{\tilde{T}} = 2^{n-1}$, the number of possible planar structures on T, as any planarization can be built by repeatedly deciding, for each vertex below the root, which of the two branches is placed to the left and which to the right. The normal subgroup $\mathrm{Aut}(\tilde{T}) \subset G_{\tilde{T}}$ consists of transformations in $G_{\tilde{T}}$ that preserve $\tilde{T}$ (this will depend on the structure of the tree T and the choice of labels); therefore, $\#(G_{\tilde{T}}/\mathrm{Aut}(\tilde{T}))$ counts the number of nonisomorphic planar trees in the orbit of $\tilde{T}$ under $G_{\tilde{T}}$ (in the fiber over the abstract tree T), by the orbit-stabilizer theorem

$$\#\mathrm{Orbit}_{G_{\tilde{T}}}(\tilde{T}) = \frac{\#G_{\tilde{T}}}{\#\mathrm{Aut}(\tilde{T})}.$$

Since the total number of planar trees on n labeled leaves is the Catalan number C_{n-1}, this gives

$$\sum_T \#(G_{\tilde{T}}/\mathrm{Aut}(\tilde{T})) = \sum_T \frac{2^{n-1}}{\#\mathrm{Aut}(\tilde{T})} = C_{n-1} = \frac{(2n-3)!!}{n!}.$$

Consider now a linearization algorithm given by a map of sets $\sigma : \mathcal{SO} \to \mathcal{SO}^{nc}$, which is a section of the projection Π. Given $T_1, T_2 \in \mathcal{SO}$, consider the abstract tree $T = \mathfrak{M}(T_1, T_2)$ and the planar trees

$$\tilde{T}^\sigma := \sigma(T) = \sigma(\mathfrak{M}(T_1, T_2)) \quad \text{and} \quad \tilde{T} := \mathfrak{M}^{nc}(\sigma(T_1), \sigma(T_2)).$$

These are both planarizations of the same T, and hence they are in the same fiber of multiplicity 2^{n-1} over T. In particular, this means that there is an element $\gamma \in G_{\tilde{T}}$ such that $\gamma(\tilde{T}) = \tilde{T}^\sigma$. The cases in which we have an incompatibility between the linearization algorithm and the asymmetric Merge $\mathfrak{M}^{nc}$ are given by pairs $(\tilde{T}^\sigma, \tilde{T})$ such that the transformation $\gamma \in G_{\tilde{T}}$ with $\gamma(\tilde{T}) = \tilde{T}^\sigma$ is not in the subgroup $\mathrm{Aut}(\tilde{T})$. Only such pairs $(\tilde{T}^\sigma, \tilde{T})$ can be possible sources of an ill-defined concatenation because any pair for which $\gamma \in \mathrm{Aut}(\tilde{T})$ will have $\tilde{T}^\sigma \simeq \tilde{T}$ as planar trees, and hence $L(\tilde{T}^\sigma) = L(\tilde{T})$ as ordered sets. So suppose that there is a pair $(\tilde{T}^\sigma, \tilde{T})$ with $\gamma \notin \mathrm{Aut}(\tilde{T})$. The condition for having $L(\tilde{T}^\sigma) = L(\tilde{T})$ as ordered sets is that $\tilde{T}^\sigma$ and $\tilde{T}$ are two different vertices among the C_{n-1} vertices of one of the $n!$ associahedra. But different vertices of the same associahedron do not belong to the same fiber over T, as the projection map $\bar{M}_{0,n+1}(\mathbb{R}) \to BHV_n^+$ is given by the origami folding of each $(n-2)$–dimensional cube (pair of cubes in the orientation double cover) in the cubical decomposition of each associahedron along each of its 2^{n-2} axes. In particular, each of these cubes contains exactly one of the associahedron vertices (giving $2C_{n-1}$ cubes in the double cover); hence, different vertices of the same associahedron are never folded together (the example of $n = 3$ is illustrated in Figure 3.18). So we conclude that every pair $(\tilde{T}^\sigma, \tilde{T})$ such that the transformation $\gamma \in G_{\tilde{T}}$ with $\gamma(\tilde{T}) = \tilde{T}^\sigma$ is not in the subgroup $\mathrm{Aut}(\tilde{T})$ is always a violation of the "well-defined concatenation" property. □

Thus, the associative, noncommutative concatenation operation of (3.6.4) *cannot be* well defined on the image of a linearization algorithm. This includes the Externalization procedure σ_L, as in (1.12.1) and (1.12.3), which is a section of the projection $\Pi : \mathcal{SO}^{nc} \to \mathcal{SO}$.

However, this problem is resolved when one restricts to the domain of a *head function*, defined as in Definitions 1.13.3 and 1.13.6.

It is shown in [139] that there are exactly $2^{\# V^o(T)}$ possible head functions h_T on an abstract binary rooted tree T, with $V^o(T)$ being the set of non-leaf vertices of T, and these correspond to the different possible choices of planarization of T. Thus, a planarization ("linearization algorithm" in linguistic terminology) is the same thing as the assignment of a head function. Among these possibilities, the actual syntactic head is a head function that is determined by the lexical items and syntactic features in $\mathcal{SO}_0$ associated with the leaves in such a way that, for each accessible term T_v of T, the head $h_T(v)$ is the item carrying the uniquely determined prominent labeling feature. We have seen in

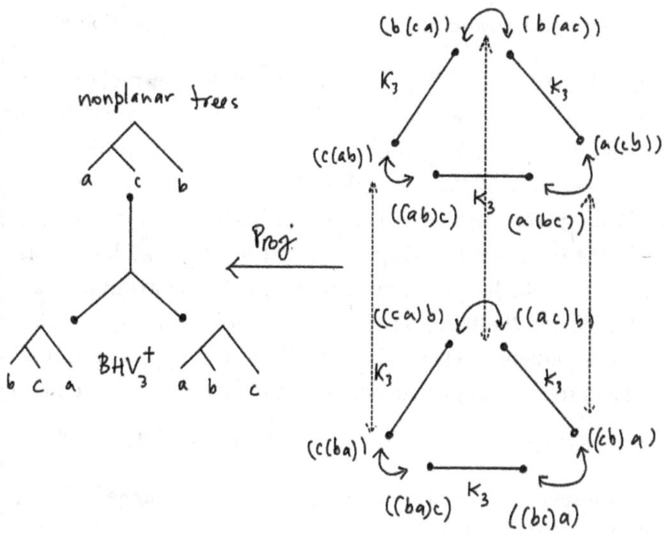

Figure 3.18
Projection from the space of planar trees to the space of nonplanar trees in the case of three leaves: planar trees at different vertices of one of the associahedra K_3 (segments) are never in the same fiber of the projection.

Lemma 1.13.7 that the properties of head functions stated in Definition 1.13.6 are in fact equivalent to the properties of the syntactic head described in [23].

As discussed in sections 1.13.2 and 1.13.3 of chapter 1, an assignment $h : T \mapsto h_T$ that would depend on the lexical items and syntactic features in $\mathcal{SO}_0$ associated with the leaves of the trees will necessarily be defined on some subdomain $\mathrm{Dom}(h) \subset \mathcal{SO}$ of the set of all syntactic objects produced by the free symmetric Merge, as these generally may include structures that don't admit good labeling. So let's assume that we have a given head function h defined on a domain $\mathrm{Dom}(h)$ with the properties of Definition 1.13.6.

We want to check that the partially defined linearization algorithm σ^h induced by the head function (i.e., the harmonic head-initial LCA planarization) has a partially defined binary associative noncommutative concatenation operation Conc as in (3.6.4), with a modified version of the partially defined property of Definition 3.6.4.

Lemma 3.6.6. *For $T_1, T_2 \in \mathrm{Dom}(h)$ such that $\mathfrak{M}(T_1, T_2) \in \mathrm{Dom}(h)$, we have*

$$L(\sigma^h(\mathfrak{M}(T_1, T_2))) = \begin{cases} \mathrm{Conc}(L(\sigma^h(T_1)), L(\sigma^h(T_2)) & h(\mathfrak{M}(T_1, T_2)) = h(T_1) \\ \mathrm{Conc}(L(\sigma^h(T_2)), L(\sigma^h(T_1)) & h(\mathfrak{M}(T_1, T_2)) = h(T_2), \end{cases}$$

where σ^h is the planar structure determined by the head function h.

Proof. First, observe that if T_1, T_2 in $\mathrm{Dom}(h)$ have the property that $T = \mathfrak{M}(T_1, T_2) \in \mathrm{Dom}(h)$, then the head $h(\mathfrak{M}(T_1, T_2))$ is necessarily either the same as $h(T_1)$ or the same as $h(T_2)$. Let σ^h denote the partially defined section $\sigma^h : \mathrm{Dom}(h) \to \mathfrak{T}^{pl}_{SO_0}$ of the projection $\Pi : \mathfrak{T}^{pl}_{SO_0} \to \mathfrak{T}_{SO_0}$ that assigns to $T \in \mathrm{Dom}(h)$ the planar structure determined by the head, by putting below each vertex v the direction pointing to the leaf $h_T(v)$ to the left in the planar embedding. (This is the harmonic head-initial LCA planarization discussed in section 1.13.2. Using the harmonic head-final choice makes no difference to the rest of the argument.) Then, given T_1, T_2 in $\mathrm{Dom}(h)$, where $\sigma^h(T_1)$ and $\sigma^h(T_2)$ are the corresponding head-initial planarizations, only one of either

$$\mathfrak{M}^{nc}(\sigma^h(T_1), \sigma^h(T_2)) \quad \text{or} \quad \mathfrak{M}^{nc}(\sigma^h(T_2), \sigma^h(T_1))$$

will be in the subset $\sigma^h(\mathrm{Dom}(h)) \subset \mathfrak{T}^{pl}_{SO_0}$, depending on whether $h(\mathfrak{M}(T_1, T_2)) = h(T_1)$ or $h(\mathfrak{M}(T_1, T_2)) = h(T_2)$, with either

$$\mathfrak{M}^{nc}(\sigma^h(T_1), \sigma^h(T_2)) = \sigma^h(\mathfrak{M}(T_1, T_2))$$

or

$$\mathfrak{M}^{nc}(\sigma^h(T_2), \sigma^h(T_1)) = \sigma^h(\mathfrak{M}(T_1, T_2))$$

in the two cases. If we consider the restriction $L : \sigma^h(\mathrm{Dom}(h)) \to \Sigma^*(SO_0)$, then we find that

$$L(\sigma^h(\mathfrak{M}(T_1, T_2))) = \begin{cases} L(\mathfrak{M}^{nc}(\sigma^h(T_1), \sigma^h(T_2))) & \text{if } h(\mathfrak{M}(T_1, T_2)) = h(T_1) \\ L(\mathfrak{M}^{nc}(\sigma^h(T_2), \sigma^h(T_1))) & \text{if } h(\mathfrak{M}(T_1, T_2)) = h(T_2). \end{cases}$$

The statement then follows from Lemma 3.6.2. □

3.6.2.1 The role of idempotents One may worry here that the *Combine* operation of Pietroski, as recalled in (3.6.2), appears to behave differently from Merge itself, while our proposed reformulation in (3.6.3) is defined directly in terms of the free symmetric Merge. A simple way in which this difference manifests itself is in the possible presence of idempotent structures. For example, one expects that *Combine*$(\alpha, \alpha) = \alpha$, while at the level of Merge,

$$\mathfrak{M}(T, T) = \overset{\frown}{T \quad T} \neq T.$$

This in itself may not constitute an example because we also need a head function, and a structure of the form $\mathfrak{M}(T, T)$ might not admit a head function. However, by the same principle, one expects cases where *Combine*$(\alpha, \beta) = \alpha$ (or β), where the head function is not an issue, and again this seems to be at odds with the fact that at the level of Merge, this never happens since $\mathcal{SO}$ is a *free* magma, and therefore, for all T, T', one has $\mathfrak{M}(T,T') \neq T$ and $\mathfrak{M}(T,T') \neq T'$. This, however, does not constitute a problem, as it is taken care of in (3.6.3) by modifying the structure of the map $s : \mathrm{Dom}(h) \to \mathcal{S}$, which extends the map $s : \mathcal{SO}_0 \to \mathcal{S}$ as in Lemma 3.2.12.

In the construction of Lemma 3.2.12, we have assumed that the semantic space that we work with has the structure of a geodesically convex Riemannian manifold and that, for a syntactic object $\mathfrak{M}(T,T')$, the image $s(\mathfrak{M}(T,T'))$ is obtained as a form of convex interpolation between the images $s(T)$ and $s(T')$. In this setting, the location of the point $s(\mathfrak{M}(T,T'))$ on the geodesic arc between $s(T)$ and $s(T')$ depends on a function $\mathbb{P}(s(T), s(T'))$ measuring syntactic relatedness. Depending on the nature of this function $\mathbb{P}$, one expects that there may be points $s, s' \in \mathcal{S}$ for which $\mathbb{P}(s,s') = 0$ or $\mathbb{P}(s,s') = 1$. When this happens, the point $s(\mathfrak{M}(T,T'))$ coincides with one of the endpoints $s(T)$ and $s(T')$. (This case was ruled out, in our previous analysis, by the assumption that $\mathbb{P}(s,s') \in (0,1)$ in Proposition 3.3.1, but here we do not require this condition.) This gives rise to precisely the type of situation where one obtains

$$Combine(\alpha,\beta) = \alpha \quad \text{or} \quad Combine(\alpha,\beta) = \beta,$$

even though $\mathfrak{M}(T,T') \neq T$ and $\mathfrak{M}(T,T') \neq T'$. Note that the function $\mathbb{P}$, which is responsible for this difference in behavior between Merge and Combine, does not implement any computational process itself, but it is only an evaluator of topological proximity in semantics. The only computational process is implemented by syntactic Merge.

This case illustrates the situation where, contrary to the case described in section 3.3 (or the possible situation discussed in section 3.10), the image of syntax inside semantics is *not* an embedding. This non-embedding situation is generally expected when one maps to a semantic model that has a discrete topology (e.g., a Boolean assignment). In the case we describe here, where the function $s : \mathrm{Dom}(h) \to \mathcal{S}$ is based on geodesic convexity, one could in principle entirely avoid idempotent cases and assume (as in section 3.3) that the semantic relatedness $\mathbb{P}(s(T), s(T'))$ may be very close to either 0 or 1, but not exactly equal to either (see the discussion in section 3.3).

3.6.2.2 An example Pietroski's *Combine* operation is designed to rule out improper inferences. In this section, we consider an example to show how this fits with the formulation of (3.6.3).[9]

[9] We thank Norbert Hornstein for this example.

Given the sentences "John ate a sandwich in the basement" and "John ate a sandwich at noon", these two sentences clearly do *not* imply that "John ate a sandwich in the basement at noon".

In our setting, consider a sentence with a series of adjuncts to a verb, such as "John ate a sandwich in the basement with a spoon at noon." We have a Merge-based inductive construction of the map $s : \mathrm{Dom}(h) \to \mathcal{S}$, starting from $s : \mathcal{SO}_0 \to \mathcal{S}$, as discussed in section 3.3. This means that, if $\tilde{T}$ is the syntactic object associated with the full sentence, we can view it as a structure of the form

$$\{T', \{T, T_{1,\ldots,k}\}\},$$

where the VP T is modified by a series of adjuncts $T_{1,\ldots,k} = \{T_1, \ldots, T_k\}$ (for simplicity, we do not draw the full tree structure).

With the construction of sections 3.2.2 and 3.3 of the extension $s : \mathrm{Dom}(h) \to \mathcal{S}$ of the map $s : \mathcal{SO}_0 \to \mathcal{S}$, we obtain points $s(T) \in \mathcal{S}$ and $s(T_i) \in \mathcal{S}$ for each $i = 1, \ldots, k$. When we consider each individual adjunct, the corresponding point

$$s_i := s(\{T, T_i\})$$

lies on the geodesic arc in $\mathcal{S}$ between $s(T)$ and $s(T_i)$, at a distance $p_i = \mathbb{P}_\sigma(s(T), s(T_i))$ from $s(T)$, where σ is the adjunct syntactic relation. In particular, there is a convex geodesic neighborhood of the point $s(T)$ in $\mathcal{S}$ that contains all the points s_i. When we consider the combinations $T_{i_1,\ldots,i_r}$ of the adjuncts T_i, this further determines a collection of points

$$s_{i_1 \ldots i_r} = s(\{T, T_{i_1 \ldots i_r}\}).$$

These points are contained in the same neighborhood of $s(T)$, and they are also contained in the intersection of neighborhoods around the points s_i with $i \in \{i_1, \ldots, i_k\}$. A sketch of this relation is illustrated in Figure 3.19, with

$$\tilde{T} := \{T, T_{12}\}.$$

Dropping the more refined metric/convexity structure, as well as the fact that the more precise location of this point depends on syntactic heads and the evaluation of semantic proximity of the lexical items involved, if we only retain the Boolean relations of these neighborhoods and their intersections, we obtain a map to the Boolean semiring that checks the fact that "John ate a sandwich in the basement at noon" implies that "John ate a sandwich in the basement"

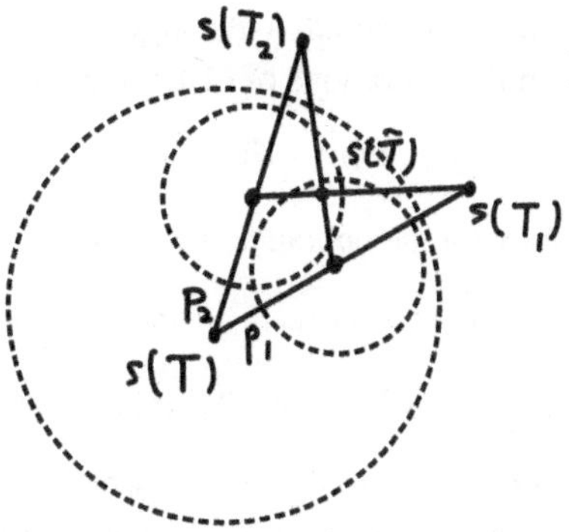

Figure 3.19
Example: adjuncts to verbs and semantic points.

and that "John ate a sandwich at noon", while the opposite implications do not hold, as desired.

Here, we can see, however, that the construction of the map $s : \mathrm{Dom}(h) \to \mathcal{S}$ that we used in sections 3.2.2 and 3.3 is only an oversimplified model, and it should be refined by directly including coverings by neighborhoods related by intersections; see the discussion in section 3.2.2.5.

3.6.3 Predicate Saturation and Operadic Structure

Other important aspects of Pietroski's semantics, in addition to the *Combine* operation discussed previously, consist of predicate saturation and existential closure; see [168]. We propose here a way to fit these aspects in our model, compatibly with the form of the *Combine* operation as in (3.6.3), using the formulation of the magma $\mathcal{SO}$ of syntactic objects in terms of *operads* (as mentioned briefly in section 2.5.1 of chapter 2).

We recall briefly the mathematical notion of an operad, introduced in [151], and we describe how to view syntactic objects as an algebra over an operad.

3.6.3.1 Syntactic objects and operads An operad in Sets is a collection $\mathcal{O} = \{\mathcal{O}(n)\}_{n\geq 1}$ of sets of n-ary operations (with n inputs and one output), with composition laws

$$\gamma : \mathcal{O}(n) \times \mathcal{O}(k_1) \times \cdots \times \mathcal{O}(k_n) \to \mathcal{O}(k_1 + \cdots + k_n) \qquad (3.6.8)$$

that plug the output of an operation in $\mathcal{O}(k_i)$ into the ith input of an operation in $\mathcal{O}(n)$. The composition of these operations γ is subject to requirements of associativity and unitarity, which we do not write explicitly here. (We will return to discuss these properties in more detail in section 3.8.1; see Figure 3.25

for the associativity relation of operads.) An algebra $\mathcal{A}$ over an operad $\mathcal{O}$ (in Sets) is a set $\mathcal{A}$ on which the operations of $\mathcal{O}$ act; namely, there are maps

$$\gamma_{\mathcal{A}} : \mathcal{O}(n) \times \mathcal{A}^n \to \mathcal{A} \tag{3.6.9}$$

that satisfy compatibility with the operad composition:

$$\begin{aligned}&\gamma_{\mathcal{A}}(\gamma_{\mathcal{O}}(T, T_1, \ldots, T_m), a_{1,1}, \ldots, a_{1,n_1}, \ldots, a_{m,1}, \ldots, a_{m,n_m}) = \\ &\gamma_{\mathcal{A}}(T, \gamma_{\mathcal{A}}(T_1, a_{1,1}, \ldots, a_{1,n_1}), \ldots, \gamma_{\mathcal{A}}(T_m, a_{m,1}, \ldots, a_{m,n_m}))\,.\end{aligned} \tag{3.6.10}$$

for $T \in \mathcal{O}(m)$, $T_i \in \mathcal{O}(n_i)$ and $\{a_{i,j}\}_{j=1}^{n_i} \subset \mathcal{A}$, where $\gamma_{\mathcal{O}}$ is the composition in the operad and $\gamma_{\mathcal{A}}$ is the operad action. This notion means that elements of the set $\mathcal{A}$ can be used as inputs for the operations in $\mathcal{O}$, resulting in an output that is again an element of $\mathcal{A}$. The category of Sets can be replaced by more general symmetric monoidal categories. In particular, we can consider cases where $\mathcal{A}$ is a topological space or a vector space that is suitable for the setting of semantic spaces. The description of the operadic composition laws in section 2.5.1, in terms of the compositions $\circ_i : \mathcal{O}(n) \times \mathcal{O}(m) \to \mathcal{O}(n+m-1)$, is equivalent, for unitary operads, to the description in terms of the compositions (3.6.8).

In particular, we are interested here in the operad $\mathcal{M}$ freely generated by a single commutative binary operation $\mathfrak{M}$, where we have $\mathcal{M}(1) = \{\mathrm{id}\}$, $\mathcal{M}(2) = \{\mathfrak{M}\}$, $\mathcal{M}(3) = \{\mathfrak{M} \circ (\mathrm{id} \times \mathfrak{M}), \mathfrak{M} \circ (\mathfrak{M} \times \mathrm{id})\}$, and so on, taking repeated compositions of $\mathfrak{M}$ operations. Consider again the set of syntactic objects $\mathcal{SO}$. The magma structure of (3.1.1) can be reformulated as the structure of algebra over this operad.

Lemma 3.6.7. *The set $\mathcal{SO}$ of syntactic objects is an algebra over the operad $\mathcal{M}$ freely generated by the single commutative binary operation $\mathfrak{M}$.*

Proof. We can identify the elements in $\mathcal{M}(n)$ with the abstract binary rooted trees with n leaves (with no labels on the leaves), where each internal (non-leaf) vertex is labeled with an $\mathfrak{M}$ operation. The maps (3.6.9) are simply given by taking $\gamma(T, T_1, \ldots, T_n)$, with $T \in \mathcal{M}(n)$ and $T_i \in \mathcal{SO}$ for $i = 1, \ldots, n$ to be the abstract binary rooted tree in $\mathfrak{T}_{\mathcal{SO}_0} = \mathcal{SO}$, obtained by grafting the root of the syntactic object T_i to the ith leaf of $T \in \mathcal{M}(n)$. If the syntactic objects T_i have n_i leaves, then the syntactic object $\gamma(T, T_1, \ldots, T_n)$ obtained in this way has $n_1 + \cdots + n_k$ leaves. Note that this operad action is just a repeated application of the product operation $\mathfrak{M}$ in the magma $\mathcal{SO}$, and hence the description as algebra over $\mathfrak{M}$ and as magma as in (3.1.1) are equivalent. □

3.6.3.2 Semantic spaces and operads The additional structure that we want to consider here, on the side of semantic spaces, is that of a partial algebra over the operad $\mathcal{M}$.

Definition 3.6.8. Let $\mathcal{M}$ be the operad freely generated by a single commutative binary operation $\mathfrak{M}$. A semantic space $\mathcal{S}$ is a *compositional semantic space* if it has the following properties:

1. There is a map $s : \mathrm{Dom}(h) \to \mathcal{S}$ extending $s : \mathcal{SO}_0 \to \mathcal{S}$.
2. There is an action of the operad $\mathcal{M}$ on $\mathcal{S}$:
$$\gamma_{\mathcal{S}} : \mathcal{M}(n) \times \mathcal{S}^n \to \mathcal{S}. \tag{3.6.11}$$
3. For $T \in \mathcal{M}(n)$ and for $T_1, \ldots, T_n \in \mathrm{Dom}(h) \subset \mathcal{SO}$ such that
$$\gamma_{\mathcal{SO}}(T, T_1, \ldots, T_n) \in \mathrm{Dom}(h),$$
we have
$$\gamma_{\mathcal{S}}(T, s(T_1), \ldots, s(T_n)) = s(\gamma_{\mathcal{SO}}(T, T_1, \ldots, T_n)). \tag{3.6.12}$$

The third condition ensures that the structure of $\mathcal{SO}$ as an algebra over the operad $\mathcal{M}$ and the structure of $\mathcal{S}$ as a partial algebra over the same operad $\mathcal{M}$ are compatible through the map $s : \mathrm{Dom}(h) \to \mathcal{S}$ from syntax to semantics.

One can more generally consider partial actions of an operad and a corresponding notion of *partial algebra over an operad* (introduced in [117]), where the operad action (3.6.9) is defined on a subdomain $\mathcal{A}_0 \subset \mathcal{A}$,
$$\gamma_{\mathcal{A}} : \mathcal{O}(n) \times \mathcal{A}_0^n \to \mathcal{A}. \tag{3.6.13}$$

For a compositional semantic space as in Definition 3.6.8, the predicate saturation operation of Pietroski's semantics, in a form compatible with syntactic Merge, can be interpreted as the operad action (3.6.11) that saturates the arguments of an n-ary operation by inputs in $s(\mathrm{Dom}(h)) \subset \mathcal{S}$ (a concept of adicity n combined with n semantic arguments). The partial compositions $\circ_i$ correspondingly give the combinations of a concept of adicity n with one semantic argument that give a concept of adicity $n-1$. Note, however, that there is an important difference here. In this model, the operations of adicity n in $\mathcal{M}(n)$ are part of the syntax core computational mechanism. They are not on the semantic side, so they cannot directly be identified with the "concept of adicity n" described in [168]. It is only through the relation (3.6.12) that they acquire that role.

3.6.3.3 Syntax-driven compositional semantics The notion of compositional semantic space that we described in Definition 3.6.8 is based on two operad actions, one (that we called γ_{SO}) on the side of syntax and one (that we called γ_S) on the side of semantics, with the compatibility (3.6.12). This is similar to the formulation of Pietroski's semantics in [168]. However, we show now that in fact, the operad action γ_{SO} on syntax suffices to completely determine its counterpart γ_S.

Proposition 3.6.9. *Let $\mathcal{S}^+ = \mathcal{S} \cup \{s_\infty\}$ be the Alexandrov one-point compactification of $\mathcal{S}$, where we denote the added point with the symbol s_∞. The action of the operad $\mathcal{M}$ on syntactic objects, described in Lemma 3.6.7, together with a function $s : \mathrm{Dom}(h) \to \mathcal{S}$, uniquely determine an action of the operad $\mathcal{M}$ on $\mathcal{S}$ by setting*

$$\gamma_S(T, s_1, \ldots, s_n) := \begin{cases} s(\gamma_{SO}(T, T_1, \ldots, T_n)) & \text{if } \begin{array}{l} s_i = s(T_i) \text{ and} \\ \gamma_{SO}(T, T_1, \ldots, T_n) \in \mathrm{Dom}(h) \end{array} \\ s_\infty & \text{otherwise.} \end{cases} \tag{3.6.14}$$

for $T \in \mathcal{M}(n)$ and $s_1, \ldots, s_n \in \mathcal{S}^+$.

Proof. We construct γ_S using γ_{SO} and the compatibility relation (3.6.12) using (3.6.14), To show that (3.6.14) does indeed define an operad action on $\mathcal{S}$, we need to check the compatibility of γ_S with the operad composition γ in $\mathcal{M}$, given by the condition (3.6.10). The left-hand-side of (3.6.10) gives

$$\gamma_S(\gamma_{\mathcal{M}}(T, T_1, \ldots, T_m), s_{1,1}, \ldots, s_{1,n_1}, \ldots, s_{m,1}, \ldots, s_{m,n_m}) . \tag{3.6.15}$$

This is equal to s_∞ unless both of the following two conditions are satisfied:

- All the $s_{i,j} = s(T_{i,j})$ for some $T_{i,j}$ in $\mathrm{Dom}(h) \subset \mathcal{SO}$
- The syntactic object

$$\gamma_{SO}(\gamma_{\mathcal{M}}(T, T_1, \ldots, T_m), T_{1,1}, \ldots, T_{1,n_1}, \ldots, T_{m,1}, \ldots, T_{m,n_m}) \tag{3.6.16}$$

  is in $\mathrm{Dom}(h)$,

in which case (3.6.15) is equal to

$$s(\gamma_{SO}(\gamma_{\mathcal{M}}(T, T_1, \ldots, T_m), T_{1,1}, \ldots, T_{1,n_1}, \ldots, T_{m,1}, \ldots, T_{m,n_m})) . \tag{3.6.17}$$

The compatibility of the action γ_{SO} with the operad composition implies that (3.6.16) is equal to

$$\gamma_{SO}(T, \gamma_{SO}(T_1, T_{1,1}, \ldots, T_{1,n_1}), \ldots, \gamma_{SO}(T_m, T_{m,1}, \ldots, T_{m,n_m})) . \tag{3.6.18}$$

Note that if the full composition in (3.6.16) is in Dom(h), then by the properties of abstract head functions, all the substructures $\gamma_{\mathcal{M}}(T_i, T_{i,1}, \ldots, T_{i,n_i})$, $i = 1, \ldots, m$ are also in Dom(h). Thus, the point (3.6.17) in $\mathcal{S}$ is the same as the point

$$s(\gamma_{SO}(T, \gamma_{SO}(T_1, T_{1,1}, \ldots, T_{1,n_1}), \ldots, \gamma_{SO}(T_m, T_{m,1}, \ldots, T_{m,n_m}))) =$$

$$\gamma_{\mathcal{S}}(T, s(\gamma_{SO}(T_1, T_{1,1}, \ldots, T_{1,n_1})), \ldots, s(\gamma_{SO}(T_m, T_{m,1}, \ldots, T_{m,n_m}))) =$$

$$\gamma_{\mathcal{S}}(T, \gamma_{\mathcal{S}}(T_1, s_{1,1}, \ldots, s_{1,n_1}), \ldots, \gamma_{\mathcal{S}}(T_m, s_{m,1}, \ldots, s_{m,n_m})),$$

which gives the right-hand-side of (3.6.10). □

The structure of algebra over the operad $\mathcal{M}$ on $\mathcal{S}^+$ makes $\mathcal{S}$ a partial algebra over $\mathcal{M}$.

Note that we are everywhere somewhat simplifying this picture, as we do not include the possibility that different syntactic objects in Dom(h) $\subset \mathcal{SO}$ may sometime map to the *same* value in $\mathcal{S}$ under s : Dom(h) $\to \mathcal{S}$, and also the possibilities of *ambiguities* of semantic assignment where s : Dom(h) $\to \mathcal{S}$ may sometimes be multivalued. These possibilities would affect the construction (3.6.14) of $\gamma_{\mathcal{S}}$ and would require a modified argument.

3.7 Adjunction, Embedded Constructions, and the Pair-Merge Problem

Our model of the map s : Dom(h) $\to \mathcal{S}$, which extends the assignment $s : \mathcal{SO}_0 \to \mathcal{S}$ of semantic values defined on lexical items, is a very simple model built using only the head function and proximity relations (and geodesic distance) in semantic space $\mathcal{S}$. In particular, since we start with the syntactic objects produced by the free symmetric Merge, the only factor that introduces asymmetry in this construction is coming from the head function.

We discuss here briefly how one can try, within the limits of such an oversimplified model, to address the question of misalignments between hierarchical syntax and compositional semantics that occur as a consequence of the particular behavior of adjunction, and in particular what is sometimes referred to as the "invisibility of adjuncts to syntax". This question was posed to us by Riny Huijbregts.

A proposal for handling this type of problem is to postulate the existence of an asymmetrical Pair-Merge operation accounting for argument-adjunct asymmetry (see [24]), in addition to free symmetric Merge. This proposal has undesirable features, as it requires the introduction of an additional form of asymmetric Merge dealing with the peculiar behavior of adjunction, while one expects that the computational mechanism of syntax should just rely entirely on free symmetric Merge. An alternative proposal (see, e.g., [159]) involves the use of "two-peaked" structures (see Figure 3.20) with $\{XP, YP\}$ being an adjunction. This proposal has the drawback that, if one considers such two-peaked structures as part of syntax, then one needs to justify them in terms of the free Merge generative process, and this is problematic because the elements of the magma $\mathcal{SO} = \mathfrak{T}_{\mathcal{SO}_0}$ do not contain such structures, nor does the action of Merge on workspaces (as can be easily seen in the formalization in chapter 1). The proposal of two-peaked structures in [159] is based on [60], but it is not formulable within the generative process of a free symmetric Merge, as formulated in [40]. We are going to discuss briefly what this means in terms of our model.

The reason why adjunction appears problematic in our setting is that adjunction can be seen as an instance of syntactic objects $\{XP, YP\}$ that do not have a well-defined head function in the sense that we have been using, $\{XP, YP\} \notin \mathrm{Dom}(h)$. This creates a problem with our simple model of mapping to semantics, which is defined only on $\mathrm{Dom}(h)$. We want to argue here that this problem can be to some extent bypassed without the need to significantly alter the construction of the mapping $s : \mathrm{Dom}(h) \to \mathcal{S}$, although of course we expect that the naive model for this map based on the datum of the head function may be replaced by some more elaborate versions, incorporating at least the full Phase Theory, as discussed in section 1.14 in chapter 1.

Consider a syntactic object of the form $\{XP, YP\} \notin \mathrm{Dom}(h)$, where both XP and YP are in $\mathrm{Dom}(h)$. In terms of our construction, the fact that the head function is not well defined on the object $\{XP, YP\}$ implies that we do not have a choice of orientation on the geodesic arc between $s(XP)$ and $s(YP)$ in $\mathcal{S}$ and a corresponding point along this arc at a distance $\mathbb{P}(s(XP), s(YP))$ from the image of the head. We do still have the geodesic arc, though, and the measurement $\mathbb{P}(s(XP), s(YP))$ of syntactic relatedness between its endpoints. So in terms of this construction, all that a hypothetical asymmetric Pair-Merge would provide is a choice of orientation on the geodesic arc. Such a datum is a geometric datum in $\mathcal{S}$ and does not necessarily require the existence of Pair-Merge as an additional part of the computational structure of syntax. One can extend $s : \mathrm{Dom}(h) \to \mathcal{S}$ to a slightly larger domain that includes adjunctions just by the requirements that geodesic arcs in $\mathcal{S}$ whose endpoints are the two terms of an adjunction come with a preferred choice of orientation.

This choice has the same effect of a Pair-Merge $\langle XP, YP\rangle$, signifying that the first element should be taken to be the "head" while the second element is to be seen as an "adjunct". Such choice of orientation then ensures that we can extend the same construction of the map $s : \mathrm{Dom}(h) \to \mathcal{S}$ to the case of adjunctions $\{XP, YP\} \notin \mathrm{Dom}(h)$. In general, one does not expect that this orientation requirement should be extendable to other types of syntactic objects $\{XP, YP\} \notin \mathrm{Dom}(h)$ that are not adjunctions. The fact that this mechanism does not require any modification of the syntactic generative process and only involves a metric property in $\mathcal{S}$ is consistent with the idea that adjuncts are on a "separate plane" (see, e.g., [24]).

While this approach can be accommodated within our setting, it leaves open the question of assigning general criteria for orientations of geodesic arcs in $\mathcal{S}$ that generalize the choice resulting from a head function, incorporating the case of adjunctions, but not the case of arbitrary $\{XP, YP\}$ objects.

Thinking in terms of two-peaked structures, on the other hand, presents another possibility for treating this problem of adjunctions in our geometric setting. Two-peaked structures are not part of the syntactic objects that can be generated by Merge. Thus, their appearance in $\mathcal{S}$ is not the result of mapping an object from syntax, and can only be realized as an overlap of structures in the image of the embedding of syntax into semantics. Given a syntactic object of the form $\{XP, YP\} \notin \mathrm{Dom}(h)$, where both XP and YP are in $\mathrm{Dom}(h)$, consider in $\mathcal{S}$ the points $s(XP)$ and $s(YP)$ and the geodesic arc between them (now without any preferred assignment of orientation). Consider also a syntactic object $T = \{Z, XP\} \in \mathrm{Dom}(h)$. Since this is in the domain of the head function, it defines a point $s(T)$ on the geodesic arc between $s(Z)$ and $s(XP)$, where the geodesic arc is oriented from the end that corresponds to the head to the other. Thus, we do indeed obtain a configuration that looks like a "two-peaked structure", as in Figure 3.20. Note that while the geodesic arc between $s(XP)$ and $s(YP)$ does not have an a priori choice of orientation, the orientation induced by the head on the geodesic arc between $s(Z)$ and $s(XP)$ induces a unique consistent orientation on the arc between $s(XP)$ and $s(YP)$.

It is important to stress here the difference between the type of two-peaked structures that we are describing and the proposed two-peaked structures in the syntactic setting, as in [159]. Here, this structure does not exist in the magma $\mathcal{SO}$; it only exists in the image of syntax under the map to semantic space $\mathcal{S}$. In other word, these "two-peaked" structures are not part of the computational process of syntax and do not need to be justified by any additional form of Merge. They exist because the images of syntactic objects inside $\mathcal{S}$ can intersect, even though the resulting configuration (like the one in Figure 3.20) is not itself the image of a syntactic object.

A different solution to the Pair-Merge problem was proposed by Riny Huijbregts [103], based only on the use of Internal Merge and External Merge, where the asymmetry of Pair-Merge is realized through the identification of an

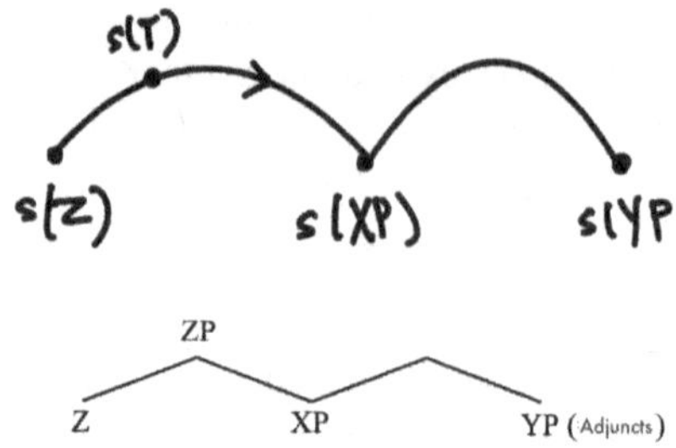

Figure 3.20
Two-peaked structures are not syntactic objects generated by Merge; rather they can arise inside $\mathcal{S}$ as overlap between images.

ordered pair (α,β) with the structure $\{\{\alpha\},\{\alpha,\beta\}\}$, where $\{\alpha\}$ is an accessible term of the structure $\{\alpha,\beta\}$.

3.7.1 Adjunctions and Hierarchical Structures

In relation to our discussion of adjunctions, we also briefly consider here the question of characterizing the region that unbounded adjunct sequences span inside the overall structure of recursivity in language; for instance, see [170] for some aspects of this question. In particular, since verbs take a finite number of arguments but an arbitrarily large number of adjuncts, it appears as if adjuncts may be able to account for a significant part of compositionality and recursive complexity in natural language.

We want to give here a mathematical argument characterizing the generative power of chains of adjunctions and the region that the resulting structures span among all the possible hierarchical structures.

It suffices to focus on the case of a verb with a sequence of adjunctions. We can assume that the overall syntactic object that contains this substructure is in the domain of a head function, and we can choose the planarization associated to that head function, so we will be dealing with *planar* binary rooted trees as hierarchical structures. One can also assume that the planar structure is induced by Externalization: the argument will not change significantly.

In a first instance, a sequence of adjunctions can be seen as appending to a given hierarchical structure a comb-tree (in the planarization associated with the head) as in Figure 3.21. Clearly, by increasing the number of such adjunctions, one can arbitrarily increase the depth of the tree by increasing the length of this comb-subtree. More generally, one can consider structures of this kind, containing a long comb-subtree, but where some of the tips of the

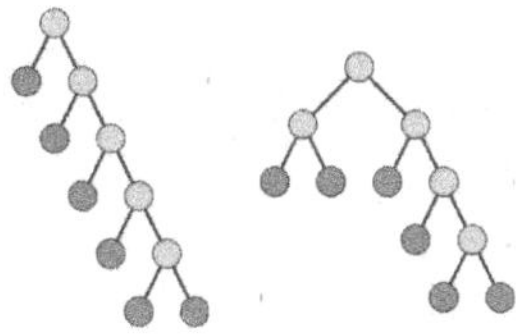

Figure 3.21
Planar trees with a sequence of adjunctions.

comb may contain additional substructures. Thus, the question is how much of the recursive structure is captured by this type of trees that contain a long comb-subtree.

We will show that we can characterize the generative power of sequences of adjunctions in terms of the Tamari order and Loday's operations of left and right sums on planar binary rooted trees; see [122] and [123].

In the case of the planar trees, consider again the associahedron that we already discussed in section 3.4. The 1-skeleton of the associahedron is the Hasse diagram of the Tamari lattice, or Tamari order (compare the Tamari order in Figure 3.22 with the associahedron in Figure 3.5). The oriented arrows in the Tamari order relate planar binary rooted trees obtained by moving one edge from right to left across a vertex (for trees drawn with the root at the top). The relation $T < T'$ if there is an oriented arrow $T \to T'$ in the Tamari lattice induces a partial order on the set $\mathfrak{T}^{pl}_{n+1}$ of planar binary rooted trees on $n + 1$ leaves (n internal vertices); hence, the set $\mathfrak{T}^{pl}_{n+1}$ is a lattice with this partial order relation. We write $n = \deg(T)$ for $T \in \mathfrak{T}^{pl}_{n+1}$ (note that for convenience, we shift here to grade by the number of internal vertices rather than by the number of leaves). We see that, in this diagram, the comb trees that correspond to the simplest chains of adjunctions occupy the extreme points of the Tamari lattice.

While we discuss here the case of *planar* binary rooted trees, observe that, when considering abstract binary rooted trees (namely, the syntactic objects $\mathcal{SO}$, without any choice of planarization), one can compare the comb-trees with other trees (with the same fixed set of lexical items at the leaves) in the link of the origin in the BHV moduli space, as in the case of Figure 3.23. Note that if the leaves had no labels, the two trees at the bottom in Figure 3.23 would also be equivalent (as nonplanar trees) to the two comb trees at the top, and this would seem to imply that abstract trees realized in a chain of adjunctions can generate most of the hierarchical structures, but in fact all four of these trees are inequivalent, when seen as trees with labeled leaves.

In Definition 2.2.3 of chapter 2, we recalled the two associative over/under operations on planar binary rooted trees, where, for $T, S \in \mathfrak{T}^{pl}$, we defined the

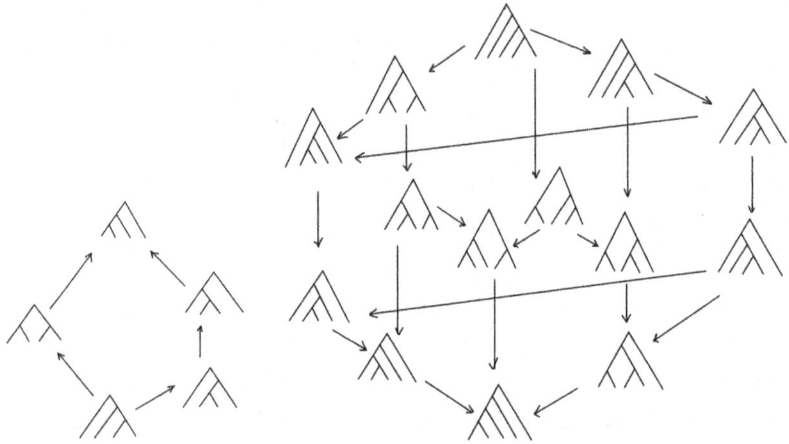

Figure 3.22
The Tamari lattice (or Tamari order) for four and five leaves.

tree $S\backslash T$ (S under T) as the planar binary rooted tree obtained by grafting the root of T to the rightmost leaf of S, while S/T (S over T) was defined as the planar binary rooted tree obtained by grafting the root of S to the leftmost leaf of T, such as

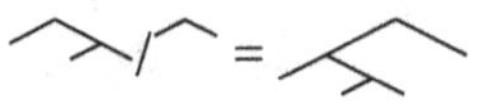 and 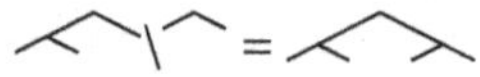 .

These two operations satisfy $S/T \leq S\backslash T$ in the Tamari order.

The *noncommutative* sum of trees (see [122] and [123]) is defined as

$$S \hat{+} T = \bigcup_{S/T \leq T' \leq S\backslash T} T' ,$$

where all $\deg(T') = \deg(S) + \deg(T)$. This sum operation is noncommutative, $S\hat{+}T \neq T\hat{+}S$, but it is associative and satisfies distributivity with respect to unions, $\cup_i T_i \hat{+} \cup_j S_j = \cup_{ij}(T_i \hat{+} S_j)$. Unlike [122], we use here the notation $\hat{+}$ for this sum of trees rather than just the notation +, as a reminder of the fact that this operation does not have the usual properties of the sum; for example, it does not satisfy commutativity. For instance, using the same example as in [123],

 ,

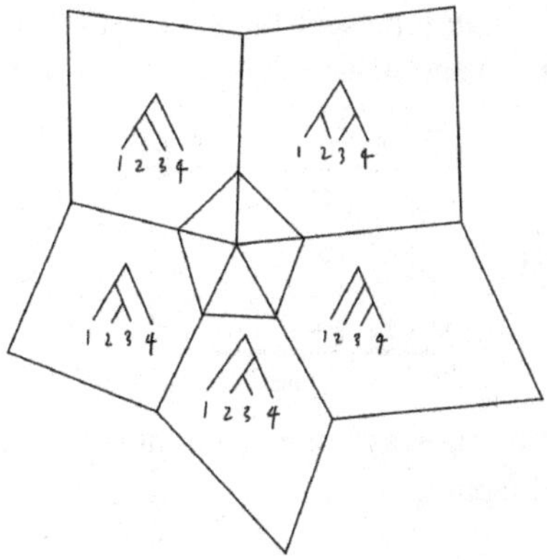

Figure 3.23
Abstract binary rooted trees with labeled leaves in the moduli space BHV_4, as shown in [11].

while

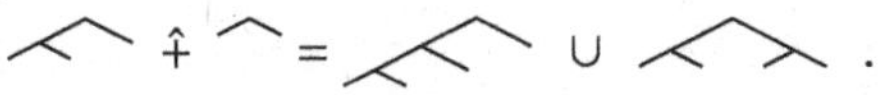

This asymmetric (noncommutative) sum of trees can be decomposed as the combination of two operations (see [122] and [123])

$$S \hat{+} T = S \dashv T \cup S \vdash T,$$

where the two operations $\dashv$ and $\vdash$ are defined recursively in the form

$$S \dashv T = \overset{\frown}{S^L \quad (S^R \hat{+} T)}$$

$$S \vdash T = \overset{\frown}{(S \hat{+} T^L) \quad T^R},$$

where

$$S = \overset{\frown}{S^L \quad S^R} \quad \text{and} \quad T = \overset{\frown}{T^L \quad T^R},$$

with L, R indicating the left and right subtrees of a planar binary rooted tree.

We denote by $\mathbb{I}$ (the analog of the integer 1 in the formulation of [122]) the tree

$\mathbb{I} :=$  .

Then we have

$\mathbb{I} \hat{+} \mathbb{I} = \mathbb{I} \dashv \mathbb{I} \cup \mathbb{I} \vdash \mathbb{I} =$  .

Thus, we see that we can represent the simplest chains of adjunctions, in head-initial form, as the planar trees

$$\underbrace{\mathbb{I} \dashv \mathbb{I} \dashv \mathbb{I} \dashv \cdots \dashv \mathbb{I} \dashv \mathbb{I}}_{N \text{ times}}$$

or in head-final form as

$$\underbrace{\mathbb{I} \vdash \mathbb{I} \vdash \mathbb{I} \vdash \cdots \vdash \mathbb{I} \vdash \mathbb{I}}_{N \text{ times}} .$$

We focus here on the first case, assuming a harmonic head-initial linearization determined by a head function.

Proposition 3.7.1. *Consider all possible chains of adjunctions that starts with a given planar binary rooted tree T and appends to its rightmost leaf a comb-tree,*

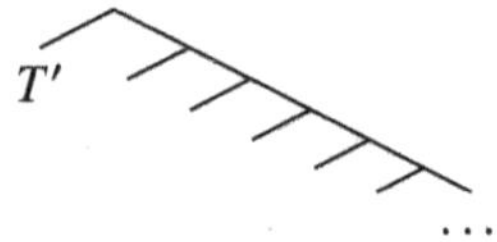

, (3.7.1)

equivalently written in the form

$$T = T' \backslash \underbrace{\mathbb{I} \backslash \mathbb{I} \backslash \cdots \backslash \mathbb{I}}_{N \text{ times}}$$

of arbitrary length N, or of the more general form

$$T = T_0 \backslash \underbrace{\mathbb{I} \backslash \mathbb{I} \backslash \cdots \backslash \mathbb{I}}_{N_0 \text{ times}} \backslash T_1 \backslash \underbrace{\mathbb{I} \backslash \mathbb{I} \backslash \cdots \backslash \mathbb{I}}_{N_1 \text{ times}} \backslash \cdots \backslash T_k \backslash \underbrace{\mathbb{I} \backslash \mathbb{I} \backslash \cdots \backslash \mathbb{I}}_{N_k \text{ times}} , \qquad (3.7.2)$$

where the trees T_i with $i = 1, \ldots, k$ are of the form

$T_i =$ 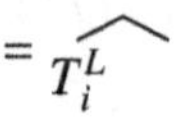

. (3.7.3)

In the case of (3.7.1), *the hierarchical structures obtained in this way span a simplicial subcomplex of the associahedron K_{n+N} of codimension N, for $T' \in K_n$. In the case of* (3.7.2), *they span a simplicial subcomplex of codimension $N_0 + \cdots + N_k$ in the associahedron $K_{n_0+\cdots+n_k+N_0+\cdots+N_k}$, where $T_i' \in K_{n_i}$.*

Proof. Consider a chain of adjunctions that starts with a given planar binary rooted tree T and appends to its rightmost leaf a comb-tree as in (3.7.1). We can see this as the tree $T =$ 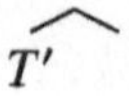

with an attached comb-tree to the rightmost

leaf of T; that is, as

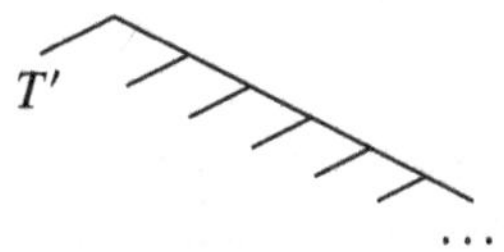

$$= T\backslash\mathbb{I}\backslash\mathbb{I}\backslash\mathbb{I}\cdots\backslash\mathbb{I}\ .$$

The operation $T\hat{+}\mathbb{I}$ gives the union of all the binary rooted tree S, with

$$T/\mathbb{I} \leq S \leq T\backslash\mathbb{I},$$

where $T\backslash\mathbb{I}$ is the tree representing an adjunction attached to the rightmost leaf of T and $T/\mathbb{I} =$  .

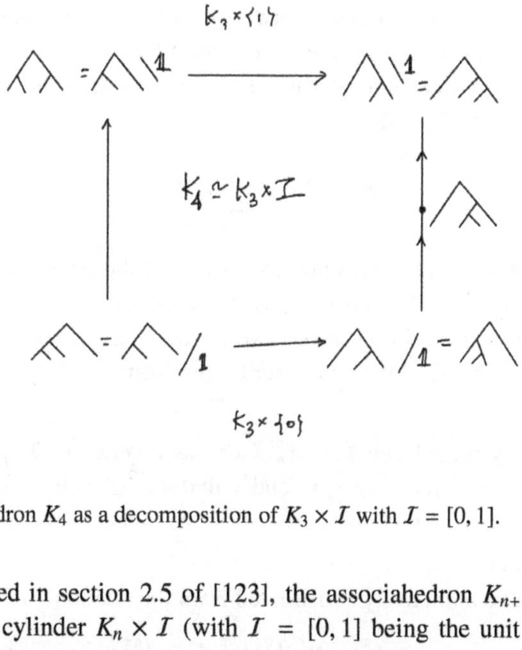

Figure 3.24
The associahedron K_4 as a decomposition of $K_3 \times I$ with $I = [0, 1]$.

As discussed in section 2.5 of [123], the associahedron K_{n+1} can be identified with a cylinder $K_n \times I$ (with $I = [0, 1]$ being the unit interval) over the associahedron K_n. For a more precise description of how to identify the n-dimensional polyhedron given by the associahedron K_{n+2} with a subdivision of the cube I^n, see [181] and [182]. The example of K_4 as a subdivision of $K_3 \times I$ is given in Figure 3.24. In this identification the boundary $K_n \times \{0\}$ has vertices given by the trees $T/\mathbb{I}$, where T is a vertex of K_n, while $K_n \times \{1\}$ has vertices $T\backslash\mathbb{I}$. The directed segment $I = [0, 1]$ between two such vertices has intermediate vertices given by the trees S, with $T/\mathbb{I} \leq S \leq T\backslash\mathbb{I}$. This means that we can identify these segments with $T\hat{+}\mathbb{I}$ with the orientation given by the Tamari ordering. The adjunction $T\backslash\mathbb{I}$ lies at the end $K_n \times \{1\}$ of the segment. In particular, the set $\mathfrak{T}_n^{pl}$ of planar binary rooted trees on n leaves is obtained as

$$\mathfrak{T}_{n+1}^{pl} = \mathfrak{T}_n^{pl} \hat{+} \mathbb{I} .$$

Indeed, this relation is just saying that the segments $T/\mathbb{I} \leq S \leq T\backslash\mathbb{I}$ cover the entire set of vertices of the associahedron K_{n+1} when T varies over the vertices of K_n.

Repeated adjunctions $T\backslash\mathbb{I}\backslash\mathbb{I}\backslash\mathbb{I}\cdots\backslash\mathbb{I}$ of length N therefore map a tree T placed at one of the vertices of the associahedron K_n to a corresponding vertex in the associahedron K_{n+N} viewed as a product of K_n with a subdivision of an N-cube, $K_{n+N} = K_n \times I^N$, where the subdivision of the cube has $T\hat{+}\mathbb{I}\hat{+}\mathbb{I}\hat{+}\cdots\hat{+}\mathbb{I}$ as its set of vertices, with oriented edges given by the Tamari order relation. The adjunctions $T\backslash\mathbb{I}\backslash\mathbb{I}\backslash\mathbb{I}\cdots\backslash\mathbb{I}$ are located at the topmost vertex $(1, 1, \ldots, 1)$ of the cube I^N. This completely characterizes the region of the associahedra that are realized by simple chains of adjunctions as in (3.7.1).

In a similar way, we can consider more general chains of the adjunctions of the form (3.7.2), with T_i with $i = 1, \ldots, k$ of the form (3.7.3). In this case, the first block

$$S_0 = T_0\backslash\underbrace{\mathbb{I}\backslash\mathbb{I}\backslash\cdots\backslash\mathbb{I}}_{N_0 \text{ times}}$$

is analyzed as before and determines a vertex S_0 in the associahedron $K_{n_0+N_0}$, given a vertex T_0 of K_{n_0}. The vertex S_0 is located at the $(1, 1, \ldots, 1)$ vertex of a cube I^{N_0}. The next step, $S_0\backslash T_1$, gives a vertex in the associahedron $K_{n_0+K_0+n_1}$, given a vertex T_1 of K_{n_1}. We can describe the location of this vertex in the following way:

- A planar binary rooted tree $T \in \mathfrak{T}_n^{pl}$ (seen as a vertex of K_n) is described uniquely by a word $w_T(\mathbb{I})$ in $\mathbb{I}$, $\vdash$, and $\dashv$ that describes the construction of T, starting from a copy of the tree $\mathbb{I} = \wedge$ as repeated applications of the operations $\vdash \mathbb{I}$ and $\dashv \mathbb{I}$.
- The word can be inductively constructed from the relation

$$w_T(\mathbb{I}) = w_{T^L}(\mathbb{I}) \vdash \mathbb{I} \dashv w_{T^R}(\mathbb{I}),$$

  see [123]. Here, (3.7.3) gives words $w_{T_i}(\mathbb{I})$ that end with $\vdash \mathbb{I}$.
- Each step $T_{i,j}$ in the construction given by the word $w_{T_i}(\mathbb{I})$ determines a vertex in the subdivision of the segment $I = [0, 1]$ in the product $T_{i,j} \times I$ of the vertex $T_{i,j}$ in K_{j+1} and the subdivided interval.
- So long as we do not impose further constraints on the choice of these T_i', each of them can range over the vertices of K_{n_i}.

- On the other hand, each step of the form $S\backslash\mathbb{I}$ in (3.7.2) determines a point that is the end $S \times \{1\}$ in $S \times \mathcal{I}$, and hence of codimension 1 with respect to the segment $\mathcal{I}$.

Thus, one sees in particular that chains of adjunctions span high codimension strata of the associahedra. This shows geometrically how the generative power of adjunctions compares to the general landscape of hierarchical structures. In the case of adjunctions (3.7.1), these occupy a codimension N locus in K_{n+N}, while in the case of (3.7.2), they determine a locus of codimension at least $N_0 + \cdots + N_k$ in the associahedron $K_{n_0+\cdots+n_k+N_0+\cdots+N_k}$, where a further increase in codimension may depend on specific requirements on the form of the T_i for $i = 1, \ldots, k$ that we do not consider explicitly. □

The argument in the case of a planarization induced by Externalization is slightly more complicated, due to the possible presence of nonharmonic head directionality, so we will not discuss it explicitly here.

3.8 Language-Specific Conditions and Theta Theory

We discuss here how to incorporate into our model some language-specific conditions that have direct linguistic reflexes that arise in the interface of the Merge-based computational model of syntax and the Conceptual–Intensional system. One should view this discussion as a further refinement of the models for semantic assignment discussed earlier in this chapter, where we mostly used only information about syntactic heads. Here, we will follow the way in which this aspect of language was organized in sections 5 and 6 of *Elements* [40]. We show how the main points outlined there can be interpreted within the mathematical formalism that we are adopting here.

In this context, the term "language specific" plays two slightly different roles. On the one hand, as in *Elements* [40], it refers to *language specific conditions* as constraints on the structures formed by free symmetric Merge, which are dictated by conditions specific to I-language. An example of such constraints is the existence of a head, its complement structure, and Phase Theory as discussed in chapter 1; another example is Theta Theory, discussed in this section. On the other hand, "language specific" can also mean "specific to one language", as when referring to the results of Externalization. The context should suffice to disambiguate these meanings.

The first topic we discuss is what is called *Theta Theory*. We follow an approach similar to the one that we adopted in the discussion of predicate saturation in section 3.6.3, based on the operad structure of syntactic objects and a structure of algebra over an operad for the space of semantic values.

3.8.1 Theta Theory

As recalled in section 5.1 of [40], Theta Theory models thematic relations between predicates and their arguments. Predicates assign *theta roles* (θ-roles) to their arguments, and a standard constraint is that there should be a one-to-one correspondence between the theta roles and the arguments to which the theta roles are assigned.

As discussed in section 5.2 of [40], there is a dichotomy in semantics[10] separating what is referred to as "argument structure" (and θ-role assignments), associated with the products of External Merge, and displacement (referred to as "non-argument structure" and pertaining to information-related properties like Focus), associated with the products of Internal Merge: in other words, a distinct between the *propositional domains* (theta-structure) and the *clausal domains* (information-related), respectively.

To see how the External Merge relates to theta positions and theta roles in our setting, and why Internal Merge does not perform the same role here as External Merge, we return to the setting discussed in §3.6.3.

In our previous discussion of the syntax–semantics interface, to keep our model as simple as possible and to better illustrate its formal algebraic properties, we used only the presence of a head function to furnish a minimal amount of semantic information. A head function h assigns a choice of a leaf to every accessible term of a syntactic object that is in Dom(h), along with a consistency condition (see Definition 1.13.6). The selected leaf is meant to represent the lexical item within that phrase that cannot be omitted (i.e., the usual notion of syntactic head). We then applied our abstract notion of head function, together with some notion of proximity in semantic space, to construct a mapping $s : \mathrm{Dom}(h) \to \mathcal{S}$ that extends the semantic assignments to lexical items $s : \mathcal{SO}_0 \to \mathcal{S}$. However, just having the head (i.e., the condition that a syntactic object is in Dom(h)) does not generally suffice to identify and remove ill-formed sentences, and one needs to use additional information on the structure of dependents (complements) of the head, in particular through the assignment of θ-roles. (For example, *John slept Bill* is not well-formed.)

Given an abstract head function as in Definition 1.13.6, the "head and complement" structure can be formulated, as we discussed in the context of Phase Theory in section 1.14, as in Definition 1.14.2 in chapter 1. We can then see how to fit θ-roles into this setting in the following way.

[10] We prefer to use the term "dichotomy" here instead of the term "duality" used in [40] to avoid confusion: in mathematics, "duality" means an involutive operation.

Consider again the magma of syntactic objects, generated by free symmetric Merge,

$$\mathcal{SO} = \text{Magma}_{na,c}(\mathcal{SO}_0, \mathfrak{M}) = \mathfrak{T}_{\mathcal{SO}_0} .$$

The magma operation $(T, T') \mapsto \mathfrak{M}(T, T')$ corresponds to the External Merge operation that combines two syntactic objects T, T' into a single structure:

$$\mathfrak{M}(T, T') = \widehat{T \quad T'}$$

If we do not assign to the leaves of the trees any lexical item labels in $\mathcal{SO}_0$, we obtain the simpler core computational structure discussed in section 1.10; namely, the free nonassociative, commutative magma $\mathfrak{T} = \text{Magma}_{na,c}(\mathfrak{M})$, whose elements are the balanced bracketed expressions in a single variable x. The elements of this magma are canonically identified with the abstract binary rooted trees (with no leaf decorations). If $\mathfrak{T}_n \subset \mathfrak{T}$ denotes the set of abstract binary rooted trees with n leaves (and no leaf decorations), we can use these sets as in section 3.6.3 to form an operad $\mathcal{M}$ with $\mathcal{M}(n) = \mathfrak{T}_n$, where the operad compositions are obtained by grafting the output root of one tree to one of the input leaves of another. As we showed in section 3.6.3 (see Lemma 3.6.7), the set of syntactic objects $\mathcal{SO}$ is an algebra over the operad $\mathcal{M}$. As described in Definition 3.6.8, one can then require that assignments of semantic values are compatible with this operad action.

We first make an observation regarding the operad $\mathcal{M}$ and its action on $\mathcal{SO}$ and $\mathcal{S}$, as in Lemma 3.6.7 and Definition 3.6.8. Note that, when we compose via the operad $\gamma(T, T_1, \ldots, T_n)$, with the grafting of the tree T_i to the leaves of T, we need to choose a bijection between the set $\{T_i\}$ and the set of leaves $L(T)$. If the corresponding trees are planar, this bijection is already assigned by the ordering of the leaves induced by the planar structure, with the root of the ith tree T_i grafted to the ith leaf of T. This is the case if we consider this operad structure and the corresponding action on $\mathcal{S}$ taking place *after* Externalization, where all trees have been assigned a planar structure through a language-dependent section σ_L of the projection from the noncommutative to the commutative magma; see section 1.12 in chapter 1 and (1.12.1).

Considering the leaves as an ordered set is fine in Externalization, with the image under a planarization σ_L of the results of the generative process of free symmetric Merge. At the level of free symmetric Merge, on the other hand, one should consider the operad composition in $\mathcal{M}$ and the operad action on $\mathcal{SO}$ in the form

$$\gamma(T, \{T_\ell\}_{\ell \in L(T)})$$

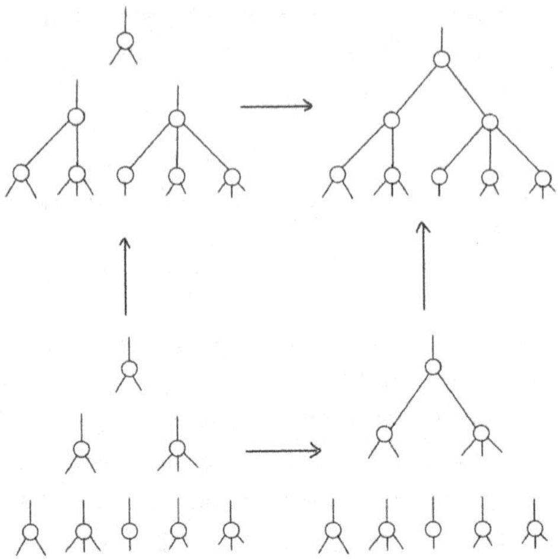

Figure 3.25
The associativity of operad composition.

for $T \in \mathfrak{T}_n$ and $T_\ell \in \mathfrak{T}_{k_\ell}$, where the root of the tree T_ℓ is grafted to the leaf $\ell \in L(T)$. The associativity of compositions is expressed as

$$\gamma(\gamma(T, \{T_\ell\}_{\ell \in L(T)}), \{T'_{\ell'}\}_{\ell' \in L(\gamma(T, \{T_\ell\}_{\ell \in L(T)}))}) = \gamma(T, \{\gamma(T_\ell, \{T'_{\ell'}\}_{\ell' \in L(T_\ell)})\}_{\ell \in L(T)});$$

see Figure 3.25. The symmetric operad structure is given by two equivariance properties for the actions of the permutation groups $\mathrm{Sym}(L(T))$ of the sets of leaves of trees. We write $T \circ \tau$, for $\tau \in \mathrm{Sym}(L(T))$ for the action that permutes the leaves of T. This means that a collection of inputs $\{a_\ell\}_{\ell \in L(T)}$ to T, with a_ℓ the input at the leaf $\ell \in L(T)$, becomes a collection with a_ℓ as the input of $\tau^{-1}(\ell)$ in $T \circ \tau$. The first equivariance condition takes the form

$$\gamma(T \circ \tau, \{T_{\tau(\ell)}\}) = \gamma(T, \{T_\ell\}) \circ \tau' \tag{3.8.1}$$

for all $\tau \in \mathrm{Sym}(L(T)) \simeq S_n$, with $n \in \#L(T)$ and S_n the symmetric group, and with $\tau' \in \mathrm{Sym}(L(\gamma(T, \{T_\ell\})) \simeq S_{\sum_\ell k_\ell}$, which permutes the n blocks of k_ℓ leaves, leaving each block unchanged. The second one takes the form

$$\gamma(T, \{T_\ell \circ \sigma_\ell\}_{\ell \in L(T)}) = \gamma(T, \{T_\ell\}_{\ell \in L(T)}) \circ \underline{\sigma}, \tag{3.8.2}$$

for $\sigma_\ell \in \mathrm{Sym}(L(T_\ell)) \simeq S_{k_\ell}$ with $\underline{\sigma} \in \mathrm{Sym}(L(\gamma(T, \{T_\ell\})) \simeq S_{\sum_\ell k_\ell}$, which permutes the leaves within each block of k_ℓ leaves, leaving the position of the blocks unchanged. These two equivariance conditions for the action of the symmetric groups are illustrated in Figure 3.26. For a formulation of symmetric operads in terms of *species* rather than sets, see [108] and [20].

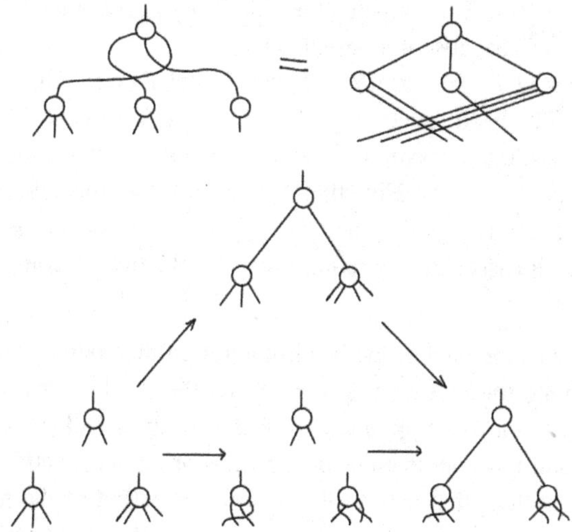

Figure 3.26
The two equivariance conditions for symmetric operads.

We argue here that this same formalism in fact accounts for Theta Theory, with the role of External Merge in argument structure, and the dichotomy between the roles of External Merge and Internal Merge. To see this, we first modify the Merge operad $\mathcal{M}$ to incorporate heads and complements.

Lemma 3.8.1. *Let $\mathcal{M}_h(n)$ denote the set of pairs (T, h_T) consisting of an abstract binary rooted tree $T \in \mathfrak{T}_n$ with n leaves (with no labeling at the leaves) and a head function $h_T : V^o(T) \to L(T)$, as in Definition 1.13.6. The collection $\mathcal{M}_h = \{\mathcal{M}_h(n)\}$ is an operad.*

Proof. To construct the operad compositiom

$$\gamma_{\mathcal{M}_h} : \mathcal{M}_h(n) \times \mathcal{M}_h(k_1) \times \cdots \times \mathcal{M}_h(k_n) \to \mathcal{M}_h(k_1 + \cdots + k_n) \ ,$$

we define the resulting

$$\gamma_{\mathcal{M}_h}((T, h_T), (T_1, h_{T_1}), \ldots, (T_n, h_{T_n})) = (T', h_{T'}) \ ,$$

where $T' = \gamma_{\mathcal{M}}(T, T_1, \ldots, T_n)$ is the operad composition in $\mathcal{M}$ that grafts the roots of the trees T_i to the leaves of T. We then need to show that the data $h_T, h_{T_1}, \ldots, h_{T_n}$ combine to define a head function, in the sense of Definition 1.13.6 in chapter 1, on the tree $T' = \gamma_{\mathcal{M}}(T, T_1, \ldots, T_n)$. We define $h_{T'}$ in the following way. For all vertices of T' that are vertices of one of the trees T_i, we set $h_{T'}(v) = h_{T_i}(v)$. This satisfies the condition on subobjects $T_w \subset T_v$, as in Definition 1.13.6, because it is satisfied by h_{T_i}.

For the vertices of T' that are non-leaf vertices of T, we define $h_{T'}(v)$ in the following way. The head function on T determines a leaf $h_T(v) \in L(T)$. Let $T_{i(\ell)}$ denote the tree that is grafted to the leaf $\ell \in L(T)$ in the resulting tree T'. Then we set $h_{T'}(v) := h_{T_{i(h_T(v))}}$. Note that this also satisfies the defining property of head functions since both h_T and $h_{T_{i(h_T(v))}}$ do. This assignment of $h_{T'}$ is compatible with the associativity requirement for the operad composition. □

This incorporates the data of the head function in the operad. We then need to also incorporate the argument structure and the θ-roles. We will assume that the head function is complemented in the sense of Definition 1.14.2, so both the head and the complement of the head are determined. The natural mathematical structure that can do this is a modification of the notion of an operad known as a *colored operad*. This notion of "colored" is nothing but another manifestation of the same principle that generalizes groups to groupoids, one that we have already encountered in our discussion of semiring parsing for Merge derivations, where we needed analogous extensions of the notion of algebras (or bialgebras and Hopf algebras) and semirings to corresponding algebroids and semiringoids. The key property of these generalizations is accounting for the constraint that not all compositions are possible—only those where the type assigned to the output of the first operation is the same as that assigned to the source of the second (the composition of arrows where the first target has to agree with the next source). This is the same idea that leads to generalizing operads to colored operads (although the terminology used is slightly different for historical reasons). We recall the definition of colored operad (see, e.g., [202]).

Definition 3.8.2. A colored operad is a collection $\mathcal{O} = \{\mathcal{O}(c, c_1, \ldots, c_n)\}$ of sets, with $c, c_i \in \Theta$ for $i = 1, \ldots, n$, where Θ is a finite set (the color set), and the c_i are color labels assigned to inputs of the n-ary operations in $\mathcal{O}(c, c_1, \ldots, c_n)$, and c is the color label assigned to the output, with composition laws

$$\begin{aligned}\gamma: \quad & \mathcal{O}(c, c_1, \ldots, c_n) \times \mathcal{O}(c_1, c_{1,1}, \ldots, c_{1,k_1}) \times \cdots \times \mathcal{O}(c_n, c_{n,1}, \ldots, c_{n,k_n}) \\ & \to \mathcal{O}(c, c_{1,1}, \ldots, c_{1,k_1}, \ldots, c_{n,1}, \ldots, c_{n,k_n})\end{aligned} \tag{3.8.3}$$

that satisfy the same associativity and unity conditions as in the noncolored case, with one unit $1_c \in \mathcal{O}(c, c)$ for each color $c \in \Theta$. In the symmetric case, one again has the two equivariance conditions of Figure 3.26, where the action $T \mapsto T \circ \tau$ maps $\mathcal{O}(c, c_1, \ldots, c_n)$ to $\mathcal{O}(c, c_{\tau(1)}, \ldots, c_{\tau(n)})$ for $\tau \in S_n$.

According to (3.8.3), the operad composition is now restricted by the requirement that the grafting of the output to one of the inputs can happen only if they are labeled by the same color.

We assume here that our model of the semantic space $\mathcal{S}$ has images of the map $s : \mathcal{SO}_0 \to \mathcal{S}$ from lexical items, which keep track of labels encoding syntactic features, including word classes. In particular, for an extension $s : \text{Dom}(h) \to \mathcal{S}$ of this map, the image $s(T)$ of a syntactic object $T \in \text{Dom}(h)$ includes information on the label assigned to the phrase by the head $s(h(T))$, according to the labeling algorithm of section 1.15 of chapter 1. The set of colors Θ that we consider here is the set of θ-roles. More precisely, we include in Θ a label "predicate" or "argument," and for the latter, various labels of familiar θ-roles such as "theme," "agent," "experiencer," "locative," "instrument," "possessor," and others.

Definition 3.8.3. Let $\text{Dom}_\Theta(h) \subset \text{Dom}(h)$ denote the set of syntactic objects $T \in \mathcal{SO}$ that are in the domain of the complemented head function h and that admit an assignment of labels in Θ to the edges of T in such a way that they are compatible, on each substructure given by an accessible term T_v of T, with the head and its complement determined by the complemented head function h.

Lemma 3.8.4. *The set* $\text{Dom}_\Theta(h) \subset \mathcal{SO}$ *of Definition 3.8.3 determines a colored operad* $\mathcal{M}_{h,\Theta} = \{\mathcal{M}_{h,\Theta}(\theta, \theta_1, \ldots, \theta_n)\}$.

Proof. Given an element $(T, h_T) \in \mathcal{M}_h$, consider all possible assignments of labels $\theta \in \Theta$ to the edges of T, in such a way that the assigned θ-roles are compatible with the "head and complement" structure determined on all the accessible terms by the head function. This means that the assignment of θ-roles (thematic role assignment) is built inductively and compatibly on each of the substructures T_v assembled to form the syntactic object T by External Merge. In other words, one can start with the leaves of T and all the "cherry trees" T_v (i.e., trees consisting of one root vertex above a pair of leaves), and assign to the two edges of T_v a predicate and a θ-position depending on the head and complement roles assigned to the leaves by the value $h(v)$ of the head function. This inductive building of substructures with the head and its complement follows the phases algorithm described in Definition 1.14.3.

One can then proceed to consider accessible terms T_w of T obtained via External Merge applied to pairs consisting of either two such cherries T_v or one of these and a leaf, and similarly consider all possible assignments of labels in Θ to the two edges below the root w, in such a way that the labeling is compatible with the labels already assigned to the two merged substructures and with the assignment of the head to T_w. Inductively proceeding in the same way, one obtains the collection of all possible compatible assignments of labels in Θ to all the edges of T. This results, then, in an object $(T, h_T, \theta_T) \in \mathcal{M}_{h,\Theta}(\theta, \theta_1, \ldots, \theta_n)$, where θ_T denotes one choice of such compatible labeling. It is then clear that, by construction, the sets $\mathcal{M}_{h,\Theta}(\theta, \theta_1, \ldots, \theta_n)$ form a colored operad with the composition law induced by the operad $\mathcal{M}_h$, where composition is possible only when it grafts outputs to inputs with matching θ-labels. □

Note that, as observed in [34], given that the core computational system of syntax is based on *binary* Merge, one should not expect n-ary operations to play a direct role in the assignment of semantic values, and hence in particular no expectation of seeing n-ary theta structures. Instances where this seems to happen (such as VPs with double object constructions) are explained by additional internal structures, as is familiar in the linguistic literature [120]: an n-ary operation in $\mathcal{M}(n)$ will be a composition of binary ones (via internal syntactic structure). In fact, this is indeed the case in the construction of Lemma 3.8.4, as all the n-ary theta structures, which are by construction the elements $(T, h_T, \theta_T) \in \mathcal{M}_{h,\Theta}(\theta, \theta_1, \ldots, \theta_n)$ are formed via the composition of binary theta structures through repeated application of binary External Merge.

Note that this also forces a dichotomy in the form referred to in section 5.2 of [40], as a segregation of External Merge and Internal Merge in semantics. Indeed, the compatible operad actions of $\mathcal{M}$ (and the modified $\mathcal{M}_h$) on $\mathcal{SO}$ and on semantic space $\mathcal{S}$, as in Definition 3.6.8 only involve External Merge and account for argument structure, while Internal Merge is not directly involved in this part of the algebraic structure and therefore maintains a separate role.

A more detailed discussion of Theta Theory and colored operads, and an interpretation of the procedure described in Lemma 3.8.4 as a "coloring algorithm" that checks correct theta-roles assignment will be presented in forthcoming work.

3.8.2 Deriving Obligatory Control

We now discuss the case of examples of obligatory control. We look in particular at the example considered in section 5.3 of *Elements* [40], in particular the sentence "the man tried to read a book." In such cases merging into the higher subject position by Internal Merge would violate the dichotomy discussed in §3.8.1 as it would move to a theta-position. As in equation (51) of [40], we consider this sentence as being derived from the syntactic object

$$\{\{\text{the}, \text{man}\}, \{\text{tried}, \{\text{to}, \{\{\text{the}, \text{man}\}, \{\text{read}, \{\text{a}, \text{book}\}\}\}\}\}\} =$$

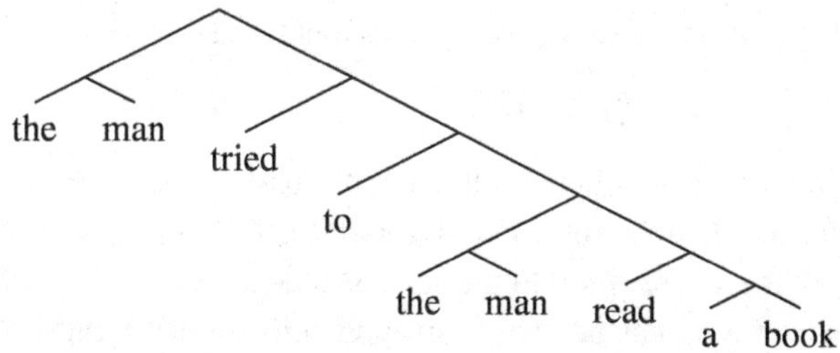

,

which should be regarded as an abstract (i.e,. nonplanar) tree, obtained via repeated applications of External Merge:

$$\mathfrak{M}(\text{the man}, \mathfrak{M}(\text{tried}, \mathfrak{M}(\text{to}, \mathfrak{M}(\text{the man}, \text{read}\ \ \text{a book})))). \quad (3.8.4)$$

In terms of the operad action described in section 3.6.3.1, we can think of this as the element $T \in \mathcal{M}(6)$ of the form

$T =$ 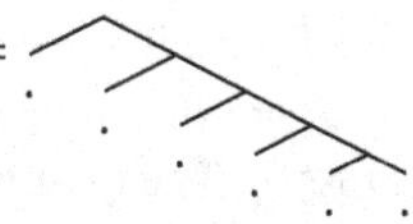 ,

where the dots stand for where inputs from $\mathcal{SO}$ are received, in this case from $s(\mathrm{Dom}(h)) \subset \mathcal{S}$. In the example discussed here, given an input $(T_1, \ldots, T_6)$ in $\mathcal{S}^6$ one obtains the output given by the syntactic object

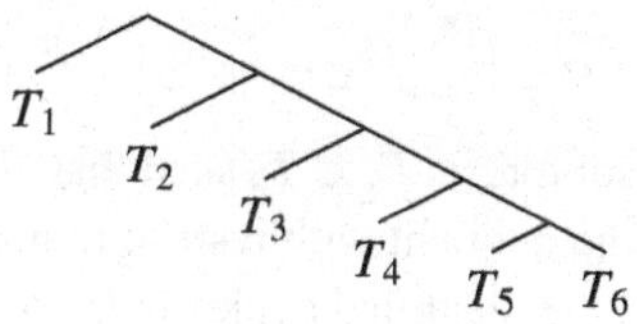

,

where

$$T_1 = T_4 = \text{the man}, \quad T_2 = \text{tried}, \quad T_3 = \text{to}, \quad T_5 = \text{read} \quad T_6 = \text{a book}.$$

Note that this description as a result of the operad action is not unique; but analyzing one such description suffices for our goals here. Since in the input we have $T_1 = T_4$, this means that we are restricting the operation $T \in \mathcal{M}(6)$ to a diagonal $\mathrm{Diag}_{1,4} \subset \mathcal{SO}^6$, with $\mathrm{Diag}_{1,4} = \{(T_1, \ldots, T_6) \in \mathcal{SO}^6 \,|\, T_1 = T_4\}$.

The fact that we are looking only at the restriction of $T \in \mathcal{M}(6)$ to this diagonal $\Delta_{1,4} \subset \mathcal{SO}^6$, has an effect on the External Merge derivation of (3.8.4). Since the action of Merge in our framework is based on the coproduct in the Hopf algebra of workspaces, we need to check the effect on the coproduct of restricting the operations of $\mathcal{M}(n)$ to a diagonal $\mathrm{Diag}_{\mathcal{I}} \subset \mathcal{SO}^n$, where the subset $\mathcal{I} \subset \{1, \dots, n\}$ is the set of coordinates identified:

$$\mathrm{Diag}_{\mathcal{I}} = \{(T_1, \dots, T_n) \in \mathcal{SO}^n \mid T_i = T_j \in \mathcal{SO}, \forall i, j \in \mathcal{I}\}. \quad (3.8.5)$$

For convenience of notation in the following discussion, we write the relations $T_i = T_j$ for all $i, j \in \mathcal{I}$ in the form $T_i = \hat{T}$ for all $i \in \mathcal{I}$, where $\hat{T}$ varies over $\hat{T} \in \mathcal{SO}$; namely, all the T_i are *equal* to the same syntactic object. The selection of a diagonal, as in (3.8.5), can be seen as an application of the FormSet operation discussed in section 1.16, which selects the subset $\{T_i\}_{i\in\mathcal{I}}$, followed by what is referred to as FormCopy in *Elements* [40], which imposes the relations $T_i = T_j$, for all $i, j \in \mathcal{I}$.

Remark 3.8.5. Consider a syntactic object

$$T = \gamma_{SO}(T', T_1, \dots, T_n), \quad \text{with } T' \in \mathcal{M}(n) \text{ and } (T_1, \dots, T_n) \in \Delta_{\mathcal{I}} \subset \mathcal{SO}^n,$$

where, as before, we write $T_i = \hat{T}$ for all $i \in \mathcal{I}$ and varying $\hat{T} \in \mathcal{SO}$. Then the fact that all the T_i with $i \in \mathcal{I}$ are *the same identical copy* of a syntactic object $\hat{T}$ modifies the coproduct $\Delta(T)$. The reason why is that now the terms of the coproduct that extract $\hat{T}$ as accessible terms (or *one* of the accessible terms $\hat{T}_v \subset \hat{T}$) will necessarily have to do that *simultaneously* in *all* the occurrences of $\hat{T}$ as accessible terms of T, since these are all identified as the same object. Thus, we write the coproduct in the form

$$\Delta_{\mathcal{I}}(T) = \sum F_{\underline{v}} \otimes T/F_{\underline{v}} + \sum (F_{\underline{v}} \sqcup \hat{T}_v) \otimes (T/F_{\underline{v}})//\hat{T}_v\,, \quad (3.8.6)$$

where both sums are over subforests $F_{\underline{v}} \subset T$, such that $\hat{T} \cap F_{\underline{v}} = \emptyset$, and where we write $T//\hat{T}_v$ to denote the quotient with respect to all occurrences of $\hat{T}_v$ in T as accessible terms of all the identified copies in $\mathrm{Diag}_{\mathcal{I}}$.

The form (3.8.6) of the coproduct $\Delta_{\mathcal{I}}(T)$ on the image of the diagonal

$$T \in \mathcal{M}(n)(\mathrm{Diag}_{\mathcal{I}}) \subset \mathcal{SO}$$

differs in the following ways from the coproduct $\Delta(T)$ of Lemma 1.2.12:

1. Accessible terms $\hat{T}_v \subseteq \hat{T}$ contained in the identified copies $T_i = \hat{T}$ for $i \in \mathcal{I}$ can only be extracted *simultaneously* in all these identified locations.
2. The extracted term occurs only once in the left channel of the coproduct output, as $\hat{T}_v$ not as $\underbrace{\hat{T}_v \sqcup \cdots \sqcup \hat{T}_v}_{\#\mathcal{I} \text{ times}}$.
3. All the occurrences of $\hat{T}_v$ in the identified copies T_i are simultaneously quotiented out in the right channel of the coproduct output, leading to

coproduct terms of the form

$$\hat{T}_v \otimes T//\hat{T}_v \,.$$

4. In terms of the coproduct that extracts subforests $F_{\underline{v}}$ with more than one component, the extracted term takes the form $F_{\underline{v}} = \hat{T}_v \sqcup \hat{F}_{\underline{v}}$, with $\hat{F}_{\underline{v}} \subset T//\hat{T}_v$ (including the cases where $\hat{T}_v = \hat{T}$ or $\hat{T}_v = 1$).
5. Accessible terms T_v of T that contain some but not all of the identified copies of $\hat{T}$ are not extracted by the coproduct, as all copies of $\hat{T}$ need to be simultaneously extracted. Such terms can be extracted only in successive steps of a derivation from a quotient $T//\hat{T}$.

We need to ensure that this modified form of the coproduct still satisfies the coassociativity property that ensures good behavior over iterations. We assume in Lemma 3.8.6 that the coproduct Δ of (1.2.10) on the Hopf algebra of workspaces is chosen as $\Delta = \Delta^c$, as discussed in section 1.2 of chapter 1, although later in this section, we will consider also the case of $\Delta = \Delta^d$.

Lemma 3.8.6. *The coproduct* $\Delta_{\mathcal{I}}$ *on the image* $\mathcal{M}(n)(\mathrm{Diag}_{\mathcal{I}}) \subset \mathcal{SO}$ *of the diagonal* $\mathrm{Diag}_{\mathcal{I}}$ *of* (3.8.5) *is coassociative,*

$$(\Delta_{\mathcal{I}} \otimes \mathrm{id}) \circ \Delta_{\mathcal{I}}(T) = (\mathrm{id} \otimes \Delta_{\mathcal{I}}) \circ \Delta_{\mathcal{I}}(T)\,,$$

for all $T \in \mathcal{M}(n)(\mathrm{Diag}_{\mathcal{I}})$.

Proof. We proceed as in Lemmas 1.2.10 and 1.2.12 to address the coassociativity question. When we compute $(\Delta_{\mathcal{I}} \otimes \mathrm{id}) \circ \Delta_{\mathcal{I}}(T)$ we obtain a sum of terms of the form

$$F_{\underline{v}} \otimes F_{\underline{w}}/F_{\underline{v}} \otimes T/F_{\underline{w}}.$$

The only case where these terms differ from the coproduct (1.2.10) is when either $F_{\underline{w}}$ or $F_{\underline{v}}$ contains some terms $\hat{T}_v \subseteq \hat{T}$. If one of the components of $F_{\underline{w}}$ is a term $\hat{T}_v$, then the forest $\hat{F}_{\underline{w}}$ of the remaining components, with $F_{\underline{w}} = \hat{T}_v \sqcup \hat{F}_{\underline{w}}$, must be a subforest of $T//\hat{T}_v$. Thus, the corresponding quotient term takes the form $(T//\hat{T}_v)/\hat{F}_{\underline{w}} = (T/\hat{F}_{\underline{w}})//\hat{T}_v$. The extracted term $F_{\underline{v}} \subset F_{\underline{w}}$ similarly may or may not contain a component equal to an accessible term $\hat{T}_u \subset \hat{T}_v$, and produces a quotient

$$F_{\underline{w}}/F_{\underline{v}} = \hat{T}_v/\hat{T}_u \sqcup \hat{F}_{\underline{w}}/\hat{F}_{\underline{u}} = \hat{T}_v/\hat{T}_u \sqcup \hat{F}_{\underline{u}}\,,$$

with $\underline{w} = (\underline{v}, \underline{u})$, so we can write the resulting coproduct term in the form

$$\hat{T}_u \sqcup \hat{F}_{\underline{v}} \otimes \hat{T}_v/\hat{T}_u \sqcup \hat{F}_{\underline{u}} \otimes (T//\hat{T}_v)/\hat{F}_{\underline{v},\underline{u}}\,.$$

We have included here the case where $\hat{T}_v = \hat{T}$; in this case, the quotient has all the copies of $\hat{T}$ completely removed. We have also the case $\hat{T}_v = 1$, where no accessible term of $\hat{T}$ is extracted from $F_{\underline{w}}$, which still contains all copies of $\hat{T}$. On the other hand, when computing the other side of the coassociativity constraint $(\mathrm{id} \otimes \Delta_{\mathcal{I}}) \circ \Delta_{\mathcal{I}}(T)$, we obtain a sum of terms of the form $F_{\underline{v}} \otimes (T/F_{\underline{v}})_{\underline{u}} \otimes (T/F_{\underline{v}})/(T/F_{\underline{v}})_{\underline{u}}$. The terms where $\hat{T}$ or accessible terms $\hat{T}_v \subset \hat{T}$ are not extracted in both $F_{\underline{v}}$ or $(T/F_{\underline{v}})_{\underline{u}}$ continue to have the same form as in the original coproduct (1.2.8), and match the corresponding terms in $(\Delta_{\mathcal{I}} \otimes \mathrm{id}) \circ \Delta_{\mathcal{I}}(T)$ by the same argument. So we only need to discuss the terms where $\hat{T}$ or accessible terms $\hat{T}_v \subset \hat{T}$ are present in either $F_{\underline{v}}$ or in $(T/F_{\underline{v}})_{\underline{u}}$. In fact, it suffices to consider the case of extraction of a single accessible term: where either $F_{\underline{v}} = \hat{T}_v$ or $(T/F_{\underline{v}})_{\underline{u}} = \hat{T}_v$, for some $\hat{T}_v \subseteq \hat{T}$. If $F_{\underline{v}} = \hat{T}_v$, then $T/F_{\underline{v}} = T//\hat{T}_v$ is the quotient of *all* occurrences of $\hat{T}_v$ in the identical copies of $\hat{T}$ inside T. Then the further extracted term $(T//\hat{T}_v)_{\underline{u}} = F_{\underline{u}}$ will not contain any $\hat{T}_v$ or any substructure of it, and the term $\hat{T}_v \otimes F_{\underline{u}} \otimes (T/F_{\underline{u}})//\hat{T}_v$ will match a corresponding term from $(\Delta_{\mathcal{I}} \otimes \mathrm{id}) \circ \Delta_{\mathcal{I}}(T)$. The case where $(T/F_{\underline{v}})_{\underline{u}} = \hat{T}_v$ is analogous. □

Remark 3.8.7. There is an equivalent way of rephrasing the associativity argument of Lemma 3.8.6, by observing that there is a bijective correspondence between syntactic objects with restrictions to diagonals and certain graphs (that are not trees) obtained by gluing together as a single copy all the substructures that are identified in the diagonal. Under this bijective correspondence, the coproduct $\Delta_{\mathcal{I}}$ that extracts accessible terms (with a single simultaneous extraction for all the terms that are identified) becomes a form of the coproduct in the Connes–Kreimer Hopf algebra of graphs. In this Hopf algebra, the terms in the coproduct have an extracted subgraph in the left-channel and a corresponding quotient graph that contracts the subgraph to a single vertex in the right-channel. Here the subgraphs used for extraction are those that are obtained from accessible terms in the tree, after the identifications performed by the restriction to diagonals. The coassociativity then follows as in the case of the Connes–Kreimer coproduct; see [45] and [46]. We do not develop this argument in full here.

We can now analyze more in detail the Merge derivation (3.8.4) of the example discussed in equation (51) of *Elements* [40]. As observed there, structures like

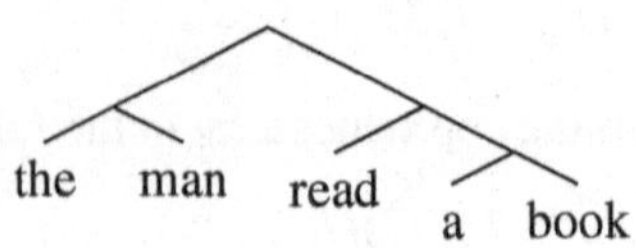

and

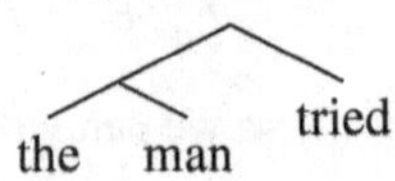

are well-formed structures produced by External Merge that performs the assignment of theta roles as discussed in section 3.8.1. The structure

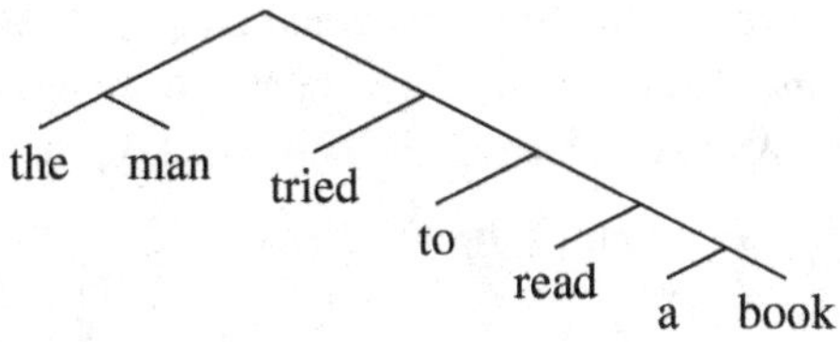

may at first be regarded as an Internal Merge $\mathfrak{M}(T_v, T/^d T_v)$ at the Externalization interface, with corresponding $\mathfrak{M}(T_v, T/^c T_v)$ of the form

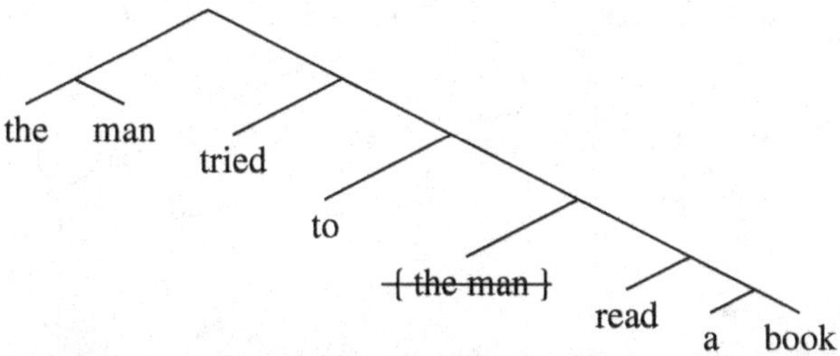

However, this reading is not compatible with the Theta Theory constraints because of the dichotomy discussed in section 3.8.1, according to which Internal Merge does not assign theta roles, while $\mathfrak{M}(T_v, T/^c T_v)$ would merge $T_v = \{\text{the}, \text{man}\}$ into a position that assigns a theta role (a theta position of the predicate *try*). Thus, this explanation is not compatible with Theta Theory.

We focus on the last External Merge operation in (3.8.4), as that is where the fact that we are restricting ourselves to the diagonal $\text{Diag}_{1,4} \subset \mathcal{SO}^6$ is relevant. For this External Merge operation, we start with a workspace given by the forest

$\hat{T} \sqcup T =$

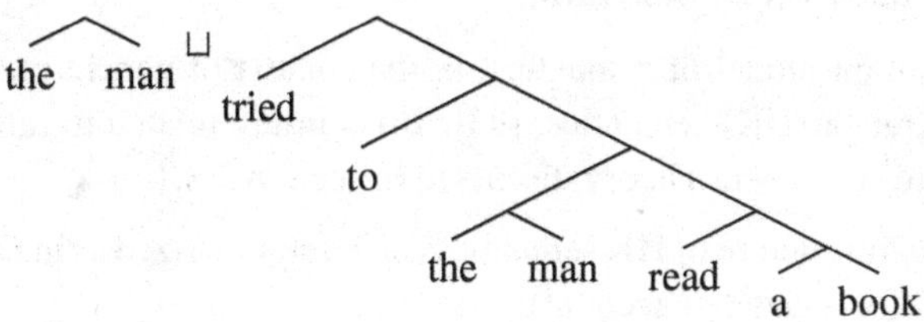

,

where

$T =$

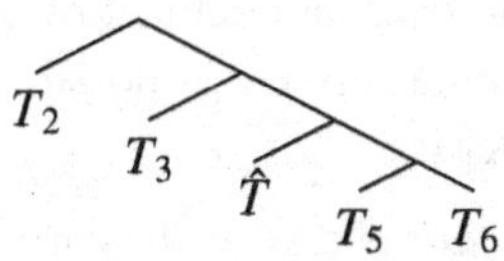

,

and obtain the structure of (3.8.4) as $\tilde{T} = \mathfrak{M}(\hat{T}, T)$. All the applications of External Merge in (3.8.4) are compatible with the assignments of theta roles. We consider then the restriction to the diagonal $\mathrm{Diag}_{1,4} \subset \mathcal{SO}^6$, which corresponds to applying the FormCopy relation to the two occurrences of $\hat{T}$, in the terminology of [40]. We have a term in the coproduct Δ^d_I that is given by

$$\hat{T} \otimes \tilde{T}//^d \hat{T}, \quad \text{with} \quad \tilde{T}//\hat{T} =$$

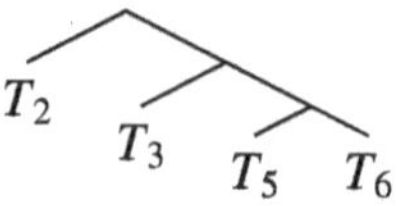

With $\mathfrak{M}^I_{S,S'} := \sqcup \circ (\mathcal{B} \otimes \mathrm{id}) \circ \delta_{S,S'} \circ \Delta^d_I$, we obtain

$$\mathfrak{M}^I_{\hat{T}, \tilde{T}//^d \hat{T}} \mathfrak{M}^I_{\hat{T},1}(\tilde{T}) =$$

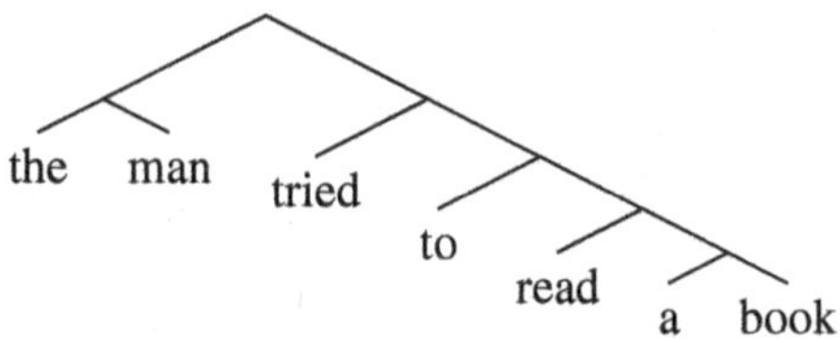

Remark 3.8.8. Here, we see another important instance of the distinction between copies and repetitions. When we consider elements $(T_1, \ldots, T_n) \in \mathrm{Diag}_I \subset \mathcal{SO}^n$ and identify all the terms $T_i = \hat{T}$ for $i \in I$ as copies, we apply the coproduct Δ_I in the Merge derivation, which keeps account of the fact that these occurrences of the term $\hat{T}$ are all identified as being *the same*. This is in contrast with the case where the elements $(T_1, \ldots, T_n)$ contain repetitions (namely, different components T_i and T_j that are isomorphic in $\mathfrak{T}_{\mathcal{SO}_0}$). In the case of repetitions, one would just apply the coproduct Δ of (1.2.8), not the coproduct Δ_I, as the different T_i and T_j, even if isomorphic, are *not* coincident and can be independently extracted by the coproduct.

3.9 Heim–Kratzer Semantics

An important model of semantics, in the context of generative linguistics, is Heim–Kratzer (HK) semantics, [93]. It is closely related to (although not reducible to) the Theta Theory discussed in section 3.8.1.

The main structure of HK semantics can be summarized as in Definition 3.9.1 (see [93] for a detailed account).

Definition 3.9.1. Semantic *types* are defined by an inductive construction of the types τ and corresponding sets D_τ of possible denotations of type τ, according to the following steps:

1. e is the type of *individuals* and D_e is the corresponding set of individuals.

2. t is the type of *truth values* and $D_t = \{0, 1\}$ is the set of possible truth values.
3. $\langle e, t\rangle$ is the type of functions $D_{\langle e,t\rangle} = \{f : D_e \to D_t\}$.
4. $\langle e, \langle e, t\rangle\rangle$ is the type of functions

$$D_{\langle e,\langle e,t\rangle\rangle} = \{f : D_e \to D_{\langle e,t\rangle}\}.$$

5. If σ and τ are types, then $\langle\sigma, \tau\rangle$ is the type of functions

$$D_{\langle\sigma,\tau\rangle} = \{f : D_\sigma \to D_\tau\}.$$

The first two types, e and t, are referred to as *saturated denotations* and the other types are *unsaturated denotations* (namely, functions that have inputs and outputs in specified types).

One assumes that there is a computational system of syntax that produces, as input for semantic interpretation, some tree structures. In the setting that we consider here, these will be the syntactic objects produced by the action of the free symmetric Merge on workspaces.

As in our setting, one can assume that the leaves of the tree are labeled by lexical items in $\mathcal{SO}_0$. As we have been doing in the previous sections with other models of semantics, one assigns a map

$$s : \mathcal{SO}_0 \to \mathcal{S}$$

to a suitable semantic space, where the lexical items are interpreted. We think of this as the map $\alpha \mapsto s(\alpha) = [[\alpha]]$ that assigns to α its denomination. Unlike our previous examples, though, the semantic space $\mathcal{S}$ considered here is some space of *functions*. This includes the case of constant (saturated) functions, which are denominations in D_e or D_τ. The whole space $\mathcal{S}$ is then given by

$$\mathcal{S} = \bigcup_\tau D_\tau, \tag{3.9.1}$$

which is the union of all denominations for all the inductively constructed types. Note also that, unlike the cases analyzed in the previous sections, this space $\mathcal{S}$ is only endowed with the discrete topology, so there is no viable notion of proximity in this case. It mostly resembles the case discussed in section 3.2.5 with the Boolean semiring.

One then proceeds to partially extend the map

$$s : \mathcal{SO}_0 \to \mathcal{S} \tag{3.9.2}$$

to a partially defined map

$$s : \mathrm{Dom}(s) \subset \mathcal{SO} \to \mathcal{S}. \tag{3.9.3}$$

The rules for extending the map (3.9.2) are as follows:

1. Consider $T = \mathfrak{M}(T_1, T_2)$. If one of the two values $s(T_i) = [[T_i]]$ in $\mathcal{S}$ (which we can take to be $s(T_1)$), is a function, $s(T_1) \in D_{\langle\sigma,\tau\rangle}$, and the other one is in the domain, $s(T_2) \in D_\sigma$, then we assign to T the value

$$s(T) = [[T]] = s(T_1)(s(T_2)) = [[T_1]]([[T_2]]) \in D_\tau \,, \qquad (3.9.4)$$

given by the evaluation of the function on the given element in its domain.
2. If neither $s(T_1)$ nor $s(T_2)$ is a function that takes the type of the other as input, then $T \notin \mathrm{Dom}(s)$ and the expression T is considered to be noninterpretable.

In case (1), we say that the "function and input" relation is satisfied at the root vertex of T, meaning that the assignments $s(T_1)$ and $s(T_2)$ at T_1 and T_2, with $T = \mathfrak{M}(T_1, T_2)$, are a function and an argument (input) of that function, respectively.

Note that, in general, if a syntactic object T is noninterpretable in HK semantics, this can be due to a mismatch of the "function and input" relation at any of the non-leaf vertices of T, while many other accessible terms $T_v \subset T$ would remain interpretable. Thus, it is useful to have a way to discern and keep track of the parts of the structure that are interpretable and the exact locations where interpretability fails. This is exactly the type of information that is built into the recursive form of the Bogolyubov preparation, built using the Hopf algebra coproduct, as introduced in section 3.1.

3.9.1 Boolean Characters and Interpretability

To see how to incorporate HK semantics in our model, we start by illustrating how Boolean characters and their Bogolyubov preparation, as discussed in section 3.2.5, are related to interpretability.

In [93], an additional rule is given for the case of nonbranching nodes, where the nonbranching node inherits the denomination assigned to the vertex immediately below it. We do not need this rule as long as we only consider binary trees. If we use the coproduct Δ^ρ in (1.2.8) in chapter 1, we also need to consider trees $\rho_C(T)$ that can have some nonbranching nodes: in this case, the rule given in [93] ensures that the result is independent of using the coproduct Δ^ρ or Δ^d. In the case of the coproduct Δ^c, one needs a rule for how to assign a value to the leaves in $T/^c F_{\underline{v}}$ labeled by $\cancel{T_{v_i}}$. Since these cancelled deeper copies are not externalized (as we discussed, Externalization uses Δ^d) but are interpreted as the syntax–semantics interface, we can take the following additional rule:

- For a tree of the form $T/^c T_v$, the value at the leaf $s(\not{T}_{\overline{v_i}})$ is taken to be

$$s(\not{T}_{\overline{v_i}}) = s(T_v)\,,$$

which is the same as the value that was assigned to the accessible term T_v of T.

The case of $T/^c F_{\underline{v}}$ is analogous.

Now observe that this can give rise to different (and incompatible) assignments of values, with

$$s(T/^d T_v) = s(T/^\rho T_v) \neq s(T/^c T_v)\,.$$

This is to be expected: as mentioned previously, in the action of Internal Merge, the cancellation of the deeper copy is *not externalized* at the interface with the Sensory–Motor system but is *interpreted* at the interface with semantics (Conceptual–Intensional system), so it is the quotient $T/^c T_v$ that carries the correct interpretation, $s(T/^c T_v) \in \mathcal{S}$. On the other hand, the use of the coproducts Δ^ρ and Δ^d has the advantage of making it possible to check the interpretability of a quotient $T/^\rho T_v$ ot $T/^d T_v$ in which the substructure T_v has been removed completely (rather than being kept as a label $\not{T}_{\overline{v}}$). This makes it possible to separately check the interpretability of the substructure T_v and of a remaining complementary structure in which T_v is entirely removed. This is the same strategy already used in the Boolean parsing in section 3.2.5.

To see how the HK semantics model fits with our Birkhoff factorization procedure, consider a character ϕ, with values in the Boolean semiring $\mathfrak{B} = (\{0,1\}, \max, \cdot)$, as in section 3.2.5, which assigns

$$\phi(T) = \begin{cases} 1 & T \text{ is interpretable in HK semantics} \\ 0 & \text{otherwise.} \end{cases}$$

Again, we consider on $\mathfrak{B}$ the Rota–Baxter operator given by the identity, so, as in section 3.2.5, the Bogolyubov preparation of this character is given by

$$\tilde{\phi}(T) = \max\{\phi(T), \phi(F_{\underline{v}_1})\phi(T/^d F_{\underline{v}_1}), \ldots, \phi(F_{\underline{v}_N})\phi(F_{\underline{v}_{N-1}}/^d F_{\underline{v}_N}) \cdots \phi(T/^d F_{\underline{v}_1})\}\,, \tag{3.9.5}$$

with the maximum over all chains of nested forests of accessible terms. As discussed in section 3.2.5, when $\phi(T) = 0$, meaning that T is noninterpretable, the maximizers of $\tilde{\phi}(T)$ are those chains of accessible terms where all the substructures and their respective quotients are themselves interpretable. Thus, $\tilde{\phi}(T)$ explicitly encodes in its recursive structure the exact locations of the points in the structure T where interpretability in HK semantics fails.

3.9.2 Topologizing Semantic Types

In addition to this kind of construction with Boolean checking of interpretability, we can consider ways of combining the setting of HK semantics with the view described in section 3.1.5 of semantic spaces embodying notions of proximity relations. In fact, we have so far assumed that in (3.9.1), the semantic space $\mathcal{S} = \cup_\tau D_\tau$ is just a set: topological proximity relations are not included in HK semantics in the formulation of [93]. However, it is also possible (and in some ways preferable) to assume that all the sets D_τ of the types τ are also topological spaces, possibly endowed with topologies that generally can differ from the discrete one. Indeed, considering $\mathcal{S}$ merely as a set does not provide a way of keeping track of when two functions are closely related (e.g., one being a small modification of the other), as these are topological, not set-theoretic notions.

For example, a topology on the set D_e of individuals specifies proximity and relatedness between certain sets of individuals, in a way similar to the semantic spaces described in section 3.1.5. We may assume that the set D_e is also a geodesically convex Riemannian manifold, a vector space, or another form of topological space with a "continuous" topology. The set $D_t = \{0, 1\}$ of truth values should necessarily be considered discrete. Then one needs to topologize the sets D_τ of functions. For that purpose, it is necessary to somewhat restrict the class of functions that one wants to consider. We will argue why such a restriction is reasonable in view of their semantic role that these functions represent. We start by assuming that D_e has a topology induced by a metric in which it is both complete and compact. (The completeness and compactness assumptions can be weakened, but for simplicity, we will restrict to discussing this case.)

We first recall some facts from point-set topology:

- The set $C(X, Y)$ of continuous functions between topological spaces X and Y is usually topologized with the *compact-open topology*, which is defined as the topology generated by the sets $\mathcal{U}_{K,U}$ for $K \subset X$ compact and $U \subset Y$ open:

$$\mathcal{U}_{K,U} = \{f \in C(X, Y) \,|\, f(K) \subset U\}.$$

In cases where X is compact and Y is metric with distance function d_Y, the *compact-open topology* is induced by a metric of the form

$$d_{C(X,Y)}(f, g) = \sup_{x \in X}\{d_Y(f(x), g(x))\}.$$

- If (X, d_X) is a metric space, the set $\mathcal{K}(X)$ of all compact subsets of X is also a metric space, with the Hausdorff metric d_H defined by

$$d_H(A, B) = \max\{\sup_a d(a, B), \sup_b d(A, b)\}\,. \tag{3.9.6}$$

  The metric space $(\mathcal{K}(X), d_H)$ is complete and compact if (X, d_X) is.
- In the more general case where X is just a topological space, the set $\mathcal{K}(X)$ can be topologized by the Vietoris topology, which is generated by the so-called *hit-and-miss sets* (i.e., sets that meet a given open set and sets that miss its complement): for U varying over the open sets of X,

$$\begin{aligned} \mathcal{V}_{+,U} &= \{K \in \mathcal{K}(X) \,|\, K \cap U \neq \emptyset\} \\ \mathcal{V}_{-,U} &= \{K \in \mathcal{K}(X) \,|\, K \subset U\}\,. \end{aligned}$$

  When the Hausdorff metric exists, it induces the Vietoris topology. The set of finite subsets of X is dense in $\mathcal{K}(X)$ with this topology.

See section 2.4 of [191] for a more detailed discussion of these topologies.

The reason why we reviewed these point-set topology properties here is in order to topologize the semantic space $\mathcal{S}$ of (3.9.1) of HK semantics in such a way that its component sets D_τ are endowed with inductively constructed topologies that maintain some common properties. The main issues involved in doing so come from two sources of difficulties in identifying the best topological properties that can be induced on the HK semantic space $\mathcal{S}$:

- Topologizing types $\langle \tau, t\rangle$ requires some care, as the set of continuous functions $f : D_\tau \to D_t = \{0, 1\}$ is too small, only detecting connected components of D_τ.
- Even if D_e is assumed to have very good topological properties, including metrizability and compactness, these do not propagate to the inductive construction of types: spaces of continuous functions are usually noncompact and their metrizability depends on compactness of the source space, so that property is also lost in the iteration process.

Proposition 3.9.2. *Let D_e be endowed with a metric topology with respect to which it is compact and complete. Consider $D_t = \{0, 1\}$ with the discrete topology. Then the subset*

$$D^c_{\langle e,t\rangle} := \{f : D_e \to D_t \,|\, f^{-1}(1) \text{ closed in } D_e\} \tag{3.9.7}$$

also has a metric topology in which it is compact and complete. Moreover, for all the types τ inductively constructed as in Definition 3.9.1, there are inductively constructed subsets $D^c_\tau \subset D_\tau$ that have compatible topologies. Thus, *the resulting space* $\mathcal{S} = \cup_\tau D_\tau$ *of* (3.9.1) *determines an associated topological space*

$$\mathcal{S}^c = \bigcup_\tau D^c_\tau . \tag{3.9.8}$$

This space is Hausdorff, although in general neither compact nor metrizable. We refer to this as the "space of topological semantic types".

Proof. We start by assuming that D_e is already a topological space that is metric, complete, and compact. Thus, we just take $D^c_e = D_e$. Similarly, $D_t = D^c_t = \{0, 1\}$ with the discrete topology. Then the set $D_{\langle e,t\rangle}$ of all functions $f : D_e \to D_t$ can be identified with the set of characteristic functions χ_A of subsets $A \subset D_e$, $\chi_A(x) = 1$ for $x \in A$ and $\chi_A(x) = 0$ for $x \notin A$; hence the set $D_{\langle e,t\rangle}$ can be identified with the *power set* 2^{D_e} of all subsets of D_e.

(Notational warning: some point-set topology literature uses the notation 2^X for the set of *closed* subsets of X rather than for the set of all subsets. We follow the more standard notation, with 2^X being the power set of X.)

We want to identify a subset of $D_{\langle e,t\rangle}$ that can be topologized nicely in terms of the topology of D_e. A common choice for this purpose consists of taking the subset $D^c_{\langle e,t\rangle} \subset D_{\langle e,t\rangle}$ of characteristic functions of closed (and hence compact, since D_e is compact) subsets of D_e, as in (3.9.7). This is then identified (by identifying the characteristic function of a set with the corresponding set) with the set $\mathcal{K}(D_e)$ of compact subsets of D_e, topologized with the Hausdorff metric d_H induced by the metric d on D_e, as we recalled previously.

This metric $(\mathcal{K}(D_e), d_H)$ is both complete and compact since D_e is. Thus, we have obtained a subset $D^c_{\langle e,t\rangle} \subset D_{\langle e,t\rangle}$, with topological properties as good as those of the space D_e.

Consider, then, the case of the set

$$D_{\langle e,\langle e,t\rangle\rangle} = \{f : D_e \to D_{\langle e,t\rangle}\} .$$

We consider the space

$$D^c_{\langle e,\langle e,t\rangle\rangle} = C(D_e, D^c_{\langle e,t\rangle}) = \{f : D_e \to D^c_{\langle e,t\rangle} \mid f \text{ continuous}\}$$

of continuous functions from the topological space D_e, with values in the topological space $D^c_{\langle e,t\rangle}$, endowed with the compact-open topology. Since D_e is compact and $D^c_{\langle e,t\rangle}$ is metric, $D^c_{\langle e,t\rangle}$ is metrizable.

We can similarly construct types $\langle e^n, t\rangle$, defined as

$$D^c_{\langle e^n,t\rangle} := D^c_{\underbrace{\langle e, \langle e, \cdots, \langle e,t\rangle \cdots \rangle}_{n\text{–times}}} = C(D_e, D^c_{\underbrace{\langle e, \langle e, \cdots, \langle e,t\rangle \cdots \rangle}_{(n-1)\text{–times}}}) ,$$

endowed with the compact-open topology. These will also be metrizable.

Similarly, since $D^c_{\langle e,t\rangle}$ is also compact, we obtain types $\langle\langle e,t\rangle^n, t\rangle$, with

$$D^c_{\langle\langle e,t\rangle^n,t\rangle} := C(D^c_{\langle e,t\rangle}, D^c_{\langle\langle e,t\rangle^{n-1},t\rangle}) = \underbrace{C(D^c_{\langle e,t\rangle}, C(D^c_{\langle e,t\rangle}, \cdots C(D^c_{\langle e,t\rangle}, D_t))\cdots)}_{n\text{-times}} .$$

These are also metrizable spaces. We also have the metrizable spaces

$$D^c_{\langle\langle e^n,t\rangle,t\rangle} = \mathcal{K}(D^c_{\langle e^n,t\rangle}) \quad \text{and} \quad D^c_{\langle\langle\langle e,t\rangle^n,t\rangle,t\rangle} = \mathcal{K}(D^c_{\langle\langle e,t\rangle^n,t\rangle})$$

with the Hausdorff distance. Similarly, for every type σ in the families constructed here, we obtain a type $\langle\sigma, t\rangle$ as before, with $D^c_{\langle\sigma,t\rangle} = \mathcal{K}(D^c_\sigma)$ endowed with the Hausdorff distance.

All these types maintain the metrizability property of the topology (although they do not maintain compactness). There are then other inductive types that do not necessarily preserve metrizability. Consider any two arbitrary choices τ_1, τ_2 of types constructed according to the procedure just described, with $\tau_2 \neq t$, and form the type $\tau = \langle\tau_1, \tau_2\rangle$, with

$$D^c_{\langle\tau_1,\tau_2\rangle} = C(D^c_{\tau_1}, D^c_{\tau_2}) = \{f : D^c_{\tau_1} \to D^c_{\tau_2} \mid f \text{ continuous}\} \subset D_{\langle\tau_1,\tau_2\rangle}, \quad (3.9.9)$$

endowed with the compact-open topology. Unlike the previous classes $\langle e^n, t\rangle$, $\langle\langle e,t\rangle^n, t\rangle$, now $D^c_{\tau_1}$ is in generally noncompact, so even though $D^c_{\tau_2}$ is metrizable, the metrizability property need not extend to $D^c_{\langle\tau_1,\tau_2\rangle}$.

Finally, we also have the case of types $\langle\tau, t\rangle$, with τ any of the types constructed here. We again take $D^c_{\langle\tau,t\rangle} = \mathcal{K}(D^c_\tau)$ in this case. However, now D^c_τ is not necessarily metrizable, so $\mathcal{K}(D^c_\tau)$ is not necessarily metrizable and we consider it endowed with the Vietoris topology. Finally, the topology on the union $\mathcal{S}^c$ can be taken to be disjoint union topology. □

Our choice to topologize the types $\langle\tau, t\rangle$ by $D^c_{\langle\tau,t\rangle} = \mathcal{K}(D^c_\tau)$ rather than by $D^c_{\langle\tau,t\rangle} = C(D^c_\tau, D_t)$ ensures a sufficiently large class of truth-values assignments. For example, a continuous function $f : D_e \to D_t = \{0, 1\}$ would have $f^{-1}(1)$ both open and closed, but if D_e is connected (which is a reasonable assumption), then it can only be $f^{-1}(1) = D_e$ or $f^{-1}(1) = \emptyset$, so the only continuous functions would be constant. On the other hand, characteristic functions of compact sets in D_e are much more rich and interesting. The fact that f is supported on a compact set implies that only knowledge of f in a bounded region of D_e is required for truth-value checking, which is a reasonable assumption. For $\tau \neq t$, we make the natural choice of topologizing the types $\langle\sigma, \tau\rangle$ as spaces of continuous functions $D^c_{\langle\sigma,\tau\rangle} = C(D^c_\sigma, D^c_\tau)$. Thus, proximity relations are compatibly preserved throughout the evaluation process, which matches functions with their arguments in the HK interpretation procedure.

3.9.3 Boolean Probes and Topological Types

With the construction of topological HK semantic types as in Proposition 3.9.2, we can revisit and extend the example of Boolean characters discussed in section 3.9.1. We present here an analog, in the setting of HK semantics, of the semantic probes Υ and the associated characters $\phi_{\Upsilon,s}$ that we discussed, for simpler models of semantic spaces, in section 3.2.1 and the following sections.

Definition 3.9.3. A probe Υ in topological HK semantics is a collection $\Upsilon = \{\Upsilon_\tau\}_\tau$ of compact subsets $\Upsilon_\tau \subset D^c_\tau$. We assign to a probe Υ a Boolean character $\phi_{\Upsilon,s}$ by setting

$$\phi_{\Upsilon,s}(T) = \begin{cases} \chi_{\Upsilon_\tau}(s(T)) & \text{if } s(T) \in D^c_\tau \\ 0 & \text{otherwise,} \end{cases} \tag{3.9.10}$$

where

- $\chi_{\Upsilon_\tau} \in D^c_{\langle \tau,t\rangle}$ is the characteristic function of the compact set Υ_τ
- $s(T) \in \mathcal{S}$ is the HK interpretation of T, when T is interpretable
- $\phi_{\Upsilon,s}(F) = \prod_a \phi_{\Upsilon,s}(T_a)$, for $F = \sqcup_a T_a$.

The Bogolyubov preparation $\tilde{\phi}_{\Upsilon,s}$ of the character $\phi_{\Upsilon,s}$ then identifies, as in (3.9.5), nested chains of substructures

$$F_{\underline{v}_N} \subset F_{\underline{v}_{N-1}} \subset \cdots F_{\underline{v_1}} \subset T\,,$$

where the probe evaluates *True* on all the $F_{\underline{v}_N}, F_{\underline{v}_{N-1}}/^d F_{\underline{v}_N}, \ldots, T/^d F_{\underline{v_1}}$.

Our choice of semantic probes is based on assignments of truth values given by characteristic functions of compact sets, according to our topological construction of Proposition 3.9.2. To see what kind of information is captured by this kind of probe, first note that we are using a family of compact sets $\Upsilon_\tau \subset D^c_\tau$. By construction, the spaces D^c_τ are typically spaces of continuous functions with the compact-open topology, and compact subsets of these are characterized by the Arzelà–Ascoli theorem, which we will not discuss here. Indeed, since the set of finite subsets of D^c_τ is dense in $\mathcal{K}(D^c_\tau)$, we can just assume for simplicity that the chosen $\Upsilon_\tau \subset D^c_\tau$ are finite sets. In that case, we have fixed a choice of a finite list of functions Υ_τ in each D^c_τ and we are probing whether the interpretation of T and of its substructures T_v and quotient structures $T/^d T_v$ is realized by functions in this given list. This can be seen as a way to reduce the complexity of the enormous spaces of functions D_τ of the original formulation with more manageable finite approximations. The density and convergence in $\mathcal{K}(D^c_\tau)$ then make it possible to check the convergence of these approximations.

3.9.4 Fuzzy Semantic Types

There is a further extension of the setting of HK semantics that we can obtain within our model, which we will call *fuzzy HK semantic types*.

Definition 3.9.4. A *fuzzy set* (X, f) is a set X together with a function $f : X \to [0, 1]$ that represent a "degree of belonging" of a point $x \in X$ to the set X. We think of a fuzzy structure $f : X \to [0, 1]$ as a "degree of confidence" replacing a $\{0, 1\}$-valued truth-value assignment. We will refer here to the interval $[0, 1]$ as "fuzzy truth-values".

Here, instead of truth-values $D_t = \{0, 1\}$ and truth-values assignments $f : D_\tau \to D_t$, we consider assignments of fuzzy truth values, where the type t with $D_t = \{0, 1\}$ is replaced by a new type $\mathfrak{f}$ with $D_\mathfrak{f} = [0, 1]$, so that $\langle\tau, \mathfrak{f}\rangle$ is the type with $D_{\langle\tau,\mathfrak{f}\rangle} = \{f : D_\tau \to D_\mathfrak{f}\}$, where a denomination $f \in D_{\langle\tau,\mathfrak{f}\rangle}$ is a fuzzy set structure $f : D_\tau \to [0, 1]$ on the set D_τ. We want to combine both the fuzziness of the truth values with the topological structures described in section 3.9.3. The fuzziness simplifies the construction of topological HK types, as we no longer need to treat in two different ways the cases $\langle\sigma, t\rangle$ and $\langle\sigma, \tau\rangle$ with $\tau \neq t$, as there are many nontrivial continuous fuzzy truth-value assignments even when continuous functions realizing genuine truth-value assignments are very scarce and only detect connected components.

Definition 3.9.5. Fuzzy topological semantic types are obtained as an inductive construction by taking:

1. e is the type of individuals, where $D_e^{\mathfrak{f},c} = D_e$ is the set of denominations of individuals, assumed to be a compact and complete metric space.
2. $\mathfrak{f}$ is the type of fuzzy truth values, with $D_\mathfrak{f}^{c,\mathfrak{f}} = [0, 1]$ also a compact and complete metric space with the Euclidean metric.
3. $\langle e, \mathfrak{f}\rangle$ is the type with $D_{\langle e,\mathfrak{f}\rangle}^{c,\mathfrak{f}} = C(D_e, [0, 1])$, with the compact-open topology.
4. For any pair of already constructed types σ, τ, the type $\langle\sigma, \tau\rangle$ has
$$D_{\langle\sigma,\tau\rangle}^{c,\mathfrak{f}} = C(D_\sigma^{c,\mathfrak{f}}, D_\tau^{c,\mathfrak{f}})$$
with the compact-open topology.
5. The semantic space of fuzzy topological semantic types is
$$S^{c,\mathfrak{f}} = \bigcup_\tau D_\tau^{c,\mathfrak{f}}, \tag{3.9.11}$$
with the disjoint union topology.

Interpretability is an assignment $s : \mathcal{SO}_0 \to \mathcal{S}^{c,\mathfrak{f}}$ that extends to a partially defined $s : \mathrm{Dom}(s) \subset \mathcal{SO} \to \mathcal{S}^{c,\mathfrak{f}}$ following the same rules as in the original HK formulation discussed previously, but with the sets D_τ replaced by the topological spaces $D_\tau^{c,\mathfrak{f}}$. Fuzzy semantic probes are collections $\upsilon = \{\upsilon_\tau\}_\tau$ of continuous fuzzy set structures $\upsilon_\tau : D_\tau^{c,\mathfrak{f}} \to [0,1]$ on $D_\tau^{c,\mathfrak{f}}$:

$$\upsilon_\tau \in C(D_\tau^{c,\mathfrak{f}}, D_{\mathfrak{f}}^{c,\mathfrak{f}}).$$

There is an associated character $\phi_{\upsilon,s,\mathfrak{f}}$ with values in the Viterbi semiring

$$\mathcal{P} = ([0,1], \max, \cdot, 0, 1)\,,$$

determined by

$$\phi_{\upsilon,s,\mathfrak{f}}(T) = \begin{cases} \upsilon_\tau(s(T)) & \text{if } s(T) \in D_\tau^{c,\mathfrak{f}} \\ 0 & \text{otherwise.} \end{cases} \tag{3.9.12}$$

With this setting, we can then consider the Rota–Baxter structure on the Viterbi semiring given by the threshold operators c_λ as in Lemma 3.2.11. We then obtain Birkhoff factorizations for the characters $\phi_{\upsilon,s,\mathfrak{f}}$ of (3.9.12) with respect to the threshold Rota–Baxter structure, with the same properties as discussed in Proposition 3.2.14, where the Birkhoff factorization identifies as maximizers those accessible terms $T_v \subset T$ with values $\phi_{\upsilon,s,\mathfrak{f}}(T_v)$ above a chosen threshold λ, meaning that their assigned fuzzy truth values have a sufficently large degree of confidence.

3.10 No, They Don't: Transformers as Characters

Recently, it has become fashionable to claim that the so-called transformer architectures underlying the functioning of many current large language models (LLMs) somehow "disprove" or undermine the theory of generative linguistics. They don't. Such claims are vacuous, not only on account that they lack any accurate description of what is allegedly being disproved, but also more specifically because one can show, as we will discuss in this section, that the functioning of the attention modules of transformer architectures fits remarkably well within the same general formalism that we have been illustrating in the previous sections, and it is consequently *fully compatible* with a generative model of syntax based on Merge and Minimalism. While this will be discussed more at length elsewhere, we will show here briefly that the weights of attention modules in transformer architectures can be regarded as another (i.e., distinct from human) way of embedding an image of syntax inside semantics, with formal properties similar to other examples that we talked about earlier in this chapter.

This does not mean, of course, that LLMs based on such architectures *necessarily* mimic the interaction between syntax and semantics as it occurs in human brains. In fact, most certainly that is *not* the case in anything close to their present form, given the well-known considerations regarding the "poverty of the stimulus," in human language acquisition (see, e.g., [10]), compared to what one may call the "overwhelming richness of the stimulus" in the training of LLMs. Some have attempted to deal with this issue by limiting the amount of training data to something argued to align more with the data available to children (e.g., as in [200] and [100], among others, including the "Baby LM-Challenge" 2023CoNLL/CMCL [3]). However, at least so far, there are still problems with such approaches with regard to both the performance on certain test-bed data sets, and in accurately mirroring the developmental trajectory of human language acquisition, with respect to training data sample sizes. This matter is discussed in more detail in [198].

This is not the main point of the discussion here, however, since several of the examples analyzed in the previous sections of this chapter are also not meant to model how syntax and semantics realistically interact in the human brain, but rather are presented simply as illustrations of the general formal algebraic properties of the mathematical model. The point that we intend to make here is that attention modules of transformer architectures can function as another choice of a Hopf algebra character that fits within the same very general algebraic formalism we illustrated throughout all the previous sections of this chapter. Therefore, transformer architectures have no intrinsic incompatibility, at this fundamental algebraic level, with generative syntax. Note also that we are not going to include here any discussion of the efficiency of computational algorithms, as we are interested only in analyzing their algebraic structure. We will only make some general comments at the end of this section, in relation to the *inverse problem* of reconstructing syntax from its image inside semantics, that we already discussed in section 3.3.

It should be pointed out that, while we can argue that the attention modules of transformer architectures in LLMs provide a form of embedding of syntax inside a semantic space, compatible with the general formalism described in this chapter, this does *not* mean that LLMs in themselves would constitute a *theory of language* (as has been occasionally claimed): such claims only reflect a profound misunderstanding of the meaning of the word "theory" in science. We will not engage here in any detailed discussion of various erroneous claims regarding implications of LLMs for the theory of generative grammar: it is difficult to take seriously any such claims when they appear to lack technical understanding of either of these topics and of the workings of scientific theory in general.

A direct comparison of LLMs with the model of the computational process of syntax as understood by generative linguistics will ultimately be provided by a comparative study of precise mathematical models of both processes. Throughout this book, we have presented a mathematical formulation of Merge and the generative process of syntax, and mathematical models of what LLMs compute are also being developed (see, e.g., [17] and [79]). In this section, we do not fully elaborate on this comparison, as the development of the appropriate mathematical description is still ongoing, but we will discuss how the architecture of LLMs can be described within our model of syntax–semantics interfaces.

3.10.1 Attention Modules

For our purposes, it suffices to consider the basic fundamental functioning of attention modules in transformers, that we recall schematically as follows.

We assume, as in our previous setting in section 3.2, a given function $s : \mathcal{SO}_0 \to \mathcal{S}$ from lexical items and syntactic features to a semantic space $\mathcal{S}$. In the case we are discussing, $\mathcal{S}$ should be a vector space model, and the map $s : \mathcal{SO}_0 \to \mathcal{S}$ is the initial token embedding. (For simplicity, we will not discuss here exactly how to relate tokens and lexical items, but see, e.g., [134] and the accompanying documentation for the handling of a related question.)

Thus, we can view elements $\ell \in \mathcal{SO}_0$ as vectors $s(\ell) \in \mathcal{S}$. In attention modules, in the case of so-called self-attention that we focus on here, one considers three linear transformations: Q (queries), K (keys), and V (values), with $Q, K \in \mathrm{Hom}(\mathcal{S}, \mathcal{S}')$ and $V \in \mathrm{Hom}(\mathcal{S}, \mathcal{S}'')$, where $\mathcal{S}'$ and $\mathcal{S}''$ are themselves vector spaces of semantic vectors (in general of dimensions not necessarily equal to that of $\mathcal{S}$; in particular, $\mathcal{S}''$ is typically a lower dimensional projection). One usually assumes given identifications $\mathcal{S} \simeq \mathbb{R}^n$, $\mathcal{S}' \simeq \mathbb{R}^m$, $\mathcal{S}'' \simeq \mathbb{R}^d$ with Euclidean vector spaces, with assigned bases, and one works with the corresponding matrix representations of $Q, K \in \mathrm{Hom}(\mathbb{R}^n, \mathbb{R}^m)$ and $V \in \mathrm{Hom}(\mathbb{R}^n, \mathbb{R}^d)$. The target Euclidean space $\mathcal{S}'$ is endowed with an inner product $\langle \cdot, \cdot \rangle$, which can be used to estimate semantic similarity, with respect to which one obtains a low-rank matrix $\langle Q(s(\ell)), K(s(\ell')) \rangle$.

The query vector $Q(s(\ell))$, for $\ell \in \mathcal{SO}_0$, can be thought of as performing a role analogous to the *semantic probes* discussed in our toy models in section 3.2. As in that case, we think of queries (or "probes" in our previous terminology) as elements $q \in \mathcal{S}^\vee$, where $\mathcal{S}^\vee$ is the dual vector space $\mathcal{S}^\vee = \mathrm{Hom}(\mathcal{S}, \mathbb{R})$; thus, a query matrix can be identified with an element in $\mathcal{S}^\vee \otimes \mathbb{R}^m \simeq \mathcal{S}^\vee \otimes \mathcal{S}' =$

$\text{Hom}(\mathcal{S}, \mathcal{S}')$, which we can regard as an m-fold probe Q evaluated on the given semantic vector $s(\ell)$.

In a similar way, we also can think of the key vector $K(s(\ell))$, for $\ell \in \mathcal{SO}_0$, as an element $K \in \text{Hom}(\mathcal{S}, \mathcal{S}')$, which we interpret in this case as a way of creating an m-fold probe out of the given vector $s(\ell)$. Thus, the space $\mathcal{S}'$ of m-fold probes plays a dual role here, as given probes to be evaluated on an input semantic vector $s(\ell)$, and as new probes generated by the semantic vector $s(\ell)$. This dual interpretation explains the use of the terminology "query" and "key" to refer to the two given linear transformations.

The values vector $V(s(\ell))$ can be viewed as a representation of the semantic vectors $s(\ell)$ *inside* $\mathcal{S}''$, such as, for example, an embedding of the set $s(L)$, for a given subset $L \subset \mathcal{SO}_0$, into a vector space $\mathcal{S}''$, of a lower dimension than $\mathcal{S}$. One refers to $d = \dim \mathcal{S}''$ as the embedding dimension.

Next, one considers a set $L \subset \mathcal{SO}_0$. Usually, this is regarded as an ordered set, a list (also called a "string"), that would correspond to an input sentence. The ordering is accounted for by what is called the "positional encoding". However, for the purpose of our discussion here, it is more convenient to consider L as the underlying unordered set. In terms of transformer models, one can focus on bi-directional architectures like BERT, [54]. To an element $\ell \in L$, one assigns an attention operator $A_\ell : L \subset \mathcal{S} \to \mathcal{S}'$, given by

$$A_\ell(s(\ell')) = \sigma(\langle Q(s(\ell)), K(s(\ell'))\rangle),$$

where σ is the softmax function $\sigma(x)_i = \exp(x_i)/\sum_j \exp(x_j)$, for $x = (x_i)$.

Note that for simplicity of notation, we are ignoring the usual rescaling factor that divides the argument of σ by the square root of the embedding dimension, since that has no influence on the algebraic structure of the model, even through it has computational significance. We write $A_{\ell,\ell'} := A_\ell(s(\ell'))$ and refer to it as the "attention matrix". The matrix entries $A_{\ell,\ell'}$ are regarded as a probability measure of how the attention from position ℓ is distributed toward the other positions ℓ' in the set L. One then assigns an output (in $\mathcal{S}''$) to the input $s(L) \subset \mathcal{S}$, as the vectors $y_\ell = \sum_{\ell'} A_{\ell,\ell'} V(s(\ell'))$, where for each $\ell \in L$, we have $y_\ell = (y_\ell)_{i=1}^d \in \mathcal{S}'' \simeq \mathbb{R}^d$.

Observe that in writing A as a matrix, one uses a choice of ordering of the set L, but the linear operator A_ℓ itself is defined independent of such an ordering. Compatible with the fact that we want to use free symmetric Merge as generator of syntactic objects, we indeed focus here on the case of bidirectional, noncausal attention, where the nontrivial entries of the attention matrix are not limited to items occurring in a specified linear order (i.e. the matrix is not necessarily lower or upper diagonal in a chosen basis/ordering). The resulting y_ℓ

is symmetric in the ordering of L, so linear ordering also does not play a role in the output.

3.10.1.1 Heads and heads In transformer architectures, one usually has several such attention modules running in parallel, and one refers to this setting as "multihead attention." In this case, the vectors $Q(s(\ell)) = \oplus_i Q(s(\ell))_i$, $K(s(\ell)) = \oplus_i K(s(\ell))_i$, and $V(s(\ell)) = \oplus_j V(s(\ell))_j$ are split into blocks, which correspond to a decomposition $\mathcal{S}' = \oplus_{i=1}^N \mathcal{S}'_i$, and similarly for $\mathcal{S}''$, with the inner product of $\mathcal{S}'$ compatible with the direct sum decomposition, so that it induces inner products $\langle\cdot,\cdot\rangle_{\mathcal{S}'_i}$. One can then compute attention matrices, for $i = 1, \ldots, N$, as

$$A^{(i)}_{\ell,\ell'} = \sigma(\langle Q(s(\ell))_i, K(s(\ell))_i\rangle_{\mathcal{S}'_i}) ,$$

which one refers to as attention distribution with *attention head i*. Multiple attention heads in the same attention layer of a transformer are independent and additive, with all their outputs additively combined and added back into the residual stream.

In this and the following sections, it is important to keep in mind that there is an unfortunate clash of notation between this meaning of "head" as "attention head" versus the usual syntactic meaning of "syntactic head", represented by the notion of "head function" in Definition 1.13.6.

For simplicity, and to avoid confusing notation, we will not consider multiple attention heads, and work only with a single attention matrix, which suffices for our illustrative purposes, while we will be referring to the term "head" only in its syntactic meaning as a head function.

3.10.2 Maximizing Attention

Since for fixed $\ell \in L$ the values $A_{\ell,\ell'}$ give a probability measure on L, we can consider characters with values in the semiring $\mathcal{R} = ([0,1], \max, \cdot)$. For example, it is natural to look for where the attention from position ℓ is maximized. Thus, we can define a character on a subdomain

$$\phi_A : \mathcal{H}^{semi} \to \mathcal{R}$$

by setting

$$\phi_A(T) = \max_{\ell \in L(T)} A_{h(T),\ell} ,$$

if $T \in \mathrm{Dom}(h)$ and zero otherwise.

Remark 3.10.1. Note that, to make ϕ well defined for all $T \in \mathfrak{T}_{SO_0}$, we need a uniform choice of the operator A_ℓ for an $\ell \in L(T)$; that is to say, we need to provide a consistent way of extracting the choice of a leaf from each tree. This corresponds to the choice of a head function h in the sense of Definition 1.13.6.

Once a head function h is chosen, the attention matrix determines an associated attention vector $A_{h(T),\ell}$ for $\ell \in L(T)$. In particular, we can choose the head function to be the same as the syntactic head, although this is not necessary; any choice of a head function will work for this purpose. Note that head functions are not everywhere defined on $\mathfrak{T}_{SO_0}$. This implies that the choice of attention vector cannot be made compatibly with substructures simultaneously across all the trees $T \in \mathfrak{T}_{SO_0}$. There is some maximal domain $\mathrm{Dom}(h) \subset \mathfrak{T}_{SO_0}$ over which such a consistent choice can be made. This issue does not arise in the construction of attention matrices from text, as sentences in text will always have a syntactic head, but it can be relevant when sentences are stochastically generated from a template (such as those used in tests of linguistic capacities of LLMs, as in [200] and [100]).

3.10.3 Attention-Detectable Syntactic Relations

Recent investigation of attention modules and syntactic relations (such as c-command; see [134]) indicate that syntactic trees and examples of specific syntactic relations such as syntactic head, prepositional object, possessive noun, and the like are embedded and detectable from the attention matrix data. The claims in [134] would require further analysis and discussion, which we do not include here. Regardless of these more specific aspects, we show that some results of this type are to be expected, given our model.

We consider the problem of detection of syntactic relations, stated as in Definition 3.10.2.

Definition 3.10.2. Consider a syntactic relation ρ, which we write as a collection $\rho = \rho_T$ of relations $\rho_T \subset L(T) \times L(T)$, with $\rho_T(\ell, \ell') = 1$ if $\ell, \ell' \in L(T)$ are in the chosen relation, and $\rho_T(\ell, \ell') = 0$ otherwise. We say that ρ is *exactly attention-detectable* if there are query/key linear maps $Q_\rho, K_\rho \in \mathrm{Hom}(\mathcal{S}, \mathcal{S}')$ and a head function h_ρ as in Definition 1.13.6, such that

$$\rho_T(h_\rho(T), \ell_{\max,h_\rho}) = 1 \tag{3.10.1}$$

for all $T \in \mathrm{Dom}(h_\rho)$, where

$$\ell_{\max,h_\rho} = \mathrm{argmax}_{\ell\in L(T)} A_{h_\rho(T),\ell}\,, \tag{3.10.2}$$

with A the attention matrix built from Q_ρ, K_ρ.

The relation ρ is *approximately attention-detectable* if there are query/key linear maps $Q_\rho, K_\rho \in \mathrm{Hom}(\mathcal{S}, \mathcal{S}')$ and a head function h_ρ as in Definition 1.13.6, such that, for some sufficiently large set $\mathcal{D} \subset \mathrm{Dom}(h_\rho)$ of trees, they satisfy

$$\frac{1}{\#\mathcal{D}} \sum_{T \in \mathcal{D}} \rho(h_\rho(T), \ell_{\max,h_\rho}) \sim 1$$

Here, the existence of query/key linear maps Q_ρ, K_ρ as in Definition 3.10.2 is considered relative to a specified context, such as a chosen corpus or data set. Moreover, since the queries and key weights are learned, the identities (3.10.1) and (3.10.2) should really be taken only up to an uncertainty estimate (see, e.g., [165]).

In the case of approximately attention-detectable syntactic relations, we think of the subset $\mathcal{D}$ as being, for instance, a sufficiently large syntactic tree-bank corpus, or a corpus of annotated syntactic dependencies (for size estimates, see [134]). Cases where the existence of query/key linear maps Q_ρ, K_ρ and a head function h_ρ with the properties required here can be ensured can be extracted from the experiments in [134], though a more detailed analysis would be required to confirm how robust are the occurrences observed in [134]. Moreover, it should be noted here that for a more significant analysis of how syntactic relations may be encoded in attention matrices, one needs to analyze multiple layers of attention modules. This fits within our formalism, if one considers compositions of attentions in subsequent layers as "virtual attention heads" in the sense discussed in [59]. In [134], the question of locating structures matching syntactic objects in data of attention heads is decoupled from the question of identifying if and to what extent the structure building operation that forms syntactic object may be, at least partially, encoded. This will need a separate discussion elsewhere, so we will not go into any details here.

3.10.4 Rota-Baxter Structures and Attention

We have seen in §3.10.2 that we can interpret attention maximization as a Hopf algebra character. We discuss here its factorization. Using a threshold Rota-Baxter operator c_λ of weight $+1$, we obtain

$$\begin{aligned} \phi_{A,-}(T) = \; & c_\lambda(\max\{\phi_A(T), c_\lambda(\phi_A(F_{\underline{v}})) \cdot \phi_A(T/F_{\underline{v}}), \ldots, \\ & c_\lambda(\phi_A(F_{\underline{v_N}})) \cdot \phi_A(F_{\underline{v}_{N-1}}/F_{\underline{v_N}}) \cdots \phi_A(T/F_{\underline{v}_1})\}) \,. \end{aligned}$$

As before, for simplicity we focus on the case of chains of subtrees $T_{v_N} \subset T_{v_{N-1}} \subset \cdots \subset T_{v_1} \subset T$ rather than more general subforests. Note that $h(T/T_v) = h(T)$ for the quotient $T/T_v = T/^c T_v$ given by contraction of the subtree, and hence the following estimate holds:

$$\max_{\ell \in L(T/T_v)} A_{h(T),\ell} \leq \max_{\ell \in L(T)} A_{h(T),\ell} \,.$$

The value $\phi_-(T)$ corresponds then to the chains of nested accessible terms of the syntactic object T for which all the values

$$\phi_A(T_{v_i}) = \max_{\ell \in L(T_{v_i})} A_{h(T_{v_i}),\ell}$$

are above the chosen threshold λ, and all the complementary quotients $T_{v_{i-1}}/T_{v_i}$ have

$$\phi_A(T_{v_{i-1}}/T_{v_i}) = \max_{\ell \in L(T_{v_{i-1}}/T_{v_i})} A_{h(T_{v_{i-1}}),\ell} = \max_{\ell \in L(T_{v_{i-1}})} A_{h(T_{v_{i-1}}),\ell} = \phi_A(T_{v_{i-1}}).$$

The first condition implies that one is selecting only chains of accessible terms inside the syntactic object T, where the maximal attention from the head of each subtree in the chain is sufficiently large, while the second condition means that, among these chains, one is selecting only those for which the recipient of maximal attention from the head of the given subtree is located outside the next subtree. This second condition guarantees that when considering the next nested subtree and trying to maximize for its attention value, one does not spoil the optimizations achieved at the previous steps for the larger subtrees.

A similar procedure can be obtained by also introducing direct implementation of some syntactic constraints. We can see this in the following way.

A syntactic relation ρ determines a character ϕ_ρ on trees $T \in \mathrm{Dom}(h) \subset \mathfrak{T}_{\mathcal{SO}_0}$ with values in the Boolean ring $\mathfrak{B} = (\{0, 1\}, \max, \cdot)$, where

$$\phi_\rho(T) = \max_{\ell \in L(T)} \rho(h(T), \ell).$$

This Boolean character detects whether the syntactic relation ρ is realized at the head $h(T)$ of a given tree T.

Using a character

$$\phi_{A,\rho}(T) = \max_{\ell \in L(T)} \rho(h(T), \ell) \cdot A_{h(T),\ell},$$

with values in $\mathcal{P} = ([0, 1], \max, \cdot)$, one maximizes the attention from the tree head over the set of $\ell \in L(T)$ that already satisfies the chosen syntactic relation with respect to the head of the tree. The corresponding Birkhoff factorization with threshold Rota–Baxter operators again identifies chains of subtrees that maximize the attention (above a fixed threshold), in a way that is recursively compatible with the larger trees as before, but now maximization is done only on the set where the relation is implemented. Subtrees with $\phi_\rho(T_v) = 0$ do not contribute even if their value of $\max_\ell A_{h(T),\ell}$ is sufficiently large.

Thus, comparison between the case with character ϕ_A and the case with character $\phi_{A,\rho}$ identifies attention-detectability of the syntactic property considered and, if detectability fails, at which level in the tree (in terms of chains of nested subtrees) the attention matrix maximum happens outside of where the syntactic relation holds.

As shown in [198], the current performance on syntactic capacities of LLMs trained on small-scale data modeling falls significantly short of the human performance when tested on LI-Adger data sets that include sufficiently diverse syntactic phenomena. Other shortcomings in the handling of syntax have been measured in [51]. This suggests that they provide a good testing ground for syntactic recoverability as outlined earlier and a possible experimental testing for aspects of the inverse problem of the syntax–semantics interface.

3.10.5 Syntax as an Inverse Problem: Physics as Metaphor

The question of reconstructing the computational process of syntax in LLMs based on transformer architectures can be seen in the same light as the situation illustrated in a simpler example in section 3.3. In that case, one views the image of syntax embedded inside a semantic space and considers the inverse problem of extracting syntax as a computational process working from these images, which live in a semantic space that is not itself endowed with the same type of computational structure. Here, the image of syntax is encoded in the key and query vectors that live in vector spaces that organize semantic proximity data, as well as in the resulting attention matrices. Inverse problems of this kind are usually expected to be computationally hard. This does not mean that the computational mechanism of syntax cannot be reconstructible, but that a significant cost in complexity, growing rapidly with the depth of the trees, may be involved, and only a partial reconstruction may be possible.

Early results showed that recurrent neural network (RNN) language models performed poorly on tests of grammaticality aimed at capturing syntactic structures on a testbed data set of pairs of sentences that differ only in their grammaticality [148], while [89] showed that language models based on RNNs can perform well at predicting long-distance number agreement, even in the absence of semantic clues (i.e., when tested on nonsensical but grammatical sentences). Results like this appear to indicate that syntax can, in principle, be extracted and disentangled from its image inside semantics. It was shown in [186] that Syntactic Ordered Memory (SOM) syntax-aware language models outperform the Chat-GPT2 LLM on syntactic generalization tests. However, this entire area remains a matter of contention, dependent in part on the testbed data set used, as described more fully in [197] and [198].

A more systematic comparison of language model architectures and their performance on syntactic tests in [99] revealed substantial differences in syntactic generalization performance by model architecture, more than by size of the data set. One can suggest that the various indicators of poor performance on syntactic tests, along with any other difficulties, might also reflect the computational difficulty involved in extracting syntax as an inverse problem from its image through the semantic interface, stored across values of the weights of attention matrices rather than in a direct syntax-first mapping.

Another possible indicator of difficulties inherent in this reconstruction problem for syntax is the dimension gap between text produced by LLM and by humans, measured in [196], for which a theoretical interpretation within our model appears possible.

In this chapter, we have used physics as a guideline for identifying mathematical structures that can be useful in modeling the relation between syntax and semantics. We conclude here by using physics again, this time only as a metaphor, for describing the relation of syntax as a generative process and the functioning of LLMs.

The generative structure underlying particle physics is given by the Feynman diagrams of quantum field theory. Disregarding epistemological issues surrounding the interpretation of such diagrams as events of particle creation and decay, we can roughly say that in a particle physics experiment, what one detects is an image of such objects embedded into the set of data collected by detectors. Detecting a particle such as the Higgs boson (the most famous recent particle physics discovery) means solving an inverse problem that identifies inside this enormous set of data the traces of the correct diagrams/processes involving the creation of a Higgs particle from an interaction of other particles (such as gluon fusion or vector-boson fusion) and its subsequent decay into other particles (such as vector-boson pairs or photons). The enormous computational difficulty implicit in this task arises from the need to solve this type of inverse problem, involving the identification of event structure (e.g., a Higgs decay into photons involving top quark loop diagrams) from the measurable data, and a search for the desired structure in a background involving a huge number of other simultaneous events. The direct map from quantum field theory consists of the Higgs boson production cross sections, which are calculated from perturbative expansions in the Feynman diagrams of quantum chromodynamics and quantum electrodynamics, involving significant higher-order quantum corrections. Such perturbative quantum field theory computations are where the algebraic formalism recalled at the beginning of this chapter plays a role. The inverse problem, instead, consists of measuring, for various possible decay channels, mass and kinematic information like decay angles of detectable particles of the expected type, which are produced either by the expected decay event or by the background of productions of the same particle type due to other events, and of searching for an actual signal in this background.

We can use this story as a metaphor and imagine the generative process of syntax embedded inside LLMs in a conceptually similar way, its image scattered across a probabilistic smear over a large number of weights and vectors trained over large data sets. This view of LLMs as the technological "particle accelerators" of linguistics, where signals of linguistic structures are detectable against a background of probabilistic noise, suggests that such models do not invalidate generative syntax any more than particle detectors would "invalidate" quantum field theory—quite the contrary, in fact.

While LLMs do not constitute a model of language in the human brain, they can still, in the sense described here, provide an apparatus for the experimental study of inverse problems in the syntax-semantic interface. Here, however, it is essential to recall again the physics metaphor. Data and technology *without theory* do not constitute science, understood as a model of the fundamental laws of nature that has both strong predictive capacity *and* a high level of concise conceptual clarity in its explanatory power. The relation between the computational process of syntax and the topological relational nature of semantics is a problem of conceptual nature. In this sense, the LLMs may contribute a technological experimental laboratory for the analysis of some aspects of this problem, but do not constitute a replacement for the necessary theoretical understanding of fundamental laws in the structure of language.

4 Summary of Mathematical Concepts

In this chapter, we present some mathematical background and fundamental mathematical definitions that are used elsewhere in the book. We will not cover in this summary basic definitions such as vector spaces, metric spaces, continuity, and manifolds, as those are already widely in use in the computational linguistics field and can easily be located in introductory mathematics textbooks. We also do not define here mathematical structures, such as formal languages, that are well known to linguists (although not always to mathematicians). We will instead focus on those mathematical structures used extensively in this book and which are not part of the usual background of linguists and are more difficult to access in the mathematical literature.

4.1 Categories

In mathematics, the formalism of category theory allows descriptions of different types of mathematical structures in terms of objects carrying the desired structures and ways of transforming them.

Definition 4.1.1. A small *category* C consists of a set of *objects* $X, Y, \cdots \in \mathrm{Obj}(C)$ and sets of *morphisms* $\mathrm{Mor}_C(X, Y)$, also written as $\mathrm{Hom}_C(X, Y)$. Morphisms have a composition rule

$$\circ : \mathrm{Mor}_C(X, Y) \times \mathrm{Mor}_C(Y, Z) \to \mathrm{Mor}_C(X, Z),$$

given by $\circ : (f, g) \mapsto g \circ f$, satisfying associativity:

$$h \circ (g \circ f) = (h \circ g) \circ f\,.$$

Every object has an identity morphism, $1_X \in \mathrm{Mor}_C(X, X)$. These are units for composition in the sense that $1_Y \circ f = f = f \circ 1_X$ for any $f \in \mathrm{Mor}_C(X, Y)$.

Morphisms are also often referred to as "arrows" and usually written with the function notation $f : X \to Y$, even though they are not necessarily functions.

(For example, when discussing Externalization in Chapter 1, we considered a case where morphisms were not functions but rather "correspondences".)

Examples of categories that the reader will come across in this book include vector spaces, with morphisms given by linear maps; magmas, with morphisms given by "magma homomorphisms" (functions compatible with the magma product operation); the category of commutative algebras (which in fact means algebras that are both associative and commutative), with "algebra homomorphisms" (functions that are compatible with the linear structure and the product of algebras); and other examples such as categories of Hopf algebras, semirings, and their generalizations, as discussed in chapter 3.

Morphisms describe how objects can transform. In turn, categories themselves can transform, and transformations of categories are described by the notion of *functors*.

Definition 4.1.2. Given categories C, C', a functor $F : C \to C'$ maps objects of the first category to objects of the second, $\mathrm{Obj}(C) \ni X \mapsto F(X) \in \mathrm{Obj}(C')$, and it also maps morphisms $f \in \mathrm{Mor}_C(X, Y)$ to morphisms $F(f) : F(X) \to F(Y)$ in $\mathrm{Mor}_{C'}(F(X), F(Y))$ compatibly with composition, $F(g \circ f) = F(g) \circ F(f)$, and identity, $F(1_X) = 1_Y$.

Functors themselves can transform into other functors. The suitable notion describing transformations of functors is given by *natural transformations*.

Definition 4.1.3. Given two functors $F_1 : C \to C'$ and $F_2 : C \to C'$ between the same pair C, C' of categories, a *natural transformation* $\eta : F_1 \to F_2$ is a collection of morphisms $\eta_X \in \mathrm{Mor}_{C'}(F_1(X), F_2(X))$ in the category C', one for every object $X \in \mathrm{Obj}(C)$. These morphisms $\eta_X : F_1(X) \to F_2(X)$ satisfy the compatibility condition $\eta_Y \circ F_1(f) = F_2(f) \circ \eta_X$, for all morphisms $f \in \mathrm{Mor}_C(X, Y)$. This compatibility is equivalently expressed as the commutativity of the diagram

$$\begin{array}{ccc} F_1(X) & \xrightarrow{F_1(f)} & F_1(Y) \\ \downarrow{\scriptstyle \eta_X} & & \downarrow{\scriptstyle \eta_Y} \\ F_2(X) & \xrightarrow{F_2(f)} & F_2(Y) \end{array}$$

for all $X, Y \in \mathrm{Obj}(C)$ and for all $f \in \mathrm{Mor}_C(X, Y)$; that is, $\eta_Y \circ F_1(f) = F_2(f) \circ \eta_X$.

In section 1.12.3 in chapter 1 we also briefly use the notion of 2-category, as in Definition 4.1.4.

Definition 4.1.4. A *2-category* is a small category where the sets of morphisms (called "1-morphisms") $\mathrm{Mor}_C(X, Y)$ are themselves the objects of a category whose morphisms (called "2-morphisms", which describe morphisms between morphisms) have two different forms of composition, vertical and horizontal. Vertical composition $\circ_1$ has a fixed pair of source and target objects, while horizontal composition $\circ_0$ follows the composition of 1-morphisms (see Figure 4.1, illustrating the two types of compositions of 2-morphisms). The compatibility between the two compositions is expressed by an identity, called the "exchange relation":

$$(\alpha \circ_0 \beta) \circ_1 (\gamma \circ_0 \delta) = (\alpha \circ_1 \gamma) \circ_0 (\beta \circ_1 \delta) .$$

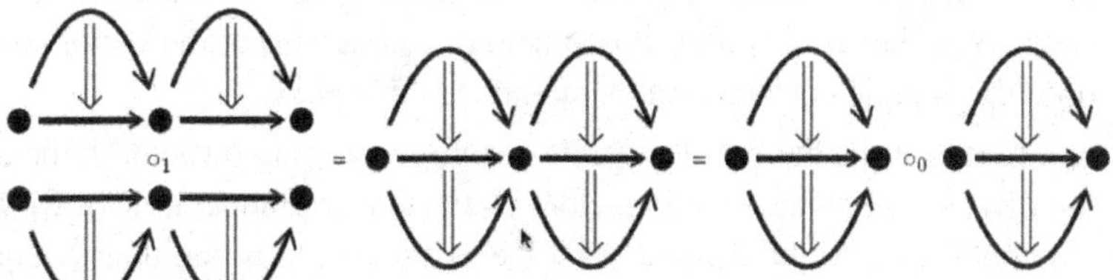

Figure 4.1
Vertical and horizontal composition of 2-morphisms in a 2-category, satisfying the exchange relation.

4.2 Hopf Algebras

One of the main mathematical structures used in this book is Hopf algebras. We present here a short summary of their properties and the mathematical definition of the concept.

4.2.1 Main Idea

The notion of Hopf algebra, along with the slightly weaker notion of bialgebra, are very useful tools in mathematics to handle problems that require accounting for all possible ways of decomposing certain objects into constituent building blocks and compatible ways to assemble the building blocks into more complex objects. Typically, they arise in many combinatorial problems, where these operations of composition/decomposition allow the appropriate enumeration of possibilities.

More precisely, the decomposition operation is called a *coproduct*: it has one input and two outputs. The input is the object that one wants to decompose, and the output is its decomposition into two parts: an extracted building block and its remaining complement. Since in general there are going to be multiple possible decompositions, corresponding to different building blocks that can be extracted, all these possibilities are listed together as a formal sum. This is the reason why one considers as the underlying structure a vector space $\mathcal{V}$ spanned by the objects that one is analyzing. This space is just a formal device where such formal sums of possibilities make sense. We assume everywhere that we are working with vector spaces over the field $\mathbb{Q}$ of rational numbers.

Note that, given a arbitrary set X, there is always a canonically defined vector space $\mathcal{V}(X)$ that is spanned by the elements of X. This does not mean that X itself has a vector space structure: the set X is the *basis* of the vector space $\mathcal{V}(X)$ and vectors in $\mathcal{V}(X)$ are formal finite sums $\sum_i a_i x_i$ of elements $x_i \in X$ weighted by scalars $a_i \in \mathbb{Q}$, the chosen field of coefficients of the vector space. We used this simple construction many times in this book.

One can then write the list of possible decompositions as a vector in the tensor product $\mathcal{V} \otimes \mathcal{V}$, where the two sides of the tensor product (which we call "left-channel" and "right-channel") list the extracted terms and their complements, respectively. It is important to note that the use of a tensor product vector space here is only signifying the fact that we are dealing with two outputs (in the case of the coproduct) or two inputs (in the case of the product). Thus, one thinks of the coproduct as an operation $\Delta : \mathcal{V} \to \mathcal{V} \otimes \mathcal{V}$, which takes an element of $\mathcal{V}$ and disassembles it into its constituents; and the product as an operation $m : \mathcal{V} \otimes \mathcal{V} \to \mathcal{V}$, which takes two elements of $\mathcal{V}$ and assembles them into a new element.

In fact, since coefficients of such enumerations of possibilities are usually integers, it is possible to work with $\mathbb{Z}$-modules rather than vector spaces in all the cases that are of interest to us. We use the vector space notation only because that is usually more familiar. Thus, the coproduct lists (as a formal sum) all the possible decompositions of a given object as a building block (or, more generally, a set of building blocks) and its complement.

The other operation, the *product*, describes a way of combining objects of the assigned type (and in particular their building blocks) to form new objects within the same class. The product and coproduct are related by consistency conditions, that make the composition/decomposition operations compatible with each other. Moreover, one usually requires a condition (associativity for the product, coassociativity for the coproduct) that guarantees that these operations behave well under iteration. (There are, however, also cases that are interesting to consider, even though these associativity and coassociativity conditions do not necessarily hold.)

An important observation is that the consistency condition imposing compatibility between product and coproduct does not mean that these are inverse operations (i.e., that one of them exactly undoes the other). Indeed, there is usually an asymmetry between the product and coproduct of a Hopf algebra, where one of them is a more complex operation than the other. In the cases discussed in this book, the product has usually a simpler form than the coproduct. As we will state more precisely in section 4.2.2, the compatibility condition compares these two operations when applied one after the other in either order.

It is customary to write the properties of operations and their compatibility in the form of "commutative diagrams". This means that following the arrows around the diagrams in two different ways yields the same result. Since following the arrows means composing the corresponding operations, this commutativity of the diagram results in an identity between different compositions of operations, which is either a compatibility condition between different operations or a property of a given operation.

The main reason why the notion of Hopf algebra is relevant in the context of the generative process of syntax is the fact that the action of Merge on workspaces uses the extraction of accessible terms from syntactic objects, and this extraction is precisely the operation that the coproduct of a Hopf algebra (or bialgebra) performs.

4.2.2 Formal Definition

The Hopf algebra $\mathcal{H}$ is a vector space over a field $\mathbb{K}$ (which we can assume to be the field $\mathbb{Q}$ of rational numbers), endowed with

- a multiplication (or product) $m : \mathcal{H} \otimes_{\mathbb{K}} \mathcal{H} \to \mathcal{H}$,
- a unit $u : \mathbb{K} \to \mathcal{H}$,
- a comultiplication (or coproduct) $\Delta : \mathcal{H} \to \mathcal{H} \otimes_{\mathbb{K}} \mathcal{H}$,
- a counit $\epsilon : \mathcal{H} \to \mathbb{K}$, and
- an antipode $S : \mathcal{H} \to \mathcal{H}$,

which satisfy the following properties. The multiplication operation is associative; namely, the following diagram commutes:

$$\begin{array}{ccc} \mathcal{H} \otimes_{\mathbb{K}} \mathcal{H} \otimes_{\mathbb{K}} \mathcal{H} & \xrightarrow{m \otimes id} & \mathcal{H} \otimes_{\mathbb{K}} \mathcal{H} \\ \downarrow{\scriptstyle id \otimes m} & & \downarrow{\scriptstyle m} \\ \mathcal{H} \otimes_{\mathbb{K}} \mathcal{H} & \xrightarrow{m} & \mathcal{H} \end{array} ,$$

which, as mentioned previously, equivalently reads as the identity $m \circ (m \otimes id) = m \circ (id \otimes m)$, which is the usual associativity condition for multiplication, $m(m(x, y), z) = (xy)z = x(yz) = m(x, m(y, z))$. The unit and multiplication are related by the following commutative diagram:

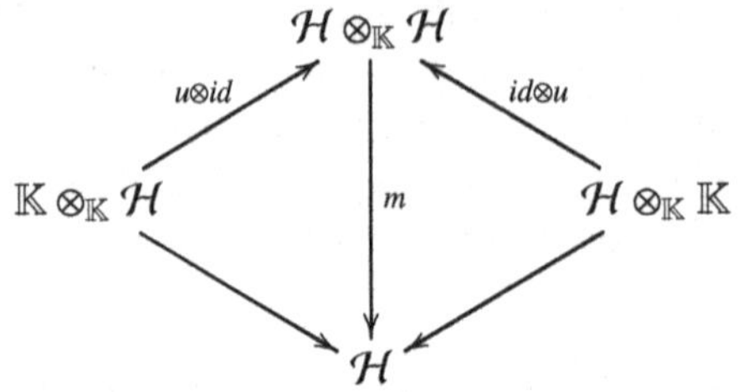

,

where the two unmarked downward arrows are the identifications induced by scalar multiplication. The unit and counit are compatible via the following commutative diagram:

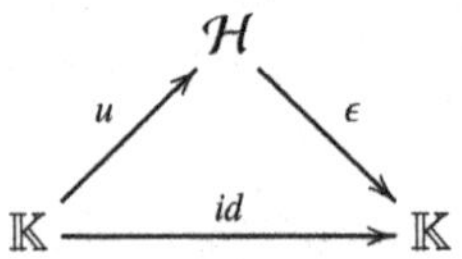

.

The coproduct is coassociative; namely, the following diagram commutes:

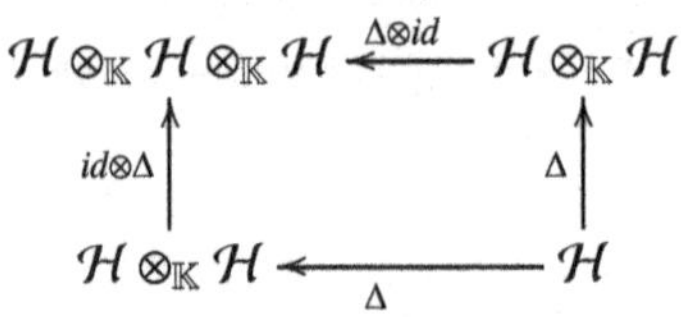

and the comultiplication and the counit are related by the following commutative diagram:

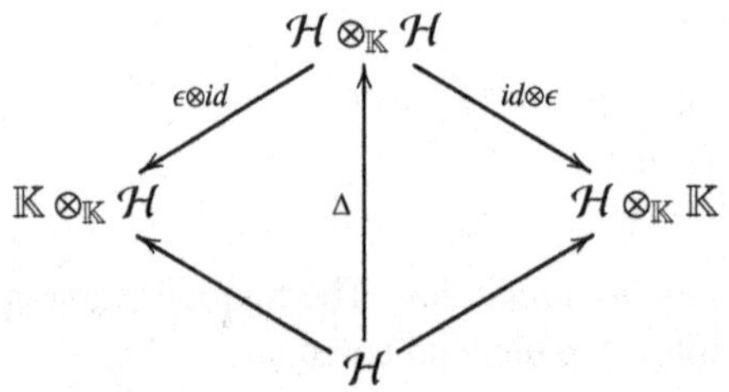

,

where the two unmarked upward arrows are the identifications given by $x \mapsto 1_{\mathbb{K}} \otimes x$ and $x \mapsto x \otimes 1_{\mathbb{K}}$, respectively, with $1_{\mathbb{K}}$ the unit of the field $\mathbb{K}$. The operations Δ and ϵ are algebra homomorphisms, and m and u are coalgebra homomorphisms; namely, one has a commutative diagram:

$$\begin{array}{ccccc} \mathcal{H}\otimes\mathcal{H} & \xrightarrow{m} & \mathcal{H} & \xrightarrow{\Delta} & \mathcal{H}\otimes\mathcal{H} \\ \downarrow{\scriptstyle \Delta\otimes\Delta} & & & & \uparrow{\scriptstyle m\otimes m} \\ \mathcal{H}\otimes\mathcal{H}\otimes\mathcal{H}\otimes\mathcal{H} & & \xrightarrow[id\otimes\tau\otimes id]{} & & \mathcal{H}\otimes\mathcal{H}\otimes\mathcal{H}\otimes\mathcal{H} \end{array},$$

where $\tau : \mathcal{H}\otimes\mathcal{H}\otimes\mathcal{H}\otimes\mathcal{H} \to \mathcal{H}\otimes\mathcal{H}\otimes\mathcal{H}\otimes\mathcal{H}$ is the permutation that exchanges the two middle factors. This diagram expresses the compatibility between the product and coproduct. The following commutative diagrams describe the behavior of unit and counit with respect to coproduct and product:

$$\begin{array}{ccc} \mathcal{H}\otimes\mathcal{H} & \xrightarrow{m} & \mathcal{H} \\ & {\scriptstyle \epsilon\otimes\epsilon}\searrow \quad \swarrow{\scriptstyle \epsilon} & \\ & \mathbb{K} & \end{array} \quad \text{and} \quad \begin{array}{ccc} \mathcal{H}\otimes\mathcal{H} & \xleftarrow{\Delta} & \mathcal{H} \\ & {\scriptstyle u\otimes u}\nwarrow \quad \nearrow{\scriptstyle u} & \\ & \mathbb{K} & \end{array},$$

using the identification $\mathbb{K}\otimes\mathbb{K} = \mathbb{K}$. The antipode $S : \mathcal{H} \to \mathcal{H}$ is a linear map such that the following diagram also commutes:

$$\begin{array}{ccccc} \mathcal{H}\otimes_{\mathbb{K}}\mathcal{H} & \xrightarrow{m} & \mathcal{H} & \xleftarrow{m} & \mathcal{H}\otimes_{\mathbb{K}}\mathcal{H} \\ \uparrow{\scriptstyle id\otimes S} & & \uparrow{\scriptstyle u\circ\epsilon} & & \uparrow{\scriptstyle S\otimes id} \\ \mathcal{H}\otimes_{\mathbb{K}}\mathcal{H} & \xleftarrow[\Delta]{} & \mathcal{H} & \xrightarrow[\Delta]{} & \mathcal{H}\otimes_{\mathbb{K}}\mathcal{H} \end{array}.$$

4.2.3 Bialgebras and Hopf Algebras

The notion of bialgebra is very similar to that of Hopf algebra recalled in section 4.2. Indeed, a bialgebra satisfies all the same properties, except for the existence of the antipode. In most cases that we are interested in, the bialgebra structure is the most important thing. However, it is convenient to think in terms of Hopf algebras because there is a class of cases, which includes those of interest to us, where the bialgebra structure already suffices to also have an antipode, and therefore the full Hopf algebra properties. This is called the "graded connected" case.

In any graded bialgebra $\mathcal{H} = \oplus_{n\geq 0}\mathcal{H}_n$ with $\mathcal{H}_0 = \mathbb{K}$ (the connected property) and the graded structure compatible with the product and coproduct, it is possible to construct an antipode map inductively by

$$S(X) = -X - \sum S(X')X'' \tag{4.2.1}$$

for any element X of the bialgebra, with coproduct $\Delta(X) = X\otimes 1 + 1\otimes X + \sum X'\otimes X''$, where the X' and X'' are terms of lower degree.

4.3 Rooted Trees

Binary rooted trees form another class of mathematical objects that plays a crucial role in our setting. A *tree* T is a finite graph whose geometric realization is simply connected (no loops), defined by a set of vertices V and a set of edges E. The *valence* of a vertex is the number of edges incident to it.

A tree is *rooted* if it has a distinguished element $v_r \in V$, the root vertex. We can assume, then, that the edges in E are oriented with the uniquely defined orientation away from the root. One can say that a vertex is below another vertex if there is an oriented edge from the second to the first. The source and target maps $s, t : E \to V$ assign to each edge of the tree its source and target vertices. The leaves of the tree are the univalent vertices. They are also the sinks (i.e., they have no outgoing edges).

A rooted tree is *binary* if below every non-leaf vertex, there are exactly two other vertices. (This property is sometimes called "full binary", but in this book we just refer to it as "binary".) Equivalently, every non-root and non-leaf vertex has valence 3, the root has valence 2, and the leaves have valence 1.

A tree is *vertex-decorated* if there is a map $L_V : V \to D_V$ to a finite set. It is edge-decorated if, similarly, there is a map $L_E : E \to D_E$ to a finite set of possible edge decorations. When we assume that the trees are vertex-decorated, we simply refer to them as "decorated trees".

A rooted tree is *planar* if it is endowed with an embedding of its geometric realization in the plane. Rooted trees without an assignment of a planar structure are referred to as *abstract* (or as *nonplanar*), binary rooted trees. Assigning a planar embedding is equivalent to assigning a linear ordering of the leaves that is compatible with the underlying abstract tree.

Consider the vector space $\mathcal{V}_k$ spanned by *decorated planar binary rooted trees* T with k internal vertices (equivalently, with $k+1$ leaves). It has the dimension

$$\dim \mathcal{V}_k = (\#D_V)^k \frac{(2k)!}{k!(k+1)!}, \tag{4.3.1}$$

where $\#D_V$ is the cardinality of the set D_V of possible vertex labels.

The difference between planar and abstract binary rooted trees plays a crucial role in the Minimalist Model of syntax, with *abstract* trees describing the *syntactic objects*, the hierarchical structures produced by free symmetric Merge. Meanwhile, the *planar* trees correspond to the result of Externalization, with the resulting linear ordering of the leaves corresponding to sentences viewed as ordered strings of words.

4.4 Other Algebraic Structures

In addition to the notion of Hopf algebra that we recalled in section 4.2, there are some other algebraic structures that play a role in our work and that we will recall here briefly. These include algebras and rings, ideals, modules, and semirings.

4.4.1 Main Idea

The notion of vector space has become widely used in computational linguistics. Encoding words as vectors makes it possible to assign to a given lexical item a collection of numbers (the vector coordinates with respect to an assigned basis), which encode, for example, the degree of relatedness of that word to a certain list of concepts, of properties, or of other words. Moreover, vectors in a vector space can be added, and formal sums in the vector space spanned by a given set are often an efficient way to encode a list of possibilities in that set. In fact, this role of sums as lists of possibilities is the one that we make extensive use of in this work, and it is our main reason for using vector spaces.

Familiar constructions in vector spaces include considering subspaces and quotient spaces, which can be seen as kernels and cokernels of linear maps between vector spaces. A quotient operation is always regarded as a way of eliminating a part of a structure, and subspaces correspond to extract substructures (where in this case the type of structure that we are considering is just that of linear space).

This simple and familiar picture can be generalized in various ways. In vector spaces, linear combinations are weighted sums where the weights (coefficients) are scalars in a field $\mathbb{K}$ (usually the field of real numbers, the smaller field of rational numbers, or the larger field of complex numbers). However, one may want weighted sums where the weights are more structured than just being numbers, while maintaining the main "linearity" properties of vector spaces and linear maps acting on them. This can be done by replacing the field $\mathbb{K}$ of scalars with a more general *ring* R of coefficients. Instead of vector spaces over $\mathbb{K}$ one has *modules* over the ring R. Examples of rings include rings of functions (e.g., polynomials, power series, Laurent series, continuous functions), so for instance if the weights that one wants to use in weighted sums depend on adjustable parameters, one is working with a module over a ring of functions (of those parameters), rather than with a vector space. Certain rings also have a simultaneous vector space structure (e.g., polynomials with real coefficients), where the addition operation as a ring and the addition operation as a vector space coincide. These are called *algebras*. We usually assume that all the rings and algebras that we consider are associative (meaning that the multiplication is associative), except when explicitly stated otherwise, as in the case of the nonassociative algebras discussed in section 1.12.4 of chapter 1.

The construction of subspaces and quotient spaces for vector spaces becomes more delicate in the case of algebras, as not all subalgebras have corresponding quotient algebras. Those that do are called *ideals*. (Things get somewhat more involved if the multiplication operation in the algebra is noncommutative, as we will review briefly later in this section.)

Semirings are another closely related algebraic structure. These are also well known in linguistics, since the work of Chomsky and Schützenberger [39]. The semiring of formal power series with coefficients in the semiring of nonnegative integers is used in the Chomsky–Schützenberger enumeration theorem, which proves that, for a language L that admits an unambiguous, context-free grammar, the power series $G_L(x) = \sum_{k=0}^{\infty} a_k x^k$, with $a_k = \#W_k(L)$ being the number of words of length k in the language, is algebraic over the field of rational functions $\mathbb{Q}(x)$; that is, it satisfies a polynomial equation over this field. Here, the difference between ring and semiring structure lies in the restriction of coefficients as nonnegative integers. More general semirings that are of familiar use in linguistics, in the context of semiring parsing, include the Boolean semiring (which just consists of 0 and 1 with the Boolean operations) or the Viterbi or probabilistic semiring, which replaces the 0 and 1 Boolean values with a probabilistic version with values $p \in [0, 1]$. We discuss various semirings and forms of semiring parsing in section 3.5 of chapter 3.

4.4.2 Rings, Algebras, Ideals, Modules, Semirings

We review here the formal definitions of the algebraic structures that we mentioned in section 4.4.1 and that we use throughout the book.

A *ring* R is a set equipped with two operations, addition and multiplication, where addition is associative, commutative, and endowed with a zero element and with additive inverses, while multiplication is associative (not necessarily commutative) and has a unit element 1. There is a relation between these two operations, given by the distributive property (multiplication distributes over sum) $x(y + z) = xy + xz$ and $(y + z)x = yx + zx$. The ring $\mathbb{Z}$ of integers, the ring $\mathbb{Z}[x]$ of polynomials with integer coefficients, the ring $\mathbb{Z}[[x]]$ of formal power series with integer coefficients, and the ring $\mathbb{Z}[x^{-1}][[x]]$ of Laurent series with integer coefficients are all examples of commutative rings. The ring $M_n(\mathbb{Z})$ of $n \times n$ matrices with integer coefficients is an example of a noncommutative ring. The difference between a ring and a field lies in the fact that a field is a commutative ring where all nonzero elements also have a multiplicative inverse.

A subring $R' \subset R$ is a subset preserved by the ring operations of R; namely, such that the ring operations of R induce a ring structure on R'. A *left-ideal* $\mathcal{I} \subset R$ is a subset with the property that combinations of the form $r_1 x_1 + \cdots + r_n x_n$, with any $x_i \in \mathcal{I}$ and any $r_i \in R$, are contained in $\mathcal{I}$; that is, $R\mathcal{I} \subset \mathcal{I}$. A *right-ideal* has the same property with respect to combinations of the form $x_1 r_1 + \cdots + x_n r_n$. A *two-sided ideal* $\mathcal{I} \subset R$ is simultaneously a left and a right ideal. In a commutative ring, there is no difference between left and right ideals. Quotient rings $R/\mathcal{I}$ can be obtained only when the substructure $\mathcal{I} \subset R$ that one quotients out is a two-sided ideal.

An *algebra* $\mathcal{A}$ is a vector space (over a field $\mathbb{K}$) endowed with a multiplication operation $m : \mathcal{A} \times \mathcal{A} \to \mathcal{A}$, as we have described in the first part of the structure or Hopf algebras and bialgebras in section 4.2—namely, with associativity of multiplication $m(m(x, y), z) = (xy)z = x(yz) = m(x, m(y, z))$ described by the commutative diagram

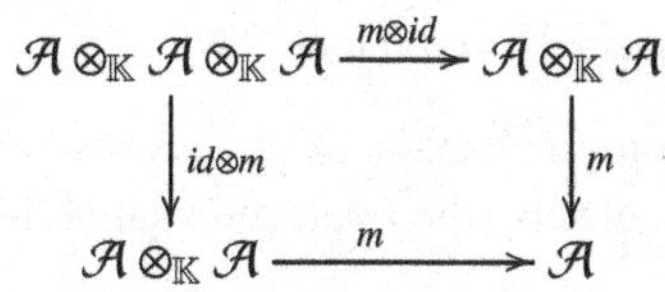

and a multiplicative unit with the commutative diagram

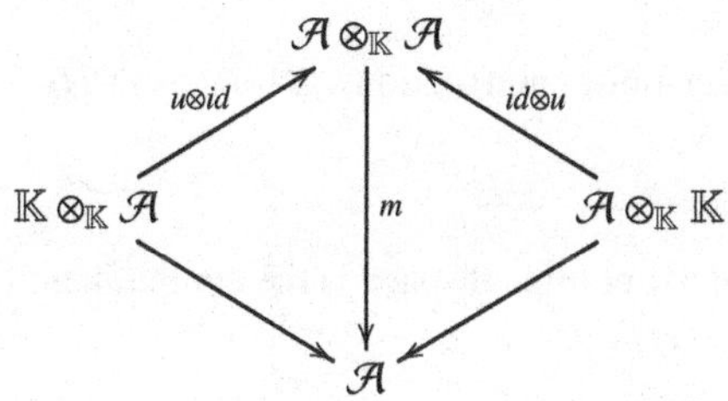

In terms of the relation to the definition of ring recalled in this section, we can think of this notion of associative algebra as a ring that is also a vector space, where addition and multiplication give the ring structure and the same addition operation, together with scalar multiplication by the field, give the vector space structure. For example, the ring of polynomials $\mathbb{Q}[x]$ is also an algebra over $\mathbb{Q}$. In this book, we also occasionally consider algebras where the multiplication operation is not required to satisfy the associativity property (nonassociative algebras), see section 1.12.4 in chapter 1.

A *left module* over a ring (or over an associative algebra) is a generalization of the notion of a vector space (over a field). Namely a module M over a ring R has an addition operation that is associative, commutative, and with the 0 element, and scalar multiplication $x \mapsto rx$ by elements r of the ring R, with $r(x + y) = rx + ry$, $(r + s)x = rx + sx$, $(rs)x = r(sx)$, $1x = x$. A *right module* has the same structure, but with scalar multiplication from the right, $x \mapsto xr$. A *bimodule* is simultaneously a left and a right module (i.e., it has a left and a right action of R that commute with each other). For commutative rings, there is no difference between left and right modules.

A *semiring* S is a set with two operations, sum $\oplus$ and product $\odot$. The sum is associative and commutative and has a zero element, but now it is no longer required to have additive inverses. The product is associative with unit element and satisfies $0 \odot s = 0$ for any $s \in S$ and distributivity $s \odot (s_1 \oplus s_2) = (s \odot s_1) \oplus (s \odot s_2)$ and $(s_1 \oplus s_2) \odot s = (s_1 \odot s) \oplus (s_2 \odot s)$.

4.5 Point Set Topology

We recall here some basic concepts and terminology regarding topological spaces that we use occasionally in the text.

Definition 4.5.1. A topological space $(X, \mathcal{T}_X)$ is a set X together with an assigned collection $\mathcal{T}_X$ of subsets (the open sets) satisfying the following properties:

1. $\mathcal{T}_X$ is closed under arbitrary (finite or infinite) unions: $\cup_\alpha \mathcal{U}_\alpha$ is in $\mathcal{T}_X$ if $\mathcal{U}_\alpha \in \mathcal{T}_X$ for all α.
2. $\mathcal{T}_X$ is closed under finite intersections: $\mathcal{U}_1 \cap \cdots \cap \mathcal{U}_N \in \mathcal{T}_X$ if $\mathcal{U}_i \in \mathcal{T}_X$ for all $i = 1, \ldots, N$.
3. X and $\emptyset$ are in $\mathcal{T}_X$.

A closed set $C \subset X$ in a topological space is the complement $C = X \smallsetminus \mathcal{U}$ of an open set, $\mathcal{U} \in \mathcal{T}_X$.

An *open covering* (often also refereed to as "a covering") of a topological space X is a collection $\{\mathcal{U}_\alpha\}_{\alpha \in \mathcal{I}}$ (either finite or infinite) of open sets $\mathcal{U}_\alpha \in \mathcal{T}_X$, with the property that every point of X is covered by at least one of the open sets; namely, such that

$$X = \bigcup_{\alpha \in \mathcal{I}} \mathcal{U}_\alpha .$$

A *finite covering* is an open covering consisting of finitely many open sets, $\{\mathcal{U}_1, \ldots, \mathcal{U}_N\}$. An open covering of a subset $A \subset X$ of a topological space is similarly defined, with open sets in the induced topology.

A warning about terminology: the word "covering" in the sense of "open covering," as intended here, should not be confused with a different use of the same word in topology; namely "covering" in the sense of "covering space", which is another topological space Y with a continuous surjection $Y \twoheadrightarrow X$, which is a locally trivial fibration with discrete fibers.

We do use the notion of covering space, as well as the more general notion of *branched covering*, when we discuss the origami folding map (3.4.1) in chapter 3, in the context of semantic spaces and Externalization, so we review it briefly here. As mentioned previously, a covering space Y of a space X is a topological space with a continuous surjection $\pi : Y \twoheadrightarrow X$, such that there is a discrete set D and open sets $U_\alpha \subset X$, such that

$$\pi^{-1}(U_\alpha) \simeq U_\alpha \times D .$$

The discrete set D is the set of branches of the covering map π. This means that locally (over the open sets U_α), the projection map consists of a discrete set of disjoint copies of U_α. The group of *covering transformations* of a covering space $\pi : Y \twoheadrightarrow X$ is the set of symmetries of Y (homeomorphisms for topological spaces, or diffeomorphisms if X and Y are smooth manifolds) that preserve the projection map π; namely, invertible transformations $\gamma : Y \to Y$ that form a commutative diagram:

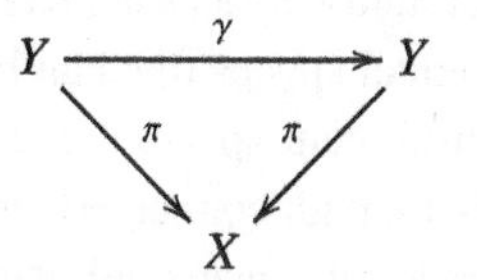

The origami folding map (3.4.1) restricted to the open set that corresponds to the binary trees is a covering space. When one includes the boundary strata consisting of trees where one of the length values goes to zero (which makes them trees with some n-ary vertices with $n \geq 3$), the map is a branched covering (the branching happening at the folds of the "origami folding"). A branched covering is a continuous surjection $\pi : Y \twoheadrightarrow X$ that is a covering on a dense open set $U \subset X$, with some of the branches colliding over the set $X \smallsetminus U$, called the "branch locus".

The notion of open coverings and finite coverings is the key to another important concept in topology: *compactness*. A subset $K \subseteq X$ in a topological space is *compact* if, from any open coverings $\{\mathcal{U}_\alpha\}_{\alpha \in I}$ of K, one can always extract a finite subcovering (i.e. there is always a way to cover K with finitely many open sets).

A topological space is *connected* if it cannot be decomposed into a *disjoint* union of open sets. The stronger notion of path connectedness requires that, for any two points in the space, there is a continuous path between them that is also entirely contained in the space.

Definition 4.5.2. A metric space (X, d_X) is a set endowed with a function $d_X : X \times X \to \mathbb{R}_{\geq 0}$ that satisfies: $d_X(x, y) = 0$ iff $x = y$, $d_X(x, y) = d_X(y, x)$ for all $x, y \in X$, and

$$d_X(x, z) \leq d_X(x, y) + d_X(y, z),$$

for all $x, y, z \in X$, the triangle inequality of distances. A Cauchy sequence $\{x_n\}_{n \in \mathbb{N}}$ of points in a metric space (X, d_X) is a sequence with the property that, for any $\epsilon > 0$, there is an $n_0 \in \mathbb{N}$ such that for all $n, m \geq n_0$, the distances satisfy $d_X(x_n, x_m) < \epsilon$. A metric space (X, d_X) is *complete* if any Cauchy sequence converges; namely, there is an $x \in X$ such that $d(x_n, x) \to 0$ as $n \to \infty$.

A metric space is also a topological space with the topology generated by the open sets $B_r(x) = \{y \in X \mid d_X(x, y) < r\}$. A topological space X is *metrizable* if the topology $\mathcal{T}_X$ is induced by a metric in this way. Not all topological spaces are metrizable: there are necessary and sufficient conditions for metrizability. We encounter some issues regarding metrizability, for example, in section 3.9 of chapter 3.

A particularly useful class of metric spaces are Riemannian manifolds. These are smooth manifolds (topological spaces that can be locally approximated by flat Euclidean spaces through tangent spaces), endowed with a metric. Connected Riemannian manifolds are path connected, and one can select paths that have minimal length, with the length measured using the Riemannian metric: these paths are called *geodesics*. A Riemannian manifold M is *geodesically complete* if the geodesic arcs $\gamma : [0, 1] \to M$ at any given point $\gamma(0) = x \in M$ extend indefinitely (i.e., to a curve $\gamma : \mathbb{R} \to M$). A Riemannian manifold M is *geodesically convex* if, for any pair $x \neq y$ of points in M, there is a unique length-minimizing geodesic arc $\gamma : [0, 1] \to M$ contained in M that connects the two points, $\gamma(0) = x$, $\gamma(1) = y$. We often assume that our models of semantic spaces are geodesically convex Riemannian manifolds (but, for example, in section 3.9, we also need to consider topological spaces that are not Riemannian manifolds).

4.6 Further Remarks on Birkhoff Factorization

We discussed in chapter 3 how to adapt a technique originally developed in theoretical physics to assign semantic values to syntactic objects in a way that recursively checks consistency over substructures. In this section, we provide further details that can help put this idea in context, both in comparison with previous approaches used in linguistics (specifically in the context-free setting) and in the context of how these ideas developed within theoretical physics.

The problem of checking the consistency of semantic values across substructures has played a significant role in linguistics. Methods used to address this question range widely: they include Knuth's semantics of context-free languages [111] and more generally attribute-grammars based on underlying context-free grammars augmented with attributes and semantic rules, as well as pregroup grammars and Lambek calculus (see, e.g., [119] and [179] for a recent version in a vector space model of semantics). Relations between earlier versions of Minimalism and Lambek calculus were analyzed in [8].

There are two main aspects in which the approach we describe here differs from these methods:

- Instead of being based on an underlying description of grammar in terms of formal languages (especially context-free), the method introduced here is designed to use a generative process encoded in a Hopf algebra structure.
- The procedure of checking consistency is packaged into a single map, which recursively modifies an initially chosen assignment of semantic values so as to incorporate the consistency checking over all substructures.

In the setting of theoretical physics, one is also considering a generative process that recursively produces hierarchical structures. These combinatorial hierarchical structures are the Feynman graphs of a given quantum field theory. The form of the action functional of the theory determines the corresponding class of graphs. As in linguistics, there are two ways in which one can model this generative process. One of them is in terms of formal languages [143], while the other is in terms of Hopf algebras, [45] and [56]. In the case of formal languages, an example of generative grammar for Feyman diagrams (for the so-called $\phi^2 A$ physical theory) is given in Figure 4.2. An example of the encoding of the generative structure of Feynman graphs through the Hopf algebra coproduct is given in Figure 4.3.

One should think of the formal language description of Feynman graphs as the analog of descriptions of older formulations of Merge and Minimalism given in terms of Merge grammars and multiple context-free grammars; see [192] and [199]. An important aspect of the formal language description of Feynman graphs is that the graph grammars involved are usually context-sensitive. Thus, the assignment of physical values to Feynman graphs in a way that is consistent over substructures cannot be directly modeled on analogs of the attribute-grammars or other forms of semantic parsing developed in the setting of context-free grammars.

The Hopf algebra description, on the other hand, is the direct analog of the formulation of Minimalism that we have presented in this book, based on

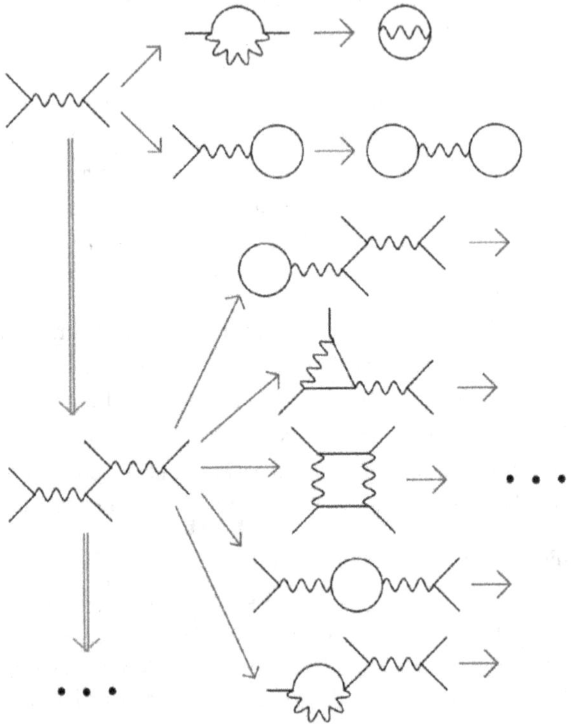

Figure 4.2
The generative grammar for the Feynman graphs of the $\phi^2 A$ physical theory, from [143].

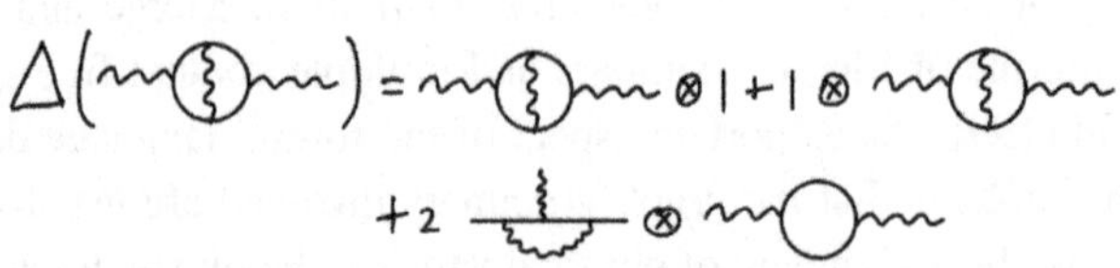

Figure 4.3
The generative structure of Feynman graphs encoded in the coproduct of the Hopf algebra.

Merge and the Strong Minimalist Thesis (SMT). In physics, this Hopf algebra formulation is crucial to bypass the problem of divergences in evaluations of Feynman integrals, and obtain a good recursive method for assigning meaningful physical values to the Feynman graphs in a way that is consistent with substructures.

The two descriptions of the generative process of Feynman graphs in terms of graph grammars and of Hopf algebras are not equivalent in terms of computational structure. This can be seen in Example 3.2 of [143], where one finds that the Lie algebra associated with the Hopf algebra of Feynman graphs is not isomorphic to the Lie algebra associated to the graph grammar. Indeed, one can see from this type of comparison that the Hopf algebra description has greater succinctness than the graph grammar description in terms of generative power. This can be seen as an analog of the result of [6] on the greater succinctness of the Merge description of Minimalism over its formulation in terms of multiple context-free grammars. For computational implementations of the generative process and the Hopf algebra calculations for Feynman graphs; see, for instance, [15].

As discussed in chapter 3, the key idea for the assignment of values consistently across substructures lies in the construction of a recursive procedure, which in physics goes under the name of "Bogolyubov preparation". It starts from a given assignment of values to the hierarchical combinatorial structures, given in the form of a map $\phi : \mathcal{H} \to \mathcal{R}$, where $\mathcal{H}$ is the Hopf algebra that describes the generative process of the Feynman graphs and $\mathcal{R}$ is a chosen target space for evaluation (such as a ring of Laurent series, or more generally a semiring, as in the setting of semiring parsing). The only requirement that is made about the map $\phi : \mathcal{H} \to \mathcal{R}$ is that formal combinations go to formal combinations (linearity) and independent objects go to independent values. In physics, this last requirement is multiplicativity over connected components of the Feynman graph. In our linguistic setting, connected components are different syntactic objects in a workspace, and this same multiplicative property is simply saying that, so long as two such objects are not combined by the action of Merge, their semantic assignments are independent and not subject to compatibility constraints. Given such an assignment of values $\phi : \mathcal{H} \to \mathcal{R}$, one proceeds to modify it recursively, in such a way that the checking of compatibility of values across substructures is built into the resulting modified function $\tilde{\phi} : \mathcal{H} \to \mathcal{R}$ (the Bogolyubov preparation of ϕ). This recursive construction uses the coproduct Δ of the Hopf algebra $\mathcal{H}$ to extract all the possible substructures and compare them with the associated quotient (i.e., what remains once the substructure is removed). It also uses a decomposition of the target algebra

$\mathcal{R}$ into two subalgebras $\mathcal{R}_\pm$ (such decomposition is called a *Rota–Baxter structure*, and the operator R that realizes the decomposition is called a *Rota–Baxter operator*).

Precise definitions of these terms are recalled in chapter 3: here, we describe their properties and their role in the construction. The two parts $\mathcal{R}_\pm$ of the target $\mathcal{R}$ represent, respectively, the cases of meaningful assignment (in physics, a finite value rather than a meaningless infinity) and the cases that one wants to disregard as meaningless. In our setting, rather than the divergences and finite values of physical theories, we are using the Rota–Baxter operator to filter the target space $\mathcal{R}$ by levels of agreement with a given semantic hypothesis (or probe). The inductive form of $\tilde{\phi} : \mathcal{H} \to \mathcal{R}$ is structured in the following way. One replaces the original map $x \mapsto \phi(x)$, for $x \in \mathcal{H}$ with a sum of inductive correction terms:

$$\tilde{\phi}(x) = \phi(x) + \sum \phi_-(x')\phi(x''),$$

where the sum is over the decompositions of x into the two terms x' and x'' according to the coproduct of the Hopf algebra:

$$\Delta(x) = 1 \otimes x + x \otimes 1 + \sum x' \otimes x'',$$

and ϕ_- is defined recursively by

$$\phi_-(x) = -T(\tilde{\phi}(x)).$$

Note that this recursive definition works if the Hopf algebra can be filtered into graded pieces (e.g., the syntactic objects are graded by the number of leaves of the tree, or equivalently the length of the sentence that they represent) and all the terms x' and x'' in the coproduct expression are in degrees lower than x, so all the $\phi_-(x')$ occurring in the definition of $\tilde{\phi}(x)$ have already been computed when $\tilde{\phi}(x)$ and $\phi_-(x)$ are computed. The recursion parameter is the degree in the Hopf algebra. The factorization then consists in separating the two parts:

$$\phi_-(x) = -R(\tilde{\phi}(x)) \quad \text{and} \quad \phi_+(x) = (1 - R)(\tilde{\phi}(x)).$$

The first map, $\phi_- : \mathcal{H} \to \mathcal{R}_-$, collects (as a formal sum) all the instances where "something goes wrong" in one of the substructures, detected by the corresponding term $\phi_-(x')$ compared with the term $\phi(x'')$ of the value assigned when that substructure is removed. The other term, $\phi_+ : \mathcal{H} \to \mathcal{R}_+$, is an assignment of values where all the instances where something goes wrong have been removed. The separation of $\tilde{\phi}$ into $\phi_\pm$ is called a "Birkhoff factorization" of ϕ because $\phi_\pm$ satisfy the product relation

$$\phi(x) = ((\phi_- \circ S) \star \phi_+)(x) = \langle \phi_- \circ S \otimes \phi_+, \Delta(x) \rangle,$$

where S is the antipode of the Hopf algebra. Again, one uses the fact that the Hopf algebra is graded and S is defined inductively in terms of the coproduct decomposition into terms of lower degree:

$$S(x) = -x - \sum S(x')x''.$$

The second line in Figure 4.3 shows this inductive form of the coproduct in the case of Feynman graphs.

In the specific examples discussed in chapter 3, we focus on computing the term $\phi_-(x)$ that detects the presence of inconsistencies. In fact, in most of these cases, we discuss explicitly the target $\mathcal{R}$ with the Rota–Baxter structure as a *semiring* rather than a ring or algebra as in the original physics setting. This means that the form in which the factorization is written is slightly changed. The main difference is that the factorization that we consider does not explicitly use the antipode (this fact is related to the absence of additive inverses in the semiring), and takes the form

$$\phi_+ = \phi_- \star \phi$$

rather than $\phi = (\phi_- \circ S) \star \phi_+$. For this reason, we do not discuss in detail the form of the antipode in our Hopf algebra of workspaces, although it is completely specified by the same recursive formula, $S(x) = -x - \sum S(x')x''$.

There are ways of relating this Hopf algebra approach to other forms of parsing. Some of them are discussed in chapter 3. Other relevant comparisons can be found in the work of Joachim Kock [112] on inductive data types and combinatorial Dyson–Schwinger (DS) equations, which can be used as a ground for comparing approaches like attribute-grammars or Lambek calculus and the Hopf algebra formulation. As we discussed briefly in chapter 1, combinatorial DS equations are a way of encoding the generative process as the solution of a fixed-point equation in a Hopf algebra, and they are known to correspond to constructions of Hopf subalgebras and Hopf ideals; see [65].

The Hopf algebra used in chapter 2 to describe the older formulation of Minimalism as in Edward P. Stabler's Computational Minimalism [192] is given by the Loday–Ronco Hopf algebra. There are several well-known and closely related Hopf algebras that we expect will play a significant role in the further study of Externalization and of the syntax–semantics interface. These include the Hopf algebras of integer binary relations of [171], which are closely related to the associahedra discussed in chapter 3 and the permutohedra, which we expect will be related to the study of the geometry of syntactic parameters; the Malvenuto–Reutenauer Hopf algebra of permutations, which is closely related to the Loday–Ronco Hopf algebra. They also include a class of Hopf algebras known as "Hopf algebras of words" [48], which also include the Malvenuto–Reutenauer Hopf algebra. We also expect that further study of the interplay between the core computational structure of syntax and the two channels of Externalization (Sensory–Motor system) and of the syntax–semantics interface (Conceptual–Intensional system) will involve several of these algebraic structures and their interplay.

Bibliography

[1] Marcelo Aguiar and Frank Sottile. "Structure of the Loday–Ronco Hopf Algebra of Trees." *Journal of Algebra*, 295, no. 2 (2006): 473–511.

[2] François Apéry and Masaaki Yoshida. "Pentagonal Structure of the Configuration Space of Five Points in the Real Projective Line." *Kyushu Journal of Mathematics*, 52, no. 1 (1998): 1–14.

[3] Association for Computational Linguistics. "BabyLM Challenge," 2023. `https://babylm.github.io/`

[4] Franz Baader and Tobias Nipkow. *Term Rewriting and All That.* Cambridge University Press, 1998.

[5] Christoph Bergbauer and Dirk Kreimer. "Hopf Algebras in Renormalization Theory: Locality and Dyson-Schwinger Equations from Hochschild Cohomology." In *Physics and Number Theory*, edited by Louise Nyssen, 133–164. IRMA Lectures in Mathematics and Theoretical Physics 10, European Mathematical Society, 2006.

[6] Robert C. Berwick. "Mind the gap." In *50 Years Later: Reflections on Chomsky's Aspects*, 1–13. MITWPL, 2015.

[7] Robert C. Berwick and Noam Chomsky. *Why only us?* MIT Press, 2017.

[8] Robert C. Berwick and Samuel David Epstein. "On the Convergence of 'Minimalist' Syntax and Categorial Grammar." In *Algebraic Methods in Language Processing 1995*, 143–148. Universiteit Twente, 1995.

[9] Robert C. Berwick, Angela D. Friederici, Noam Chomsky, and Johan J. Bolhuis. "Evolution, Brain, and the Nature of Language." *Trends in Cognitive Sciences*, 17, no. 2 (2013): 89–98.

[10] Robert C. Berwick, Paul Pietroski, Beracah Yankama, and Noam Chomsky. "Poverty of the Stimulus Revisited." *Cognitive Science*, 35, no. 7 (2011): 1207–1242.

[11] Louis J. Billera, Susan P. Holmes, and Karen Vogtmann. "Geometry of the Space of Phylogenetic Trees." *Advances in Applied Mathematics*, 27, no. 4 (2001): 733–767.

[12] Pawel Blasiak, "Combinatorial Route to Algebra: The Art of Composition and Decomposition." *Discrete Mathematics & Theoretical Computer Science*, 12, no. 2 (2010): 381–400.

[13] John Michael Boardman and Rainer M. Vogt. *Homotopy Invariant Algebraic Structures on Topological Spaces*, Vol. 347. Lecture Notes in Mathematics. Springer, 1973.

[14] Nikolay Nikolayevich Bogoliubov and Dmitry Vasil'evich Shirkov. *The Theory of Quantized Fields*. Interscience Publishers, 1959.

[15] Michael Borinsky. "Feynman Graph Generation and Calculations in the Hopf Algebra of Feynman Graphs." *Computer Physics Communications*, 185, no. 12 (2014): 3317–3330.

[16] Andrzej Borowiec, Wieslaw A. Dudek, and Steven Duplij. "Basic Concepts of Ternary Hopf Algebras.", *Journal of Kharkov National University*, ser. Nuclei, Particles and Fields, 529, no. 3 (15) (2001): 21–29.

[17] Tae-Danae Bradley, John Terilla, and Yannis Vlassopoulos. "An Enriched Category Theory of Language: From Syntax to Semantics." *Matematica*, 1, no. 2 (2022): 551–580.

[18] Christian Brouder and Alessandra Frabetti. "QED Hopf Algebras on Planar Binary Trees." *Journal of Algebra*, 267, no. 1 (2003): 298–322.

[19] Damien Calaque, Kurusch Ebrahimi-Fard, and Dominique Manchon. "Two Interacting Hopf Algebras of Trees: A Hopf-Algebraic Approach to Composition and Substitution of B-Series." *Advances in Applied Mathematics*, 47, no.2 (2011): 282–308.

[20] Frédéric Chapoton. "Operads and Algebraic Combinatorics of Trees." *Séminaire Lotharingien de Combinatoire*, 58 (2008): B58c [27 pages].

[21] Noam Chomsky. *Lectures on Government and Binding*. Foris Publications, 1982.

[22] Noam Chomsky. *The Minimalist Program*. MIT Press, 1995.

[23] Noam Chomsky. "Bare Phrase Structure." In *Evolution and Revolution in Linguistic Theory*, edited by H. Campos and P. Kempchinsky, 51–109. Georgetown University Press, 1995.

[24] Noam Chomsky. "Beyond Explanatory Adequacy." In *Structures and Beyond: The Cartography of Syntactic Structures*, edited by Adriana Bellett, Vol. 3, 104–131. Oxford University Press, 2004.

[25] Noam Chomsky. "On Phases." In *Foundational Issues in Linguistic Theory: Essays in Honor of Jean-Roger Vergnaud*, edited by R. Freidin, C.P. Otero, and M.L. Zubizarreta, 133–166. MIT Press, 2008.

[26] Noam Chomsky. "Problems of Projection." *Lingua*, 130 (June 2013): 33–49.

[27] Noam Chomsky. "Problems of Projection: Extension." In *Structures, Strategies and Beyond: Studies in Honour of Adriana Belletti*, edited by E. Di Domenico, C. Hamann, and S. Matteini, 1–16. John Benjamins Publishing, 2015.

[28] Noam Chomsky. "Some Puzzling Foundational Issues: The Reading Program." *Catalan Journal of Linguistics Special Issue* (2019): 263–285.

[29] Noam Chomsky. *The UCLA lectures*. Preprint 2019, lingbuzz/005485

[30] Noam Chomsky. "Simplicity and the Form of Grammars." *Journal of Language Modeling*, 9, no. 1 (2021): 5–15.

[31] Noam Chomsky. "Minimalism: Where Are We Now, and Where Can We Hope to Go." *Gengo Kenkyu*, 160 (2021): 1–41.

[32] Noam Chomsky. "Genuine Explanation and the Strong Minimalist Thesis." *Cognitive Semantics*, 8, no. 3 (2022): 347–365.

[33] Noam Chomsky. *Working toward the Strong Interpretation of SMT.* Lecture Series Theoretical Linguistics at Keio-EMU, 2023.

[34] Noam Chomsky. "The Miracle Creed and the Strong Minimalist Thesis." In *A Cartesian Dream: A Geometrical Account of Syntax. In Honor of Andrea Moro,* edited by Matteo Greco and Davide Mocci, 17–40. Research in Generative Grammar Monographs. Lingbuzz Press, 2023.

[35] Noam Chomsky. "Language and the Miracle Creed." *Researching and Teaching Chinese As a Foreign Language*, 4, no. 1 (2023): 1–12.

[36] Noam Chomsky. "Genuine Explanation." In *Rich Descriptions and Simple Explanations in Morphosyntax and Language Acquisition*, edited by Giuliano Bocci, Daniele Botteri, Claudia Manetti, and Vicenzo Moscati, 15–44. Oxford University Press, 2024.

[37] Noam Chomsky. "Displacement." Preprint, 2023. To appear in *Structured Sentences and the Computational Theory of Mind,* edited by M. Beslin, K. Howitt, A. Krauska, L. Seguin, and J. Uriagereka. Cambridge University Press.

[38] Noam Chomsky and Howard Lasnik. "The Theory of Principles and Parameters." In *Syntax: An International Handbook of Contemporary Research*, edited by Joachim Jacobs, Arnim von Stechow, Wolfgang Sternefeld, and Theo Vennemann, 506–569, de Gruyter, 1993.

[39] Noam Chomsky and Marcel P. Schützenberger. "The Algebraic Theory of Context-Free Languages." In *Computer Programming and Formal Systems*, edited by P. Braffort and D. Hirschberg, 118–161. North-Holland, 1963.

[40] Noam Chomsky, T. Daniel Seely, Robert C. Berwick, Sandiway Fong, Marinus A. C. Huybregts, Hisatsugu Kitahara, Andrew McInnerney, and Yushi Sugimoto. *Merge and the Strong Minimalist Thesis.* Elements in Generative Syntax. Cambridge University Press, 2023.

[41] Sue Yeon Chung, Daniel D. Lee, and Haim Sompolinsky. "Classification and geometry of general perceptual manifolds." *Physical Review X,* 8, no. 3 (2008): 031003.

[42] Guglielmo Cinque, "Cognition, Universal Grammar, and Typological Generalizations." *Lingua*, 130 (June 2013): 50–65.

[43] Guglielmo Cinque. *On Linearization: Toward a Restrictive Theory.* MIT Press, 2023.

[44] Chris Collins and Edward Stabler. "A Formalization of Minimalist Syntax." *Syntax,* 19, no. 1 (2016): 43–78.

[45] Alain Connes and Dirk Kreimer. "Hopf Algebras, Renormalization and Noncommutative Geometry." *Communication in Mathematics Physics*, 199, no. 1 (1998): 203–242.

[46] Alain Connes and Matilde Marcolli. *Noncommutative Geometry, Quantum Fields, and Motives*, Colloquium Publications, Vol.55. American Mathematical Society, 2008.

[47] Cas W. Coopmans, Karthikeya Kaushik, and Andrea E. Martin. "Hierarchical Structure in Language and Action: A Formal Comparison." *Psychological Review*, 130, no. 4 (2023): 935–952.

[48] Martin Duncan Crossley. "Some Hopf Algebras of Words." *Glasgow Mathematics Journal*, 48, no. 3 (2006): 575–582.

[49] Carina Curto and Vladimir Itskov. "Cell Groups Reveal Structure of Stimulus Space." *PLoS Computational Biology*, 4, no. 10 (2008): e1000205 [13 pages].

[50] Colleen Delaney and Matilde Marcolli. "Dyson-Schwinger Equations in the Theory of Computation." In *Feynman Amplitudes, Periods and Motives*, edited by Luis Álvarez-Cónsul, José Ignacio Burgos-Gil, and Kurusch Ebrahimi-Fard, 79–107, Contemporary Mathematics, Vol. 648. American Mathematical Society, 2015.

[51] Vittoria Dentella, Fritz Günther, and Evelina Leivada. "Systematic Testing of Three Language Models Reveals Low Language Accuracy, Absence of Response Stability, and a Yes-Response Bias." *Proceedings of the National Academy of Sciences (PNAS)*, 120, no. 51 (2023): e2309583120 [10 pages].

[52] Satyan Linus Devadoss. "Tessellations of Moduli Spaces and the Mosaic Operad." In *Homotopy Invariant Algebraic Structures: A Conference in Honor of J. Michael Boardman*, edited by Jean-Pierre Meyer, Jack Morava, and W. Stephen Wilson, 91–114, Contemporary Mathematics, Vol. 239. American Mathematical Society, 1999.

[53] Satyan Linus Devadoss and Jack Morava. "Navigation in Tree Spaces." *Advances in Applied Mathematics*, 67 (June 2015): 75–95.

[54] Jacob Devlin, Ming-Wei Chang, Kenton Lee, Kristina Toutanova. "BERT: Pre-Training of Deep Bidirectional Transformers for Language Understanding." In *Proceedings of NAACL-HLT 2019*, edited by Jill Burstein, Christy Doran, and Thamar Solorio, 4171–4186. Association for Computational Linguistics, 2019.

[55] Persi Diaconis, C. Y. Amy Pang, and Arun Ram. "Hopf Algebras and Markov Chains: Two Examples and a Theory." *Journal of Algebraic Combinatorics,* 39, no. 3 (2014): 527–585.

[56] Kurusch Ebrahimi-Fard and Dirk Kreimer. "The Hopf Algebra Approach to Feynman Diagram Calculations." *Journal of Physics A: Mathematical and General,* 38, no. 50 (2005): R385–R407.

[57] Kurusch Ebrahimi-Fard and Dominique Manchon, "The Combinatorics of Bogoliubov's Recursion in Renormalization." In *Renormalization and Galois Theories*, edited by Alain Connes, Frédéric Fauvet, and Jean-Pierre Ramis, 179–207, IRMA Lectures in Mathematics and Theoretical Physics, Vol. 15. European Mathematical Society, 2009.

[58] Herbert Edelsbrunner and John L. Harer. *Computational Topology: An Introduction.* American Mathematical Society, 2010.

[59] Nelson Elhage, Neel Nanda, Catherine Olsson, Tom Henighan, Nicholas Joseph, Ben Mann, Amanda Askell, Yuntao Bai, Anna Chen, Tom Conerly, Nova DasSarma, Dawn Drain, Deep Ganguli, Zac Hatfield-Dodds, Danny Hernandez, Andy Jones, Jackson Kernion, Liane Lovitt, Kamal Ndousse, Dario Amodei, Tom Brown, Jack Clark, Jared Kaplan, Sam McCandlish, and Chris Olah. "A Mathematical Framework for Transformer Circuits." In *Transformer Circuits Thread*, 2021. `https://transformer-circuits.pub/2021/framework/index.html`

[60] Samuel David Epstein, Hisatsugu Kitahara, and T. Daniel Seely. "Structure Building That Can't Be." In *Ways of Structure Building*, edited by M. Uribe-Etxebarria and V. Valmala, 253–270. Oxford University Press, 2012.

[61] Ezra Van Everbroeck, Maria Polinsky, and Garrison W. Cottrell. "Does English Need Its Pronouns? Simulating the Effect of Pro-Drop on SVO Languages." In *AAAI Spring Symposium*, Technical Report 5 (2004): SS-04-05-013 [4 pages]

[62] Martin B. H. Everaert, Marinus A. C. Huybregts, Noam Chomsky, Robert C. Berwick, and Johan J. Bolhuis. "Structures, Not Strings: Linguistics as Part of the Cognitive Sciences." *Trends in Cognitive Sciences*, 19, no. 12 (2015): 729–743.

[63] Evelina Fedorenko, Po-Jang Hsieh, Alfonso Nieto-Castañón, Susan Whitfield-Gabrieli, and Nancy Kanwisher. "New Method for fMRI Investigations of Language: Defining ROIs Functionally in Individual Subjects." *Journal of Neurophysiology*, 104, no. 2 (2010): 1177–1194.

[64] Richard P. Feynman, Robert B. Leighton, and Matthew Sands. *The Feynman Lectures on Physics; New Millennium Edition.* Basic Books, New York, 2010.

[65] Loïc Foissy. "Classification of Systems of Dyson–Schwinger Equations in the Hopf Algebra of Decorated Rooted Trees." *Advances in Mathematics*, 224, no. 5 (2010): 2094–2150.

[66] Loïc Foissy. "Les algèbres de Hopf des arbres enracinés, I." *Bulletin des Sciences Mathématiques*, 126, no. 3 (2002): 193–239.

[67] Sandiway Fong, Robert C. Berwick, and Jason Ginsburg. *The Combinatorics of Merge and Workspace Right-Sizing*. Lecture at Evolinguistics Workshop, 2019.

[68] Sandiway Fong and Jason Ginsburg. *On Constraining Free Merge.* The 43rd Meeting of the Kansai Linguistics Society. Konan University, 2018.

[69] Sandiway Fong and Jason Ginsburg. "On the Computational Modeling of English Relative Clauses." *Open Linguistics*, 9, no. 1 (2023): 20220246 [35 pages].

[70] Sandiway Fong and Masayuki Oishi, "On the nature of FormSet", preprint, FIND-Workshop, 2023.

[71] Angela D. Friederici. *Language in Our Brain: The Origins of a Uniquely Human Capacity*. MIT Press, 2017.

[72] Angela D. Friederici, Noam Chomsky, Robert C. Berwick, Andrea Moro, and Johan J. Bolhuis. "Language, Mind and Brain." *Nature Human Behaviour*, 1, no. 10 (2017): 713–722.

[73] Sitanshu Gakkhar and Matilde Marcolli. "Syntactic Structures and the General Markov Models." *Mathematics in Computer Science,* 18, no. 1 (2024): Paper 4 [38 pages].

[74] Ángel J. Gallego and Román Orús. "Language Design as Information Renormalization." *SN Computer Science*, 3, no. 2 (2022) Article 140.

[75] Xing Gao, Li Guo, and Huhu Zhang. "Rota's Program on Algebraic Operators, Rewriting Systems and Gröbner-Shirshov Bases." *Advances in Mathematics (China)*, 51, no. 1 (2022): 1–31.

[76] Peter Gärdenfors. *Conceptual Spaces: The Geometry of Thought.* MIT Press, 2000.

[77] Peter Gärdenfors. *The Geometry of Meaning: Semantics Based on Conceptual Spaces.* MIT Press, 2014.

[78] Hans-Martin Gärtner and Jens Michaelis. "A Note on Countercyclicity and Minimalist Grammars." In *Proceedings of FGVienna: The 8th Conference on Formal Grammar,* edited by Gerald Penn, 95–109. CSLI Publications, 2008.

[79] Stéphane Gaubert and Yiannis Vlassopoulos. "Directed Metric Structures arising in Large Language Models." Preprint 2024, arXiv:2405.12264.

[80] Lothar Gerritzen and Ralf Holtkamp. "Hopf Co-addition for Free Magma Algebras and the Nonassociative Hausdorff Series." *Journal of Algebra*, 265, no. 1 (2003): 264–284.

[81] Sabrina Gerth and Peter beim Graben. "Unifying Syntactic Theory and Sentence Processing Difficulty through a Connectionist Minimalist Parser." *Cognitive Neurodynamics*, 3, no. 4 (2009): 297–316.

[82] Joshua Goodman. "Semiring Parsing." *Computational Linguistics*, 25, no. 4 (1999): 573–605.

[83] Peter beim Graben and Sabrina Gerth. "Geometric Representations for Minimalist Grammars." *Journal of Logic, Language and Information*, 21, no. 4 (2012): 393–432.

[84] Joseph Greenberg. "Some Universals of Grammar with Particular Reference to the Order of Meaningful Elements." In *Universals of Language*, edited by J.Greenberg, 73–113. MIT Press, 1963.

[85] Ulf Grenander. *Elements of Pattern Theory*. Johns Hopkins University Press, 1996.

[86] Ulf Grenander and Michael I. Miller. *Pattern Theory: From Representation to Inference*. Oxford University Press, 2007.

[87] Ulf Grenander. "Patterns in Mathematical Semantics." In *Regular Structures: Lectures in Pattern Theory, Volume III.* Chap. 9, 451–538. Springer, 1981.

[88] Cristina Guardiano and Giuseppe Longobardi. "Parameter Theory and Parametric Comparison." In *Oxford Handbook of Universal Grammar,* edited by I.G. Roberts, 377–398. Oxford University Press, 2017.

[89] Kristina Gulordava, Piotr Bojanowski, Edouard Grave, Tal Linzen, and Marco Baroni. "Colorless Green Recurrent Networks Dream Hierarchically." In *Proceedings of the 2018 Conference of the North American Chapter of the Association for Computational Linguistics: Human Language Technologies*, edited by Marilyn Walker, Heng Ji, and Amanda Stent, Volume 1, 1195–1205, 2018

[90] Li Guo. "Operated Semigroups, Motzkin Paths and Rooted Trees." *Journal of Algebraic Combinatorics,* 29, no. 1 (2009): 35–62.

[91] Hubert Haider. *On Minimalist Theorizing and Scientific Ideology in Grammar Theory.* ResearchGate, 2018. `https://doi.org/10.13140/RG.2.2.15886.82242`

[92] Boris Hanin and David Rolnick. "Complexity of Linear Regions in Deep Networks." *Proceedings of Machine Learning Research,* 97 (2019): 2596–2604.

[93] Irene R. Heim and Angelika Kratzer. *Semantics in Generative Grammar*. Blackwell Publishing, 1998.

[94] Ralf Holtkamp. "Comparison of Hopf Algebras on Trees." *Archiv der Mathematik,* 80, no. 4 (2003): 368–383.

[95] Ralf Holtkamp. "Rooted Trees Appearing in Products and Co-Products." In *Combinatorics and Physics,* edited by Kurusch Ebrahimi-Fard, Matilde Marcolli, and Walter D. van Suijlekom, 153–169, Contemporary Mathematics, Vol. 539, American Mathematical Society, 2011.

[96] Ralf Holtkamp. "A Pseudo-Analyzer Approach to Formal Group Laws Not of Operad Type." *Journal of Algebra,* 237, no. 1 (2001): 382–405.

[97] Norbert Hornstein. *Move!: A Minimalist Theory of Construal.* Wiley-Blackwell, 2001.

[98] Norbert Hornstein, Jairo Nunes, and Kleanthes K. Grohmann. Understanding Minimalism. Cambridge University Press, 2005.

[99] Jennifer Hu, Jon Gauthier, Peng Qian, Ethan Wilcox, and Roger P. Levy. "A Systematic Assessment of Syntactic Generalization in Neural Language Models." *Proceedings of the 58th Annual Meeting of the Association for Computational Linguistics*, edited by Dan Jurafsky, Joyce Chai, Natalie Schluter, and Joel Tetreault, 1725–1744. Association for Computational Linguistics, 2020.

[100] Philip A. Huebner, Elior Sulem, Fisher Cynthia, and Dan Roth. "BabyBERTa: Learning More Grammar With Small-Scale Child-Directed Language." In *Proceedings of the 25th Conference on Computational Natural Language Learning*, edited by Arianna Bisazza and Omri Abend, 624–646. Association for Computational Linguistics, 2021.

[101] Marinus A. C. Huijbregts. *Empirical Cases That Rule Out Ternary Merge.* Notes, April 19, 2021.

[102] Marinus A. C. Huijbregts. *Special and General Theory of Merge With Special and General Thanks to Noam.* Syntax Interface Lectures, Utrecht University, 2019.

[103] Marinus A. C. Huijbregts. *Merge Based Morphology and Phonology*. Preprint, 2024.

[104] Sagar Indurkhya. "Inferring Minimalist Grammars with an SMT-Solver." In *Proceedings of the Society for Computation in Linguistics 2020*, 457–460. Association for Computational Linguistics, 2020.

[105] Sagar Indurkhya. "Parsing as Deduction Revisited: Using an Automatic Theorem Prover to Solve an SMT Model of a Minimalist Parser." In *Proceedings of the 26th Conference on Computational Natural Language Learning (CoNLL)*, edited by Antske Fokkens and Vivek Srikumar, 157–175. Association for Computational Linguistics, 2023.

[106] S. A. Joni and Gian-Carlo Rota. "Coalgebras and Bialgebras in Combinatorics." *Studies in Applied Mathematics*, 61, no. 2 (1979): 93–139.

[107] David E. Johnson and Shalom Lappin. "A Critique of the Minimalist Program." *Linguistics and Philosophy*, 20, no. 3 (1997): 273–333.

[108] André Joyal. "Foncteurs analytiques et espèces de structures." In *Combinatoire énumérative*, edited by G. Labelle and P. Leroux, 126–159. Lecture Notes in Mathematics, Vol. 1234. Springer, 1986.

[109] Richard S. Kayne. *The Asymmetry of Syntax*. MIT Press, 1994.

[110] Richard S. Kayne. "Temporal/Linear Order, Antisymmetry and Externalization." *Research in Generative Grammar*, 45, no. 1 (2023): 1–22.

[111] Donald E. Knuth. "Semantics of Context-Free Languages." *Mathematical Systems Theory*, 2 (1968): 127–145.

[112] Joachim Kock. "Combinatorial Dyson-Schwinger Equations and Inductive Data Types." *Frontiers of Physics*, 11 (2016): Article 111205 [15 pages].

[113] Karen Thompson Kohl. "An Analysis of Finite Parameter Learning in Linguistic Spaces." S.M. thesis, Massachusetts Institute of Technology, 1999.

[114] Masayuki Komachi, Hisatsugu Kitahara, Asako Uchibori, and Kensuke Takita. "Generative Procedure Revisited." *Reports of the Keio Institute of Cultural and Linguistic Studies*, 50 (2019): 269–283.

[115] Dirk Kreimer. "On the Hopf Algebra Structure of Perturbative Quantum Field Theories." *Advances in Theoretical and Mathematical Physics*, 2, no. 2 (1998): 303–334.

[116] Dirk Kreimer and Walter D. van Suijlekom. "Recursive Relations in the Core Hopf Algebra." *Nuclear Physics B*, 820, no. 3 (2009): 682–693.

[117] Igor Križ and J. Peter May. *Operads, Algebras, Modules and Motives*. Astérisque No. 233. Société Mathématique de France, 1995.

[118] Nicholas LaCara. The Linear Correspondence Axiom. LIN331 Syntactic Theory, course at the University of Toronto, 2018.

[119] Joachim Lambek. "Pregroup Grammars and Chomsky's Earliest Examples." *Journal of Logic, Language and Information*, 17, no. 2 (2008): 141–160.

[120] Richard K. Larson. "On the Double Object Construction." *Linguistic Inquiry*, 19, no. 3 (1988): 335–391.

[121] Jean-Louis Loday. Cyclic Homology. Grundlehren der Mathematischen Wissenschaften, Vol. 301. Springer, 1998.

[122] Jean-Louis Loday. "Arithmetree." *Journal of Algebra*, 258, no. 1 (2002): 275–309.

[123] Jean-Louis Loday. "Dichotomy of the Addition of Natural Numbers." In *Associahedra, Tamari lattices and Related Structures*, edited by Folkert Mïller-Hoissen, Jean Marcel Pallo, and Jim Stasheff, 65–79, Progress in Mathematics, Vol. 299. Birkhäuser/Springer, 2012.

[124] Jean-Louis Loday and María O. Ronco. "Hopf Algebra of the Planar Binary Trees." *Advances in Mathematics*, 139, no. 2 (1998): 293–309.

[125] Jean-Louis Loday and María O. Ronco. "Order Structure on the Algebra of Permutations and of Planar Binary Trees." *Journal of Algebraic Combinatorics*, 15. no. 3 (2002): 253–270.

[126] Jean-Louis Loday and María O. Ronco. "Combinatorial Hopf algebras." In *Quanta of Maths*, edited by Étienne Blanchard, David Ellwood, Masoud Khalkhali, Matilde Marcolli, Henri Moscovici, and Sorin Popa, 347–383. Clay Mathematics Proceedings, Vol. 11. American Mathematical Society, 2010.

[127] Giuseppe Longobardi and Cristina Guardiano. "Evidence for Syntax as a Signal of Historical Relatedness." *Lingua*, 119, no. 11 (2009): 1679–1706.

[128] Giuseppe Longobardi and Alessandro Treves. *Grammatical Parameters From a Gene-Like Code To Selforganizing Attractors: A Research Program.* Preprint, 2023.

[129] Yuri I. Manin. "Renormalization and Computation I: Motivation and Background." In *OPERADS 2009*, edited by Jean-Louis Loday and Bruno Vallette, 181–222. SÉminaires et Congrès, Vol. 26. Société Mathématique de France, 2013.

[130] Yuri I. Manin. "Renormalisation and Computation II: Time Cut-off and the Halting Problem." *Mathematics Structures in Computer Science*, 22, no. 5 (2012): 729–751.

[131] Yuri I. Manin. "Complexity vs Energy: Theory of Computation and Theoretical Physics." *3Quantum: Algebra Geometry Information (QQQ Conference 2012), Journal of Physics: Conference Series*, 532 (2014): 012018 [15 pages].

[132] Yu.I. Manin. "Neural Codes and Homotopy Types: Mathematical Models of Place Field Recognition." *Moscow Mathematical Journal*, 15, no. 4 (2015): 741–748.

[133] Yuri I. Manin and Matilde Marcolli. "Semantic Spaces." *Mathematics in Computer Science*, 10, no. 4 (2016): 459–477.

[134] Christopher D. Manning, Kevin Clark, John Hewitt, Urvashi Khandelwal, and Omer Levy. "Emergent Linguistic Structure in Artificial Neural Networks Trained by Self-Supervision." *Proceedings of the National Academy of Sciences (PNAS)*, 117, no. 48 (2020): 30046–30054.

[135] Matilde Marcolli. "Information Algebras and Their Applications." In *Geometric Science of Information*, edited by Frank Nielsen and Frédéric Barbaresco, 271–276. Lecture Notes in Computer Science, Vol. 9389. Springer, 2015.

[136] Matilde Marcolli. *Pareto Optimization in Categories*. Preprint 2022, arXiv:2204.11931.

[137] Matilde Marcolli. "Syntactic Parameters and a Coding Theory Perspective on Entropy and Complexity of Language Families." *Entropy*, 18, no. 4 (2016): Paper 110 [17 pages].

[138] Matilde Marcolli, Noam Chomsky, and Robert C. Berwick. *Mathematical Structure of Syntactic Merge*. Preprint 2023, arXiv:2305.18278.

[139] Matilde Marcolli, Robert C. Berwick, and Noam Chomsky. *Old and New Minimalism: A Hopf Algebra Comparison*. Preprint 2023, arXiv:2306.10270.

[140] Matilde Marcolli, Robert C. Berwick, and Noam Chomsky. *Syntax–Semantics Interface: An Algebraic Model*. Preprint 2023, arXiv:2311.06189.

[141] Matilde Marcolli, Riny Huijbregts, and Robert C. Berwick. *Modeling Externalization*. In preparation.

[142] Matilde Marcolli, Richard K. Larson, and Riny Huijbregts. *Extension Condition "Violations" and Merge Optimality Constraints*. In preparation.

[143] Matilde Marcolli and Alexander Port. "Graph Grammars, Insertion Lie Algebras, and Quantum Field Theory." *Mathematics in Computer Science*, 9, no. 4 (2015): 391–408.

[144] Matilde Marcolli and Nicolas Tedeschi. "Entropy Algebras and Birkhoff Factorization." *Journal of Geometry and Physics*, 97 (2015): 243–265.

[145] Matilde Marcolli and Ryan Thorngren. "Thermodynamic Semirings." *Journal of Noncommutative Geometry*, 8, no. 2 (2014): 337–392.

[146] Matilde Marcolli and Chelsea Walton. *Generalized Quasi-Hopf Algebras and Merge*. In preparation.

[147] Andrea E. Martin and Leonidas A. A. Doumas. "Tensors and Compositionality in Neural Systems." *Philosophical Transaction of the Royal Society B*, 375, no. 1791 (2019): 20190306 [7 pages].

[148] Rebecca Marvin and Tal Linzen. "Targeted Syntactic Evaluation of Language Models." *Proceedings of the 2018 Conference on Empirical Methods in Natural Language Processing*, edited by Ellen Riloff, David Chiang, Julia Hockenmaier, and Jun'ichi Tsujii, 1192–1202. Association for Computational Linguistics, 2018.

[149] Akira Masuoka. "Quotient Theory of Hopf Algebras." In *Advances in Hopf Algebras*, edited by J. Bergen and S. Montgomery, 107–133. Marcel Dekker, 1994

[150] David W. Matula. "A Natural Rooted Tree Enumeration by Prime Factorization." *SIAM Review*, 10 (1968): 273.

[151] J. Peter May. *The Geometry of Iterated Loop Spaces*, Lecture Notes in Mathematics. Vol. 271. Springer 1972.

[152] Jens Michaelis. "Transforming Linear Context Free Rewriting Systems into Minimalist Grammars." In *Logical Aspects of Computational Linguistics (NY, 2001)*, edited by P. de Groote, G. Morrill, and C. Retoré, 228–244. Lecture Notes in Artificial Intelligence, Vol. 2099. Springer, 2001.

[153] John W. Milnor and John C. Moore. "On the Structure of Hopf Algebras." *Annals of Mathematics (2)*, 81 (1965): 211–264.

[154] A. Moro. *The Boundaries of Babel*. MIT Press, 2015.

[155] A. Moro. *Dynamic Antisymmetry*. Linguistic Inquiry Monographs, No. 38. MIT Press, 2000.

[156] David Mumford and Agnès Desolneux. *Pattern Theory: The Stochastic Analysis of Real-World Signals*. CRC Press, 2010.

[157] Antônio Francisco Neto. "A Bijection Between Rooted Trees and Fermionic Fock States: Grafting and Growth Operators in Fock Space and Fermionic Operators for Rooted Trees." *Journal of Physics A*, 46, no. 43 (2013): Paper 435205 [19 pages].

[158] Andrew Ortegaray, Robert C. Berwick, and Matilde Marcolli. "Heat Kernel Analysis of Syntactic Structures." *Mathematics in Computer Science*, 15, no. 4 (2021): 643–660.

[159] Yohei Oseki. "Eliminating Pair-Merge." In *Proceedings of the 32nd West Coast Conference on Formal Linguistics*, edited by Ulrike Steindl, Thomas Borer, Huilin Fang, Alfredo García Pardo, Peter Guekguezian, Brian Hsu, Charlie O'Hara, and Iris Chuoying Ouyang, 303–312. Cascadilla Proceedings Project, 2015.

[160] C. Y. Amy Pang. *Markov Chains from Descent Operators on Combinatorial Hopf Algebras*. Preprint 2016, arXiv:1609.04312.

[161] C. Y. Amy Pang. "The Eigenvalues of Hyperoctahedral Descent Operators and Applications to Card-Shuffling." *Electronic Journal of Combinatorics*, 29, no.1 (2022): Paper No. 1.32 [50 pages].

[162] Jeong Joon Park, Ronnel Boettcher, Andrew Zhao, Alex Mun, Kevin Yuh, Vibhor Kumar, and Matilde Marcolli. "Prevalence and Recoverability of Syntactic Parameters in Sparse Distributed Memories." In *GSI 2017: Geometric Science of Information*, edited by Frank Nielsen and Frédéric Barbaresco, 265–272, Lecture Notes in Computer Science, Vol 10589. Springer, 2017.

[163] Frédéric Patras. "L'algèbre des descentes d'une bigèbre graduée." *Journal of Algebra*, 170, no. 2 (1994): 547–566.

[164] Massimo Piattelli-Palmarini and Giuseppe Vitiello. "Linguistics and Some Aspects of Its Underlying Dynamics." *Biolinguistics*, 9 (2015): 96–115.

[165] Jiahuan Pei, Cheng Wang, and György Szarvas. "Transformer Uncertainty Estimation with Hierarchical Stochastic Attention." *Proceedings of the AAAI Conference on Artificial Intelligence*, 36, no. 10 (2022): 11147-11155.

[166] Michael Edward Peskin and Daniel V. Schroeder. *An Introduction to Quantum Field Theory*. ABP, 1995.

[167] Vasily Pestun and Yiannis Vlassopoulos. *Tensor Network Language Model*. Preprint, 2017, arXiv:1710.10248.

[168] Paul M. Pietroski. "Minimalist Meaning: Internalist Interpretation." *Biolinguistics*, 2, no. 4 (2008): 317–341.

[169] Paul M. Pietroski. *Conjoining Meanings. Semantics Without Truth Values*. Oxford University Press, 2018.

[170] Paul M. Pietroski. "Function and Concatenation." In *Logical Form and Language*, edited by Gerhard Preyer and Georg Peter, 91–117. Oxford University Press, 2002.

[171] Vincent Pilaud and Viviane Pons. *The Hopf Algebra of Integer Binary Relations*. Preprint, 2018, arXiv:1807.03277.

[172] Alexander Port, Taelin Karidi, and Matilde Marcolli. "Topological Analysis of Syntactic Structures." *Mathematics in Computr Science*, 16, no. 1 (2022): Paper No. 2 [68 pages]

[173] Douglas C. Ravenel. "Complex Cobordism and Stable Homotopy Groups of Spheres." Pure and Applied Mathematics, Vol. 121. Academic Press, 1986.

[174] Luigi Rizzi. "Labeling, Maximality and the Head – Phrase Distinction." *The Linguistic Review*, 33, no. 1 (2016): 103–127.

[175] Brian Roark and Richard Sproat. *Computational Approaches to Morphology and Syntax*. Oxford University Press, 2007.

[176] Ian Roberts. *Parameter Hierarchies and Universal Grammar*. Oxford University Press, 2019.

[177] Gian-Carlo Rota. "Baxter Operators, an Introduction." In *Gian-Carlo Rota on Combinatorics, Introductory Papers and Commentaries*, edited by Joseph P.S.Kung, 504–512. Birkhäuser, 1995.

[178] Gian-Carlo Rota. "Hopf Algebra Methods in Combinatorics." In *Problèmes combinatoires et théorie des graphes (Colloq. Internat. CNRS, Univ. Orsay, Orsay, 1976)*, pp. 363–365, Colloq. Internat. CNRS 260, CNRS, Paris 1978.

[179] Mehrnoosh Sadrzadeh. *Pregroup Grammars, Their Syntax and Semantics*. Preprint, 2019, arXiv:2109.11237.

[180] Saima Samchuck-Schnarch. *An Introduction to Operad Theory*. Preprint, 2020.

[181] Samson Saneblidze and Ronald Umble. *A Diagonal on the Associahedra*. Preprint, 2000, arXiv:math/0011065v2

[182] Samson Saneblidze and Ronald Umble. "Diagonals on the Permutahedra, Multiplihedra and Associahedra." *Homology Homotopy and Applications*, 6, no. 1 (2004): 363–411.

[183] Kevin Scharp. *Replacing Truth*. Oxford University Press, 2013.

[184] Peter Schauenburg and Hans-Jürgen Schneider. "On Generalized Hopf Galois extensions." *Journal of Pure and Applied Algebra*, 202, no. 1-3 (2005): 168–194.

[185] William R. Schmitt. "Hopf Algebras of Combinatorial Structures." *Canadian Journal of Mathematics*, 45, no. 2 (1993): 412–428.

[186] Yikang Shen, Shawn Tan, Alessandro Sordoni, Siva Reddy, and Aaron Courville. "Explicitly Modeling Syntax in Language Models With Incremental Parsing and

a Dynamic Oracle." In *Proceedings of the 2021 Conference of the North American Chapter of the Association for Computational Linguistics: Human Language Technologies*, edited by Kristina Toutanova, Anna Rumshisky, Luke Zettlemoyer, Dilek Hakkani-Tur, Iz Beltagy, Steven Bethard, Ryan Cotterell, Tanmoy Chakraborty, and Yichao Zhou, 1660–1672. Association for Computational Linguistics, 2021.

[187] Kevin Shu and Matilde Marcolli. "Syntactic Structures and Code Parameters." *Mathematics in Computer Science*, 11, no. 1 (2017): 79–90.

[188] Paul Smolensky. "Tensor Product Variable Binding and the Representation of Symbolic Structures in Connectionist Systems." *Artificial Intelligence*, 46, no. 1-2 (1990): 159–216.

[189] Paul Smolensky. "Harmony in Linguistic Cognition." *Cognitive Science*, 30, no. 5 (2006): 779-801.

[190] Donald Spector. "Supersymmetry and the Möbius Inversion Function." *Communications in Mathematical Physics*, 127, no. 2 (1990): 239–252.

[191] Sashi Mohan Srivastava. *A Course on Borel Sets*. Springer, 1998.

[192] Edward P. Stabler. "Computational Perspectives on Minimalism." In *Oxford Handbook of Linguistic Minimalism*, edited by C. Boeckx, 616–641. Oxford University Press, 2010.

[193] Yuji Takano. "Exploring Merge: A New Form of Sideward Movement." *The Linguistic Review*, 37, no. 1 (2020): 7–45.

[194] Mitsuhiro Takeuchi. "Quotient Spaces for Hopf Algebras." *Communications in Algebra*, 22, no. 7 (1994): 2503–2523.

[195] Neil Tennant. "A New Unified Account of Truth and Paradox." *Mind*, 124, no. 494 (2015): 571–605.

[196] Eduard Tulchinskii, Kristian Kuznetsov, Laida Kushnareva, Daniil Cherniavskii, Sergey Nikolenko, Evgeny Burnaev, Serguei Barannikov, and Irina Piontkovskaya. "Intrinsic Dimension Estimation for Robust Detection of AI-Generated Texts." In *Proceedings of the 37th International Conference on Neural Information Processing Systems*, edited by A. Oh, T. Naumann, A. Globerson, K. Saenko, M. Hardt, and S. Levine, Article No. 1706, 39257–39276. Curran Associates, 2024.

[197] Héctor Vázquez. "The Acceptability Delta Criterion: Testing Knowledge of Language Using the Gradience of Sentence Acceptability." In *Proceedings of the Fourth BlackboxNLP Workshop on Analyzing and Interpreting Neural Networks for NLP*, edited by Jasmijn Bastings, Yonatan Belinkov, Emmanuel Dupoux, Mario Giulianelli, Dieuwke Hupkes, Yuval Pinter, and Hassan Sajjad, 479–495, Association for Computational Linguistics, 2021.

[198] Héctor Vázquez, Annika Lea Heuser, Charles Yang, and Jordan Kodner. "Evaluating Neural Language Models As Cognitive Models of Language Acquisition." In *Proceedings of the 1st GenBench Workshop on (Benchmarking) Generalisation in NLP*, edited by Dieuwke Hupkes, Verna Dankers, Khuyagbaatar Batsuren, Koustuv Sinha, Amirhossein Kazemnejad, Christos Christodoulopoulos, Ryan Cotterell, and Elia Bruni, 48–64. Association for Computational Linguistics, 2023.

[199] Krishnamurti Vijay-Shanker and David Weir. "The Equivalence of Four Extensions of Context Free Grammar Formalisms." *Mathematical Systems Theory*, 27, no. 6 (1994): 511–545.

[200] Alex Warstadt and Samuel R. Bowman. "What Artificial Neural Networks Can Tell Us about Human Language Acquisition." In *Algebraic Structures in Natural Language*, edited by Shalom Lappin and Jean-Philippe Bernardy, 17–60. CRC Press, 2022.

[201] Stefan Weinzierl. *Hopf Algebras and Dyson-Schwinger Equations*. Preprint, 2015, arXiv:1506.09119.

[202] Donald Yau. *Colored Operads*. American Mathematical Society, 2016.

[203] Karen Yeats. *Rearranging Dyson-Schwinger Equations*, Memoirs of the American Mathematical Society, Vol. 211. American Mathematical Society, 2011.

[204] Elia Zardini. "Truth without Contra(di)ction." *Review of Symbolic Logic*, 4, no. 4 (2011): 498–535.

[205] Yi Zhang and Xing Gao. "Hopf Algebras of Planar Binary Trees: An Operated Algebra Approach." *Journal of Algebraic Combinatorics*, 51, no. 4 (2020): 567–588.

Index

Linguistic Inquiry Monographs

Samuel Jay Keyser, general editor

1. *Word Formation in Generative Grammar*, Mark Aronoff
2. *$\bar{X}$ Syntax: A Study of Phrase Structure*, Ray Jackendoff
3. *Recent Transformational Studies in European Languages*, S. Jay Keyser, editor
4. *Studies in Abstract Phonology*, Edmund Gussmann
5. *An Encyclopedia of AUX: A Study of Cross-Linguistic Equivalence*, Susan Steele
6. *Some Concepts and Consequences of the Theory of Government and Binding*, Noam Chomsky
7. *The Syntax of Words*, Elisabeth O. Selkirk
8. *Syllable Structure and Stress in Spanish: A Nonlinear Analysis*, James W. Harris
9. *CV Phonology: A Generative Theory of the Syllable*, George N. Clements and Samuel Jay Keyser
10. *On the Nature of Grammatical Relations*, Alec P. Marantz
11. *A Grammar of Anaphora*, Joseph Aoun
12. *Logical Form: Its Structure and Derivation*, Robert May
13. *Barriers*, Noam Chomsky
14. *On the Definition of Word*, Anna-Maria Di Sciullo and Edwin Williams
15. *Japanese Tone Structure*, Janet Pierrehumbert and Mary E. Beckman
16. *Relativized Minimality*, Luigi Rizzi
17. *Types of $\bar{A}$-Dependencies*, Guglielmo Cinque
18. *Argument Structure*, Jane Grimshaw
19. *Locality: A Theory and Some of Its Empirical Consequences*, Maria Rita Manzini
20. *Indefinites*, Molly Diesing
21. *Syntax of Scope*, Joseph Aoun and Yen-hui Audrey Li
22. *Morphology by Itself: Stems and Inflectional Classes*, Mark Aronoff
23. *Thematic Structure in Syntax*, Edwin Williams
24. *Indices and Identity*, Robert Fiengo and Robert May
25. *The Antisymmetry of Syntax*, Richard S. Kayne
26. *Unaccusativity: At the Syntax–Lexical Semantics Interface*, Beth Levin and Malka Rappaport Hovav
27. *Lexico-Logical Form: A Radically Minimalist Theory*, Michael Brody
28. *The Architecture of the Language Faculty*, Ray Jackendoff
29. *Local Economy*, Chris Collins
30. *Surface Structure and Interpretation*, Mark Steedman

31. *Elementary Operations and Optimal Derivations*, Hisatsugu Kitahara
32. *The Syntax of Nonfinite Complementation: An Economy Approach*, Željko Bošković
33. *Prosody, Focus, and Word Order*, Maria Luisa Zubizarreta
34. *The Dependencies of Objects*, Esther Torrego
35. *Economy and Semantic Interpretation*, Danny Fox
36. *What Counts: Focus and Quantification*, Elena Herburger
37. *Phrasal Movement and Its Kin*, David Pesetsky
38. *Dynamic Antisymmetry*, Andrea Moro
39. *Prolegomenon to a Theory of Argument Structure*, Ken Hale and Samuel Jay Keyser
40. *Essays on the Representational and Derivational Nature of Grammar: The Diversity of Wh-Constructions*, Joseph Aoun and Yen-hui Audrey Li
41. *Japanese Morphophonemics: Markedness and Word Structure*, Junko Ito and Armin Mester
42. *Restriction and Saturation*, Sandra Chung and William A. Ladusaw
43. *Linearization of Chains and Sideward Movement*, Jairo Nunes
44. *The Syntax of (In)dependence*, Ken Safir
45. *Interface Strategies: Optimal and Costly Computations*, Tanya Reinhart
46. *Asymmetry in Morphology*, Anna Maria Di Sciullo
47. *Relators and Linkers: The Syntax of Predication, Predicate Inversion, and Copulas*, Marcel den Dikken
48. *On the Syntactic Composition of Manner and Motion*, Maria Luisa Zubizarreta and Eunjeong Oh
49. *Introducing Arguments*, Liina Pylkkänen
50. *Where Does Binding Theory Apply?* David Lebeaux
51. *Locality in Minimalist Syntax*, Thomas S. Stroik
52. *Distributed Reduplication*, John Frampton
53. *The Locative Syntax of Experiencers*, Idan Landau
54. *Why Agree? Why Move? Unifying Agreement-Based and Discourse-Configurational Languages*, Shigeru Miyagawa
55. *Locality in Vowel Harmony*, Andrew Nevins
56. *Uttering Trees*, Norvin Richards
57. *The Syntax of Adjectives*, Guglielmo Cinque
58. *Arguments as Relations*, John Bowers
59. *Agreement and Head Movement*, Ian Roberts
60. *Localism versus Globalism in Morphology and Phonology*, David Embick
61. *Provocative Syntax*, Phil Branigan

62. *Anaphora and Language Design*, Eric J. Reuland
63. *Indefinite Objects: Scrambling, Choice Functions, and Differential Marking*, Luis López
64. *A Syntax of Substance*, David Adger
65. *Subjunctive Conditionals*, Michela Ippolito
66. *Russian Case Morphology and the Syntactic Categories*, David Pesetsky
67. *Classical NEG Raising: An Essay on the Syntax of Negation*, Chris Collins and Paul M. Postal
68. *Agreement and Its Failures*, Omer Preminger
69. *Voice and v: Lessons from Acehnese*, Julie Anne Legate
70. *(Re)labeling*, Carlo Cecchetto and Caterina Donati
71. *A Two-Tiered Theory of Control*, Idan Landau
72. *Concepts, Syntax, and Their Interface: Tanya Reinhart's Theta System*, Martin Everaert, Marijana Marelj, and Eric Reuland, editors
73. *Contiguity Theory*, Norvin Richards
74. *Impossible Persons*, Daniel Harbour
75. *Agreement beyond Phi*, Shigeru Miyagawa
76. *The Final-over-Final Condition: A Syntactic Universal*, Michelle Sheehan, Theresa Biberauer, Ian Roberts, and Anders Holmberg
77. *Situations and Syntactic Structures: Rethinking Auxiliaries and Order in English*, Gillian Ramchand
78. *Features of Person: From the Inventory of Persons to Their Morphological Realization*, Peter Ackema and Ad Neeleman
79. *Cardinals: The Syntax and Semantics of Cardinal-Containing Expressions*, Tania Ionin and Ora Matushansky
80. *Composing Questions*, Hadas Kotek
81. *Probes and Their Horizons*, Stefan Keine
82. *A Theory of Indexical Shift: Meaning, Grammar, and Crosslinguistic Variation*, Amy Rose Deal
83. *Merge: Binarity in (Multidominant) Syntax*, Barbara Citko and Martina Gračanin-Yüksek
84. *A Selectional Theory of Adjunct Control*, Idan Landau
85. *Syntax in the Treetops*, Shigeru Miyagawa
86. *On Linearization*, Guglielmo Cinque
87. *Principles of Argument Structure: A Merge-Based Approach*, Chris Collins
88. *When Arguments Merge*, Elise Newman
89. *Mathematical Structure of Syntactic Merge: An Algebraic Model for Generative Linguistics*, Matilde Marcolli, Noam Chomsky, and Robert C. Berwick

Publisher contact:
The MIT Press
Massachusetts Institute of Technology
77 Massachusetts Avenue, Cambridge, MA 02139
mitpress.mit.edu

EU Authorised Representative:
Easy Access System Europe, Mustamäe tee 50,
10621 Tallinn, Estonia
gpsr.requests@easproject.com

Printed by Integrated Books International,
United States of America